WASTES: SOLUTIONS, TREATMENTS AND OPPORTUNITIES

T0239588

SELECTED PAPERS FROM THE 3RD EDITION OF THE INTERNATIONAL CONFERENCE ON WASTES: SOLUTIONS, TREATMENTS AND OPPORTUNITIES, VIANA DO CASTELO, PORTUGAL, 14–16 SEPTEMBER 2015

Wastes: Solutions, Treatments and Opportunities

Editors

Cândida Vilarinho & Fernando Castro
University of Minho, Guimarães, Portugal

Mário Russo
Polytechnic Institute of Viana do Castelo, Viana do Castelo, Portugal

CRC Press
Taylor & Francis Group
Boca Raton London New York

CRC Press is an imprint of the
Taylor & Francis Group, an **informa** business

A BALKEMA BOOK

Published by:
CRC Press/Balkema
P.O. Box 447, 2300 AK Leiden, The Netherlands
e-mail: Pub.NL@taylorandfrancis.com
www.crcpress.com – www.taylorandfrancis.com

First issued in paperback 2020

© 2015 by Taylor & Francis Group, LLC
CRC Press/Balkema is an imprint of the Taylor & Francis Group, an informa business

No claim to original U.S. Government works

Typeset by MPS Limited, Chennai, India

ISBN 13: 978-0-367-73785-6 (pbk)
ISBN 13: 978-1-138-02882-1 (hbk)

This book contains information obtained from authentic and highly regarded sources. Reasonable efforts have been made to publish reliable data and information, but the author and publisher cannot assume responsibility for the validity of all materials or the consequences of their use. The authors and publishers have attempted to trace the copyright holders of all material reproduced in this publication and apologize to copyright holders if permission to publish in this form has not been obtained. If any copyright material has not been acknowledged please write and let us know so we may rectify in any future reprint.

Except as permitted under U.S. Copyright Law, no part of this book may be reprinted, reproduced, transmitted, or utilized in any form by any electronic, mechanical, or other means, now known or hereafter invented, including photocopying, microfilming, and recording, or in any information storage or retrieval system, without written permission from the publishers.

For permission to photocopy or use material electronically from this work, please access www.copyright.com (http://www.copyright.com/) or contact the Copyright Clearance Center, Inc. (CCC), 222 Rosewood Drive, Danvers, MA 01923, 978-750-8400. CCC is a not-for-profit organization that provides licenses and registration for a variety of users. For organizations that have been granted a photocopy license by the CCC, a separate system of payment has been arranged.

Trademark Notice: Product or corporate names may be trademarks or registered trademarks, and are used only for identification and explanation without intent to infringe.

Visit the Taylor & Francis Web site at
http://www.taylorandfrancis.com

and the CRC Press Web site at
http://www.crcpress.com

Table of contents

Wastes: Solutions, Treatments and Opportunities – Vilarinho, Castro & Russo (eds)
© 2015 Taylor & Francis Group, London, ISBN 978-1-138-02882-1

Preface

Dear colleagues,

It is with great pleasure that we bring to you the book "Selected papers from the 3rd edition of the International Conference Wastes: Solutions, Treatments and Opportunities".

The papers published in this book were submitted, revised and approved by the Scientific Committee in a full peer review process to be presented at Wastes 2015 that took place in the Polytechnic Institute of Viana do Castelo, between 14 and 16 of September.

The Wastes conferences, happening every two years, are a platform for the scientists and industries from the waste management and recycling sectors from around the world, to share experiences and knowledge with all who attend. Discussions regarding the balance between economic, environmental and social outcomes are carefully addressed. The development of innovative techniques, tools and strategies on how wastes can be transformed into good ideas, improving both the overall environmental performance and the understanding of the industry impact on the environment, as well as the options analysis for its improvement are key objectives of this conference.

With this publication we expect to take the scope of this event beyond the limits of its physical occurrence by providing both attendants and general public with an instrument that is the materialization of the main contributions to Wastes 2015.

The editors wish to thank all the reviewers that, taking part of the Scientific Committee of the Conference, gave a fundamental input to the process of reviewing the papers included in this book, namely:

Ana Luísa Fernando	João Labrincha
Ana Maria Segadães	Joel Oliveira
André Mota	Jorge Araújo
André Ribeiro	José Barroso de Aguiar
Anje Nzhiou	José Teixeira
António Brito	Luís Marinheiro
António Roque	Madalena Alves
Benilde Mendes	Manuel Fonseca Almeida
Carlos Nogueira	Margarida Gonçalves
Felipe Macias	Margarida Quina
Fernanda Margarido	Maria Alcina Pereira
Gerasimus Lyberatos	Mário Costa
Hugo Silva	Nuno Cristelo
Javier Escudero	Nuno Lapa
Javier Viguri	Rosa M. Quinta-Ferreira
Joana Carvalho	Tiago Miranda
Joana Dias	

Wastes: Solutions, Treatments and Opportunities – Vilarinho, Castro & Russo (eds)
© 2015 Taylor & Francis Group, London, ISBN 978-1-138-02882-1

Evaluation of foamed bitumen efficiency in warm asphalt mixtures recycling

L. Abreu, J. Oliveira & H. Silva
CTAC – Territory, Environment and Construction Centre, University of Minho, Guimarães, Portugal

D. Palha & P. Fonseca
Elevo Group, Porto, Portugal

ABSTRACT: The recycling of pavements is nowadays a very important question to the road paving industry. With the objective of incorporating higher percentages of reclaimed asphalt (RA) materials in recycled asphalt mixtures, new techniques have been developed in the last years. The use of foamed bitumen is normally associated with the production of cold asphalt mixtures, which usually show lower quality standards. However, the objective of the work presented in this paper is to assess the use of foamed bitumen as the binder of warm asphalt mixtures incorporating 30% RA, which have quality standards similar to those of conventional mixtures. Thus, five mixtures have been produced with 30% RA, one of them with a conventional bitumen (control mix) and the others with foamed bitumen at different production temperatures. The mixtures were tested for compactability and water sensitivity and the results show a possible reduction of 25°C in the production temperatures, while the water sensitivity test results were kept close to 90%.

1 INTRODUCTION

Nowadays the constant concern of saving natural resources and reduce waste production is promoting an increase on the recycling of used materials. Industries and the Society are very focused on finding the right treatment for waste materials. On the asphalt paving industries, the effort has been concentrated in the incorporation of reclaimed asphalt (RA) material in new mixtures.

The incorporation of RA have been studied by a significant number of authors with the principal objective of increasing the percentage of RA in new mixtures (Celauro et al., 2010, Shu et al., 2012, Su et al., 2009, Abreu et al., 2015). The incorporation of RA have numerous advantages, namely, the reduction of costs when compared with the production of new mixtures (Chiu et al., 2008), the reduction of space needed to dispose of the waste from pavement rehabilitation (saving money and the environment) and the reduction of the environmental impact (lower consumption of natural resources and its associated emissions).

However some disadvantages can be also pointed out to the incorporation of high percentages of RA (above 20%). The fact that the material may be heterogeneous implies the need for a correct treatment before the incorporation in new mixtures. Some studies mention different ways to treat the RA before the incorporation in order to mitigate the effect of this issue (Aravind & Das, 2007, Oliveira et al., 2012, Abreu et al., 2013). Other disadvantage is related with the aged bitumen present in that material that normally implies the use of a virgin bitumen softer than the conventional bitumens normally used and, in some cases, the use of rejuvenators.

Another problem that results from the incorporation of high percentages of RA is the increase on the heating temperature of the virgin aggregates. This is due to the fact that the virgin aggregates have to be overheated in order to transfer part of the temperature to the RA material that is normally introduced in the mixing chamber at ambient temperature. Using this procedure, it is possible to mitigate the ageing effect of the temperature on the bitumen (which is already significantly aged by the time it has been exposed to oxygen and sun light in the road pavement). In recent studies, the RA material has been divided into two fractions (coarse and fine, separated in the 10 mm sieve) and the coarse fraction is heated together with the virgin aggregates while the fine fraction is kept at ambient temperature up to the moment it is introduced in the mixer (Abreu et al., 2015, Oliveira et al., 2012, Palha et al., 2013). The main reason that justifies this procedure is the fact that the fine fraction contains a significantly higher percentage of bitumen, which can thus be preserved from unwanted ageing, and the production rates can be increased by reducing the mixing time (needed for the temperature transfer to occur), since a higher proportion of material is heated up.

To reduce the production temperature new techniques have been recently used. The mixtures produced at lower temperatures are known as warm mix asphalt (WMA). These techniques have been pointed out as a solution to reduce energy consumption and greenhouse gas emissions, to improve the working conditions, to improve the workability of the mixtures, and to increase time available to apply and compact the mixtures (Zaumanis et al., 2012, D'Angelo et al., 2008, Rubio et al., 2012). These techniques can be classified, according to the production process, into: i) utilization of organic additives; ii) utilization of chemical additives or iii) foaming processes (Sengoz et al., 2013, Zaumanis et al., 2012, He & Wong, 2007). In terms of foaming processes they can be differentiated between water-containing technologies, e.g., by using a synthetic zeolite, and water-based technologies where the water is directly injected into the hot bitumen (Rubio et al., 2012).

The process of injecting water into the hot bitumen originates an expansion on the bitumen known as foamed bitumen (FB). This technique can be used as a stabilizing and recycling agent with a variety of materials, which can be a good quality material or with not so appropriated materials (He & Wong, 2006). When producing FB, the characteristics that are important to control are the expansion ratio (ER) and the half-life (HL) (Jenkins, 2000). The ER is a relation between the volume of the bitumen after expansion and the initial volume of bitumen. Normally the foamed bitumen is characterized by the maximum expansion ratio (ERmax) that represents the maximum volume obtained. The half-life (HL) is the elapsed time between the moment that the foam was at its maximum volume and the time when this volume reduces to a half of that value. It gives an idea of the time available to produce the mixture (Brennen et al., 1983). Other index that are referred to as important to analyze the results of the foaming process is the foam index (FI), that was defined by Jenkins (2000) as the area below the decay curve of a bitumen with some restrictions. The highest FI is normally associated to a better bitumen in terms of foaming characteristics. Foam promoting additives may be necessary to increase the half-life time in order to obtain an appropriate mixture (aggregates fully covered by the bitumen).

In the present paper, the possibility of using FB as a binder to produce asphalt mixtures with incorporation of 30% RA at temperatures lower than those normally used, has been tested. The expansion characteristics of the bitumen with and without foaming additive and its basic characteristics have been evaluated and the effect of the production temperature has been assessed by testing different asphalt specimens for compactability and water sensitivity.

2 MATERIALS AND METHODS

2.1 *Materials*

A virgin 70/100 bitumen has been chosen because of the need of soften the aged bitumen present on the RA. The new aggregates are granite igneous rocks and the filler is limestone; the use of those aggregates was justified by the proximity of their sources to the laboratory and to the asphalt plant.

In terms of RA, the material results from milling off the surface layer of a highway pavement and was subsequently separated between a fine and a coarse fraction by a classifier with a mesh of 10 mm.

Table 1. Reference and real temperatures of the mixtures produced.

Mixture	Ref. Temp. °C	Real Temperatures (°C)				
		New aggregates	Coarse fraction	Fine fraction	Production	Compaction
FB160	160.0	212.3	206.8	20.5	160.0	146.8
FB150	150.0	195.6	179.7	20.1	145.0	133.0
FB140	140.0	181.8	178.8	19.5	138.0	122.5
FB130	130.0	167.5	166.8	17.9	125.0	108.6
Control	160.0	211.8	211.0	25.3	165.8	146.6

2.2 Methods

2.2.1 Foamed bitumen production process

In this study, a Wirtgen WLB 10 S lab scale plant was used to produce the foamed bitumen, which injects water and air into hot bitumen in an expansion chamber, promoting the formation of bitumen foam. The equipment manufacturer mentioned that it has been developed to a laboratory scale with the objective of making the analysis of the FB characteristics in small scale possible, but it is similar to the equipment used in a normal scale (Wirtgen, 2008).

In order to increase the stability (measured by the half-life) of the foamed bitumen, a specific additive, already tested by other authors like Hailesilassie et al. (2015), was used. The air pressure was maintained at 5.5 bar (default value), the percentage of water used was 3% (by mass of bitumen) and the bitumen temperature was 150°C.

2.2.2 Selection of the mixing temperature

Taking the importance of the mixing temperature to the production of WMAs into consideration and the need to better understand the role of foamed bitumen in the behavior of the mixtures, compactability and water sensitivity tests have been carried out. Thus, four mixtures were produced (with 30% RA and foamed bitumen) at different temperatures. A control hot mix asphalt (HMA) mixture (with 30% RA and a 70/100 pen virgin bitumen without being foamed) was also produced for comparison.

The production of the mixtures with incorporation of RA was carried out taking into account that further ageing of the bitumen present in the RA should be avoided. Thus, the fine RA fraction was introduced in the mixture at ambient temperature and only the new aggregates and the coarse RA fraction have been heated (Table 1). To mitigate the ageing effect of heating, the time during which the coarse RA fraction was submitted to high temperatures was limited to two hours.

In order to maintain the homogeneity of the specimens produced at each studied temperature (the bitumen content of the RA material and the amount of foamed bitumen injected by the foaming equipment may vary slightly amongst different samples), all specimens of a specific temperature were manufactured from the same batch (3 specimens for compactability tests and the other 6 for water sensitivity tests).

The compactability tests (EN 12697-10) were carried out using a Marshall Impact compactor (EN 12697-30) with a measuring device for automatically recording the thickness of the specimen after each compacting blow. The compaction of the specimens comprised the application of a total of 200 blows.

For the water sensitivity tests, 6 specimens were used in accordance with the EN 12697-12 standard, according to which, two groups of three specimens are tested for the indirect tensile strength (ITS) after a different conditioning period (three specimens are kept dry and the others three are kept in water). In this particular study, a curing period of 36 hours was used for all specimens prior to the tests, during which they were at room temperature. With this test it is possible to assess the loss of strength of the mixtures when in contact with water, by determining the ratio between the indirect tensile strength of the wet specimens and the dry specimens (ITSR), which is an indicator of the durability of the mixtures when applied in a real context.

Table 2. Expansion characteristics of foamed bitumen.

Bitumen	ERmax	Half-life sec	Foam Index sec
70/100	20.0	10.5	392.1
70/100 + 0.1% Additive	17.0	47.3	671.5

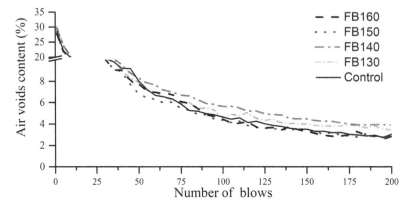

Figure 1. Compactability test results of the studied mixtures.

3 RESULTS AND DISCUSSION

3.1 *Foamed bitumen characteristics*

As previously mentioned, a foam promoting additive was used in this study. The main characteristics of the foamed bitumen produced with and without the additive are presented in Table 2. It can be observed that the expansion ratio (ERmax) is not significantly affected by the incorporation of the additive, but half-life (HL) has increased more than 4 times, which is a good advantage to the production of mixtures with this technique. The foam index (FI) obtained show the importance of the HL on the analysis of the expansion process, and prove the importance of the use of a foam promoting additive.

Both bitumens were also tested in terms of rheological characteristics and the higher limit PG grade of both bitumens was 64°C, confirming that the additive does not affect the properties of the virgin bitumen.

3.2 *Compactability test results*

The compactability test results show the evolution of the air void content of a specimen during the compaction process for a specific temperature. Repeating this procedure for a series of different temperatures is important to understand how the temperature may influence the workability of the mixture and, consequently, the porosity of the mixture. Based on the results obtained (Figure 1), it is possible to see that the difference in terms of air voids content is not significant for the mixtures produced with FB at 160 and 150°C. In fact the results are very similar to those of the control mixture. However, for the mixtures produced at 140 and 130°C, a higher air void content was obtained. In order to compare the results to those of the specimens prepared for the water sensitivity tests, the air void content should be measured after 150 blows, which is equivalent to the compaction energy used in the Marshall compaction by applying 75 blows in each face of the specimen, as presented in Table 3.

Table 3. Air voids content of the specimens after 150 compacting blows.

	FB160	FB150	FB140	FB130	Control
Air voids content (%)	3.3	3.4	4.4	3.9	3.5

Table 4. Air voids content of the specimens used in the water sensitivity tests.

	FB160	FB150	FB140	FB130	Control
Air voids content (%)	3.5	3.1	4.7	4.9	3.2

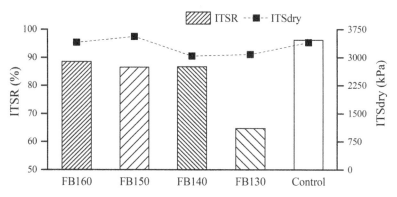

Figure 2. Water sensitivity test results.

3.3 *Water sensitivity test results*

The value of air voids content is closely related to the water sensitivity of asphalt mixtures. Thus, all specimens used for this test were also volumetrically characterized (Table 4). As can be observed, the results obtained are similar to those of the compactability test specimens. The main differences are observed for the FB130 mixture, as a result of the lower production temperature.

The water sensitivity test results are shown in Figure 2, where it is possible to see that the ITSR values of the mixtures produced with foamed bitumen are slightly more influenced by the presence of water than the control mixture. Nevertheless, the results obtained for the mixtures produced at 140°C and above showed very interesting ITSR values (above 85%). The lower ITSdry values of the mixtures produced at 140 and 130°C may be related to the higher air void contents of those mixtures, even though it did not significantly affect the ITSR of the FB140 mixture. Thus it can be concluded that it is possible to produce recycled asphalt mixtures with 30% RA with foamed bitumen, reducing the production temperature by 20°C without significantly affecting the volumetric properties of the mixtures and their water susceptibility.

4 CONCLUSIONS

The use of foamed bitumen in the production of asphalt mixtures with incorporation of 30% RA material have been tested and the results show that this technique could be a good option to reduce the production temperatures of this type of mixtures.

Up to a reduction of 20°C on the production temperatures, the mixtures with foamed bitumen showed similar properties to those of the control mixture produced with conventional hot bitumen. Thus it can be concluded that this type of binder can be efficient in reducing the production temperature of mixtures with RA incorporation, reducing the possibility of inducing undesirable ageing on the bitumen present in the RA material which is already hard and brittle.

ACKNOWLEDGMENT

The authors would like to acknowledge the contribution of the companies Galp and Elevo Group, who provided the binders and the RAP used in this study. This work is being funded by ERDF funds through the Operational Competitiveness Program – COMPETE in the scope of Project "Energy Efficiency and Environmental Design of Bituminous Mixtures and Reducing Emissions of Greenhouse Gases" (SI Innovation Project 7603). Thanks are also due to the Foundation for Science and Technology (FCT) for funding allocated through the PhD grant SFRH/BD/85448/2012.

REFERENCES

Abreu, L. P. F., Oliveira, J. R. M. & Silva, H. M. R. D. 2013. Formulação e caracterização de ligantes numa mistura betuminosa com uma taxa de reciclagem elevada. *7° Congresso Rodoviário Português*. Lisboa.

Abreu, L. P. F., Oliveira, J. R. M., Silva, H. M. R. D. & Fonseca, P. V. 2015. Recycled asphalt mixtures produced with high percentage of different waste materials. *Construction and Building Materials*, 84, 230–238.

Aravind, K. & Das, A. 2007. Pavement design with central plant hot-mix recycled asphalt mixes. *Construction and Building Materials*, 21, 928–936.

Brennen, M., Tia, M., Altschaeffl, A. G. & Wood, L. E. 1983. Laboratory investigation of the use of foamed asphalt for recycled bituminous pavements. *Transportation Research Record*, p. 80–87.

Celauro, C., Bernardo, C. & Gabriele, B. 2010. Production of innovative, recycled and high-performance asphalt for road pavements. *Resources, Conservation and Recycling*, 54, 337–347.

Chiu, C.-T., Hsu, T.-H. & Yang, W.-F. 2008. Life cycle assessment on using recycled materials for rehabilitating asphalt pavements. *Resources, Conservation and Recycling*, 52, 545–556.

D'angelo, J., Harm, E., Bartoszek, J., Baumgardner, G., Corrigan, M., Cowsert, J., Harman, T., Jamshidi, M., Jones, W., Newcomb, D., Prowell, B., Sines, R. & Yeaton, B. 2008. Warm-Mix Asphalt: European Practice. Alexandria.

Hailesilassie, B., Schuetz, P., Jerjen, I., Hugener, M. & Partl, M. 2015. Dynamic X-ray radiography for the determination of foamed bitumen bubble area distribution. *Journal of Materials Science*, 50, 79–92.

He, G. & Wong, W. 2006. Decay properties of the foamed bitumens. *Construction and Building Materials*, 20, 866–877.

He, G. & Wong, W. 2007. Laboratory study on permanent deformation of foamed asphalt mix incorporating reclaimed asphalt pavement materials. *Construction and Building Materials*, 21, 1809–1819.

Jenkins, K. J. 2000. *Mix Design Considerations for Cold and Half-warm Bitumen Mixes with Emphasis on Foamed Bitumen* Doctor of Philosophy (Engineering), University of Stellenbosch.

Oliveira, J. R. M., Silva, H. M. R. D., Abreu, L. P. F. & Pereira, P. a. A. 2012. Effect of Different Production Conditions on the Quality of Hot Recycled Asphalt Mixtures. *Procedia – Social and Behavioral Sciences*, 53, 266–275.

Palha, D., Fonseca, P., Abreu, L. P. F., Oliveira, J. R. M. & Silva, H. M. R. D. 2013. Solutions to improve the recycling rate and quality of plant produced hot mix asphalt. *2nd International Conference – WASTES: solutions, treatments, opportunities*. Braga.

Rubio, M. C., Martínez, G., Baena, L. & Moreno, F. 2012. Warm mix asphalt: an overview. *Journal of Cleaner Production*, 24, 76–84.

Sengoz, B., Topal, A. & Gorkem, C. 2013. Evaluation of natural zeolite as warm mix asphalt additive and its comparison with other warm mix additives. *Construction and Building Materials*, 43, 242–252.

Shu, X., Huang, B., Shrum, E. D. & Jia, X. 2012. Laboratory evaluation of moisture susceptibility of foamed warm mix asphalt containing high percentages of RAP. *Construction and Building Materials*, 35, 125–130.

Su, K., Hachiya, Y. & Maekawa, R. 2009. Study on recycled asphalt concrete for use in surface course in airport pavement. *Resources, Conservation and Recycling*, 54, 37–44.

Wirtgen 2008. Suitability test procedures of foam bitumen using Wirtgen WLB 10 S. Germany: Wirtgen, GmbH.

Zaumanis, M., Haritonovs, V., Brencis, G. & Smirnovs, J. 2012. Assessing the Potential and Possibilities for the Use of Warm Mix Asphalt in Latvia. *Construction Science*, 13, pp. 6–59.

Wastes: Solutions, Treatments and Opportunities – Vilarinho, Castro & Russo (eds)
© *2015 Taylor & Francis Group, London, ISBN 978-1-138-02882-1*

Building an integrated perception and attitude towards municipal solid waste management in Nigeria

O.M. Aderoju, A. Guerner Dias & R. Guimarães
Faculty of Sciences University of Porto, Porto, Portugal

ABSTRACT: Municipal Solid Waste Management (MSWM) as an integral part of the urban environment that require proper planning to ensure the safety of human and its environment. Nigeria became an increasingly urbanized society because of the oil boom in the 1970s. Rapid population growth simultaneously increases the quantity of waste generation in any society. Again, environmental indiscipline, poor management practices by the authorities, non implementation and enforcement of the environmental laws attributes to indiscriminate dumpsites in our cities. Hence this study aimed at the behavior of Nigerians towards municipal solid waste disposal using an integrated approach. The methodology adopted was a conceptual framework which has four stages (Initiators, Channel, procedure, and recipient). In conclusion, incorporating the conceptual framework it into the formal solid waste management structure in Nigeria will encourage self discipline, educate the people, promote pollution free and sustainable environment.

1 INTRODUCTION

Managing municipal solid wastes (MSW) is obviously becoming a major challenge in major cities of developing nations due to rapid urbanization and increasing population growth. Saheed et al. (2008) reported that the generation of solid waste is directly proportional to population, industrialization, urbanization and the changing lifestyle, food, habits and living standards of the people. According to recent estimates, the waste sector contributes about one-fifth of global anthropogenic methane emissions (IEA 2005). Various activities of the society lead to the generation of waste materials that include both organic and inorganic wastes (Naveen et al., 2014). Municipal Solid Waste (MSW) are commonly known as trash or garbage that consist of everyday item like food item, metal scrap, polythene, and paper among others (Cunningham and Saigo,1997). Waste management is an important objective of planning to ensure that the future generations inherit an environment that is as pollution free as possible given the present scientific, economic, social and political constraints (United States Environmental Protection Agency, 2005). Municipal Solid Waste (MSW) management involves various steps, namely collection, transportation, processing and disposal and perhaps open land disposal is the most common method adopted in most developing nation like Nigeria. Babayemi and Dauda (2009) described the solid waste management in Nigeria as being challenged with ineffective collection technique, inadequate coverage of the collection routine and unlawful dumping of solid waste. In Portugal, we observed in our little experience that the populaces have been able to consent that the reduction of environmental pollution contributes immensely to the environmental sustainability of their cities. This is probably the same standard of perception of most people all over Europe due to the high level of awareness on solid waste management.

In Nigeria, municipal waste disposal management has received a lot of attention with little done. Pronouncements of waste management programmes were made in the past but lacks political will and implementation. With little awareness on integrated waste management by its people, the pattern of disposal of refuse and poor management practice by the authorities has led to an increasing number of indiscriminate dumpsites the urban cities. Hence this study is aimed at the behavior of

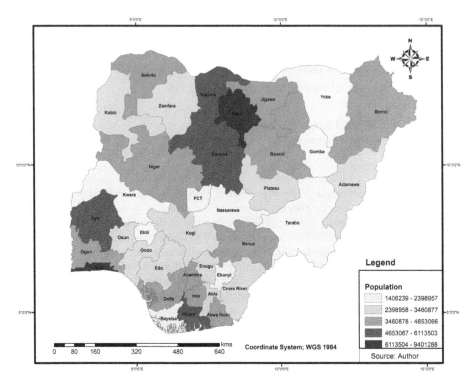

Figure 1. Nigeria population distribution map.

Nigerian towards municipal solid waste disposal and management using an integrated approach. The objective is to educate the people on environmental ethics and discipline and to promote the involvement of institutions and citizen participation in the waste management sensitization programmes.

2   AN OVERVIEW OF SOLID WASTE MANAGEMENT IN NIGERIA

Nigeria is a country located in the western part of the African continent, covering a landmass of approximately 924,000 km^2. It lies between the latitudes 5°N and 15°N and longitudes 5°E and 15°E. Nigeria as a nation has 36 states and the Federal Capital Territory (FCT) which are collectively constituted of 774 Local Government Areas; Solid Waste Management has been identified as one of the major environmental challenges being faced in the country (Adeyinka et al., 2005). Nigeria has a population of about 170 million, an estimated growth rate of 2.6% and is placed 7th only behind China, India, the United States, Indonesia, Brazil and Pakistan on a global population ranking (CIA, 2012, Agbesola, 2013). It is blessed with so many natural resources like crude oil, natural gas, limestone, gold and as many among others. Nigeria became an increasingly urbanized and urban-oriented society because of the oil boom in the 1970s. During the 1970s, Nigeria had possibly the fastest urbanization growth rate in the world (Ochuo 1986; ICMPD 2010). The Figure 1 shows the Nigeria population distribution map of the 2006 census leaving Lagos and Kano to be densely populated among other states.

The impact of urban growth has led to an increasing volume and variety of solid waste, resulting from increased flow of goods and services, and changed lifestyle and consumption pattern. (Solomon 2009, Nabegu 2011) in their research quoted the Federal Ministry of Environment and Housing that the municipal solid waste generation for average Nigerian communities with

8

Table 1. Waste generation in some urban cities in Nigeria.

City	Population	Agency	Tonnage/ month	Density (kg/m^3)	kg/ capita/day
Lagos	8,029,200	Lagos state waste management authority	255,556	294	0.63
Kano	3,348,700	Kano state environmental protection agency	156,676	290	0.56
Ibadan	307,840	Oyo state environmental protection commission	135,391	330	0.51
Kaduna	1,458,900	Kaduna state environmental protection agency	114,443	320	0.58
Port Harcourt	1,053,900	Rivers state environmental protection agency	117,825	300	0.60
Makurdi	249,000	Urban development board	24,242	340	0.48
Onitsha	509,500	Anambra state environmental protection agency	84,137	310	0.53
Nsukka	100,700	Enugu state environmental protection agency	12,000	370	0.44
Abuja	159,000	Abuja environmental protection board	14,785	280	0.66

Source: All sites engineering Ltd. (Ogwueleka, 2009).

household and commercial centres is 0.49 kg/capita/day. Table 1 shows the waste generation in the some major Nigerian cities.

As shown in Table 1, waste density in Nigeria ranges from 280–370 kg/m³ with waste generation rate of 25 million tons annually and at a daily rate of 0.44–0.66 kg/capital/day (Ogwueleka, 2009). Nabegu in 2011 reported that in most cities across Nigeria, solid waste is disposed by transporting and discharging in open dumps location almost close to residential areas, which are environmentally unsafe. The collection from dump sites is the function of state and local government agencies. Informal solid waste collection operations exist in parallel with official agencies in most major cities in Nigeria. There is inadequate service coverage as only limited areas of the cities are covered (Ogwueleka 2003; Nabegu 2009). Ogbonna et al. (2002) also observed that little or no attention is given to some traditional suburban settlements for provision of waste collection and disposal services. The sources of solid waste generation in Nigeria among others are basically from commercial, industrial, household, agricultural and institution of any kind. In 2005, Adewunmi et al. deduced that in Ibadan the capital of Oyo, the total solid waste generated was 66.1% for domestic, 20.3% commercial and 11.4% industrial. Several states in the country are coming up with various means of waste collection procedures which is initiated by both public and private sectors, although the effectiveness of this is largely a function of location; and where the collection is done by private sectors, it is a function of income of the owner of the waste to be able to pay the amount charged (Babayemi and Dauda, 2009). With regards environmental legislation, the unfortunate incident of dumping of toxic waste from Italy at Koko port in the then Bendel state in the 1988 prompted to establish Federal Environmental Protection Agency (FEPA) Act of 1988. FEPA vested with the statutory responsibility for overall protection of the environment. It is in the light of this reality that very interesting strides were taken by Nigeria government decided to establish the Federal Environmental Protection Agency and its inspectorate and enforcement department in 1991 (Nwufo, 2010). In 1999, Nigeria embraced democracy after a long reign of military rule. The need for an environmental policy in the country was adopted in the 1999 constitution of the Federal Republic of Nigeria which states" In accordance with section 20 of the constitution, the state is empowered to protect and improve the environment and safeguard the water, air, land, forest and wildlife of Nigeria. Among all other environmental laws are Waste Management Regulations, National Environmental Protection Regulation, and the Environmental Impact Assessment (EIA) Act.

3 CONCEPTUAL FRAMEWORK FOR SUSTAINABLE MUNICIPAL SOLID WASTE MANAGEMENT

The conceptual framework (see Figure 2) is to remodel the attitude of the Nigerian people towards a successful and sustainable waste management program. In every waste management program,

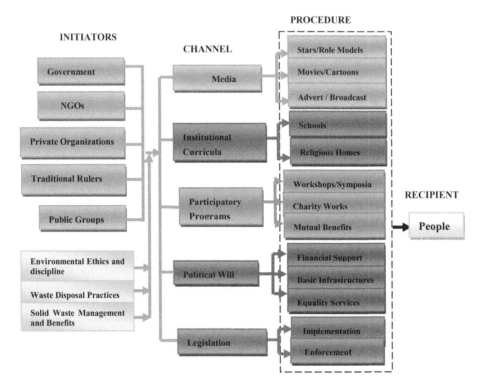

Figure 2.   Concept of waste management sensitization program.

the evacuation phase is very significant. It is imperative to educate the people on the hierarchy of Integrated Waste management program, the health risk and possible benefits that is attainable. This simply the 3Rs refer to the reduction in the amount of waste being generated, the reuse of items prior to their being commissioned as waste, and the recycling of items once they become waste. In the light of this, the framework focuses on educating the people on environmental ethics and discipline, waste management practices and solid waste management and its benefits.

4   DISCUSSION

This concept comprises of 4 major stages (initiators, channel, procedure and the recipient) with the initiators as the principal actors educating the people on environmental ethics and discipline, waste management practices and solid waste management and its benefits. The Figure 2 describes the waste management sensitization program as follows; the initiators comprises of the government, NGOs, Private Organizations, traditional rulers as well as the public groups. The initiators carry out their responsibilities through a set of channel. This set of channel includes the media, institutional curricula, participatory programs, political will and legislation.

The media is one of the best ways of reaching out to the people/public to pass on information because people spend much time using one form of the media service harnessing information and for pleasure. This is done using sport/movie stars and role models to educate people on environmental ethics and discipline on solid waste management through movies, television series, comedy shows, cartoons, animation, video games, social media, music and radio adverts and broadcast, billboards and many more.

Institutional curricula are another smart way of educating the kids, youths and adults in schools (elementary, high and tertiary), orphanage homes, and religion homes on environmental ethics and discipline regarding solid waste and more. This is simply because kids/youths believe strongly in their teachers and tend to reflect whatever they are taught in their daily lives. Again in religious homes, the fear and believe in the words of God has made people practice whatever that is taught for the benefit of man.

Participatory program involves incorporating the people of the society into the sensitization program from the start of the program/project which automatically craves their indulgence on the importance of solid waste management to their environment. In doing this, organizing workshop programs, symposia and talk show, charity work among the recipient (people) goes a long way educating them. Another aspect of this participatory program is the mutual benefit logic which involves companies recovering recyclables materials from homes through institutional arrangement. Schools instruct students to bring certain quantity of plastic bottles in from their homes which attract remuneration on the part of the schools and students.

Political will simply means the government should be committed providing support to the campaign on solid waste management program by providing financial support, employ dedicated youths to educate the people, provide basis infrastructures and assure services provided would be of equal interest irrespective of the individual or set of elite.

Legislation is basically the implementation of the available environmental laws and enforcement with respect to waste disposal and management. People should be aware of such laws and if defaulted, it attracts heavy penalty

Recipient's (people's) mind will be enlightened and refined in terms of perception and attitude towards in environment and particularly solid waste because of constant and continuous education and understanding on environmental discipline and solid waste management in general. With the media disseminating environmental education to the people during official and leisure times, schools and religious homes teaching environmental discipline and norms, citizens are practically involved in environmental participatory programs and projects, and the government fulfilling its goals towards pursuing an environmental friendly society for its people, and finally proper implementation and enforcement of environmental legislation, the people (recipient) will adhere and adjust to the new environmental consciousness of the society.

Since the youths of today are the ambassadors of the future, it is expected that environmental discipline, perception and attitude of the Nigerian people and the world at large on municipal solid waste disposal and management will be passed on to the generations to come.

5 CONCLUSION

Nigeria has a long way to go in the area of environmental education and awareness for the citizens to put off the habit of indiscriminate waste disposal. This problem is also a function of non-implementation and lack of enforcement of the existing environmental laws. This study gives an insight that by incorporating this conceptual framework it into the formal solid waste management structure in Nigeria, a safer, healthier working condition is assured. Public education on solid waste disposal and segregation is of great importance. Municipal solid waste disposal and management is not solely the responsibility of the government but also requires self discipline, change in perception and attitude of the people in order build a sustainable environment. The researcher recommends that government should give priority to solid waste management to prevent environmental disaster in our society. Solid waste management sensitization and environmental education programs should be extended to the rural communities and should be continuous and not to be abolished by any government so that a good solid waste management culture is built and passed on to the coming generation. Solid waste management policies and enforcement of sanitation laws in various arms of the government should be implemented, and the researcher urges all stakeholders to more until a utopian environment in Nigeria becomes a reality.

REFERENCES

Adewumi, I.K., Ogedengbe, M.O., Adepetu, J.A., Fabiyi, Y.L. 2005. Planning Organic Fertilizer Industries for Municipal Solid Wastes Management. *Journal of Applied Sciences Research* 1(3): 285–291.

Agbesola, Y.O. 2013. *Sustainability of Municipal Solid Waste Management in Nigeria: A Case Study of Lagos.* A Published Master's Thesis, Water and Environmental Studies, Department of Thematic Studies, Linkoping University, Sweden.

Agunwamba, J.C., Ukpai, O.K., Onyebuenyi, I.C. 1998. Solid Waste Management in Onitsha, Nigeria *Waste Management and Research* 16(1): 23–31.

Babayemi, J.O., Dauda, K.T. 2009. Evaluation of Solid Waste Generation, Categories and Disposal options in Developing Countries: A case study of Nigeria. *J. Appl. Sci. Environ. Management* 13(3): 83–88.

CIA. 2012. Central Intelligence Agency. *Hämtat från Central Intelligence Agency World Fact Book.* www.cia.gov den 10 May 2012.

Cointreau, S. 1982. *Environmental Management of Urban Solid Wastes in Developing Countries: A Project Guide.* Washington, DC: Urban Development Department, World Bank.

Cunningham, W.P., Saigo, B.W. 1997. *Environmental Science a Global Concern.* McGraw-Hill Company, New York.

International Center for Migration Policy Development (ICMPD) 2010. *Introduction to the Inventory: Mediterranean Transit Migration (MTM): A Dialogue in Action Linking Emigrant Communities for More Development, Inventory of Institutional Capacities and Practices.* Funded by France, Italy, The Netherlands and Switzerland.

IEA, 2005. *CO_2 Emissions from Fuel Combustion 1971.2003.* 2005 Edition. Paris, France: IEA.

Kadafa, A.A., Latifah, A.M., Abdullah, H.S., Sulaiman, W.N.A. 2013. Current Status of Municipal Solid Waste Practise in FCT, Abuja. *Research Journal of Environmental and Earth Sciences* 5(6): 295–304.

Nabegu, A.B. 2008. An Assessment of Refuse Management and Sanitation Board (REMASAB) Solid waste management in Kano metropolis. *Techno–Science Africana Journal* 1: 101–108.

Nabegu, A.B. 2009. Municipal Solid Waste Management in Kano and Katsina cities: A comparative Analysis. *Journal of Geography and Development* 2(2): 97–209.

Nabegu, A.B. 2011. Solid Waste and Its Implications for Climate Change in Nigeria. *Journal for human Ecology* 34(2): 67–73.

Naveen, B.P., Sivapullaiha, P.V., Sitharam, T.G. 2014. Compressibility and Shear Strength of dumped Municipal Solid Waste. *The Journal of Solid Waste Technology and Management*, Issue 4.

NPC, 2012. National Population Commission, Nigeria. Retrieved from www.population.gov.ng. Assessed on the 17th of April, 2015.

Nwufo, C.C. 2010. Legal framework for the Regulation of Waste in Nigeria. *An International Multi-Disciplinary Journal, Ethiopia* 4(2): 491–501.

Ogbonna, D., Ekweozor, K., Ig, E. 2002. Waste management: a tool for environmental protection in Nigeria. *Journal of Environmental Science* 3(1): 55–57.

Ogwueleka, T.C. 2003. Planning Model for Refuse Management. *Journal of Science and Technology* 3(2): 71–76.

Ogwueleka, T.C. 2009. Municipal Solid Waste Characteristics and Management in Nigeria. *Iranian Journal of Environmental Health Science & Engineering* 6(3): 173–180.

Oucho, J. 1986. *Population Studies and Research Institute, Our Common Future. Report of the World Commission on Environment, Public Hearing, Nairobi, 23 Sept. 1986 A/42/427.*

United States Environmental Protection Agency 2005. *Report of the municipal solid waste in the United States: 2005 Facts and Figures.* 5–15.

Saeed M.O., Nasir M.H., Mujeebu, M.A. 2008. Development of Municipal Solid Waste generation and Recyclable components rate of Kualalumpur: Perspective Study Paper presented in *International Conference on Environment (ICENV) Penang, Malaysia.*

Solomon, U.U. 2009. *The State of Solid Waste Management in Nigeria*, Wuhan, Hubei, Department of Environmental Engineering, China University of Geosciences, Wuhan, Hubei.

The Constitution of the Federal Republic of Nigeria 1999. Retrieved from www.vanguardngr.com. Assessed on the 14th of April, 2015.

Wastes: Solutions, Treatments and Opportunities – Vilarinho, Castro & Russo (eds)
© 2015 Taylor & Francis Group, London, ISBN 978-1-138-02882-1

Kinetic study of the pyrolysis of the wastes tire components

N. Akkouche & M. Balistrou
Laboratoire d'Energétique, Mécanique et Ingénieries, Université M'hamed Bougara, Boumerdes, Algérie

K. Loubar & M. Tazerout
Laboratoire GEPEA, UMR 6144 CNRS, Ecole des Mines de Nantes, Nantes, France

ABSTRACT: Waste tire pyrolysis is considered sustainable waste management sector. The main problem of this process is the choice of optimal operating conditions such as temperature and heating rate. During pyrolysis, the tire rubbers are converted into three main products such as liquid, gas and carbon black. In order to study the thermal behavior of used tire, it is useful to know well the behavior of rubbers that constitute it (NR, BR and SBR). In this study, a model with several parallels reactions is used to simulate the conversion of rubber and to determine his degradation kinetic. A software code was developed to determine the kinetic parameters by modeling the conversion peak by a symmetric normal distribution. The chemical kinetics of the pyrolysis of each type of rubber is modeled by two or three parallel reactions. The results obtained have a good agreement with the experimental results (TG/DTG).

1 INTRODUCTION

In waste management, recovery does not involve the production of quality materials or higher utility. The ethics of sustainable development promises the use of organic waste for energy production where the process eliminates waste and saves a fossil resource simultaneously. Indeed, energy waste is a set of methods that reduce waste and transform it into by-products of hydrocarbon. In recent years, the deposit of unserviceable tires is a major environmental problem. Annually, they estimated that 2.5 Million ton are generated in the European Community, 2.5 Million ton in North America, and one Million ton per year in Japan (Isabel (2001)).

Different technologies can be used for energy recovery from tire waste. However, pyrolysis technology is a more sustainable process. This because, by varying the operating conditions, the quantity and/or quality of the product can be directed to a product or to another (Adetoyese (2012), Bridgid (2014), Edwin (2013)).

When submitting the waste tire to heat flow, several thermo-chemical steps take place. The thermally degradation of the tyre starts at around 350°C and therefore pyrolysis experiments are usually in the range of 450–700°C (Williams (2013)). During this temperature range, the organic materials constituting the tire, especially rubbers, are converted into three main products which are oils, gas and carbon fiber. The yields and the chemical compositions of each product depend on the operating conditions, such as temperature, heating rate and residence time of the material at high temperatures, but they also depend on the type of tire and his constituent rubbers.

Williams (1995) studied thermogravimetry (TG/DTG) of three types of tires where the weight fractions of their rubbers are different. They showed that pyrolysis of tire has two steps conversion: the first at low temperatures, due primarily to the conversion of the natural rubber (BR); the second at high temperatures, due to the conversion of the synthetic rubber (SBR and BR). However, Berrueco (2005) note that the thermal conversion of tire is done into three steps:

- Conversion between 200 and 325°C, due to the conversion of the plasticizers, additives and other oils;
- Conversion between 325 and 400°C due to the conversion of natural rubber (NR);
- Conversion between 400 and 500°C due to the conversion of the synthetic rubber (SBR and BR).

To control the thermal behavior of used tires, it is useful to know well the behavior of rubbers that constitute it. There are three main types of rubbers: Natural Rubber (NR), Styrene Butadiene Rubber (SBR) and Butadiene Rubber (BR).

Rubber pyrolysis is the subject of several studies. Bhowmick (1987) found that the pyrolysis of natural rubber (NR) under a heating rate of 10 K.min^{-1} starts from 330°C and ends at 485°C, with a maximum degradation peak at 400°C. Macaione (1988) have shown that the natural rubber (NR) decomposes at temperatures below compared to the synthetics rubbers (SBR and BR). Brazier (1978) have shown that the butadiene rubber (BR) is divided into two stages. Wiliams (1995) found similar results during the heating to 720°C under various heating rate (5, 20, 40 and 80 K.min^{-1}). They show that the pyrolysis of rubber creates a low formation of carbon black (4%). Natural rubber (NR) decomposes at temperatures below 325 to 350°C compared to synthetic rubbers that decompose between 350 and 375°C. Butadiene rubber (BR) has two stages of degradation of which the second is slower than the first one.

This work is interested in the thermogravimetric study of the main rubbers constituting the tires. The kinetic study is based on the TG/DTG results carried out in the laboratory of School of Mines of Nantes. The approach to model the conversion rate peak with a Gaussian distribution has been validated. The DTG peak is assimilated to a sum of several parallel peaks which appear simultaneously. A software to determine the kinetic parameters is developed and the found results are modeled and compared with experimental results.

2 KINETIC STUDY

The kinetic of net rubbers conversion during pyrolysis is modeled by two or three parallel reactions. The approach of the kinetic parameters estimation is based on the TG-DTG results where the DTG curve is divided and treated as a sum of several peaks. Each peak is modeled by a normal (Gaussian) distribution inspired on the theory of probability.

The overall conversion of the sample is considered as a parallel conversion of three components. Each conversion generates only the degradation of a weight fraction of the sample.

$$\alpha^j = \sum_{i=1}^{2or3} \omega_i \alpha_i^j \tag{1}$$

where $\alpha^j =$ overall conversion of the sample, $\alpha_i^j =$ conversion of the component, $\omega_i =$ weight fraction of the component. $i = i$th component, $j = j$th recorded value in the time.

$$\alpha^j = \frac{w^j - w^\infty}{w^1 - w^\infty} \tag{2}$$

(w^j, w^1, w^∞) respectively are instantaneous weight, initial weight and final weight of the sample.

2.1 Peaks separation

The principle of the method is to divide the DTG peaks and to assimilate it as a sum of several (two or three) peaks. Each peak corresponds to a chemical reaction. The derivative of equation (1) gives:

$$\frac{\partial \alpha^j}{\partial t} = \sum_{i=1}^{2or3} \omega_i \frac{\partial \alpha_i^j}{\partial t} \tag{3}$$

To model the conversion rate by a Gaussian distribution, the conversion rate which is given versus time, is rewritten as a function of temperature. When the heating rate β is constant, Equation (4) may be given by:

$$\frac{\partial \alpha_i^j}{\partial t} = \frac{\partial \alpha_i^j}{\partial T} \frac{\partial T}{\partial t} = \beta \frac{\partial \alpha_i^j}{\partial T} \tag{4}$$

where $\beta = 10$°C.min^{-1}.

The modeling of the conversion peak is inspired from the probability and statistics theory where the peak is modeled by a symmetric normal distribution, given by the following equation:

$$f_i(T^j) = \frac{1}{\sigma_i \sqrt{2\pi}} \exp\left(-\frac{(T^j - \mu_i)^2}{2\sigma_i^2}\right) \tag{5}$$

It models the probability density with a position parameter (μ) and a scale parameter (σ). In this case, the position parameter is the temperature of the maximum of degradation while the scale parameter is the temperature interval of degradation.

The fact that the rate conversion is negative and the Gaussian distribution is positive, the distribution function is multiplied by a negative adjustment coefficient ($-H$).

$$\left(\frac{\partial \alpha^j}{\partial t}\right)^{CAL} = -\sum_{i=1}^{2or3} \frac{\omega_i \beta H_i}{\sigma_i \sqrt{2\pi}} \exp\left(-\left(\frac{T^j - \mu_i}{\sigma_i \sqrt{2}}\right)^2\right) \tag{6}$$

The terms of the equation (6), namely ($\omega_i H_i$, μ_i, σ_i) are estimated by the least squares method. The estimation is optimized by minimizing the sum of the squared differences between both experimental $(d\alpha^j/dt)^{EXP}$ and calculated $(d\alpha^j/dt)^{CAL}$ conversion rates.

Using Eq. 3 and Eq. 6, the estimated terms are used to determine the value of $(\omega_i * d\alpha_i^j/dT)$ which models the conversion of component.

$$\omega_i \frac{\partial \alpha_i^j}{\partial T} = -\frac{\omega_i \beta H_i}{\sigma_i \sqrt{2\pi}} \exp\left(-\frac{(T^j - \mu_i)^2}{2\sigma_i^2}\right) \tag{7}$$

To determine the weight conversions (α_i^j) of the components, it is sufficient to integrate the equation (7) and to use the experimental data of the overall conversion $(\alpha^j)^{EXP}$.

$$\omega_i \alpha_i^j = -\frac{\omega_i \beta H_i}{2} erf\left(\frac{T^j - \mu_i}{\sigma_i \sqrt{2}}\right) \tag{8}$$

Knowing that the Gauss Error function (erf) is an even function, the curve of $(\omega_i \alpha_i^j)$ versus temperature is canceled in the abscissa ($T^j = \mu_i$). The change of the variable (scale), given by equation (9), is carried out so that the weight conversion (α_i^j) varies between 1 and 0;

$$\alpha_i^j = \frac{\omega_i \alpha_i^j - \omega_i \alpha_i^\infty}{\omega_i \alpha_i^1 - \omega_i \alpha_i^\infty} \tag{9}$$

where $(\omega_i \alpha_i^j)$ = calculated value at considered temperature, $(\omega_i \alpha_i)$ = calculated value at final temperature, $(\omega_i \alpha_i^1)$ = calculated value at initial temperature.

To proceed to the estimation of kinetic parameters, it is necessary to determine the weight fractions (ω_i) to subtract the conversion rates given by Eq. 7.

2.2 Estimation of the weight fractions

Kknowing that the terms ($\omega_i H_i$) of Eq. 6 are already determined, it is necessary to determine the constants (H_i) to be able to deduct the weight fractions (ω_i).

Using Eq. 8, the conversions (α_i^j) are given by the following equation:

$$\alpha_i^j = -\frac{\beta H_i}{2} erf\left(\frac{T^j - \mu_i}{\sigma_i \sqrt{2}}\right) \tag{10}$$

15

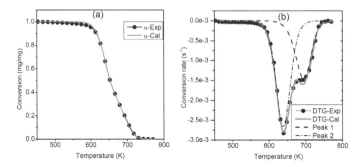

Figure 1. Experimental and calculated results of Natural Rubber conversion.

The constants (H_i) of the Eq. 10 will be estimated by the least square method. The optimization is ensured by minimizing the sum of square differences between the conversions (α_i^j) calculated using Eq. 9 and (α_i^j) calculated using Eq. 10. The estimated values are used to determine the weight fractions (ω_i) and to subtract the conversion rates ($d\alpha_i^j/dT$) given by Eq. 7.

2.3 *Calculation of kinetic parameters*

The conversion rate of component is modeled by the Arrhenius law. It is based on a fundamental hypothesis where the conversion of the component depends only on the current concentration of the component itself. Mathematically, it is translated by Eq. 11.

$$\frac{d\alpha_i^j}{dT} = -\frac{1}{\beta} A_i \exp\left(\frac{-E_i}{RT^j}\right)\left(\alpha_i^j\right)^{n_i} \tag{11}$$

where A_i = Pre-exponential factor, E_i = Energy activation, n_i = reaction order.

This model is used by several authors (Aylon (2005), Sorum (2001), Kim (1995), Leung (1998), Adetoyese (2012)). It is used in the case where the sample consists of one neat component (such as NR, SBR, BR, cellulose, hemicelluloses, lignin, ..., etc.) and is degraded into a single chemical reaction.

The kinetic parameters are estimated using the least square method. The optimization is ensured by minimizing the sum of square differences between the conversion rates ($d\alpha_i^j/dT$) calculated using Eq. 7 and ($d\alpha_i^j/dT$) calculated using Eq. 12.

3 RESULTS AND DISCUSSION

The kinetic parameters determination, which model the conversion of the tire rubbers (NR, SBR, BR), is based on TG/DTG results. Pyrolysis tests were carried out in a thermobalance (SETARAM SETSYS Evolution-1750) equipped with 2000 SETSOFT software. Each sample of about 200 mg, is heated to 900°C in a heating rate of 10°C.min^{-1} and under an inert atmosphere in which it is swept by a flow of 25 ml min^{-1} of nitrogen.

Fig. 1, Fig. 2 and Fig. 3 respectively show the experimental and calculated results of NR, SBR and BR. Figures (1-a, 2-a, 3-a) show the results of the weight conversions. Figures (1-b, 2-b, 3-b) show the results of the conversion rates. The curves of the calculated results are determined using the calculated kinetic parameters. For NR and SBR, the peak of overall conversion is divided into two peaks, which means that the degradation is performed into two parallel reactions. The peak of overall conversion of BR is divided into three peaks, which means that the degradation of the BR is performed into three parallel reactions.

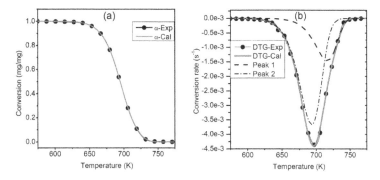

Figure 2.   Experimental and calculated results of Styrene Butadiene Rubber conversion.

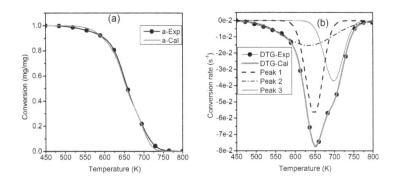

Figure 3.   Experimental and calculated results of Butadiene Rubber conversion.

Table 1.   Kinetic parameters of the tire rubbers pyrolysis.

Rubber	Reaction	A s^{-1}	E kJ·mol^{-1}	n –	ω –
NR	1st	9.59×10^{15}	238	1.5	0.370
	2nd	1.06×10^{17}	231	1.7	0.630
SBR	1st	2.35×10^{17}	264	1.04	0.280
	2nd	6.40×10^{17}	261	1.16	0.720
BR	1st	4.85×10^{8}	140	1.7	0.109
	2nd	2.35×10^{17}	270	1.4	0.802
	3rd	6.94×10^{24}	310	1.1	0.089

Table 1 summarizes the determined kinetic parameters of the reactions that govern the pyrolysis of each type of rubber. It also illustrates the weight fraction generated by each reaction.

The modeling of the rubbers pyrolysis using the determined kinetic parameters allows to calculate the evolution of the weight conversion during pyrolysis under the heating rate of $10°C.min^{-1}$. The numerical results presented graphically and superimposed on the experimental results (Figures above) show a good agreement between theory (simulation) and practice (experimental). This can be strengthened by the fact that the convergence criterion, given by the sum of square differences between the conversion rates, is $5.84 \cdot 10^{-6}$; $2.73 \cdot 10^{-6}$; and $3.84 \cdot 10^{-5}$ respectively for NR, SBR and BR. The obtained parameters are the same range than other published works (Williams (1995), Danon (2015), Conesa (1996)).

4 CONCLUSION

The complexity of the tires composition rejects all the simplifying assumptions that encourage the use of non-parallel reaction models to model the pyrolysis of tires and its constituents. To move forward and propose coherent models, it is very useful to know the reactions mechanism of pyrolysis of each component. Thermogravimetric analysis and mastery of his interpretation leads to predict at the outset the number of reactions and the reaction mechanism that governs the weight conversion.

The reaction orders governing the pyrolysis of tire rubbers are greater than unity and are between 1.04 and 1.7. They are of the same order as the values in the literature where the conversion of the tire rubbers into several parallel reactions is considered. The activation energies are in good agreement with the literature and physics. They are lower than the binding energies of the rubber components. They are between 230 and 240 kJ mol^{-1} for the NR, between 260 and 270 kJ mol^{-1} for the SBR and between 140 and 310 kJ mol^{-1} for the BR.

The superposition of the simulation results on the experimental results shows good agreement between theory (simulation) and practice (experimental). This supports the use of these results to model the pyrolysis of tires constituents. However, the modeling of used tires need to take into account all the components of the tire, including sulfur which acts as a liaison between the polymer chains NR, SBR and BR. The good agreement of the results leads to conclude that the modeling of the conversion peak, modeled by a symmetric normal distribution, is very acceptable.

REFERENCES

Adetoyese, O., Ka-Leung. L., Malte, F. & Chi-Wai, H. 2012. Optimisation of particle size in waste tyre pyrolysis. *Fuel*. 95: 417–424.

Aylon, E., Callen, M.S., Lopez, J.M., Mastral, A.M., Murillo, R., Navarro, M.V. & Stelmach, S. 2005. Assessment of tire devolatilization kinetics. *Journal of Analytical and Applied Pyrolysis*. 74: 259–264.

Berrueco, C., Esperanza, E., Mastral, F.J., Ceamanos, J. & Garcia-Bacaicoa, P. 2005. Pyrolysis of waste tyres in an atmospheric static-bed batch reactor: Analysis of the gases obtained. *Journal of Analytical and Applied Pyrolysis*. 74: 245–253

Bhowmick, A.K., Rampalli, S., Gallagher, K., Seeger, R. & McIntyre, D. 1987. The degradation of guayule rubber and the effect of resin components on degradation at high temperature. *Journal of Applied Polymer Science*. 33(4): 1125–1139.

Brazier, D. W. & Schwartz, N. V. 1978. The effect of heating rate on the thermal degradation of polybutadiene. *Journal of Applied Polymer Science*. 22(1): 113–124.

Fui Chin, B.L., Yusup, S., Al Shoaibi, A., Kannan, P., Srinivasakannan, C. & Sulaiman, S.A. 2014. Kinetic studies of co-pyrolysis of rubber seed shell with high density polyethylene. *Energy Conversion and Management*. 87:746–753.

Conesa, J.A. & Marcilla, A. 1996. Kinetic study of the thermogravimetric behavior of different rubbers; *Journal of Analytical and Applied Pyrolysis*. 37: 95–110.

Danon, B., Mkhize, N.M., Van Der Gryp, P. & Görgens, J.F. 2015. Combined model-free and model-based devolatilisation kinetics of tyre rubbers. *Thermochimica Acta*. 601: 45–53.

Edwin Raj, R., Robert Kennedy, Z. & Pillai, B.C. 2013. Optimization of process parameters in flash pyrolysis of waste tyres to liquid and gaseous fuel; *Energy Conversion and Management*. 67:145–151.

Isabel, D.M.R, Laresgoiti, M.F, Cabrero, M.A., Torres, A., Chomon, M.J. & Caballero, B. 2001. Pyrolysis of scrap tyres, *Fuel Processing Technology*. 72: 9–22.

Kim, B.S., Park, J.K. & Chun, H.D. 1995. Pyrolysis kinetics of scrap tire rubbers. I: using dtg and tga. *Journal of Environmental Engineering*. 121: 507–514.

Leung, D.Y.C. & Wang, C.L. 1998. Kinetic study of scrap tyre pyrolysis and combustion, *Journal of Analytical and Applied Pyrolysis*. 45: 153–169.

Macaione, D.P., Sacher, R.E. & Singler, R.E. 1988. In 'Compositional Analysis by Thermogravimetry' (Ed. C. M. Earnest), *American Society for Testing and Materials*. Philadelphia, PA.

Sorum, L., Gonli, M.G. & Hustad, J.E. 2001. Pyrolysis characteristics and kinetics of municipal solid wastes. *Fuel*. 80: 1217–1227.

Williams, P.T. 2013. Pyrolysis of waste tyres : A review, *Waste Management*. 33: 1714–1728.

Williams, P.T. & Besler, S. 1995. Pyrolysis-thermogravimetric analysis of tyres and tyre components. *Fuel*. 74(9): 1277–1283.

Wastes: Solutions, Treatments and Opportunities – Vilarinho, Castro & Russo (eds)
© 2015 Taylor & Francis Group, London, ISBN 978-1-138-02882-1

Valorization of inorganic waste, end of waste and by-product for ceramics

L. Barbieri, F. Andreola, R. Taurino, I. Lancellotti & C. Leonelli
Department of Engineering "Enzo Ferrari", University of Modena and Reggio Emilia, Modena, Italy

ABSTRACT: Recycling and re-use of waste are economically attractive options for public and private actors due to widespread separate collection and the development of functional markets for secondary raw materials. The ceramic sector is particularly sensitive to this setting through re-entry into the production cycle of all by-products and partially amounts of residues derived from depuration treatment process. Besides, thanks to academic and industrial research results, there is the tendency to see different inorganic wastes, "end of wastes" and by-products as new good candidates. In the present work, two case studies, one describing ceramics with high amount of glass waste and one ceramics with different kind of wastes, end-of-wastes and by-products, are reported with a commented list of technological and environmental benefits.

1 INTRODUCTION

Wastes European Directives, in addition to reporting specific criteria and targets, gave the impression of a new cultural attitude that offers the combination of sustainable use of resources coupled with sustainable management of waste. Raw material recovery at local level is also supported because only a few countries have economically viable stocks of raw materials. It becomes so important the concept of Urban Mining that, in its broadest sense, represents the actions and technologies that can be put in place to recover the resources produced by urban catabolism (civil residues, industrial, agricultural, ...) in terms of secondary raw materials and energy.

The latest figures from Eurostat (ISPRA, 2013) on the production of hazardous and non-hazardous waste in the European Union in 2010 estimated a total quantity of approximately 2,515 million tons of which about 102 million are dangerous. The valorization of such wastes as well as end-of-waste and by-products must therefore become a prime target.

End-of-waste criteria specify when a certain waste ceases to be waste and obtains a status of a product (or a secondary raw material). According to Article 6 (1) and (2) of the Waste Framework Directive 2008/98/EC, certain specified wastes shall cease to be waste when they have undergone to a recovery (including recycling) operation and complies with specific criteria to be developed in line with certain legal conditions, in particular:

- the substance or object is commonly used for specific purposes;
- there is an existing market or demand for the substance or object;
- the use is lawful (substance or object fulfils the technical requirements for the specific purposes and meets the existing legislation and standards applicable to products);
- the use will not lead to overall adverse environmental or human health impacts.

By-product is a substance or object which results from a production process and its main aim is not the production of that item. By-products can come from a wide range of business sectors, and can have very different environmental impacts.

A lot of inorganic recovery materials can become useful candidates for clay-based ceramic products which require massive amount of natural raw materials (based mainly on the traditional clay-silica-feldspar system). They show a wide range of compositional variations and the resulting products are very heterogeneous.

Regarding ceramic tiles processing, several studies made in the last decades are related to the substitution of conventional raw materials by other natural resources as zeolites (de Gennaro et al., 2007; de Gennaro et al, 2003), volcanic rocks (Carbonchi et al., 1999; Ergul et al., 2007) and nepheline syenite (Salem et al., 2009). Besides, the quartz substitution by Si-rich waste as rice husk ash and silica fume was studied in whiteware compositions with good results concerning the reduction of sintering temperature (~50°C) (Prasad et al., 2003). Other works have investigated the use of different alternative fluxing agents (soda–lime glass cullet (Tucci et al., 2004), cathode ray tube of TV or PC monitor (CRT glass) (Andreola et al., 2008; Andreola et al., 2010) and granite cutting sludge (Torres et al., 2004)) substituting partially the feldspars to obtain sintering behaviour and mechanical characteristics similar to the industrial compositions. In general, the addition of higher amount of fluxes leads to an increase of amorphous phase in the final product, which can have negative effect on the mechanical properties.

In the present work, two case studies of recovery and valorization of matter will be presented.

2 CASE STUDIES

2.1 *Case study 1: Use of huge amount of glass waste (>60%) in new ceramic products*

The present study describes the feasibility of formulating novel ceramic body compositions containing high percentage (>60 wt%) of glass waste deriving from different chains of collection. Material was prepared by using a specific particle sizes of glass waste and other components as inorganic plastifier, and organic binder. The interesting results obtained permitted to patent the application (Leonelli et al., 2013). Some advantages of the invention can be listed below:

- Thanks to the very high amount of recycled glass used and the low percentage of water added (less than the amount usually necessary for conventional ceramic slurries) it is not necessary to use high energy mixing equipment, such as turbo mixer, while simple mechanical mixers are sufficient.
- The plastic behaviour of the final mixture gives rise to a paste easy to work not only by pressing, but also by lamination and also by hand. Such workability allows the production of both sheets – with even very thin thicknesses (down to 5 mm) – and artistic furnishing products.
- The consolidation step reaches temperatures (≤1000°C) significantly below glass melting (1450°C) and even below the average temperature adopted for highly sintered ceramic tiles treatments (up to 1250°C). In addition, the firing times can be reduced till being comparable with those used in industrial fast firing cycles of tiles (≤60 min).
- The benefits in terms of safety and environmental compatibility are evident and numerous, thanks to presence of a high amount of recycled glass in place of raw materials and the elimination of other traditional components (flux and inert materials), whose use involves environmental impacts associated with their extraction and transport.

All the advantages listed above allow the achievement of an environmentally sustainable material that can contribute to fabricate finished products with easy access to both ECOLABEL and LEED certifications. Besides, the final products can be inserted in the policy of Green Public Procurement.

The investigation was realized using packaging glass waste (PGW) and end-of-life fluorescent lamp glass (FLG). The packaging glass waste received after the separation process is polluted, contains around 2–3 wt% of impurities determined by sieving and gravimetry. Its average chemical composition (Table 1) is similar to that of typical common soda-lime glasses. The difference in terms of chemical composition between the common glass (CM) and the packaging glass waste (PGW) is mainly due to the presence of impurities such as plastic, ceramics/porcelain, rubber, paper, etc.

Table 1. Chemical compositions (wt%) of the glass wastes compared to common glasses.

Oxide wt%	Common glasses (CM)	Packaging glass waste (PGW)	Fluorescent lamp glass waste (FLG)
SiO_2	68–74	71.70	67.9
Al_2O_3	0–4	2.25	2.26
Na_2O	10–16	12.50	17.50
K_2O	0–4	1.00	1.60
CaO	9–14	9.50	5.09
MgO	0–4	2.00	2.96
BaO	–	0.04	0.94
SrO	–	0.37	0.07
PbO	–	–	0.78
Fe_2O_3	0–0.45	0.43	0.08
TiO_2	–	0.07	0.002
others	–	–	0.086
L.O.I (1100°C)	–	0.2–2.5	–

Also the chemical composition of FLG, coming from a Waste of Electrical and Electronic Equipment treatment plant, is similar to a soda-lime glass, but compared to PGW contains a higher amount of alkali oxides (~19%) and lower amount of silica and alkaline earth oxides, in particular, calcium oxide and only traces of Fe_2O_3 chromophore. Besides, small amount of PbO derived from coating and low-melting frit used to seal the lamp to electric connector. The predominance of alkali oxides compared to the alkaline-earth and a low alumina content, suggest a low tendency to crystallization for this glass and characteristic temperatures of low-melting glass.

From the thermal properties recorded by the optical heating microscopy, it appears evident a longer stability range, from sintering to softening point, for FLG (682–826°C) with respect to PGW (730–837°C). Then, within this range the material sinters without melting and preserves its shape. Beyond this point the glass powder begins to soften and to behave like a liquid and the surface tension decreases, the surface area goes to a minimum, causing the formation of a sphere and then hemisphere and finally the melting or flow point. All these characteristic temperatures for PGW are higher than 50–80°C with respect to FLG, in full agreement with the different chemical composition. These considerations, together with sintering curves obtained by optical dilatometry on the green batch, permits to predicte different behavior of these glasses during the firing process and identify the temperatures where the sintering rate is maximum in order to use a suitable firing cycle.

The obtained products have been characterized in accordance with ISO technical rules. The sintering parameters of fired samples, obtained by lamination and pressing (40 MPa) forming, were evaluated. Linear shrinkage and water adsorption indicate that the higher densification degree depends both on glass particle size and pressure applied. From the data reported in Table 2 it is possible to observe that the materials have mechanical properties comparable to those of traditional tile ceramics. These results are consistent with tests of porosity, density and water absorption (WA%).

In particular, when in the ceramic batch, fluorescent lamp glass (<400 μm) and packaging glass waste (<100 μm) are used the obtained material could be classified into the BIa group (WA% < 0.5) according to ISO rules, which corresponds to high sintered floor ceramic tiles, demonstrating the positive role of the glass in the sintering process. Besides, the mechanical properties of FLG (<400 μm) sample are higher than the minimum imposed by ISO rule (35 MPa) for BIa group.

Concerning the material surface properties, the sheets were covered with a crystalline glaze created specifically for the product in order to obtain a good dilatometric agreement. This coating allowed to reach the highest classes for both acid and stain resistances. In addition, the product has

Table 2. Sintering parameters and mechanical properties for fired samples (1000°C, 60 min) obtained by pressing (40 MPa).

Glass waste	Linear srinkage (LS%)	Water absorption (WA%) (EN ISO 10545-3)	Bending strenght (MPa) (EN-ISO 10545-4)
FLG (<400 μm)	8.79	0.02	40.93
PGW (<400 μm)	4.87	1.83	23.37
PGW (<100 μm)	5.46	0.29	30.60

the capability to develop very interesting colours and chromatic effects when added with inorganic pigments, without alter its aesthetic surface properties.

The final products obtained with recycled glasses formulation can be proposed in different fields: building materials, artistic ceramics, furniture industry and ceramic tiles. The products are distinguished from those already existing for the highest percentage of recycled glass used and for the versatility forming shaping processes. Changing the organic binder, solid content and other minor components specific of each forming technique, it is possible to obtain these ceramic materials by uniaxial-pressing, lamination or slip casting.

2.2 Case study 2: Use of different kinds of wastes, end of wastes and by-products

A first category of waste that can be recognized as candidate for end-of-waste are the Municipal Solid Waste Incinerator (MSWI) bottom ash, codified in Italy as non-hazardous waste. When appropriately treated by authorized plants, MSWI acquires an economical value and becomes an "artificial aggregate". A commercial ceramic glaze composed by both olivine (magnesium iron silicate, $((Mg,Fe)_2SiO_4)$ and commercial frits, rich in lead (about 30 wt%), was reformulated by using secondary raw materials (CRT cone glass and MSWI post-treatment bottom ash before and after vitrification). The waste-based products were characterized and, compared to the standard glaze, showed better acid resistance, comparable aesthetic characteristics and slightly lower stainless resistance. Environmental benefits were obtained by saving natural raw material (olivine), by reducing lead percentage in the proposed formulations (from around 30 to 5 wt%), by energy saving (for the avoided use of commercial frits) and by reducing lead content in the new compositions (Schabbach et al., 2011). Another ceramic glaze was successfully obtained by substituting more than 80 wt% of the glassy component (frit) of a commercial double firing ceramic glaze. A decrease in the environmental impact was also demonstrated by LCA method with respect to the commercial glazed obtained starting by virgin raw materials (Andreola et al., 2007a).

Even glazing and polishing ceramic sludges were used as raw materials in commercial formulations of bricks and tiles using traditional thermal cycles. 0–20 wt% polishing sludges-added bricks showed aesthetical (effluorescences, colour) and technological (water absorption, weight loss, shrinkage, compressive resistance) properties comparable to those of the standard, within the limits of industrial tolerance. The partial substitution of Na-feldspars with polishing sludge in porcelain stoneware ceramic body produces a decrease of the sintering temperature maintaining unchanged the final product properties. Moreover, the addition of the same waste into a single firing ceramic body, by decreasing the porosity improves the mechanical resistance of the tiles (Andreola et al., 2006).

Fluorescent lamps or cathode ray tubes glass and packaging scrap glass has been used to produce glass foams. Other wastes or by-products, egg shells and cutting sludge from treatment of glass polishing (containing $CaCO_3$ in percentage of 95–97 and 7, respectively) were used as foaming agents. These last, introduced in the matrix in media from 1 to 5 wt%, permitted the obtainment of glass foams with density values from 0.24 to 0.69 g/cm^3, porosity (81–87%) and presenting interesting mechanical properties such as compressive strength up to around 15 MPa for samples obtained at temperatures from 650 to 800°C for 45, 30 or 15 minutes (Andreola et al., 2012; Fernandes et al., 2013).

Lastly, rice husk ash (RHA) ranging from 5 to 20 wt% was used as silica resource mixed with an industrial clay mixture and 20% of water to produce bricks. Laboratory brick samples were obtained by extrusion, and were fired in an industrial kiln with a thermal cycle (24 hrs total time, Tmax: 960°C and 6 hrs soaking time). The dried and fired specimens were characterized following the technical rules and compared to the standard one (clayed materials). This research has revealed that RHA behaves as a raw materials with high silicate content which have mainly plasticity reducing effect on the brick bodies. RHA is compatible in the formulation of bricks up to 5 wt% while not worsening the mechanical and structural properties compared to the standard one. For that percentage, the addition of RHA results satisfactory from mechanical point of view and the data obtained are in accordance to the recommended values for floor (10 MPa). Bricks containing higher amounts of RHA could be used in building manufacturing (light weighted faced load bearing walls) where moderate strengths and penetration protection (porosity/permeability) are required (Andreola et al., 2007; Bondioli et al., 2010).

REFERENCES

Andreola, F., Barbieri, L., Lancellotti, I, Piccagliani, V., Rabitti, D., Esposito, L., Rambaldi, E., Beneventi, M. C., Medici, C. & Rompianesi, P. 2006. Treatment and valorisation of polishing and glazing ceramic sludge. *Innovative technologies and Environmental Impacts in Waste Management*, pp. 152–155, Ed. L. Morselli et al., Maggioli Editore. ISBN:88-387-3645-6.

Andreola, F., Barbieri, L., Corradi, A., Ferrari, A.M., Lancellotti, I. & Neri, P. 2007a. Recycling of EOL CRT glass into ceramic glaze formulations and its environmental impact by LCA approach. *International Journal of LCA*, 12(6), 448–454.

Andreola F., Barbieri L., Bondioli, F., Ferrari A.M. & Manfredini, T. 2007b. Valorization of Rice Husk Ash as Secondary Raw Material in the Ceramic Industry. *Proc. 10th ECerS Conf.*, Göller Verlag, Baden-Baden, 1794–1798.

Andreola, F., Barbieri, L., Karamanova, E., Lancellotti, I. & Pelino, M. 2008. Recycling of CRT panel glass as fluxing agent in the porcelain stoneware tile production. *Ceramic International*, 34(5), 1289–1295.

Andreola, F., Barbieri, L., Bondioli, F., Ferrari, A. M., Lancellotti, I. & Miselli, P. 2010. Recycling screen glass into new traditional ceramic materials. *International Journal of Applied Ceramic Technology*, 7(6), 909–917.

Andreola, F., Barbieri, L., Giuranna, D. & Lancellotti, I. 2012. New eco-compatible materials obtained from WEEE glass, in Atti The ISWA world solid waste congress.

Bondioli, F., Barbieri, L., Andreola, N.M. & Bonvicini, M. 2010. Agri-food waste: an opportunity for the heavy clay sector. *Brick World Review*, 1, 34–40.

Carbonchi, G., Dondi, M., Morandi, N. & Tateo, F. 1999. Possible use of altered volcanic ash in ceramic tile production. *Industrial Ceramic*, 19, 67–74.

de Gennaro, R., Dondi, M., Cappelletti, P. & Cerri, G. 2007. Zeolite-feldspar epiclastic rocks as flux in ceramic tile manufacturing. *Mesoporous and Microporous Materials*, 105, 273–278.

de Gennaro, R., Dondi, M., Cappelletti, P. & Cerri, G. 2003. Influence of zeolites on the sintering and technological properties of porcelain stoneware tiles. *Journal of European Ceramic Society*, 23, 2237–2245.

Ergul, S., Akyildiz, M. & Karamanov, A. 2007. Ceramic material from basaltic tuffs. *Industrial Ceramic*, 37, 75–80.

Fernandes, H.R., Andreola, F., Barbieri, L., Lancellotti, I., Pascual, M.J. & Ferreira, J.M.F. 2013. The use of egg shells to produce Cathode Ray Tube (CRT) glass foams. *Ceramic International*, 39(8), 9071–9078.

ISPRA – National Report of Municipal Waste 2013. Roma, Italy.

Leonelli, C., Barbieri, L., Andreola, F., Reggiani, E. & Ingrami, M. 2013. Materiale a base vetrosa per la produzione di manufatti ceramici e metodo per la sua preparazione. Industrial Patent 0001404410. Italy.

Prasad, C.S., Maiti, K.N. & Venugopal, R. 2003. Effect of the substitution of quartz by RHA and silica fume on the properties of Whiteware compositions. *Ceramic International*, 29, 907–1014.

Salem, A., Jazayeri, S.H., Rastelli, E. & Timellini, G. 2009. Dilatometric study of shrinkage during sintering process for porcelain stoneware body in presence of nepheline syenite. *Journal of Material Processing Technology*, 20, 1240–1246.

Schabbach, L.M., Andreola, F., Lancellotti, I. & Barbieri, L. 2011. Minimization of Pb content in a ceramic glaze by reformulation the composition with secondary raw materials. *Ceramic International*, 37, 1367–1375.

Torres, P., Fernandes, H.R., Agathopoulos, S., Tulyaganov, D.U. & Ferreira, J.M.F. 2004. Incorporation of granite cutting sludge in industrial porcelain tile formulations. *Journal of European Ceramic Society*, 24(2004), 3177–3185.

Tucci, A., Esposito, L., Rastelli, E., Palmonari, C. & Rambaldi, E. 2004. Use of soda-lime scrap-glass as a fluxing agent in a porcelain stoneware tile mix. *Journal of European Ceramic Society*, 24, 83–92.

Wastes: Solutions, Treatments and Opportunities – Vilarinho, Castro & Russo (eds)
© 2015 Taylor & Francis Group, London, ISBN 978-1-138-02882-1

Use of different inorganic solid wastes to produce glass foams

A.R. Barbosa, A.A.S. Lopes & R.C.C. Monteiro
Department of Materials Science, CENIMAT/I3N, Faculty of Sciences and Technology,
Universidade Nova de Lisboa, Caparica, Portugal

F. Castro
Department of Mechanical Engineering, University of Minho, Guimarães, Portugal

ABSTRACT: Cathode Ray Tube (CRT) waste glasses produced from dismantling TV sets were used to prepare glass foams by a simple and economic processing route, consisting of a direct sintering process of mixtures of CRT waste as glass powder with different foaming agents (coal fly ash and limestone quarrying residues). The influence of firing temperature, amount and type of foaming agent on the apparent density, pore size distribution and compressive strength have been studied. The aim of the work was to investigate the possibility to obtain glass foams using exclusively wastes as starting materials, and therefore replacing the conventional raw materials.

1 INTRODUCTION

In Europe, about 7.5 million tons of waste of electrical and electronic equipment (WEEE) are produced annually, which corresponds to about 4% of the urban solid waste flow (Andreola et al. 2007). Nowadays, Cathode Ray Tube (CRT) glasses represent about one third of electronics waste tonnage. From the total weight of a computer monitor or a television set about 65% corresponds to CRT that contains hazardous and heavy elements such as lead, cadmium or mercury (Bernardo & Albertini 2006, Fernandes et al. 2013). The collected monitors are usually dismantled and treated in order to achieve correctly the removal of any hazardous materials or components and to recycle to the maximum (Mear et al. 2006). However, recycling of CRT glasses remains limited and a large proportion of CRT glasses is still being landfilled (Fernandes et al. 2013). Hence, existing recycling methods have to be improved or a new recovery method has to be developed (Petersen & Yue 2014).

Foam glass production, supply, sales, market status and forecast at international level (Asian, European and North American regions) have deserved a great attention, and information is provided in specialized literature (Market Publishers 2014, Orbis Research 2015). The present work is focused on glass foam production as it is one of the most promising recycling technologies for CRT glasses. Glass foams exhibit excellent fire resistance (Petersen et al. 2014), thermal and acoustic insulation properties (Bernardo & Albertini 2006), and may be produced with low densities (Wu et al. 2006). Commercial glass foams exhibit porosity, apparent density and compressive strength values of about 45–85 vol.%, 0.1–1.2 g cm^{-3} and 0.4–6 MPa, respectively (Vancea & Lazău 2014). In general, the amount of added foaming agent is much higher than theoretically required, $\sim$1 wt% (Chen et al. 2011). Several studies have been reported in literature regarding the preparation of glass foams from CRT glasses with addition of different amounts of foaming agents (Francis & Rahman 2014, Blengini et al. 2012). Silicon carbide (SiC) is the most commercially used foaming agent, but calcium sulphate and calcium and sodium carbonates have also been tried (Vancea & Lazău 2014). In the present investigation, two different types of solid wastes have been used as foaming agents to produce glass foams from CRT glasses: fly ashes from coal-fired power plants (FA), and calcite (CaCO$_3$) from limestone quarrying residues (CA). It was aimed to investigate the feasibility of using such solid wastes as foaming agents, to evaluate the evolution of the foaming process during the thermal treatment and to characterize the mechanical strength of glass foams produced from CRT glass. Moreover, it was aimed that the obtained results can contribute to the optimization of processing conditions, to the understanding of the evolution of the porosity and of the pore size distribution and to establish a relationship between these microstructural characteristics and the compressive strength of the developed materials.

2 MATERIALS AND METHODS

Glass powder from CRT waste was obtained by wet milling with distilled water, using an agate ball mill (Pulverizette; Fritsch) rotating during 3 h at a constant speed of 400 rpm. After drying, the powder was sieved to obtain a glass powder fraction smaller than 63 μm that was used along the experimental work. This particle size is comparable to that used in the foam glass industry (Petersen 2014). Both foaming agents, calcite ($CaCO_3$) from limestone quarrying (CA) and fly ashes from coal-fired power plant (FA), were also milled and sieved under the same conditions, and the size fractions below 63 μm were collected. The particle size was determined by a laser diffraction analyzer (Malvern, Mastersizer Hydro 2000MU). The powder density (ρ_{as}) was measured by a helium pycnometer (Micrometrics, Accupyc 1330).

The oxides present in CRT waste glass were (wt%): SiO_2 (46), PbO (27), K_2O (8), Na_2O (46), CaO (6), Al_2O_3 (2.8), BaO (2.8), SrO (1.6), MgO (1.5) and some minor oxides, the content of each one being less than 0.5 wt%. The glass transition temperature (T_g) of CRT waste glass was determined by differential thermal analysis (DTA), using a DTA equipment (Linseis, STA PT1600), and a value of about 550°C was determined at a heating rate of 5°C/min. CA was mainly composed by $CaCO_3$ and exhibited a weight loss on ignition of about 50% due to carbonate dissociation. FA was mainly constituted by (wt%): SiO_2 (44), Al_2O_3 (19), Fe_2O_3 (6), K_2O (1.7), other minor oxides present as less than 1 wt%, and exhibited a weight loss on ignition of about 16% due to the presence of unburnt coal.

The CRT glass powder was mixed for 3h with CA or with FA (Turbula WAB, T2F). The proportions of foaming agent were 2, 5 and 8 wt%, based on previous studies (Bernardo & Albertini 2006, Francis & Rahman 2014). Cylindrical powder compacts ($\approx$6 mm height, 20 mm diameter) were prepared from the different CRT waste glass/foaming agent compositions by uniaxial pressing the powders under a compressive stress of 80 MPa. The samples were treated during 30 min at different temperatures, in the range 600 to 800°C, using a heating rate of 5°C/min in an electric muffle furnace.

Apparent density (ρ_a) was calculated according to the standard procedure (ASTM C20-83, vol.15.01, 1985), using the dry weight of the sample and the suspended weight in water. The total porosity (P) of each specimen has been calculated from the apparent density (ρ_a) and absolute density (ρ_{as}) according to Eq. 1:

$$P\ (\%) = \left(1 - \frac{\rho_a}{\rho_{as}}\right) 100 \qquad (1)$$

The morphology of the pores in the produced glass foams was investigated by optical microscopy (Olympus-BX51 TRF), and the pore size distribution was evaluated using ImageJ software analysis. Compression tests were performed in specimens with an average dimension of 10 mm × 10 mm × 10 mm, using a universal testing machine (Shimadzu AG) with a load cell of 50 kN and a crosshead speed of 0.5 mm/min.

3 RESULTS AND DISCUSSION

3.1 *Apparent density and porosity*

For a successful foaming process it is essential that the gases are released in the sintered glass body when the porous structure is already closed. Subsequently, the gases can be confined in the glass

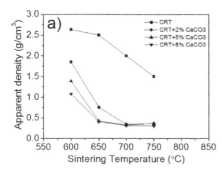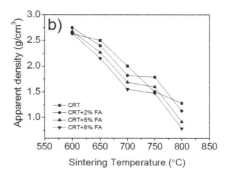

Figure 1. Apparent density versus sintering temperature for CRT waste-based glass foams produced with different amounts of added foaming agent; (a) calcite, (b) coal fly ash.

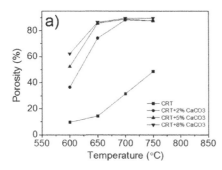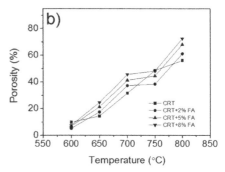

Figure 2. Total porosity versus sintering temperature for CRT waste-based glass foams produced with different amounts of added foaming agent; (a) calcite, (b) coal fly ash.

matrix. In the present study, although both foaming agents are able to produce glass foam by the release of CO_2, two different ways of foaming are compared. FA is an oxidative foaming agent, as the residual non-burnt coal reacts during the firing of the samples according to the following reaction C (s) $+$ O_2 (g) $\rightarrow$ CO_2 (g), while CA is a dissociative foaming agent, reacting according to $CaCO_3$ (s) $\rightarrow$ CaO (s) $+ CO_2$ (g). Both chemical reactions are activated by temperature. The decomposition of $CaCO_3$ in an air atmosphere starts close to 630°C, and the oxidation of the carbon from fly ashes begins about 450°C. The sintering temperatures (in the range 600–800°C), applied to the samples prepared from CRT glass/CA and CRT glass/FA mixtures, have been selected taking into account the temperatures needed for the occurrence of the above reactions and the glass transition temperature of the CRT glass ($\sim$550°C). The influence of sintering temperature and of foaming agent type and concentration on the apparent density (ρ_a) and total porosity (P) of the produced glass foams is shown in Figure 1 and Figure 2.

For glass foams produced with $CaCO_3$ addition, the apparent density decreased rapidly from 600 to 650°C, and for these materials the density could be reduced from 1.80 to 0.30 g/cm^3 with increasing treatment temperature. CRT glass-CA compositions heated at 650°C exhibited a porosity from $\sim$70 up to $\sim$85%, while samples made from CRT glass alone when heated at 650°C showed a porosity of only $\sim$15%. Comparatively, glass foams produced with added fly ash showed a slower tendency to decrease in apparent density with sintering temperature, reaching an apparent density smaller than 1.5 g/cm^3 when heated at 800°C, and only after heating at this temperature, a porosity of $\sim$70% could be attained.

Figure 3 presents a photograph of CRT glass foam samples obtained with calcite and with fly ash addition (5 wt%) after sintering at different temperatures. Figure 4 presents a photograph of

Figure 3. Photograph of the CRT glass foams produced with 5 wt% addition of calcite (above) and fly ash (below), sintered at different temperatures: (a) 650°C; (b) 700°C; (c) 750°C; (d) 800°C.

Figure 4. Foam glass structures resulting from treatments at different temperatures; (a) 650°C–5 wt.% CaCO$_3$ (b) 700°C–5 wt.% CaCO$_3$ (c) 750°C–5 wt.% fly ash, (d) 800°C–5 wt.% fly ash.

some samples after cutting with a diamond saw in order to show the internal surfaces. For identical processing conditions, a higher porosity in CRT-CA samples is observed, which is attributed to the larger amount of CO$_2$ produced during the dissociative reaction of CaCO$_3$ comparatively to the amount of CO$_2$ resulting from the oxidative reaction of the residual carbon in the fly ashes.

One of the objectives is to obtain the maximum porosity for a higher compressive strength. The glass foams produced with the various CA additions (2–8 wt%) exhibited a porosity higher than 70% when sintered at T $\geq$ 650°C. At temperatures between 650–750°C, CRT-FA samples presented a porosity varying between 20 and 50%, and the apparent density was approximately 1.5 g/cm^3. Porosity values higher than 60% were observed for glass foams produced with 2–8 wt% of fly ash, only when sintered at the highest temperature (800°C).

3.2 Mechanical strength

The mechanical resistance of the various produced glass foams was evaluated, and the mechanical strength values for samples obtained under specific experimental conditions (sintering temperature and foaming agent content) are presented in Figure 5 and Figure 6 for CRT-CA and CRT-FA compositions, respectively. In general, the foams produced with calcite addition (Figure 5) have a compressive strength lower than the foams produced with fly ash (Figure 6). This behavior can be associated to the lower apparent density and consequently the higher total porosity of CRT glass-CA compositions than CRT glass-FA compositions sintered under similar conditions.

According to the obtained results, for the foams produced with calcite addition the highest compressive strength values (160–180 kPa) correspond to the samples containing 2 wt% CaCO$_3$ sintered at 700°C and 750°C and to samples containing 5 wt% CaCO$_3$ sintered at 600°C, followed by samples with 8 wt% CaCO$_3$ ($\sim$150 kPa) sintered at 750°C. These processing conditions resulted in CRT glass-CA samples with a homogeneous porous structure, as can be observed in the photograph

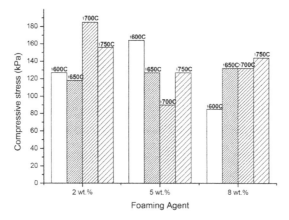

Figure 5. Compressive strength of CRT glass-CaCO$_3$ samples sintered in the range 600–750°C.

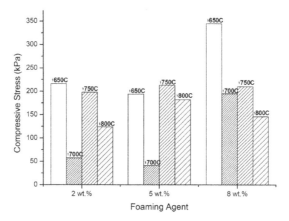

Figure 6. Compressive strength of CRT glass-fly ash samples sintered in the range 650–800°C.

presented in Figure 4(a). According to the microstructural image analysis (ImageJ software), the pore size was reasonably homogenous in CRT glass-5 wt% CA, where the average pore size was about 60 µm. Samples containing 8 wt% CaCO$_3$ sintered at 650 and 700°C exhibited high porosity (80–90%), with a wide pore size distribution, and very scattered compressive strength values were obtained.

Samples containing fly ash as foaming agent, when sintered at 750°C showed a higher mechanical resistance than those sintered at 800°C that exhibited a higher porosity. The highest compressive strength value (~213 kPa) corresponded to the sample with 5 wt% added fly ash, sintered at 750°C, having an apparent density of 1.60 g/cm^3 and a total porosity of ~45%. For this sample, the average pore size was very small, about 12 µm, and the foam structure was very homogenous, see Figure 4(c). For a constant processing temperature (750°C or 800°C), the resultant compressive strength of CRT glass-FA samples appeared to be independent of FA content, within the selected range of 2–8 wt%.

4 CONCLUSIONS

The possibility of producing glass foams by recycling CRT waste glasses along with calcite quarrying residues or coal fly ashes as foaming agent has been presented. Both foaming agents revealed to be suitable precursors when mixed in small amount (<8 wt%) with waste glass in order to produce low density glass foams. The expansion ability of the compositions, the pore size distribution and the mechanical properties of the resulting foams depend on the type and amount of added foaming agent and on the heat treatment temperature. CRT waste glass, calcite ($CaCO_3$) and coal fly ashes proved to be promisor alternatives to conventional raw materials for glass foam production, offering interesting advantages: (i) all of them are wastes and then potentially cost-free; (ii) their use in glass foam production contributes to saving natural resources and reusing of waste; (iii) the use of CRT glass provides a reusing of landfilled glass; (iv) the relatively low temperature for foam processing, due to CRT glass powder sintering ability and ease of gas release from $CaCO_3$ and fly ashes, enables energy saving; and (v) the optimized combination of processing temperature and foaming agent amount led to foams with a good microstructural homogeneity and satisfactory compressive strength (180 kPa and 213 kPa) suggesting that the produced glass foams are potential materials for construction applications.

ACKNOWLEDGEMENTS

This work was funded by FEDER funds through the COMPETE 2020 Programme and National Funds through FCT – Portuguese Foundation for Science and Technology under the project UID/CTM/50025/2013 and funded by M-ERA.NET/0010/2012 project.

REFERENCES

Andreola, F., Barbieri, L., Corradi, A. & Lancellotti I. 2007. CRT glass state of the art – A case study: Recycling in ceramic glazes, Journal of the European Ceramic Society 27: 1623–1629.

Bernardo, E. & Albertini, F. 2006. *Glass foams from dismantled cathode ray tubes, Ceramics International* 32: 603–608.

Blengini, G.A., Busto, M., Fantoni, M. & Fino, D. 2012. Eco-efficient waste glass recycling: Integrated waste management and green product development through LCA, *Waste Management* 32: 1000–1008.

Chen, M.J. Zhang, F.S. & Zhu, J.X. 2011. A novel process for foam glass preparation from waste CRT panel glass. In: The Minerals, Metals & Materials Society (ed.). *Proceedings of the Recycling of Electronic Waste II*: 97–102.

Fernandes, H.R., Andreola, F., Barbieri, L., Lancellotti, I., Pascual, M.J. & Ferreira, J.M.F. 2013. The use of egg shells to produce Cathode Ray Tube (CRT) glass foams. *Ceramics International* 39: 9071–9078.

Francis, A.A. & Rahman, M.K.A. 2014. Structure characterization and optimization of process parameters on compressive properties of glass-based foam composites. *Environmental Progress Sustainable Energy*, 33: 800–807.

König, J., Petersen, R.R. & Yue, Y. 2014. Influence of the glass–calcium carbonate mixture's characteristics on the foaming process and the properties of the foam glass, *Journal of the European Ceramic Society* 34: 1591–1598.

Market Publisher 2014. 2014 Market Research Report on Global Foam glass Industry, https://marketpublishers. com/r/22E653A81BCEN.html (accessed on 12 June 2015).

Mear, F., Yot, P., Cambon, M. & Ribes, M. 2006. The characterization of waste cathode-ray tube glass. *Waste Management* 26: 1468–1476.

Orbis Research 2015. Global Foam Glass Market 2015. Trend, Analysis and Forecast 2020. http://www. orbisresearch.com/reports/index/Global-Foam-Glass-Market-2015-Trend-Analysis-and-Forecast-2020 (accessed on 12 June 2015).

Petersen, R.R., König, J., Smedskjaer, M.M. & Yue, Y. 2014. Effect of Na_2CO_3 as foaming agent on dynamics and structure of foam glass melts. *Journal of Non-Crystalline Solids* 400: 1–5.

Scarinci, G., Brusatin, G. & Bernardo, E. 2005. Glass foams. In: M. Scheffler & P. Colombo (eds). *Cellular ceramics: structure, manufacturing, properties and applications*: 158–176. Weinheim: Wiley-VCH.

Wu, J.P., Boccaccini, A.R., Lee, P.D., Kershaw, M.J. & Rawlings, R.D. 2006. Glass ceramic foams from coal ash and waste glass: production and characterization. *Advances in Applied Ceramics* 105(1): 32–39.

Vancea, C.C. & Lazău, I. 2014. Glass foam from window panes and bottle glass wastes. *Central European Journal of Chemistry* 12(4): 804–811.

Zegowitz, A. 2010. Cellular Glass Aggregate Serving as Thermal Insulation and a Drainage Layer. *Buildings* XI: 1–8.

Wastes: Solutions, Treatments and Opportunities – Vilarinho, Castro & Russo (eds)
© 2015 Taylor & Francis Group, London, ISBN 978-1-138-02882-1

Improving selective collection of urban waste using a business intelligence system

O. Belo, B. Oliveira, M. Medeiros, M. Leite & P. Faria
Algoritmi R&D Centre, University of Minho, Guimarães, Portugal

ABSTRACT: Today recovering urban waste requires effective management services, which usually imply sophisticated monitoring and analysis mechanisms. This is essential for the smooth running of the entire recycling process as well as for planning and control urban waste recovering. In this paper we present a business intelligence system especially designed and implemented to support regular decision-making tasks on urban waste management processes. The system provides a set of domain-oriented analytical tools for studying and characterizing potential scenarios of collection processes of urban waste, as well as for supporting waste management in urban areas, allowing for the organization and optimization of collection services. In order to clarify the way the system was developed and the how it operates, particularly in process visualization and data analysis, we also present the organization model of the system, the services it disposes, and the interface platforms for exploring data.

1 INTRODUCTION

Today, the management of natural and environmental resources is a very recognized need, which figures clearly in our consciousness as a citizen and as an inhabitant of a given urban zone. In practice, this is reflected day after day in all we do related to urban waste management, trying to reduce waste production, both globally and regionally (André & Cerda, 2006). As such, collecting and processing municipal waste is now a very common activity in any urban context. In general, today people recognize the importance of making the separation of solid waste produced daily in their homes. Waste is no longer seen as simple garbage but as a resource that can be used in the production of goods for daily consumption. However, to be valued urban waste need to be collected and recycled. Waste valuation processes are not simple. They require means and resources, covering different systems, infrastructures and personnel. In this process, all begins in a domestic environment. It is there where the first process's task happens with the individual separation of each type of waste produced by domestic activities. Therefore, this is real the primary action of any selective collection process whose end is reached with the placement of waste in recycle bin points. In fact, this simple action is a very important step in waste recovery process, helping to minimize the waste of natural resources and reduce environmental contamination.

In the last decades solid waste management has changed quite positively in Portugal, largely due to the emergence of specific legislation and to the creation of regional entities especially-oriented for this purpose, being possible already in 2002 to quantify the production of municipal solid waste (Magrinho et al., 2006). Mainland Portugal, for example, in 2013 produced about 4.4 million tons of municipal waste, which corresponds to a production of 440 kg per capita (Fernandes et al., 2014). If we consider the size of the country and its population, this figure is significant, revealing the potential occurrence of some environmental and public health problems. So, during the last years we have been witnessing numerous initiatives and alert campaigns about this problem, in which were addressed frequently people's lifestyle and consumption habits, highlighting the importance of domestic separation and of selective collection of waste, as an essential process for recycling of materials and its valuation.

Usually, a recycling process involves collecting, transporting and processing urban waste. Besides allowing for the valuation of municipal waste, this process also allows for reducing the amount of waste placed in landfills and the waste of energy, raw materials and natural resources. In any developing society, daily needs of inhabitants are growing systematically. Thus, it is expected that year after year the production of municipal waste will increase. This is a very serious problem for all national or regional entities involved with. The definition of a sustainable solution to this problem is urgent and requires effective solid waste management policies (Abrelpe, 2015). However, such policies cannot be defined merely based on simple updates of the law. It must be complemented with the implementation of concrete measures addressed for governing how to manage urban solid waste and to promote the implementation of concrete recycling services, from its initial collection stage to its processing stage. To be effective and meet their goals, these policies and services need to be supported by tools for planning and monitoring the actions involved with urban waste recycling life cycle. As such, it is important to have also some sort of decision support system with the ability for providing relevant information for improving the quality of analysis and decision of managers, as well as helping to plan and manage the tasks involved in urban waste recycling process. In this sense, we addressed the recycling process of urban waste using a particular perspective of the selective production and municipal waste collection processes, developing a business intelligence system for monitoring and control waste collection in urban areas. The system was conceived based on the various aspects of management usually involved in these processes, taking particular attention to the specificities of identifying the "when, the "where" and the "how" an urban selective collection process was carried out, considering the various types of waste involved, the seasonality of waste collection, with the subsequent analysis of the respective production, the mapping of collection processes at certain times, or the efforts expended in collection processes in distinct urban areas, recycling bins and respective containers.

Thus, in this paper we present the general characteristics, functional architecture and services of a business intelligent system we built for supporting decision-making activities in all the domains related to municipal waste collection. The basic idea was to provide a computational tool having the ability to manage and assess performance of waste collection processes of a predefined set of urban zones. The paper is structured as follows: section 2 presents and discusses a few references about related work; section 3 describes how the business intelligence system was organized and implemented, what kind of services we have access, and a general overview of system's data visualization structures and mechanisms; and, finally, section 4 exposes some conclusions and a few guidelines for future work.

2 RELATED WORK

In the last decades a lot of research efforts have been involved with the design and development of decision-support systems for the management of waste in urban areas. Apparently it may be seen as one more area of management, involving a set of very common tasks, ranging from collecting to recycling. However, this is not true. Waste management is a very singular and complex activity, involving a large diversity of multidisciplinary problems in many areas considering economic and technical aspects about recycling and sustainable development issues. So, it is not a surprise the appearance of decision support systems on this area, as a way to help and support waste managers in their daily activities, especially on tasks involving decision-making and comparative analysis.

Since 2006, research activities in the field increased a lot (Beliën et al., 2011), revealing the emergent importance of all environment issues related to solid waste. However, even before, many information systems applications appeared approaching relevant issues on field of waste management. Lets see some of them. Already in 1998, Haastrupa et al. (1998) presented a very specific urban waste decision support system, especially oriented for evaluating waste collection services and for identifying suitable areas for waste treatment and disposal, taking into consideration several evaluation aspects, including environmental consequences. Later, Harrison et al. (2001) presented a new decision-making framework for the quantification of the life-cycle inventory of a range of pollutants and costs for an extensive municipal solid waste system, discussing some typical waste management scenarios using a case study for examining waste management strategies. Then, Fiorucci et al. (2003) developed another type of decision support system to assist municipal solid waste management activities, providing means for planning landfills and treatment plants. Similarly, Massukado & Zanta (2005) conceived and developed a simulation decision-support tool to help municipal administrations for analyzing potential scenarios related to solid waste management, in order to anticipate the effect of different management perspectives. Other kind of tools was also developed for waste management tasks. See for example, the application of reverse logistics for locating collection areas (Bautista & Pereira, 2006), the use of geographical information systems for modeling waste collection and transportation (Tavares et al., 2009), or the utilization of designing management strategies in spatial characterization of waste based on urban areas (Katpatal & Rao, 2011). Most recently, Oliveira & Lima (2012) develop an interesting work studying selective collection and transportation of waste in a specific city, applying as well geographic information systems technology to demonstrate how to plan a selective collection system. In fact, this kind of technology has been strongly adopted on a large variety of waste management scenarios in many countries – e.g. Ghose et al. (2006), Rada et al. (2013) or Gallardo et al. (2015).

3 THE BUSINESS INTELLIGENCE SYSTEM

3.1 *The application case*

Solid urban waste collection is recognized as an essential task in any city's daily life. Day after day it assumes even more relevance as cities are expanding and their population increasing. So it is not surprising that the entities responsible for urban zones are more and more worry about the effectiveness of their waste collection processes implementation, in order to maximize, within the possibilities, the usefulness of the solutions they adopted. However, the establishment of a good solution is not an easy thing to achieve. It requires the use of different problem solving methods, in several multidisciplinary domains, usually involving a large diversity of aspects that are related to the routes of waste collection, personnel and vehicle usage, planning and implementation services costs, among other things.

Many approaches can be use to help managers to implement better and more efficient waste collection processes and services. One of the most recent approaches is the use of business intelligence techniques especially oriented to the implementation of decision support systems with the abilities to monitor, control, plan, predict any kind of event related to waste collection. Some application cases were quite successful – e.g. Finnveden & Moberg (2005), Oliveira & Lima (2012) or Verge & Rowe (2013) – and demonstrated the utility and effectiveness of business intelligence applications on this field. In order to cover some of the most relevant information needs that managers usually have on waste collection management, we designed and implemented a business intelligence system especially oriented to monitor, control and predict waste management services as well as to characterize and profile urban zones, taking into consideration their type, number of inhabitants, population density, human-built features, or social and economic organizations, just to name a few. The system has the ability to combine all this information with the one related to waste collection, providing relevant management indexes to implement, control, revise or optimize waste collection services in pre-defined urban zones. Currently, the system acts based exclusively on a package of information related to waste collection processes implemented in a part of the district of Braga, in Portugal. However, the extension of the influence and competence of the system it easily acquired. To do that, we only need to load a new package of data into the system's data warehouse related to one or more urban zones for a given period of time, and process it using system's data integration mechanisms.

3.2 *System's architecture and services*

The system works on a temporal basis. It aims to provide a general picture about waste collection processes done in a specific period of time, between now and a date in the past. Of course, for

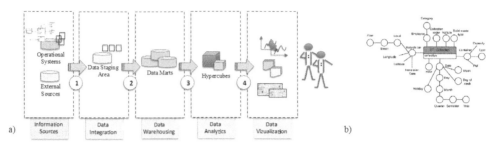

a)

b)

Figure 1.   a) Global system's architecture, and b) a the conceptual model of a system's data mart.

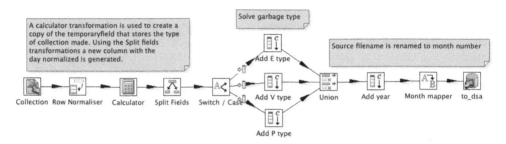

Figure 2.   A data integration service implemented in Kettle.

this to be possible it is necessary that the system have information about the period required for analysis. Thus, it is not difficult to see that the more information the system contains, the greater is its ability to represent and analyze waste collection scenarios. To support such ability, the system was organized into four different scalable functional platforms (Figure 1), namely: 1) information sources, where predefined packages of data are selected and collected in order to populate system's data marts; 2) data integration, which is an intermediate system's area (a data staging area) that supports the execution of all data preparation processes – auditing, cleansing, transformation, etc. – accordingly the requirements of the data marts integrated in the data warehousing platform; 3) data warehousing, which receives and maintains system's data marts; each data mart supports specific activities of a particular decision-making area – waste collection (Figure 1b)), waste profiling, etc.; 4) data analytics, where are located on-line analytical processing mechanisms and structures, providing the means to explore system's data based on the different perspectives of analyses (dimensions) of each decision-maker; and finally 5) data visualization, which integrates a set of specific data visualization applications that presents the results of the analysis processes, using dashboards, custom reports, geo charts or gauges. This is a typical architecture of a regular business intelligence system. The difference resides in the application area, obviously. Thus, all system's mechanisms and structures were designed and implemented especially to reflect the needs of an urban waste management entity and in particular to provide selective waste collection information to its managers.

All services implemented communicate directly with others located in neighborhood layers, exchanging data and keeping intermediary results whenever necessary. Inside the system, information flows always from information sources to analytical structures. To guarantee the correctness of all data elements located in each functional layer, the system uses several software mechanisms to ensure data extraction, transformation and loading into the data warehouse that comes from operational systems or other external information sources. Figure 2 presents a schematic representation of a data integration service, implemented in Kettle (Pentaho, 2015), which is responsible to collect and prepare waste collection data accordingly the requisites established for data warehousing and data analytics platforms.

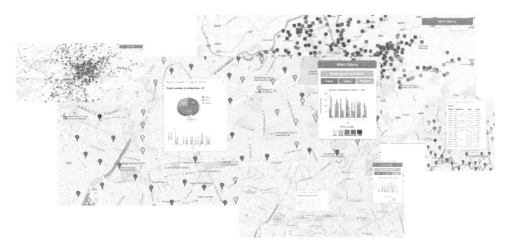

Figure 3.   System's data visualization interfaces – dashboards and geo charts.

3.3 *Data visualization mechanisms*

The data visualization platform is the system's top layer module. It is responsible for ensuring process interaction with system's users, providing a set of data visualization mechanisms that show in a glance sophisticated analysis elements using dashboards, reports or geo charts. Using the system's information catalog users can explore collection processes' data across all the dimensions of analysis – e.g. calendar, time, urban zones, recycling bins, type of waste or containers – that were defined and processed inside the data analytics platform, accordingly some predefined business models. Crossing waste collection data, using for example the calendar, time, recycling bins and type of waste dimensions we can get some collection scenarios through specific geo charts snapshots that can be used for comparative analysis. As any business intelligence system, users can explore data easily, drilling down or rolling up data following their own perspectives of analysis without having the need to know advanced technological knowledge. System's dashboards and geo charts are quite easy to configure and understand, providing rich visual collection processes insights, as we can see in Figure 3. The different types of waste are represented using specific gradient colors – green, blue, yellow and red, represent glass, paper, plastic and batteries, respectively –, which assumes a different color shape according to the number of collections made.

In order to provide a best characterization of urban zones and consequently the correspondent waste collection processes, system's data was enhanced with some demographic data elements from third parties. This provided the necessary means to implement some additional business analytics components in a next version of the system, which will have the ability to profile urban zones and predict future waste collection processes.

4 CONCLUSIONS AND FUTURE WORK

In this paper it was presented a business intelligence system especially conceived for helping waste collection managers and decision makers on daily management activities. Basically, this is a complementary tool in the point of view that add to current operational systems some very interesting application features, especially the ones concerning with comparative analysis, visual processes insights, or dynamic multidimensional data exploration in some pertinent business perspectives involved with urban waste collection. The high level systems' interaction interfaces take users to another level of experience on waste collection management, mostly when we deal with dashboards and geo charts, which guarantee very powerful and user-friendly platforms for data visualization and analysis. In the development of this first version of the system, we focused essentially in the structures and services necessary to support data integration mechanisms, data warehousing structures and services, and data visualization interface platforms. Regarding information source integration, we started working with a small set of urban zones, which were quite enough to prove the system's concept and demonstrate its viability as decision support system for waste collection management. Shortly, we will extend the system's intervention area, incorporating news means to receive waste collection data from other urban zones of the district of Braga, and access to other external information sources to get and enrich system's demographic data. At short term, this last issue will make possible a better urban zone characterization and a more effective profiling of urban waste production.

REFERENCES

Abrelpe. *Resíduos Sólidos: Manual de boas práticas no planejamento*. Available at: http://www.abrelpe.org.br/manual_apresentacao.cfm [Accessed April 27, 2015].

André-García, F.J. & Cerdá Tena, E. 2006 Gestión de residuos sólidos urbanos: análisis económico y políticas públicas. *Cuad. Econ. ICE*. 71: 71–91.

Bautista, J. & Pereira, J. 2006 Modeling the problem of locating collection areas for urban waste management. An application to the metropolitan area of Barcelon, *Omega*, Volume 34, Issue 6: 617–629. Elsevier.

Beliën, J., De Boeck, L. & Van Ackere, J. 2011. Municipal Solid Waste Collection Problems: A Literature Review, *Hub Research Papers 2011/34, Economics & Management*.

Fioruccia, P., Minciardia, R., Robbaa, M. & Sacilea, R. 2003. Solid waste management in urban areas Development and application of a decision support system, Resources, *Conservation and Recycling*, 37: 301–328. Elsevier.

Finnveden & Gand Moberg, Å. 2005. Environmental systems analysis tools – an overview. *Journal of Cleaner Production* 13:1165–1173.

Gallardo, A., Carlos, M., Peris, M. & Colomer, F. 2015. Methodology to design a municipal solid waste pre-collection system. A case study, *Waste Management*, Vol. 36: 1–11. Elsevier.

Ghose, M.K., Dikshit, A.K. & Sharma, S.K. 2006. A GIS based transportation model for solid waste disposal – A case study on Asansol municipality. *Waste Management* 26(11): 1287–1293.

Haastrupa, P., Maniezzob, V., Mattarellia, M., Mazzeo Rinaldia, F., Mendesa, I. & Paruccinia, M. 1998. A decision support system for urban waste management, *European Journal of Operational Research*, Vol. 109, Issue 2: 330–341. Elsevier.

Harrison, K., Dumas, R., Solano, E., Barlaz, M., Brill Jr., E. & Ranjithan, S. 2011. Decision Support Tool for Life-Cycle-Based Solid Waste Management. *J. Comput. Civ. Eng. 15, special issue: information technology for life-cycle infrastructure management*: 44–58.

Katpatal, Y. & Rao, B. 2011. Urban Spatial Decision Support System for Municipal Solid Waste Management of Nagpur Urban Area Using High-Resolution Satellite Data and Geographic Information System, *Journal of Urban Planning and Development*, Vol. 137, Issue 1.

Magrinho, A., Didelet, F. & Semiao, V. 2006. Municipal solid waste disposal in Portugal, *Waste Management*, Vol. 26:1477–1489.

Massukado, L. & Zanta, V. 2006. SIMGERE – Software Para Avaliação de Cenários de Gestão Integrada de Resíduos Sólidos Domiciliares, *Eng. sanit. ambient.*, Vol. 11, Nr. 2:133–142.

Oliveira, R. & Lima, R. 2012. Using a Geographic Information System in the Selective Collection of Recyclable Materials, In *Proceedings of ISWA World Solid Waste Congress*, Florence, Italy.

Pentaho. Pentaho Data Integration. Available at: http://www.pentaho.com/product/data-integration [Accessed March 16, 2015].

Rada, E.C., Ragazzi, M. & Fedrizzi, P. 2013. Web-GIS oriented systems viability for municipal solid waste selective collection optimization in developed and transient economies, *Waste Management*, 33(4): 785–792.

Fernandes, A., Teixeira, A., Guerra, M., Ribeiro, R., Rodrigues, S. & Alvarenga, A. 2014. *Relatório do Estado do Ambiente REA 2014*, Agência Portuguesa do Ambiente, Departamento de Estratégias e Análise Económica.

Tavares, G., Zsigraiova, Z., Semiao, V. & Carvalho, M.G. 2009. Optimisation of MSW collection routes for minimum fuel consumption using 3D GIS modeling, *Waste Management*, 29 (3):1176–1185.

Verge, A. & Rowe, R. 2013. A framework for a decision support system for municipal solid waste landfill design, *Waste Management & Research*, 31(12): 1217–1227.

Wastes: Solutions, Treatments and Opportunities – Vilarinho, Castro & Russo (eds)
© 2015 Taylor & Francis Group, London, ISBN 978-1-138-02882-1

Remediation of soils contaminated with Zinc by *Miscanthus*

S. Boléo, A.L. Fernando, B. Barbosa, J. Costa, M.P. Duarte & B. Mendes
Departamento de Ciências e Tecnologia da Biomassa/MEtRiCS, Faculdade de Ciências e Tecnologia (FCT), Universidade Nova de Lisboa, Caparica, Portugal

ABSTRACT: *Miscanthus* is a perennial grass, characterized by high yields. Yet, the cultivation of *Miscanthus* for bioenergy may generate land-use conflicts which might be avoided through the establishment of energy crops on marginal land. Moreover, in the Mediterranean region, water is a valuable resource and irrigation can represent a negative impact. In this context, this work aims to study the potentiality of *Miscanthus* (*Miscanthus sinensis*, *Miscanthus floridulus* and *Miscanthus* × *giganteus*) in Zn contaminated soils under a low irrigation regime. Results showed that biomass productivity of *M*. × *giganteus*, but not of *M*. *sinensis* and *M*. × *floridulus* were negatively affected by the contamination. *M*. × *giganteus* was the most productive genotype. Although the zinc removal percentages by *Miscanthus* accumulation represent 6% maximum, after three consecutive years, towards the Zn soil bioavailable fraction, the establishment of a *Miscanthus* cover represent an approach to attenuate and stabilize contaminated sites with additional revenue to owners.

1 INTRODUCTION

Miscanthus is a woody rhizomatous C4 grass originated in South-East Asia and was initially imported to Europe as an ornamental plant. It is a perennial plant, related to sugarcane, with an estimated productive lifetime of at least 10–15 years, and both the stems and leaves of the crop can be harvested annually (El Bassam 2010). It is characterized by relatively high yields, low moisture content at harvest, high water and nitrogen efficiencies and an apparently low susceptibility to pests and diseases (Oliveira et al. 2001). Its robustness and physiological characteristics as a deep, dense and extensive root system, allows it to easily adapt to different types of soils and ecological conditions, being indicated for reducing soil erosion, minimize nutrient leaching and sequester more C in soils (Fernando et al. 2010, Fernando 2013). *Miscanthus* is a multi-use crop, being mostly use as a renewable energy crop. Other uses are: thatching, animal bedding, high quality paper pulp, fiberboards production and inclusion in composites (Fernando 2005).

Recent approaches utilize this crop in the removal of nutrients from wastewaters (Bandarra et al. 2013, Lino et al. 2014, Barbosa et al. 2015) and in the phytoremediation of contaminated soils with heavy metals (Fernando et al. 2004, Boléo et al. 2013, Nsanganwimana et al. 2014). Human activities as industry, localized agriculture and mining, contribute to undesirable accumulations of heavy metals in the environment, namely on soils and water bodies. In this line, this work tries to apply the advantages of phytoremediation technology, an "*in situ*" and solar energy driven remediation approach for the restoration of contaminated soils. This could be an approach to restore or attenuate and stabilize contaminated sites while bringing additional revenue to owners, and simultaneously to avoid land use conflicts with food crops (Dauber et al. 2013, Fernando et al. 2014). Moreover, in the Mediterranean region, water is a valuable resource and irrigation can represent a negative impact. In this context, this research work aims to study the potentiality of *Miscanthus* production in Zn contaminated soils under a low irrigation regime.

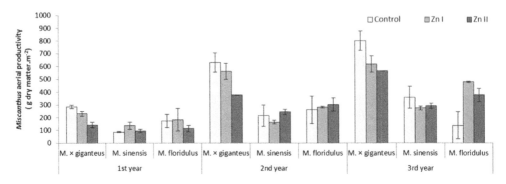

Figure 1. *Miscanthus* aerial productivity ($g \cdot m^{-2}$ dry matter) during three growing cycles.

2 MATERIALS AND METHODS

The trials were established in April 2012 in pots, with three different *Miscanthus* genotypes, M. *sinensis*, M. *floridulus*, and a hybrid – *Miscanthus* × *giganteus*. Rhizomes were provided by the University of Catania. In each pot ($0.06154\,m^2$, $12\,kg$ of soil) 2 rhizomes were established (a pot for each genotype with replicates). After the establishment of the rhizomes, pots were fertilized: $3\,g\,N/m^2$ (urea, 46% N); $3\,g\,N/m^2$ (nitrolusal, mixture of $NH_4NO_3 + CaCO_3$, 27% N); $17\,g\,K_2O/m^2$ (potassium sulphate, 51% K_2O); $23\,g\,P_2O_5/m^2$ (superphosphate, 18% P_2O_5). Two concentrations of zinc in contaminated soils were tested (450 and $900\,mg\,Zn.kg^{-1}$ dry matter, corresponding to maximum allowable and to twice as maximum, respectively, ZnI and ZnII) (Decreto-Lei n° 276/09, 2009). Simultaneously, a low irrigating regime was applied: 475 mm.

At the end of each growing season (December-January), during three consecutive years (2012–2014), the plants were harvested and the aerial productivity and zinc accumulation was monitored. Total below-ground dry weight and its zinc accumulation were also determined in the third year. Zn content was determined by atomic absorption following calcination of biomass at 550°C for two hours, in a muffler furnace and nitric acid digestion of the ash material. Zn released by percolated waters was also evaluated along the three growing seasons. Bioavailable Zn in the soils at the beginning of the experiment was also evaluated by EDTA extraction (Iqbal et al. 2013).

3 RESULTS AND DISCUSSION

3.1 *Biomass productivity*

Figure 1 presents the aerial biomass productivity obtained in the trials, corresponding to stems leaves, panicules and litter produced during three consecutive years. According to the results obtained, there was an increase of productivity from the 1st to the 2nd year and from the 2nd to the 3rd year: this behavior was observed in all of the studied genotypes. This reflects the energy spent by the plant, on the first years, to develop the extensive under-ground rhizome system (El Bassam 2010).

M. × *giganteus* is significantly more productive than M. *floridulus*, and this genotype more productive than *M* × *sinensis*. This relation was consistent in all of the three years of *Miscanthus* experiment. The contamination with zinc affected negatively the production of *M*.× textitgiganteus, but not of *M. sinensis* and *M. floridulus*. *M* × *giganteus* and M. *sinensis* produced inflorescence, but not *M. floridulus*. Leaves fraction represent the highest share in *M*. × *giganteus* and *M. sinensis*. Stems represent the highest fraction in *M. floridulus* (data not shown).

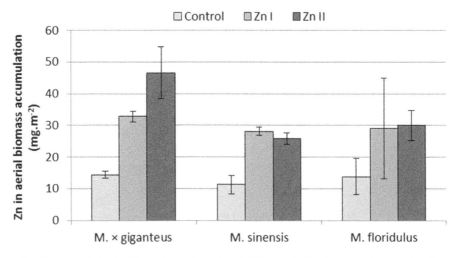

Figure 2. Zn accumulation in *Miscanthus* and weeds aerial biomass during 3 consecutive years (mg Zn·m^{-2}).

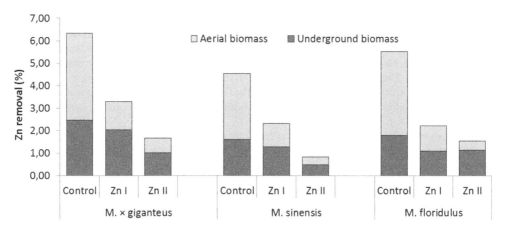

Figure 3. *Miscanthus* and weeds aerial biomass bioavailable Zn removal (%) from soil, after three consecutive years.

3.2 *Accumulation of Zn in the biomass*

Figure 2 present the Zn accumulated by the biomass of *Miscanthus* in three growing seasons. Results show that biomass obtained in Zn contaminated soils presented significantly higher zinc accumulation then biomass from non-contaminated soils, thus showing phytoextraction capacity. It was observed (data not shown) that the zinc accumulation between stems and leaves is not significantly different. *M. × giganteus* was the genotype which accumulated most.

3.3 *Zn remediation*

Figure 3 shows the Zn removal (%) by the aerial biomass of *Miscanthus* towards the bioavailable Zn content in the soil, after 3 consecutive years.

As it is observed by Figure 3, the removal percentage of bioavailable Zn in soil by the aerial biomass is very limited: 6.35% the maximum obtained and 3.15% the average. Zn mostly accumulates in the below ground biomass. In terms of remediation, if we consider the phytoextraction, than the results are not optimal. But if considering that the aerial biomass does not contain Zn in high amounts, that the aerial biomass can be economically valorized, and that the plant is tolerant to the contamination, than, remediation occurs because a vegetative crop was established in a contaminated soil, with the environmental benefits associated with.

3.4 *Zn in percolated waters*

During the three years of trials, percolated waters were monitored. Results were always below $0.74\,\text{mg Zn.L}^{-1}$, being $0.11\,\text{mg Zn.L}^{-1}$ the mean value obtained in all experiments (including control). This means that the amount of Zn in the percolated waters do not represent risk to ground/surface waters. Differences among genotypes did not follow a pattern.

4 CONCLUSIONS

The three studied *Miscanthus* genotypes showed zinc phytoextraction and accumulation capacity. The higher the Zn contamination in the soil, the higher the accumulation of Zn in aerial and below ground biomass. *M. × giganteus* was the genotype which accumulated most Zn. The contamination with zinc also affected negatively and significantly the production of *M. × giganteus*, but not of *M. sinensis* and *M. floridulus*. Among genotypes, *M. × giganteus* was the most productive, followed by *M. floridulus*. *M. sinensis* was the lesser productive genotype.

Further studies are needed to assess other parameters of the biomass quality with a view to its end use; to check if the studied features change during the life time of the planted *Miscanthus* rhizomes; and to clarify the interaction among zinc and other nutrients in their uptake and translocation by *Miscanthus* plants. The prospect of the valorisation of the *Miscanthus* aerial biomass, for bioenergy or bio-products production purposes, could lessen the financial costs of soil remediation, compared to the traditional physical – chemical processes with the associated revenue of environmental benefits.

Results showed that *Miscanthus* have potential to simultaneously deliver high yields, restore soil properties. Factoring in issues such as yields, inputs and biomass quality, as well as potential environmental impacts shows that the production of *Miscanthus* may present advantages. Due to the extensive radicular system, *Miscanthus* is associated with control of soil erosion, carbon sequestration and minimization of nutrient leaching as well as with the restoration of soil properties (fertility, structure, organic matter). The biomass being produced if balanced may contribute to a positive energy balance and to greenhouse gases emissions reduction. However, the removal of metals from the soil by *Miscanthus* is a slow process. Yet, it represents an opportunity to produce sustainable biomass in a resource constrained World.

ACKNOWLEDGEMENTS

This work was supported by the European Union (Project Optimization of perennial grasses for biomass production (OPTIMA), Grant Agreement No: 289642, Collaborative project, FP7-KBBE-2011.3.1-02).

REFERENCES

Bandarra, V., Fernando, A.L., Boléo, S., Barbosa, B., Costa, J., Sidella, S., Duarte, M.P. & Mendes, B. 2013. Growth, Productivity and Biomass Quality of Three Miscanthus Genotypes Irrigated with Zn and Cu Contaminated Wastewaters. In: Eldrup A., Baxter D., Grassi A., Helm P. (Eds.), *Proceedings of the 21th European Biomass Conference and Exhibition, Setting the course for a Biobased Economy*, 3–7 June 2013, Copenhagen, Denmark, 147–150.

Barbosa, B., Costa, J., Fernando, A.L. & Papazoglou, E.G. 2015. Wastewater reuse for fiber crops cultivation as a strategy to mitigate desertification, *Industrial Crops and Products*, 68, 17–23.

Boléo, S., Fernando, A.L., Borda d'Água, R., Bandarra, V., Barbosa, B., Costa, J., Sidella, S., Duarte, M.P. & Mendes, B. 2013. Phytoremediation response of three Miscanthus genotypes to soils contaminated with zinc. In: Eldrup A., Baxter D., Grassi A., Helm P. (Eds.), *Proceedings of the 21th European Biomass Conference and Exhibition, Setting the course for a Biobased Economy*, 3–7 June 2013, Copenhagen, Denmark, ETA-Renewable Energies and WIP-Renewable Energies, 311–314.

Dauber, J., Brown, C., Fernando, A.L., Finnan, J., Krasuska, E., Ponitka, J., Styles, D., Thrän, D., Groenigen, K.J.V., Weih, M. & Zah, R. 2012. Bioenergy from "surplus" land: environmental and social-economic implications. *BioRisk*, 7, 5–50.

Decreto-Lei nº 276/09 2009. Anexo I, Valores limite de concentração relativos a metais pesados, compostos orgânicos e dioxinas e microrganismos, *Diário da República*, nº 192, I Série, 2 de Outubro 2009, 7154–7165 (in Portuguese).

El Bassam, N. 2010. *Handbook of Bioenergy Crops. A complete reference to species, development and applications*. Earthscan Ltd. London, UK.

Fernando, A.L. 2013. Miscanthus for a Sustainable Development: How Much Carbon Is Captured in the Soil? In: Eldrup A, Baxter D, Grassi A, Helm P (Eds.), *Proceedings of the 21st European Biomass Conference and Exhibition, Setting the course for a Biobased Economy*, 3–7 June 2013, Copenhagen, Denmark, ETA-Renewable Energies and WIP-Renewable Energies, 1842–1843.

Fernando, A.L., Boléo, S., Barbosa, B., Costa, J., Lino, J., Tavares, C., Sidella, S., Duarte, M.P. & Mendes, B. 2014. How Sustainable is the Production of Energy Crops in Heavy Metal Contaminated Soils? In: Hoffmann C., Baxter D., Maniatis K., Grassi A., Helm P. (Eds.), *Proceedings of the 22th European Biomass Conference and Exhibition, Setting the course for a Biobased Economy*, Hamburg, Germany, 23–26 June 2014, ETA-Renewable Energies, 1593–1596.

Fernando, A.L., Duarte, M.P., Almeida, J., Boléo, S. & Mendes, B. 2010. Environmental impact assessment of energy crops cultivation in Europe. *Biofuels, Bioproducts & Biorefining*, 4, 594–604.

Fernando, A.L., Godovikova, V. & Santos Oliveira, J.F. 2004. Miscanthus × giganteus: contribution to a sustainable agriculture of a future/present – oriented biomaterial. Materials Science Forum, *Advanced Materials Forum II*, 455–456, 437–441.

Fernando, A.L.A.C. 2005. Fitorremediação por Miscanthus × giganteus de solos contaminados com metais pesados, *PhD Thesis*, FCT/UNL, Lisbon, Portugal.

Iqbal, M., Bermond, A. & Lamy, I. 2013. Impact of miscanthus cultivation on trace metal availability in contaminated agricultural soils: Complementary insights from kinetic extraction and physical fractionation. *Chemosphere*, 91, 287–294.

Lino, J., Fernando, A.L., Barbosa, B., Boléo, S., Costa, J., Duarte, M.P. & Mendes, B. 2014. Phytoremediation of Cd and Ni Contaminated Wastewaters by Miscanthus. In: Hoffmann C., Baxter D., Maniatis K., Grassi A., Helm P. (Eds.), *Proceedings of the 22th European Biomass Conference and Exhibition, Setting the course for a Biobased Economy*, Hamburg, Germany, 23–26 June 2014, ETA-Renewable Energies, 303–307.

Nsanganwimana, F., Pourrut, B., Mench, M. & Douay, F. 2014. Suitability of Miscanthus species for managing inorganic and organic contaminated land and restoring ecosystem services. A review. *Journal of Environmental Management*, 143, 123–134.

Oliveira, J.S., Duarte, M.P., Christian, D.G., Eppel-Hotz, A. & Fernando, A.L. 2001. Environmental aspects of Miscanthus production. In: Jones M.B., Walsh M. (eds.) *Miscanthus for energy and fibre*. James & James (Science Publishers) Ltd, London, 172–178.

Wastes: Solutions, Treatments and Opportunities – Vilarinho, Castro & Russo (eds)
© 2015 Taylor & Francis Group, London, ISBN 978-1-138-02882-1

Anaerobic co-digestion of cork based oil sorbent and cow manure or sludge

A.J. Cavaleiro, T.M. Neves, A.P. Guedes & M.M. Alves
Centre of Biological Engineering, University of Minho, Braga, Portugal

P. Pinto & S.P. Silva
Corticeira Amorim, S.G.P.S., S.A., S. Paio de Oleiros, Santa Maria da Feira, Portugal

D.Z. Sousa
Laboratory of Microbiology, Wageningen University, Wageningen, The Netherlands
Centre of Biological Engineering, University of Minho, Braga, Portugal

ABSTRACT: Cork, a material with great economic, social and environmental importance in Portugal, is also a good oil sorbent that can be used in the remediation of oil spills. The oil-impregnated cork can be easily removed, but requires further treatment. In the case of vegetable oil spills, anaerobic digestion may be a potential solution. This study aims to evaluate the effect of adding cork contaminated with sunflower oil as co-substrate in anaerobic digestion processes. Biodegradability assays were prepared with cow manure or sludge from a wastewater treatment plant, in the presence of five concentrations of oil-contaminated cork, between 200 and 1000 mg·L^{-1} as COD. Maximum cumulative methane production increased with the amount of oily cork up to 41% and 101% in the assays with manure and sludge, respectively. Sporadic addition of cork contaminated with vegetable oil during anaerobic digestion of manure or sludge increases significantly the methane production of these processes.

1 INTRODUCTION

Vegetable oil spills, although less perceived than mineral oil spills, cause deleterious effects on ecosystems and present serious environmental problems (Mudge 1995, Li et al. 2007). Acute and chronic contamination of marine environments has been reported, causing depletion of dissolved oxygen from the water column, and mortality of fish, birds and sessile animals (Mudge 1995, Li et al. 2007, EPA 2011). While poorly documented, these accidents occur frequently during storage and transportation of oils (Mudge 1995, EPA 2011), and may also represent an important source of contamination in oil refineries.

The use of sorbent materials is a fast and effective strategy for oil removal from contaminated sites, rapidly decreasing the environmental damage of oil spills. Maximum sorption capacity, prolonged oil retention, biodegradability or potential reuse of the sorbent material are factors that determine the choice of a sorbent. Sorbents may be inorganic (e.g. clay, vermiculite, diatomite) or organic, from natural or synthetic origin. Organic synthetic products, such as polypropylene or polyurethane foam, have high sorption capacity, but are not renewable or biodegradable (Teas et al. 2001).

The use of cork as natural sorbent has been promoted by Corticeira Amorim, with its commercial product CorkSorb. This product has a maximum oil absorption capacity of 9.43 L·kg^{-1}, which is higher than that of mineral sorbents, and similar or even greater than other organic sorbents such as peat, cellulose or polypropylene. Absorption occurs by capillarity in approximately 15 s, and the oil is retained in the cork cells for months. The hydrophobic characteristics of the CorkSorb products are advantageous when spills occur in aquatic environments or wet floors (Silva & Reis 2007, Corticeira Amorim 2009).

The oil-impregnated cork is easily removed from the oil spill site, but has to be treated later. One of the possible treatments for the oily cork waste, and currently most used, is incineration. Another possible alternative is anaerobic digestion, which has the advantage of coupling waste treatment with energy production in the form of biogas (Esposito et al. 2012). High biogas production can be expected from oils and fats, and the addition of lipid-rich wastes to anaerobic digestion processes was shown to significantly increase the net energy balance of these systems (Alves et al. 2009, Neves et al. 2009a). The anaerobic treatment of two or more wastes is called co-digestion, and presents several technological and economic advantages, provided the mixture of wastes is carefully controlled to avoid inhibition (Neves et al. 2009a, Álvarez et al. 2010). The digestate produced in the co-digestion processes may be used in agriculture, after stabilization or composting (Álvarez et al. 2010). The presence of cork in the digestate may improve its characteristics as a soil conditioner.

In this work, cork contaminated with sunflower oil was added as co-substrate in anaerobic digestion processes, and the effect on biogas production was assessed.

2 MATERIALS AND METHODS

2.1 Oil-contaminated cork waste

Thermal treated hydrophobic cork granules (0.3-1 mm) were provided by Corticeira Amorim. Cork was contaminated in the laboratory by adding 5 kg of commercial sunflower oil per kg of cork, which corresponds to approximately half of its maximum absorption capacity (Pintor et al. 2012). Total and volatile solids of the non-contaminated cork were 96 and 92%, respectively. Chemical oxygen demand (COD) of the commercial oil was $1.6 \pm 0.1 \text{ kg} \cdot \text{kg}^{-1}$. Long chain fatty acids (LCFA) were analyzed after promoting hydrolysis of the oil, based on the methods of Brandl et al. (1988) and Neves et al. (2009b). LCFA content of the commercial sunflower oil was 32% linoleic acid (C18:2), 23% oleic acid (C18:1), 22% stearic acid (18:0), 21% palmitic acid (C16:0) and 3% other LCFA.

2.2 Specific methanogenic activity of manure and sewage sludge

Manure from a cattle production unit, or secondary sludge from a wastewater treatment plant in northern Portugal, were used in the biodegradability assays. The specific methanogenic activity of these organic wastes was evaluated in closed bottles prepared with a volatile solids content of approximately $3 \text{ g} \cdot \text{L}^{-1}$. Acetate ($30 \text{ mmol} \cdot \text{L}^{-1}$) or H_2/CO_2 (80:20%, total pressure of 2 bar) were added as direct substrates for the methanogens. A pressure transducer was used to measure changes in the pressure over time. Simultaneously, control tests were conducted without substrate addition, or pressurized with N_2/CO_2 (80:20%, total pressure of 2 bar) (Colleran et al. 1992). Bicarbonate buffered medium was used, with sodium sulfide ($0.8 \text{ mmol} \cdot \text{L}^{-1}$) as reducing agent and resazurin as redox indicator. All the assays were performed in triplicate, and were incubated in the dark, at 37°C under a rotation speed of 120 min^{-1}. The specific methanogenic activity was calculated by dividing the initial slope of the methane production curve by the amount of volatile solids (VS) in the bottle at the end of the assay ($\text{mL g}^{-1} \cdot \text{day}^{-1}$). Relatively low activity values were obtained for the manure, i.e. 29 ± 4 and $279 \pm 12 \text{ mL g}^{-1} \cdot \text{day}^{-1}$ in the presence of acetate or H_2/CO_2, respectively. For the sewage sludge, acetoclastic and hydrogenotrophic activity were 13 ± 1 and $0 \text{ mL g}^{-1} \cdot \text{day}^{-1}$, respectively.

2.3 Biodegradability assays

Biodegradability assays were performed under strict anaerobic conditions in closed vials in which the oil-contaminated cork waste was incubated in the presence of cow manure or sewage sludge

Table 1. Experimental conditions applied in the biodegradability assays.

Code	Cork $(mg \cdot L^{-1})$	Oil COD concentration $(mg \cdot L^{-1})$	Oil concentration $(mg \cdot L^{-1})$
Blk	0	0	0
CO-200	26	200	128
CO-400	51	400	256
CO-600	77	600	385
CO-800	103	800	513
CO-1000	128	1000	641
C-200	26	0	0
C-400	51	0	0
C-600	77	0	0
C-800	103	0	0
C-1000	128	0	0
O-200	0	200	128
O-400	0	400	256
O-600	0	600	385
O-800		800	513
O-1000	0	1000	641

Blk – blank; CO – cork + oil; C – cork; O – oil.

(mass concentration, expressed as SV, of 3 g $\cdot$ L^{-1}). The conditions applied are summarized in Table 1.

Five different concentrations of oil-contaminated cork waste were tested (CO code in Table 1). As described in section 2.1, the amount of oil and cork are proportional, and thus CO tests were prepared with increasing concentrations of both oil and cork. In parallel, blank assays were performed in the absence of contaminated cork residue, i.e. containing only manure or sludge (Blk in Table 1). Control experiments were also prepared with (i) non-contaminated cork granules (C in Table 1), and (ii) commercial vegetable oil without cork (O in Table 1). These two sets of controls were prepared to evaluate potential inhibitory effects or stimulation of biogas production, due to the individual presence of cork or oil. The five control assays performed with non-contaminated cork were prepared with the same amounts of cork added to the different biodegradability tests (CO assays). In the controls with oil, the same five concentrations of oil used in CO assays were tested.

The assays were prepared with bicarbonate buffered basal medium supplemented with salts and vitamins, as described by Angelidaki et al. (2009). All assays were performed in triplicate, and the flasks were incubated at 37°C without stirring. Methane production was quantified during the biodegradation tests by gas chromatography (GC). Methane yields (in %) were calculated after discounting the value of the blanks, through the ratio between the maximum cumulative methane production obtained in each test and the expected theoretical value, which was calculated from the LCFA composition of the added oil. The statistical significance of differences detected in the maximum cumulative methane production values was assessed using analysis of variance (ANOVA) single factor. Statistical significance was set at $P < 0.05$. Volatile fatty acids were analyzed at the beginning and end of the tests. At the end, medium chain fatty acids (MCFA) and LCFA were also quantified after freeze-drying the content of the bottles.

3 RESULTS AND DISCUSSION

The use of cork as oil absorbent in oil spills is expected to generate modest amounts of oil-contaminated cork waste, at a relatively low frequency. Therefore, this waste can be added as a

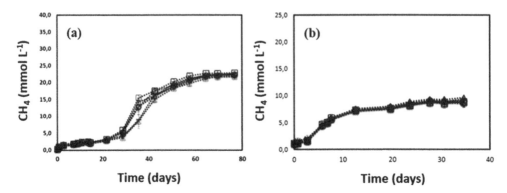

Figure 1. Cumulative methane production in the controls where non-contaminated cork was co-digested with manure (a) or sludge (b). Blk (-), C-200 (■), C-400 (○), C-600 (△), C-800 (◇) e C-1000 (□).

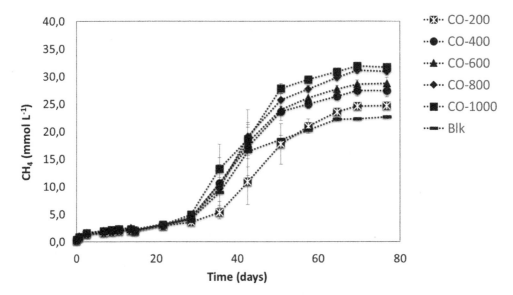

Figure 2. Cumulative methane production in the assays with manure and oil-contaminated cork.

co-substrate to ongoing anaerobic digestion processes. Biodegradation of the oil may improve the energy yield of the process, or oil can be inhibitory to the anaerobic microorganisms. Specifically the methanogens have been reported as very sensitive to LCFA toxicity (Lalman & Bagley 2002). Moreover, the potential presence of aromatic compounds in the cork granules can also inhibit the microbial activity. Control trials (C and O) were set to assess these effects.

Comparing to the blank assays (Blk), no significant differences were observed in methane production when non-contaminated cork was added to both experiments with manure or sludge (Fig. 1). These results indicate that cork did not influenced the anaerobic degradation process.

The ability of the complex microbial communities, present in the manure or sludge, to convert the sunflower oil to methane was confirmed in the control tests performed with oil (O). Methane yields varied between 32–55% and no inhibition was observed with the increase of oil concentrations (data not shown). The absence of free fatty acids at the end of the tests suggest that the relatively low methane yields are probably related with limitations in the hydrolysis step.

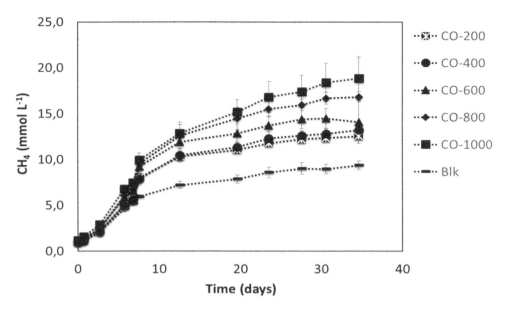

Figure 3.   Cumulative methane production in the assays with sewage sludge and oil-contaminated cork.

In the assays performed with oil-contaminated cork, maximum cumulative methane production increased significantly ($p < 0.01$) in comparison with the blanks (Blk). This increase was proportional to the amount of waste added (Figs. 2–3) and reached a maximum value of 41% and 101% in CO-1000 assays with manure and sludge, respectively. Thus, it can be concluded that sporadic addition of oil-contaminated cork during the anaerobic treatment of manure or sludge improves the methane production of these processes.

4   CONCLUSIONS

Addition of oil-contaminated cork had a positive effect on the maximum cumulative methane production obtained in the anaerobic digestion of manure or sewage sludge. This effect was proportional to the amount of oil-impregnated cork added, and reached maximum values of 41% in the test with manure and 101% in the assay with sewage sludge. The use of this approach allows the recovery of the energetic potential of the waste lipids present in the oil-contaminated cork obtained from the remediation of oil spills.

ACKNOWLEDGMENTS

The European Regional Development Fund – ERDF, through the Operational Program Thematic Factors of Competitiveness – COMPETE, and the Portuguese Foundation for Science and Technology (FCT), in the frame of projects FCOMP01-0124-FEDER-014784 (FCT: PTDC/EBB-EBI/114364/2009) e FCOMP-010124-FEDER-027917 (FCT: PTDC/AAG-TEC/3428/2012), are acknowledged. The authors also thank to the FCT Strategic Project PEst-OE/EQB/LA0023/2013 and the Project "BioEnv – Biotechnology and Bioengineering for a sustainable world", REF. NORTE-07-0124-FEDER-000048", co-funded by the Programa Operacional Regional do Norte (ON.2 – O Novo Norte), QREN, FEDER.

REFERENCES

Álvarez, J.A, Otero, L. & Lema, J.M. 2010. A methodology for optimising feed composition for anaerobic co-digestion of agro-industrial wastes. *Bioresource Technology* 101: 1153–1158.

Alves, M.M., Pereira, M.A., Sousa, D.Z., Cavaleiro, A.J., Picavet, M., Smidt, H. & Stams, A.J. 2009. Waste lipids to energy: how to optimize methane production from long-chain fatty acids (LCFA). *Microbial Biotechnology* 2: 538–550.

Angelidaki, I., Alves, M., Bolzonella, D., Borzacconi, L., Campos, J.L., Guwy, A.J., Kalyuzhnyi, S., Jenicek, P. & van Lier, J.B. 2009. Defining the biomethane potential (BMP) of solid organic wastes and energy crops: a proposed protocol for batch assays. *Water Science & Technology* 59: 927–934.

Brandl, H., Gross, R.A., Lenz, R.W. & Fuller, R.C. 1988. *Pseudomonas oleovorans* as a source of poly (β-hydroxyalkanoates) for potential applications as biodegradable polyesters. *Applied and Environmental Microbiology* 54: 1977–1982.

Colleran, E., Concannon, F., Golden, T., Geoghegan, F., Crumlish, B., Killilea, E., Henry, M. & Coates, J. 1992. Use of methanogenic activity tests to characterize anaerobic sludges, screen for anaerobic biodegradability and determine toxicity thresholds against individual anaerobic trophic groups and species. *Water Science and Technology* 25: 31–40.

Corticeira Amorim, S.G.P.S., S.A. 2009. CorkSorb, sustainable absorbents. http://www.corkSorb.com/index.php, accessed on 8 April 2015.

U.S. Environmental Protection Agency 2011. *Oil spills*. htpp://www.epa.gov/oilspill/, accessed on 8 April 2015.

Esposito, G., Frunzo, L., Giordano, A., Liotta, F., Panico, A. & Pirozzi, F. 2012. Anaerobic co-digestion of organic wastes. *Reviews in Environmental Science and Biotechnology* 11: 325–341.

Lalman, J. & Bagley, D.M. 2002. Effects of C18 long chain fatty acids on glucose, butyrate and hydrogen degradation. *Water Research* 36: 3307–3313.

Li, Z., Lee, K., Cobanli, S.E., King, T., Wrenn, B.A., Doe, K.G., Jackman, P.M. & Venosa, A.D. 2007. Assessment of sediment toxicity during anaerobic biodegradation of vegetable oil using Microtox® and *Hyalella azteca* bioassays. *Environmental Toxicology* 22: 1–8.

Mudge, S.M. 1995. Deleterious effects from accidental spillages of vegetable oils. *Spill Science & Technology Bulletin* 2: 187–191.

Neves, L., Oliveira, R. & Alves, M.M. 2009a. Co-digestion of cow manure, food waste and intermittent input of fat. *Bioresource Technology* 100: 1957–1962.

Neves, L., Pereira, M.A., Mota, M. & Alves, M.M. 2009b. Detection and quantification of long chain fatty acids in liquid and solid samples and its relevance to understand anaerobic digestion of lipids. *Bioresource Technology* 100: 91–96.

Pintor, A.M.A., Ferreira, C.I.A., Pereira, J.C., Correia, P., Silva, S.P., Vilar, V.J.P., Botelho, C.M.S. & Boaventura, R.A.S. 2012. Use of cork powder and granules for the adsorption of pollutants: a review. *Water Research* 46: 3152–3166.

Silva, S.P. & Reis, R.L. 2007. *Meio de absorção/adsorção à base de derivados de cortiça para absorção/adsorção de óleos. PT 103492.*

Teas, Ch., Kalligeros, S., Zanikos, F., Stournas, S., Lois, E. & Anastopoulos, G. 2001. Investigation of the effectiveness of absorbent materials in oil spills clean up. *Desalination* 140: 259–264.

Wastes: Solutions, Treatments and Opportunities – Vilarinho, Castro & Russo (eds)
© 2015 Taylor & Francis Group, London, ISBN 978-1-138-02882-1

Altering process conditions to enhance amine loading of clinoptilolite

B. Cene, Y.A. Aydın & N. Deveci Aksoy
Istanbul Technical University, Istanbul, Turkey

ABSTRACT: The influence of process parameters on preparation of hexadecylamine (HA) and dodecylamine (DA) modified natural clinoptilolite was studied by one factor at a time approach. The presence of amine groups on modified samples was verified by thermogravimetric analysis, in addition to determination of amine concentration in aqueous medium. As a result of the study, it was demonstrated that chloroform and deionized water were the most favorable solvents for modification and washing, respectively. The uptake of HA and DA was maximized at clinoptilolite to solvent ratio of 1/100 with clinoptilolite samples sized within 0.106–0.300 mm. It was shown that, HA and DA modified clinoptilolites were 1.87 and 8.41 times more effective than natural Bigadic clinoptilolite for the remediation of waste streams polluted with Cr(VI).

1 INTRODUCTION

Clinoptilolite is a member of heulandite group of naturally occurring zeolites that exhibit unique physicochemical properties in addition to their availability at low cost. Due to these properties, clinoptilolites have previously been applied as molecular sieves, ion-exchangers, adsorbents and catalysts in industry and environmental protection (Mozgawa & Bajda 2005). The tetrahedral framework of zeolites, composed of SiO_4 and AlO_4, possesses a stable negative charge, which hinders the adsorption of anions and organic pollutants (Zeng et al. 2010; Asgari et al. 2013; Xie et al. 2013). For improvement of anion exchange capacity, charge reversal is necessary. Positive surface charge can be successfully attained via preparation of organo-clinoptilolites through surface modification by long chain organic cations such as surfactants or H^+ clinoptilolites by either acid treatment or deammonization. In the latter pathway, compositional, textural and structural changes accompany the shift in surface charge, which are also preferential for organic modification (Vujakovic et al. 2003). Thus, such treatments are most often applied as a preliminary step of surfactant loading (Vujakovic et al. 2000; Wang and Peng 2010).

Amine-clinoptilolite complexes have previously been proposed for the remediation of wastes including anionic pollutants. However, the majority of the work is focused on modification via ion exchange with a quaternary ammonium salt, most frequently hexadecyltrimethylammonium bromide (HDTMA-Br) (Chutia et al. 2009, Rozic et al. 2009, Bajda & Klapyta 2013). Modification can be realized through an alternative route in which long-chained primary amines are adsorbed on natural clinoptilolite (Vujakovic et al. 2001). Vujakovic et al. (2001, 2003) have prepared modified clinoptilolite samples with oleylamine, which exhibited very high affinity, i.e. more than 90% removal was achieved within the studied concentration range, for SO_4^{2-}, $Cr_2O_7^{2-}$ and $H_2PO_4^-$ anions.

In this study, we prepared organo-clinoptilolites using hexadecylamine (HA) and dodecylamine (DA) as primary amines. The parameters in the modification procedure were optimized to attain maximum amine adsorption. The resulting organo-clinoptilolites were suggested as adsorbents for remediation of wastewaters polluted with Cr(VI) anions.

2 MATERIALS AND METHODS

2.1 *Pretreatment of clinoptilolite*

Natural Bigadic zeolite was ground and classified to particle diameter ranges of 0.106–0.300, 0.300–0.425 and 0.425–0.600 mm. Soluble impurities and dusts were removed by Soxhlet extraction with distilled water for 4 hours. Samples were dried at 105°C for 24 h and stored in a desiccator.

H$^+$ clinoptilolite was prepared in attempt to increase the available sites for amine loading. Accordingly, raw clinoptilolite was contacted with 1M HCl in 5% solids ratio for 2 hours at 60°C under an agitation rate of 100 rpm. The samples were further suspended in this solution for 24 hours at room temperature. Cl$^-$ ions were removed by washing with distilled water and samples were dried at 105°C prior to amine modification.

2.2 *Amine modification*

1-dodecylamine (Alfa Aesar, ≥98%) and 1-hexadecylamine (Alfa Aesar, ≥90%) were used as surfactants. Surfactant solutions were prepared either in hexane, ethanol, isopropyl alcohol (IPA) or chloroform to discuss the effects of solvent on amine grafting. Surfactant solutions were added individually to the suspension of H$^+$-clinoptilolite in water. The system was contacted for 2 hours at 60°C under an agitation rate of 100 rpm. Solid ratio was applied as 1/100, 1/20 and 1/10 and the effect of particle diameter was investigated for ranges 0.106-0.300, 0.300-0.425 and 0.425-0.600 mm. Modified clinoptilolite samples were filtrated and the filtrate was analyzed for amine concentration according to ASTM D-1026 standard method. Solid fraction composed of organo-clinoptilolites and free amine residues was washed with different eluent solutions, i.e ethanol-water (50% v/v) or water, to investigate its effect on the removal of free amines. Finally, amine grafted clinoptilolite samples were dried at 60°C to until constant mass was recorded. The amine concentration in the eluent solution

2.3 *Characterization*

Thermal gravimetric analysis of raw clinoptilolite and amine grafted samples were carried out by Perkin Elmer, Diamond TG/DTA under nitrogen atmosphere within the temperature range of 20–550°C with a heating rate of 5°C/min. Amine loading was calculated from the relative mass loss recorded within the temperature ranges of 200–220°C and 320–330°C.

2.4 *Chromium adsorption*

Stock solution of 100 ppm was prepared by dissolution of extra pure K$_2$CrO$_4$ (Merck) in distilled water. The stock was further diluted to 10 ppm and used in batch adsorption experiments. All trials were run in triplicates under room temperature. Initial pH was adjusted to 3.0, adsorbent dose was 10 g.l^{-1} and the contact time was 10 hours. Agitation rate was held constant at 100 rpm. The concentration of the samples was analyzed spectrophotometrically (Shimadzu UV 1240) at 540 nm using 1,5-diphenyl carbazide reagent. Chromium removal was calculated according to Equation 1:

$$removal\% = (10 - C_t)/10*100 \qquad (1)$$

where C_t = Cr(VI) concentration at the end of 10 hours.

3 RESULTS AND DISCUSSION

3.1 *Solvent selection*

Organo clinoptilolites were prepared by the contact of clinoptilolite sample with a surfactant solution. When modification is realized by ion exchange with quartenary ammonium salts such as

Table 1. The concentration of amines adsorbed upon clinoptilolite.

Solvent	C_{Amines} (mmol.kg^{-1})	
	HA-C	DA-C
Hexane	68.40	71.75
IPA	64.00	68.00
Ethanol	69.00	92.00
Chloroform	117.0	147.00

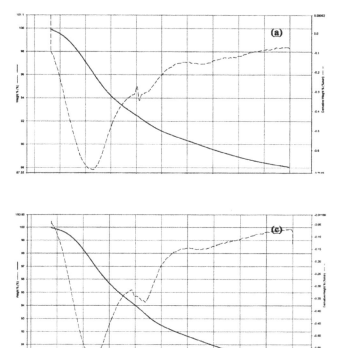

Figure 1. Thermogravimetric curve of DA-C prepared in (a) hexane, (b) ethanol, (c) chloroform.

HDTMA-Br, water is involved as solvent since the salts are soluble in water (Bajda & Klapyta 2013). Long chained primary amines such as HA and DA are insoluble in water but readily soluble in polar and non-polar organic solvents. Therefore, the experiments were designed to include both solvent types, i.e. hexane and chloroform to represent polar solvents and IPA and ethanol to represent non-polar ones. The results of amine loading calculated form the difference between initial and final amine concentration in solution were summarized in Table 1.

According to the results, for both HA and DA, chloroform was far more favorable as a solvent and IPA was the least efficient one. In contrast to the previous literature on clinoptilolite-amine interactions (Li et al. 2002, Xie et al. 2013), the surfactant with lower chain length provided higher surface coverage. For confirmation of these findings, calculations were repeated using thermogravimetric curves of which some are shown in Figure 1a–c.

Table 2. The effect of washing solution on amine adsorption upon clinoptilolite.

Solution	C_{Amines} (mmol.kg^{-1})	
	HA-C	DA-C
Ethanol-water	116.00	140.40
Distilled water	120.00	149.00

Table 3. The effect of solid/solvent ratio on amine loading of Bigadic clinoptilolite.

Solid/solvent ratio	C_{Amines} (mmol.kg^{-1})	
	HA-C	DA-C
1/100	151.50	164.00
1/20	116.00	149.00
1/10	103.50	130.50

The TGA curves shown in Figures 1a–c showed three major peaks between 20–200°C, 220–250°C and 320–330°C. The first region was ascribed to the physical and zeolitic water loss and the other regions were related to the oxidation of organic fraction, with the former representing physisorbed material (Vujakovic et al. 2003). For ethanol, major fraction of organic mass loss occurred in the region of physisorption, which indicated that, the interaction between ethanol and amines were superior over the interaction between amines and clinoptilolite surface.

The amine loadings calculated according to the relative mass losses of the total of latter regions were 70.14 mmol.kg^{-1}, 86.32 mmol.kg^{-1} and 124.09 mmol.kg^{-1} for DA-C prepared in hexane, ethanol and chloroform, respectively. The differences were explained with the removal of free amines, which were initially present on the filtrated clinoptilolite but were removed with washing prior to thermal analysis.

3.2 Selection of the washing solution

The previous works by Vujakovic et al. (2001, 2003) involved a binary solution of ethanol-water of equal volume as washing solution. However, our work showed that there exist strong interactions between HA, DA and ethanol. Therefore, it was suspected that the ethanol present in the washing solution might have removed some fraction of the weakly adsorbed amines from the clinoptilolite surface. Distilled water was utilized as an alternative washing solution and comparative results were shown in Table 2.

According to the results, amine loadings increased by 3.5% and 6.1% as distilled water was used instead of ethanol water mixture. Though the values were slightly lower, increments we also confirmed with TGA as 3.1% and 5.7%. Therefore, distilled water was suggested as washing solution in the modification of clinoptilolites with primary amines.

3.3 The effect of solid ratio

The efficiency of contact between the surfactant and clinoptilolite surface is prone to be affected by the ratio of solids to liquids as such is valid for adsorption studies in which the effect is reflected by either adsorbent dose or shaking rate. Experiments were conducted at 1/100, 1/20 and 1/10 and results were shown in Table 3.

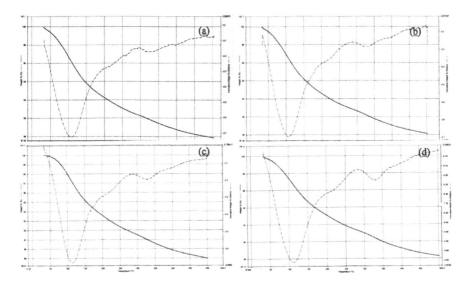

Figure 2.   TGA of modified clinoptilolites. (a) 5/100 (b) 1/100 DA-C, (c) 5/100 HA-C, (d) 1/100 HA-C.

Table 4.   The effect of particle size on amine loading of Bigadic clinoptilolite.

Particle size (nm)	Amines (mmol.kg^{-1})	
	HA-C	DA-C
0.106-0.300	160.70	172.40
0.300-0.425	158.00	165.80
0.425-0.600	151.50	149.00

The results proved that solid ratio was highly influential on the extent of amine loading on Bigadic clinoptilolite. For both HA and DA, amine loading improved by gradually as solid content was lowered. Enhancements of 46.4% and 25.7% were attained as a solid ratio of 1/100 was preferred instead of 1/10. Thermal analyses confirmed the significant enhancements in the amine loads, as shown with the increased relative mass losses within 300–320°C in Figure 2a–d.

3.4   The effect of particle size

Zeolites are renown with their high surface area. However, as a general rule, smaller particle sized adsorbents increase the efficiency of adsorption due to the fact that smaller particles give large surface areas (Amuda et al. 2007). With this regard, we suggested lowering the particle size and obtained the results summarized in Table 4.

As expected, amine loading increased with decreasing particle size. The enhancement was much more significant in the case of DA-C. Thermogravimetric analysis provided quite similar values with differences being in the order of 1.2–5.0% (data not shown).

3.5   Chromate adsorption

Organo clinoptilolites have previously been suggested for use in chromate removal (Zeng et al. 2010, Bajda & Klapyta 2013, Asgari et al. 2013). The uptake capacity was determined as 10.4 mg.g^{-1}, uttermost (Bajda & Klapyta 2013). For comparative purposes, DA-C and HA-C were utilized as adsorbent in Cr(VI) removal and the results were listed in Table 5.

Table 5. The adsorption of chromate by synthesized organo-clinoptilolites.

Adsorbent	Cr(VI) uptake (mg.g^{-1})	Removal (%)
HA-C	60.35	61.00
DA-C	12.38	13.60

The uptake capacities were superior over the value reported by Bajda and Klapyta (2013) for Pohang clinoptilolite modified with HDTMA-Br. Despite the higher amine load of DA-C with respect to HA-C, the latter showed higher efficiency in Cr(VI) removal. This result contradicted with previous literature that explained enhanced removal efficiency with increased surface coverage (Li et al. 2002, Xie et al. 2013).

4 CONCLUSIONS

With this study, Bigadic clinoptilolite with cation exchange capacity and external cation exchange capacity of 0.43 meq.g^{-1} and 0.18 meq.g^{-1} was modified with quaternary amines of 12 carbon and 16 carbon chain lengths. Through determination of ideal modification conditions, amine loading was increased from 117.0 and 147.0 mmol.kg^{-1} to 160.7 and 172.4 mmol.kg^{-1} for HA and DA, respectively. Adsorption experiments proved that the resulting adsorbent could efficiently remove Cr(VI) from synthetic wastewater.

REFERENCES

Amuda O.S., Giwa, A.A. & Bello I.A. 2007. Removal of heavy metal from industrial wastewater using modified activated coconut shell carbon. *Biochemical Engineering Journal* (36): 174–181.

Asgari, G., Ramavandi, B., Rasuli, L. & Ahmadi, M. 2013. Cr (VI) adsorption from aqueous solution using a surfactant-modified Iranian zeolite: characterization, optimization, and kinetic approach. *Desalination and Water Treatment* (51): 6009–6020.

Bajda, T. & Klapyta, Z. 2013. Adsorption of chromate from aqueous solutions by HDTMA-modified clinoptilolite, glauconite and montmorillonite. *Applied Clay Science* (86): 169–173.

Chutia, P., Kato, S., Kojima, T. & Satokawa, S.: Adsorption of As(V) on surfactant-modified natural zeolites. *Journal of Hazardous Materials* (162): 204–211.

Li, Z., Alessi, D. & Allen, L. 2002. Influence of quaternary ammonium on sorption of selected metal cations onto clinoptilolite zeolite. *Journal of Environmental Quality* (31):1106–1114.

Mozgawa W. & Bajda T. 2005. Spectroscopic study of heavy metals sorption on clinoptilolite. *Physical Chemistry of Minerals* (31): 706–713.

Rozic, M., Sipusic, D.I., Sekovanic, L., Miljanic, S., Curkovic, L. & Hrenovic, J. 2009. Sorption phenomena of modification of clinoptilolite tuffs by surfactant cations. *Journal of Colloids Interface Science* (331): 295–301.

Vujakovic, A., Djuricic, M.A. & Tomasevic, E. 2001. Thermal study of surfactant and anion adsorption on clinoptilolite. *Journal of Thermal Analysis and Calorimetry* (63): 161–172.

Vujakovic, A., Dakovic, A., Lemic, J., Mihajlovic, A. & Canovic, T. 2003 Adsorption of inorganic anionic contaminants on surfactant modified minerals. *Journal of Serbian Chemical Society* (68): 833–841.

Wang, S. & Peng, Y. 2010. Natural zeolites as effective adsorbents in water and wastewater treatment. *Chemical Engineering Journal* (56), 11–24.

Xie, Q., Xie, J., Wang, Z., Wu, D., Zhang, Z. & Kong, H. 2013. Adsorption of organic pollutants by surfactant modified zeolite as controlled by surfactant chain length. *Microporous and Mesoporous Materials* (179): 144–150.

Zeng, Y., Woo, H., Lee, G. & Park, J. 2010. Removal of chromate from water using surfactant modified Pohang clinoptilolite and Haruna chabazite. *Desalination* (257): 102–109.

Wastes: Solutions, Treatments and Opportunities – Vilarinho, Castro & Russo (eds)
© *2015 Taylor & Francis Group, London, ISBN 978-1-138-02882-1*

Glycerolysis of two high free fatty acid waste materials for biodiesel production

E.T. Costa, M.F. Almeida & J.M. Dias
LEPABE, DEMM, Faculdade de Engenharia da Universidade do Porto, Porto, Portugal

A. Matos
ENC Energy SGPS SA, S.Félix da Marinha, Portugal

ABSTRACT: Biodiesel is generally produced from edible vegetable oils; however, some waste materials with high free fatty acids (FFA) content might be used if an adequate pretreatment technology is employed. The present study evaluated the use of glycerolysis for the pre-treatment of two high FFA waste materials, for further biodiesel production. The raw-materials were sludges, one from an urban wastewater treatment (sample one – acid value around 140 mg KOH g^{-1}) and one from a wastewater treatment of the dairy industry (sample two – acid value around 4.5 mg KOH g^{-1}). Biodiesel was further produced by alkaline transesterification. The glycerolysis (1:2 glycerol:oil (w/w), 200°C, 100 rpm, 2 h) allowed to decrease between 85% to 96% the FFA content of the raw-materials. The best results were obtained for sample two; however, although with effective pretreatment, transesterification led to a maximum product purity of 67% and relatively low product yields.

1 INTRODUCTION

1.1 *Energy paradigm and waste management*

In order to avoid the deterioration of the ecosystems quality, and aiming to promote the independence of society from fossil fuels (which may soon run low), it is imperative to develop efforts to find substitutes for these conventional fuels. Biofuels can play a key role in this paradigm shift.

Biodiesel is a biofuel generally produced from edible oils and fats (mostly vegetable oils such as rapeseed and soybean). Currently, the efforts are concentrated in increasing the use of alternative (non-edible) raw materials to allow a greater sustainability of its production, taking into account the environmental, social and economic pillars.

The change from a traditional fuel (fossil) to an alternative fuel (biofuel) might be associated with the waste management technologies, with significant environmental and economic benefits (Dias et al., 2009). The question that now arises is whether it is possible to simultaneously change the energy policy and value certain specific wastes streams.

It is possible to foresee that certain types of wastes might present significant oil/fat contents, such as those originated in the activities of: agriculture; aquaculture; forestry; hunting and fishing; processing of animals and food products; and, industrial or municipal wastewater treatment. Such wastes are difficult to process, namely due to the high levels of free fatty acids (FFA) usually found and many times end up being landfilled (last priority regarding waste management strategies). The use of such waste materials might promote the energy paradigm shift and contribute for the reduction of the generated waste and the economic and environmental costs associated with their disposal.

1.2 *Pretreatment of waste by glycerolysis*

Oils and fats with high FFA content are unsuitable to be used directly in the conventional biodiesel production process (alkali transesterification). It is generally considered that less than 1 wt.% of FFA (around 2 mg KOH g^{-1}) is required for an effective reaction, although some studies report higher values (Canakci, 2007; Dias et al., 2013). It is therefore necessary to make a pre-treatment of these oils, since they may present FFA contents between 20–50 wt.% (Gerpen, 2004).

The glycerolysis is an esterification reaction that can be used for that purpose and by which the fatty acids react with glycerol being converted into mono, di and triglycerides and water (Felizardo et al., 2011; Gole and Gogate, 2014). In order to shift the reaction towards the products it is necessary to remove water from the reaction vessel as it is being generated, and, therefore, the reactional systems should take into account of this issue.

This reaction is highly endothermic being generally conducted at high temperature and low pressure, being also dependent upon the glycerol:FFA ratio established (Gole and Gogate, 2014). No clear information exists concerning the relative distribution of the obtained products at the different reaction conditions used.

In comparison with other pretreatment processes (acid esterification and saponification), glycerolysis is recognized as very effective when raw materials present very high acid values. An initial FFA content up to 60 wt.% is reported (Felizardo et al., 2011).

An additional advantage of using glycerolysis is that, although conducted at relatively high temperature (ex. 200°C), it can be performed in the absence of catalyst and therefore it does not require additional purification processes of the product (as for instances when using acid esterification) as well as loss of product, therefore reducing the complexity of the overall process and the associated economic and environmental costs (Costa et al., 2013). To speed up the reaction, it is however to possible to use heterogeneous acid catalysts, such as zinc chloride (Gerpen, 2004).

At the end of the glycerolysis reaction, a product of low acidity is obtained, which is easily separated from the glycerol present (in excess), through decantation.

2 MATERIALS AND METHODS

2.1 *Raw materials – Waste samples*

Sludge from an urban wastewater treatment plant (Sample 1) and from a wastewater treatment plant of a dairy industry (Sample 2) were used. Both samples were collected at facilities located in the North of Portugal.

2.2 *Raw materials – Characterization*

In order to assess the feasibility of using the selected waste materials, considering the glycerolysis pretreatment, the samples from the wastewater treatment were characterized in terms of density (NP 83:1965), dry matter content (EN 12880:2000) and oil and fats content (Method 9071 A (standard methods)). For each parameter and for each sample, determinations were carried out in triplicate (results are presented as mean values). After oil and fats separation, the acid value of the raw materials was determined (EN ISO 660:2000). Reagents used were of analytical grade.

2.3 *Separation of oils and fats*

Given the source of the samples, it is expected that they have a heterogeneous composition, containing, in addition to oils and fats, other substances, such as water and debris. In agreement, a simple process was established to separate the oils and fats, aiming future application at an industrial level. The process consisted on decanting at 80°C, to separate oils, fats and water from other substances, and after centrifuging during 4 min at 4000 rpm, to separate the lipid fraction from the water. The lipid fraction obtained was further pre-treated by glycerolysis.

Figure 1. Laboratory setup used for the glycerolysis reaction.

2.4 *Pretreatment of oils and fats by glycerolysis*

Figure 1 shows the laboratory equipment used for the glycerolysis reaction. The setup included a small electric furnace, to ensure the required temperature of the reaction, a reaction vessel of stainless steel with 250 mL (where the reaction occurred), a condenser connected to a vacuum pump (to condense and extract water from the reaction) and a mechanical stirrer (to ensure homogenization and appropriate contact between the reactants). The process started by inserting the required amount of oil/fat inside the reactor (50 g), which was heated to the reaction temperature (200°C). After, 25 g of glycerol (commercial, Higilin brand), at the same temperature, were added and the reaction started. The reaction mixture was stirred a speed of 100 rpm. In order to evaluate the evolution of the glycerolysis reaction over time, samples (1 mL each time) were extracted at different periods of time (0 min, 15 min, 30 min, 60 min, 90 min and 120 min) through a septum using a syringe. The acid value of each withdrawn sample was determined (EN ISO 660:2000). After the reaction, the reactor content was transferred into a separatory funnel and settling occurred during 2 hours to ensure total phase separation. The glycerolysis reaction was performed in triplicate for each sample.

2.5 *Transesterification of the glycerolysis product*

The product obtained from 2.4 was subjected to an alkali transesterification reaction for biodiesel production. The laboratory setup used included a glass reaction vessel equipped with a thermometer and connected to a condenser, immersed in a heating bath; magnetic stirring was performed. For biodiesel synthesis, 30 g of the pre-treated sample were used. The reaction was performed, in duplicate, at the conventional reaction conditions, which were: 65°C; 6:1 molar ratio of methanol (99% purity) to oil/fat; 1%wt. of NaOH (97% purity) as catalyst; and 1 h (Dias et al., 2008).

After the end of the reaction, the contents of the reaction vessel were cooled, transferred to a sep-aratory funnel and left to decant for one hour in order to ensure the complete separation of biodiesel and glycerol. Biodiesel was purified by washing, first with 50% (V/V) of an acid solution (0.5% HCl) and after repeatedly with 50% (V/V) of distilled water until the pH of the washing water was

Table 1. Characterization of the sludge samples.

	Sample 1	Sample 2
Density (kg/m^3)	969.65	918.03
*CV (%)	1.78	0.38
Dry Matter (wt.%)	78.37	52.12
CV (%)	5.37	7.76
Oil and Fat Content (wt.%)	89.63	41.65
CV (%)	11.38	1.69

*CV – coefficient of variation.

the same as the distilled water and after dried. Finally, the product was characterized by gas chromatography regarding total methyl esters and linolenic methyl ester content (EN 14103 – 2003) as well as iodine number (EN 14214 – 2008). Composition was also accessed by gas chromatography.

3 RESULTS AND DISCUSSION

3.1 *Raw-materials characterization*

Table 1 summarizes the results obtained for the different evaluated parameters.

The results obtained show that the main differences between the samples relate to the dry matter content as well as the oil and fats content. The urban wastewater treatment sludge (sample 1) presented significantly higher dry matter content and also higher content of oils and fats. In both cases, the oil and fats content is very significant. The variation observed in the results for the oil and fats content of sample 1 (CV = 11.38%) might indicate a higher heterogeneity of the sample.

3.2 *Pretreatment of oils and fats by glycerolysis*

After oil/fats separation, the acid value of sample 1 was 139.69 mg KOH g^{-1} whereas for sample 2, the acid value, although high, was low in comparison, being 8.99 mg KOH g^{-1}.

Figure 2 shows the evolution of the acid value along the reaction of glycerolysis, for sample 1.

It was found that during the first 15 min there was a reduction in the acid value of 92%, reaching the final value of 8.21 mg KOH/g. Despite the rapid decrease of acid value verified in the first 15 minutes, the value at the end of 120 minutes (6.13 mg KOH g^{-1}) was still above the threshold for use in transesterification. In fact, under the studied conditions, it was not possible to decrease further the acid value of the raw material. Considering that at the end of the reaction only monoglycerides and triglycerides exist, the yield was 95%.

Figure 3 shows that for sample 2 a drastic reduction of the acidity occurs immediately after the addition of glycerol to the reaction vessel (initial acid value less than 4.5 mg KOH g^{-1}), suggesting that the reaction occurs at high speed, with immediate conversion of mono, di and triglycerides. After 90 min, the decrease of the acid value is low. The final value of 1.37 mg KOH g^{-1} was obtained indicating the adequacy of this raw-material for biodiesel production. For this sample, the yield was 127%, also considering that at the end of the reaction only monoglycerides and triglycerides exist (the result indicates that such an assumption presents an associated error).

3.3 *Transesterification of glycerolysis product*

The pre-treated oil was subjected to the alkali transesterification reaction. For the case of sample 1, a yield (wt.product/wt.raw material × 100) of 80 wt.% was obtained while for sample 2, the yield was 48 wt.%. The low yield obtained for sample 2 should be related to the significant formation of emulsions observed during the wash phase (Dias et al., 2009).

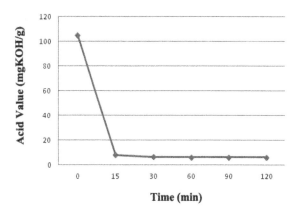

Figure 2. Evolution of the acid value along the glycerolysis reaction for sample 1.

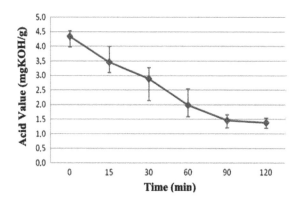

Figure 3. Evolution of the acid value along the glycerolysis reaction for sample 2.

Table 2 shows the methyl ester composition of both samples. For biodiesel obtained from the pretreated sample 1, the composition shows as most significant esters the methyl ole-ate, linoleate and palmitate, whereas for sample 2, the most significant esters were the methyl palmitate, oleate, stearate and myristate.

The composition of the biodiesel reflects the composition of the original oil/fat (Dias et al., 2009). The significant differences in terms of unsaturation degree might affect the stability (related to the presence of double bonds) and the cold flow properties of the products.

In terms of methyl esters content (purity), despite the high product yield obtained for sample 1, most of that product contains other substances which are not methyl esters, since only a 16.5 wt.% purity was obtained for this sample. In the case of sample 2, a purity of 67 wt.% was obtained showing that most of the product was in fact methyl esters. However, even in the case of sample 2, the product is far from meeting the minimum quality requirements for this parameter (>96.5 wt.%). The linolenic methyl ester content was within the limit (<12 wt.%); in fact, a very small content was found for sample 1 and this ester was not detected at sample 2.

Regarding the iodine value, the results found are significantly lower than the maximum of 120 g $I_2/100$ g imposed by EN14214. For sample 1 the value obtained was 83 g $I_2/100$ g whereas for sample 2 the result was 27 g $I_2/100$ g. Although within the limit, the low iodine values obtained indicate poor cold flow properties of the products.

Table 2. Composition of biodiesel produced from the two pre-treated samples.

Ester of the following fatty acids	Sample 1 (wt.%)	Sample 2 (wt.%)
Capric (C10:0)	–	2.7
Lauric (C12:0)	–	3.5
Myristic (C14:0)	1.1	11.9
Palmitic (C16:0)	20.0	35.6
Stearic (C18:0)	6.9	11.7
Oleic (C18:1)	41.2	24.6
Linoleic (C18:2)	26.0	2.6
Linolenic (C18:3)	0.8	–
Others	4.0	7.4

4 CONCLUSIONS

The feasibility of using wastewater treatment sludge as alternative raw material for biodiesel production was accessed and glycerolysis was studied as pre-treatment to reduce the acid value of the oil/fats present.

The glycerolysis performed using a 1:2 glycerol:oil mass ratio, at 200°C, 100 rpm and during 2 h, allowed to decrease significantly the FFA of the raw-materials (between 85 and 96%); however, in the case of the sludge from the urban waste water treatment, due to the extremely high initial acid value of the separated oils/fats (139.69 mg KOH g^{-1}), it was not possible to achieve the required acid value after pre-treatment. The best final results were obtained for the industrial wastewater treatment sludge (dairy), but the transesterification of the pre-treated sample led to a maximum product purity of 67% and relatively low product yield. The methyl ester profiles obtained are significantly different for both samples and the iodine values, although in agreement with the standard, are very low, which indicate poor cold flow properties.

Overall, the results show that glycerolysis is an effective pre-treatment and that, to ensure the feasibility of using such wastes for biodiesel production, there is the need to conduct optimization experiments for the glycerolysis pre-treatment as well as for the transesterification process in order to obtain products with higher quality, to fulfil the quality requirements imposed by Biodiesel European Quality Standard.

REFERENCES

Canakci, M. 2007. The potential of restaurant waste lipids as biodiesel feedstocks. *Bioresource Technology* 98: 183–190.

Costa, J. F., Almeida, M. F., Alvim-Ferraz, M. C. M., and Dias, J. M. 2013. Biodiesel production using oil from fish canning industry wastes. *Energy Conversion and Management* 74: 17–23.

Dias, J. M., Alvim-Ferraz, M. C. M., and Almeida, M. F. 2008. Comparison of the performance of different homogeneous alkali catalysts during transesterification of waste and virgin oils and evaluation of biodiesel quality. *Fuel* 87: 3572–3578.

Dias, J. M., Alvim-Ferraz, M. C. M., and Almeida, M. F. 2009. Production of biodiesel from acid waste lard. *Bioresource Technology* 100: 6355–6361.

Dias, J. M., Araújo, J. M., Costa, J. F., Alvim-Ferraz, M. C. M., and Almeida, M. F. 2013. Biodiesel production from raw castor oil. *Energy* 53: 58–66.

Felizardo, P., Machado, J., Vergueiro, D., Correia, M. J. N., Gomes, J. P., and Bordado, J. M. 2011. Study on the glycerolysis reaction of high free fatty acid oils for use as biodiesel feedstock. *Fuel Processing Technology* 92: 1225–1229.

Gerpen, J. V., Shanks, B., Pruszko, R., Clements, D., and Knothe, G. 2004. Biodiesel Production Technology. *National Renewable Energy Laboratory*.

Gole, V. L., and Gogate, P. R. 2014. Intensification of glycerolysis reaction of higher free fatty acid containing sustainable feedstock using microwave irradiation. *Fuel Processing Technology* 118: 110–116.

Wastes: Solutions, Treatments and Opportunities – Vilarinho, Castro & Russo (eds)
© 2015 Taylor & Francis Group, London, ISBN 978-1-138-02882-1

The effect of a joint clay-microorganism system to treat Ni and diethylketone solutions

F. Costa, B. Silva & T. Tavares

Centre of Biological Engineering, University of Minho, Braga, Portugal

ABSTRACT: The capacity of a combined system using a *Streptococcus equisimilis* biofilm supported in vermiculite to treat aqueous solutions polluted with diethylketone and nickel was accessed. In order to evaluate the interaction between the sorbent matrices and the two adsorbates several batch experiments were performed exposing 1) different amounts of vermiculite to Ni and diethylketone (Singular Sorbent Systems – SSS) or 2) a biofilm supported in different amounts of vermiculite to both pollutants (Binary Sorbent Systems – BSS). Fed batch pilot scale experiments were also conducted. For the SSS experiments, the removal of 3 g/L of DEK was complete for all the assays, whereas the removal of 0.45 g/L of Ni ranged between 31% and 100%. For the BSS experiments, the removal percentages of DEK and Ni decreased (77% to 97%, 23% to 97% respectively). Maximum removal percentages of 87.3% of DEK and 57.6% of Ni were reached in the pilot scale experiments.

1 INTRODUCTION

Water and soil contamination with hazardous compounds has triggered and attracted increasing attention by researchers worldwide. Heavy metals, polycyclic aromatic hydrocarbons, dyes and ketones are examples of such hazardous substances. Although heavy metals and ketones exist in nature, their overuse in several anthropogenic activities and their subsequent release into the environment, lead to their accumulation in several environmental matrices.

Among the several metals known to exert several hazardous effects on the environment and on living beings, Ni stands out, not only due to its extensive use in numerous industries (mining operations, electronics, batteries and petrochemical industries), but also due to its carcinogenic, embryotoxic and nephrotoxic properties as well as its ability to cause several types of acute and chronic health disorders (Suazo-Madrid et al., 2011). Diethylketone, also known as DEK, is a simple, symmetrical dialkyl ketone that has been used in several activities (Quintelas et al., 2013) and it is known to cause dizziness, tachycardia, fainting and even cause coma or death in cases of prolonged exposure. Thus, taking into consideration the negative impact on the environment, it is urgent the development of green methods able to remove those substances from the environment and thus minimize their impact, as well as the development and implementation of stricter environmental legislation.

2 OBJECTIVES

The main objective of this work is the development of an environment-friendly technology able to treat aqueous solutions containing both Ni and DEK. This two pollutants are usually found together in wastewaters from different industries such as electronics, metal extraction and paint production. The simultaneous treatment of aqueous solutions contaminated with heavy metals and VOCs is still in a very preliminary state and in order to tackle with this specific pollution problem this work was divided in two stages. In a first stage two sets of batch adsorptions assays were conducted: one combining different doses of vermiculite and a specific concentration Ni and DEK. These assays aimed to determine the adsorption capacity of vermiculite towards these two pollutants.

In a second stage, batch adsorption experiments were conducted combining the use of concentrated biomass of *S. equisimilis* and different concentrations of sterilized vermiculite to remove and/or biodegrade, respectively, Ni and DEK from aqueous solutions. This experiments aimed 1) to determine the ability of this joint clay-microorganism system to simultaneously remove both pollutants from aqueous solutions, 2) to determine whether these two substances compete for the available adsorption sites either on the biomass or on the vermiculite and 3) to infer about the role of the biomass on the sorption and degradation process. The kinetic parameters were estimated using five growth kinetic models and four biodegradation kinetic models all reported in literature (Costa et al., 2012; Raghuvanshi and Babu 2010). The experimental equilibria data were analyzed using the BET, Dubinin-Radushkevich (D-R), Langmuir and Freundlich adsorption isotherms, also reported in literature (Khamis et al., 2009; Saravanan et al., 2009).

3 MATERIALS AND METHODS

Streptococcus equisimilis (CECT 926) obtained from the Spanish Type Culture Collection – University of Valencia was used in this work. Brain Heart Infusion (BHI, OXOID CM1135) was used for bacteria growth. The vermiculite was obtained from Sigma-Aldrich, and presented an average particle diameter of 8.45 mm, BET surface area of 39 m^2/g and porosity of 10%. DEK was purchased from Acros Organics (98% pure) and it was diluted in sterilized distilled water. Stock solutions of nickel (1 g/L) were prepared by dissolving an appropriate amount of $NiCl_2 6H_2O$ (Carlo Erba Reagents, CAS number 7791-20-0) in 1 L of sterilized distilled water. The range of nickel concentrations (5 mg/L to 450 mg/L) was obtained by dilution of the stock solution. All glassware used for experimental purposes was washed in 60% HNO_3 and subsequently rinsed with deionised water to remove any possible interference by other metals.

3.1 *Adsorption assays – single sorbent systems (SSS)*

The singular sorbents systems adsorption assays were conducted with 1 L Erlenmeyer flasks containing 0.425 L of Ni (0.45 g/L) and DEK (3 g/L) solution and different amounts of sterilized vermiculite (5 g and 7.5 g). The flasks were kept at 150 rpm and 37°C until equilibrium was reached. Previous assays were made in order to determine the time required for equilibrium to be reached (circa 20 hours for Ni and DEK). Samples of 1 mL were collected, centrifuged at 13400 rpm for 10 minutes, and the supernatant was analyzed by GC or acidified with nitric acid to be analyzed by ICP-OES, respectively for the determination of DEK and Ni concentrations. The control consisted of aqueous solutions with Ni and DEK to determine any kind of Ni or DEK adsorption by the Erlenmeyer flasks.

3.2 *Adsorption assays – binary sorbent systems (BSS)*

The binary sorbents assays were performed with 2 L Erlenmeyer flasks containing 0.850 L of Ni (0.45 g/L), DEK (3 g/L) solution and biomass supported in vermiculite (0.1 g to 10 g). All the flasks were rotated at a constant rate of 150 rpm until equilibrium was reached. Previous assays were made to determine the time needed for equilibrium to be reached (5 to 7 days). Samples of 1 mL were collected, centrifuged at 13400 rpm for 10 minutes, and the supernatant was used for the determination of DEK and Ni concentrations. The samples were analyzed by GC and by ICP-OES, respectively for DEK and for Ni. A control with Ni, DEK and vermiculite was used to infer about the influence of the biomass on the sorption of both pollutants.

All the assays were conducted in duplicate and the results are an average of both duplicates. The relative standard deviation and relative error of the experiments measurements were less than 2 and 5%, respectively.

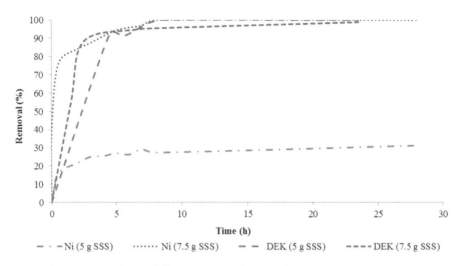

Figure 1. Ni and DEK sorption on different amounts of vermiculite on singular sorbent system (SSS).

3.3 *Adsorption assays – pilot scale*

The pilot adsorption assays were performed with a column with $16\,dm^3$ and can be divided in two stages. The first stage aimed the growth and development of the *S. equisimilis* biofilm on the vermiculite (700 g) and had the following conditions: 40 L of HBI culture medium with a flow of 125 mL/min at 25°C during 7 days. The second stage consisted in exposing the biofilm formed previously to an aqueous solution of DEK (7.5 g/L) and Ni (0.1 g/L) with a flow of 25 mL/min. This aqueous solution (S1) was subsequently replaced by a new solution (S2 and S3) containing DEK and Ni. At the end of the experiment only DEK was added to the S3 aqueous solution (S3 – DEK). Samples were collected and analyzed by GC and by ICP-OES, respectively for DEK and Ni concentration.

4 RESULTS AND DISCUSSION

4.1 *Adsorption assays – single sorbent systems (SSS)*

It is possible to conclude that for the SSS assays, the removal of DEK was always higher than 95%, reaching complete removal after about 8 hours for 5 g of vermiculite. For Ni, the removal percentage ranged between 31.13% and 100%, respectively for 5 g and 7.5 g of vermiculite (Figure 1).

Costa et al. (2015 a, b) studied the adsorption of DEK and/or Ni into vermiculite (single sorbate assays). According to these authors, the sorption of DEK (3 g/L) increase as the amount of vermiculite employed increased, reaching values of almost 100% for the assays using vermiculite doses equal and higher than 5 g. According to several authors this behavior can be explained as a result of an increased surface area, which is translated into a greater number of sites available for sorption (Quintelas et al., 2011). Regarding the sorption of Ni, these authors conclude that the percentage of initial Ni sorbed is similar (71% to 84%) for all assays conducted with different doses of vermiculite. When comparing these results it is possible to infer that the presence of Ni has a synergetic effect on the DEK's removal from aqueous solutions.

For both Ni and DEK, and for all the vermiculite doses tested, the kinetic model that best describes the obtained results is the pseudo-second order ($0.975 \leq R^2 \leq 0.999$ and $0.977 \leq R^2 \leq 0.999$, respectively).

The relationship between the pseudo-second order parameters and the adsorption performance was studied by several authors (Wu et al., 2009; Ofomaja, 2010). Wu et al. (2009) established

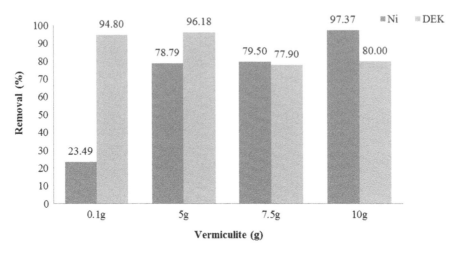

Figure 2. Removal percentage of Ni and of DEK using a biofilme of *S. equisimilis* supported into different doses of vermiculite.

an approaching equilibrium factor (R_w) which represents the characteristics of kinetic curve of an adsorption system using the pseudo-second order model. This approaching equilibrium factor (R_w) is defined as:

$$R_w = 1 / (1 + K_2 \cdot q_e \cdot t_{ref}) \tag{1}$$

In here t_{ref} is the longest operation time (based on kinetic experiments), q_e and K_2 correspond respectively to the uptake and to the pseudo-second order constant. For all the assays and for both pollutants the R_w factor was found to range between 0.006 and 0.120 which indicates that the kinetic curve is largely curved with a good approach to equilibrium, confirming the good performance of the system.

4.2 Adsorption assays – Binary Sorbent Systems (BSS)

Regarding the biosorption experiments it was possible to conclude that as the amount of vermiculite doses increases, the removal percentage of Ni also increases reaching its maximum removal value with 10 g of vermiculite (97.37%). For DEK, the maximum removal percentage was obtained for 5 g of vermiculite (96.18%) (Figure 2).

For vermiculite amounts higher than 5 g the removal percentage of DEK decreased significantly (about 16%) and may be justified by the decrease of the biomass concentration through time, which leads to the formation of a weaker and not fully developed biofilm.

For both Ni and DEK and for all the vermiculite amounts tested, the kinetic model that best describes the obtained results is the pseudo-second order ($R^2 > 0.986$ and $R^2 > 0.873$, respectively). The adsorption isotherm that best describe the results for the biosorption equilibrium of Ni was the BET isotherm ($R^2 = 0.997$). The high value obtained for Q_{max} (3.50×10^8) for the BET model and the fact that C_s is smaller than C_e suggest that the BET approaches the Langmuir isotherm.

The best fit for the biosorption of DEK was obtained with the Langmuir adsorption isotherm ($R^2 = 0.871$), followed by the Freundlich adsorption isotherm ($R^2 = 0.801$). These results suggest that the adsorption of DEK occurs in a monolayer and that all adsorption sites are equally probable (Costa et al., 2015a).

During the analyses of BSS samples experiments, it was possible to observe the appearance of several intermediates. These compounds were subsequently identified as 2-pentanone,

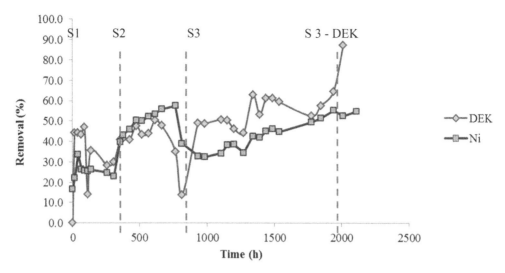

Figure 3. Removal percentage of DEK and Ni in the pilot scale assays, using a biofilme of *S. equisimilis* supported into different doses of vermiculite. S1 – Solution with Ni and DEK at time zero; S2 – replacement of the S1 solition, for a new Ni and DEK solution; S2 – replacement of the S2 solition, for a new Ni and DEK solution; S3 – DEK – addition of DEK to the S3 solution.

methyl-acetate and ethyl-acetate and did not appear at the end of the experiments, which means that they were also subjected to degradation by the *S. equisimilis* biofilm.

4.3 *Adsorption assays – pilot scale*

It was possible to infer that the removal of DEK was continuous through time, whereas the removal of Ni decreased significantly since the replacement of the S1 solution for the S2 solution (Figure 3).

These results can be explained by the increasing number of molecules competing for the active sites of the sorbent (not only Ni and DEK, but also DEK's metabolites), which can be translated by the decrease of removal percentages. Maximum removal percentages of 57.6 and 87.3% were obtained for Ni and DEK respectively.

Meena et al. (2005) studied the removal of several heavy metals from aqueous solutions using carbon aerogel as an adsorbent. These authors obtained Ni removal percentages of about 50%, 60%, 65% and 80% for carbon aerogel adsorbent doses of 5 g/L, 8 g/L, 10 g/L and 12 g/L respectively and an initial concentration of 3 mg/L. When comparing the results obtained by these authors, with the results obtained herein it seems that apart from the superior removal percentages, the initial concentration of Ni is also superior (150 times superior) obtained, which reveals and corroborate the good performance and capacity of these systems to decontaminate aqueous solutions conclusions.

5 CONCLUSIONS

From the results presented in this work it is possible to concluded that when Ni and DEK are mixed together in singular sorbent systems the maximum removal percentage of Ni decreases significantly to 30% for the assays conducted with 5 g of vermiculite, whereas the maximum removal percentages obtained for DEK were always higher than 71%, reaching a maximum value of almost 100%. It was also demonstrated that a joint system composed by a biofilm of *S. equisimilis* supported on vermiculite is able to efficiently remove Ni from aqueous solution and degrade and/or adsorb DEK. The presence of DEK does not affect negatively the removal of Ni and the presence of biofilm is a benefit in this type of systems, since it increases the removal of Ni and it degrades not only DEK but also its intermediates. The removal percentages obtained in the pilot scale experiments are lower than the ones obtained for the BSS system assays, but still it is possible to infer that the system can remove high concentration of Ni and DEK from aqueous solutions.

ACKNOWLEDGMENTS

The authors would like to gratefully acknowledge the financial support of this project by the Fundação para a Ciência e Tecnologia, Ministério da Ciência e Tecnologia, Portugal and co-funding by FSE (programme QREN–POPH), by the "FCT Strategic Project PEst-OE/EQB/LA0023/2013" and by the "BioEnv - Biotechnology and Bioengineering for a sustainable world", REF. NORTE-07-0124-FEDER-000048, co-funded by the Programa Operacional Regional do Norte (ON.2 – O Novo Norte), QREN, FEDER. Filomena Costa thanks FCT for a PhD Grant (SFRH/BD/77666/2011).

REFERENCES

Costa, F., Quintelas, C. & Tavares, T. 2012. Kinetics of biodegradation of diethylketone by *Arthrobacter viscosus*. *Biodegradation* 23: 81–92.

Costa, F., Silva, B. & Tavares, T. 2015a. Simultaneous sorption of cadmium and diethylketone by *Streptococcus equisimilis* supported on vermiculite: kinetics and equilibrium studies. (*Unpublished results*).

Costa, F., Silva, B. & Tavares, T. 2015b. Nickel and diethylketone simultaneous biosorption by *Streptococcus equisimilis* biofilm supported on vermiculite. (*Unpublished results*).

Khamis, M., Jumea, F. & Abdo, N. 2009. Speciation and removal of chromium from aqueous solution by white, yellow and red UAE sand. *Journal of Hazardous Materials* 169: 948–952.

Meena, A. K., Mishra, G. K., Rai, P. K., Rajagopal, C. & Naga, P. N. 2005. Removal of heavy metals from aqueous solutins usingcarbon aerogel as na adsorbent. *Journal of hazardous materials* 122 (1–2): 161–70.

Ofomaja, A. E. 2010. Biosorption studies of Cu(II) onto Mansonia sawdust: process design to minimize biosorbent dose and contact time. *Reactive and Functional Polymers* 70: 879–889.

Quintelas, C., Figueiredo, H. & Tavares, T. 2011. The effect of clay treatment on remediation of diethylketone contaminated wastewater: Uptake, equilibrium and kinetic studies. *Journal of Hazardous Materials* 186: 1241–1248.

Quintelas, C., Costa, F. & Tavares, T. 2013. Bioremoval of diethylketone by the synergistic combination of microorganism and clays: uptake, removal and kinetic studies. *Environmental Science and Pollution Research* 20: 1374–1383.

Raghuvanshi, S. & Babu, B.V. 2010. Biodegradation kinetics of methyl iso-butyl ketone by acclimated mixed culture. *Biodegradation* 21: 31–42.

Saravanan, P., Pakshirajan, K. & Saha, P. 2009. Batch growth kinetics of an indigenous mixed microbial culture utilizing m-cresol as the sole carbon source. *Journal of Hazardous Materials* 162: 476–481.

Suazo-Madrid, A., Morales-Barrera, L., Aranda-Garcia, E. & Cristiani-Urbina, E. 2011. Nickel (II) biosorption by *Rhodotorulaglutinis*. *Journal of industrial Microbiology & Biotechnology* 38: 51–64.

Volesky, B. 1990. Biosorption of heavy metals, Boca Raton, Florida, CRC press.

Wu, F.C., Tseng, R.L., Huang, S.C. & Juang, R.S. 2009. Characteristics of pseudo-second-order kinetic model for liquid-phase adsorption: a mini-review. *Journal of Chemical Engineering* 151: 1–9.

Wastes: Solutions, Treatments and Opportunities – Vilarinho, Castro & Russo (eds)
© 2015 Taylor & Francis Group, London, ISBN 978-1-138-02882-1

Plastic waste use as aggregate and binder modifier in open-graded asphalts

L. Costa, H. Silva, J. Oliveira, S. Fernandes & E. Freitas
CTAC – Territory, Environment and Construction Centre, University of Minho, Guimarães, Portugal

L. Hilliou
IPC/I3N – Institute of Polymers and Composites, University of Minho, Guimarães, Portugal

ABSTRACT: Road pavements are very important infrastructures for the Society, but they can cause serious environmental impacts during construction, operation and rehabilitation phases. Thus, it is essential to develop surface paving solutions that promote not only the durability but also a comfortable and safe use. In fact, this work aims to study the properties of new open-graded mixtures for surface layers produced with plastic wastes. First, HDPE and EVA wastes were used as bitumen modifiers, and then another plastic waste (PEX) replaced part of the aggregates. After studying the modified binders, the open-graded mixtures were designed, and then they were tested concerning their particle loss, rutting resistance, surface texture and damping effect. It was concluded that both ways of using the plastic wastes can improve the mechanical and functional properties of the open-graded mixtures related to the pavement performance.

1 INTRODUCTION

The rapid urbanization growth results in higher traffic levels creating the need for roads with improved performance. However, building a road pavement requires large amounts of materials, and their extraction can lead to the devastation of natural resources and cause negative impacts on the environment.

The asphalt mixture, a matrix of aggregates linked with an asphalt binder, is one of the most important materials used for road paving. The aggregates represent approximately 90% of asphalt mixtures, and the alternative of their partial substitution by plastic wastes can be considered as a sustainable technology, since an equivalent performance can be assured (Robinson, 2004). Regarding the asphalt binder, it is the most valuable constituent and largely responsible for the asphalt mixture performance (Becker et al., 2001).

One method often used to improve the asphalt mixtures mechanical performance is the addition of polymers (Becker et al., 2001), which can be applied in asphalt mixtures by using the wet method (as asphalt binder modifier) or the dry method (when the polymers are added to the aggregates). However, the use of virgin polymers can even double the final price of asphalt binders (Kalantar et al., 2012). The use of plastic wastes, instead of virgin polymers, can be a good answer for that economic concern, being also a better environmental solution.

Besides the road mechanical performance, related to the durability and structural design of asphalt road pavements, the pavement must also provide a safe, comfortable and noiseless surface. These functional characteristics are related with the tire/pavement interaction, adhesion and noise, and the projection of water in wet weather, which in turn are related with the porosity, the surface texture, and the incorporation of polymers (usually elastomers) (Biligiri, 2013).

Open-graded and porous asphalts mixtures are frequently applied in surface layers due to their ability to reduce water splash, aquaplaning and noise, promoting the adhesion between the tires to and the pavement mainly in wet conditions. Due to the high porosity of these types of asphalt

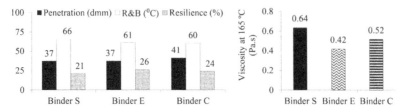

Figure 1. Basic properties of the control binder and of the binders produced with plastic wastes.

mixtures, they need to be carefully designed using modified binders with better rheological properties in order to assure a good disaggregation resistance (Arrieta & Maquilón, 2014).

Because these types of mixtures are so challenging, the objective of this study is to evaluate the performance of open-graded mixtures produced with new waste plastic modified binders, and in addition the possibility of using another plastic waste to substitute part of the aggregates.

For open-graded asphalt mixtures, the main problems caused by their high porosity are typically the low cohesion or resistance to disaggregation and the low rutting resistance, and these problems are deeply dependent on the content or type of binder used. Thus, the particles loss and the rutting resistance in the wheel tracking test will be assessed in order to evaluate the suitability of the new solution with plastic wastes for production of open-graded asphalt mixtures.

A method based on the spectra of hammer impacts on the pavement (Sandberg, 1987, Nils, 2009) may also be used to indirectly measure the asphalt pavement stiffness.

2 MATERIALS AND METHODS

2.1 *Materials*

A 70/100 penetration grade bitumen was used as a base bitumen for the production the new plastic waste modified binders. Another commercial polymer modified binder Elaster 13/60 (S) was used as control binder for comparison reasons (also during asphalt mixtures production).

The waste plastics used for bitumen modification were recycled EVA and HDPE. These two plastic wastes were selected as being those more viable for bitumen modification among several plastic wastes tested in previous studies (Costa et al., 2013a, Costa et al., 2013b).

A modified binder with elastic properties should be used to produce open-graded (OG) mixtures (Biligiri, 2013), thus justifying the inclusion of EVA wastes in both modified binders produced in this study, namely:

– a 70/100 penetration grade bitumen modified with 5% EVA (E);
– a 70/100 penetration grade bitumen modified with 4% EVA and 2% HDPE (C).

The binder modification was performed in a high speed mixer (IKA T65 ULTRA-TURRAX disperser) during 20 minutes at 160°C, and at 5000 rpm.

The basic asphalt binder's properties were assessed, namely the penetration value (EN 1426), the softening point or R&B temperature (EN 1427), the recovery after penetration (resilience) (EN 13880-3) and the dynamic viscosity (EN 13 302). Those properties obtained for the commercial binder and for both binders produced with plastic wastes are presented in the Figure 1.

It is a valuable fact that all modified binders are in the range of a 35/50 penetration grade bitumen typically used in Portugal. Furthermore, it is expected that the commercial binder (S) would have an improved high temperature performance due to its better R&B results. The resilience of all binders is similar. Mixtures also need to be workable at high temperatures, and here the binder with EVA (E) presents best results due to its lower viscosity at 165°C.

Granite aggregates and limestone filler were used to obtain the aggregate matrix of the OG mixture. In addition, a crosslinked polyethylene (PEX) plastic waste partly (5.5% in volume) also substituted the aggregates used in the production of another open-graded mixture (OGP).

Table 1. Percentage by volume of each fraction of aggregates used in the studied OG and OGP mixtures.

Mixture	Fraction 0/4	Fraction 4/6	Fraction 6/14	Fraction 6/12.5	PEX 0.5/4	Filler
OG	14.0%	20.0%	25.0%	40.0%	0.0%	1.0%
OGP	11.5%	17.0%	25.0%	40.0%	5.5%	1.0%

According to EN 12697-5, the density of PEX is 936 kg/m^3, while the density of the stone aggregates is 2.8 times higher (nearly 2650 kg/m^3). Taking into account that the dry methods of polymer modification typically use rates of 2% to 4% of the asphalt mixtures weight, this work was slightly conservative and selected the lowest of these values. Thus, nearly 2% of the total aggregates of OGP mixtures were substituted by PEX, corresponding to 5.5% in volume.

The possible interaction between the PEX and the binder was evaluated through the gel content of PEX (ASTM D2765-11), and it was 54.0% (cross-linked material that does not melt).

Table 1 shows the percentage of each fraction of aggregates (stone fractions, PEX fraction and filler) used for asphalt mixtures' production, in order to produce both open-graded mixtures (OG and OGP) with a maximum nominal size of 12.5.

The distribution of mineral aggregates for the OGP mixtures was adjusted in the corresponding PEX fraction dimensions in order to have similar aggregate gradations for OG and OGP. Summing up, the asphalt mixtures selected for performance evaluation are the following:

– OG_S – Open-graded asphalt with commercial modified binder;
– OG_E – Open-graded asphalt with waste EVA (5%) modified binder;
– OG_C – Open-graded asphalt with combined modified binder (4% EVA and 2% HDPE);
– OGP_S – Open-graded asphalt with commercial modified binder and 5.5% PEX in volume;
– OGP_E – Open-graded asphalt with waste EVA modified binder and 5.5% PEX in volume;
– OGP_C – Open-graded asphalt with combined modified binder and 5.5% PEX in volume.

2.2 Methods

After defining the constituent materials, the next step of the study was the mix design of the OG mixtures, i.e. definition of the binder content. The binder content was determined by using the particle loss dry test (EN 12697-17), as suggested in the Portuguese specifications, testing five different binder contents in order to define the best binder content. The test specimens were moulded according to EN 12697-30, and tested at a temperature of 25°C.

As the aggregates matrix of the open-graded mixtures with PEX (OGP) is similar to that of OG mixtures, its binder content was defined as being the same previously obtained for OG mixtures. The particle loss dry test was also performed to assess the disaggregation resistance performance of the OGP mixture, but only for that optimum binder content.

The rutting resistance of the asphalt mixtures was assessed by means of the Wheel Tracking Test (WTT), according to the EN 12697-22, using the procedure B (in air), with a standard wheel load of approximately 700 N. The test was carried at the temperature of 50°C, as being representative of the hotter summer days in Portugal.

In order to evaluate some surface properties of the asphalt mixtures, the texture of the WTT samples was measured with laser equipment (average value per stretches of 1 meter). The results obtained were the mean profile depth (MPD) according to NP ISO13473-1.

Finally, in an attempt to assess additional mechanical properties, the mechanical impedance was measured. The mechanical impedance is a different way to evaluate the stiffness of the asphalt mixtures. In this test, the damping factor (ξ) is determined, which is directly related to the phase angle (ϕ) of the asphalt mixture (Nils, 2009), as expressed by Equation 1.

$$\xi = \tan(\phi)/2 \qquad (1)$$

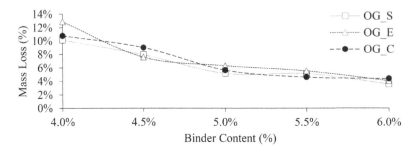

Figure 2. Influence of the binder content in the particle loss of the OG mixtures.

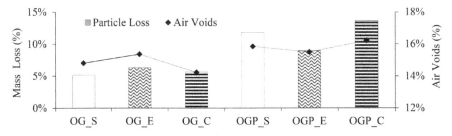

Figure 3. Particle loss results and air voids contents of the OG and OGP mixtures.

The test consists in exciting the material to be studied (in this case the asphalt mixtures of the WTT samples), by means of an impact hammer. Since the response of each material (damping and resonance frequency) is dependent on its shape and support conditions, it was decided to suspend the slabs and glue an accelerometer in the center of the sample with beeswax. The excitation was also produced in the center of the sample, but in the side of the slab opposite to the accelerometer. The mechanical impedance tests were performed both at 20°C and 30°C.

3 RESULTS

The binder content determination (mix design) of the OG mixtures was based on the mechanical characterization of the disaggregation resistance obtained through the Cantabro dry test. The mass or particle loss results for each OG mixture, for five different binder contents, are present in Figure 2. The binder content was selected according to the evolution of the mass loss.

For all mixtures (the control mixture OG_S with commercial modified binder and both mixtures OG_E and OG_C with the new binders modified with plastic waste) it can be observed that the mass loss quickly decreases for binder contents below 5.0%. The reduction of the mass loss is much lower for higher binder contents. Thus, the value of 5.0% was selected as being the optimum binder content obtained during the mix of the studied open-graded mixtures, since it meets the mechanical and economic needs of these road materials.

The same binder content (5.0%) was adopted for the corresponding mixtures using PEX as partial substitute of the aggregates (OGP). The mass loss results of the OG and OGP mixtures, for the optimum binder content of 5.0%, are compared in Figure 3.

Regarding the effect of the binder on the particle loss, the new binders with EVA and HDPE presented a performance similar to that of the commercial modified binder, both in OG and OGP mixtures. In fact, the particle loss of the OGP mixture with the EVA modified binder was even lower than that of the OGP mixture produced with the commercial modified binder. Besides, when comparing the global results of OG and OGP mixtures, it can be observed that the use of PEX as aggregate increases the particle loss of the open-graded asphalt mixtures. The higher particle

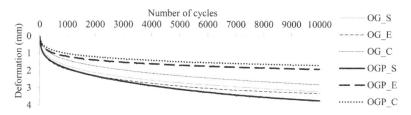

Figure 4. Wheel tracking test results for the OG and OGP mixtures.

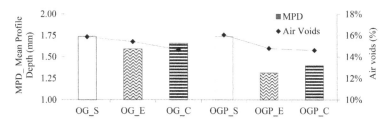

Figure 5. Surface texture evaluation through the MPD values for the OG and OGP mixtures.

loss of the OGP mixtures with PEX can be related with their higher air voids contents, which in turn can be associated with the higher difficulty to compact mixtures with plastics. In fact, PEX have elastic recovery and will work like a small spring that brings additional challenges during the compaction of these mixtures. Moreover, although the particle size of OG and OGP mixtures is identical, the OGP mixtures (with PEX) may need more binder due to some interaction with PEX aggregates that are not fully cross-linked. It should also be mentioned that the lower density of PEX also reduces the volume of binder available cover the aggregates, even if the binder contents by weight of OG and OGP mixtures are the same.

Then, the WTT rutting test was used to determine the susceptibility of the asphalt mixture to deform under repeated loads at high service temperatures Figure 4.

Concerning the modified binders' rutting performance, the combined binder (EVA and HDPE) has the best deformation resistance. Besides, the use of PEX as aggregate usually improves the rutting resistance, except for the mixture with commercial binder (OGP_S). The higher deformation resistance apparently indicates that the addition of PEX increases the stiffness of the asphalt mixture at high temperatures. This result, as well as the lower density of this type of mixtures with PEX could be a very interesting advantage in some situations, allowing the production of light but rut resistant asphalt mixtures for road pavements.

The surface properties influence the tire/pavement noise production, the adhesion and the permeability. The voids content of slab samples presented in Figure 5 was similar for all the mixtures produced, but in terms of superficial properties it is also important to evaluate the surface texture of the mixtures produced, namely by computing their mean profile depth (MPD).

The Portuguese specifications establish that the MPD should be higher than 1.25, and it was observed that all the results are similar among them and higher than that specified value.

Finally, the mechanical impedance was assessed as being a different way to evaluate the stiffness of the asphalt mixtures. In this work, only the first resonant frequency was evaluated. As the damping factor related to the phase angle, it also provides information about the stiffness of the asphalt mixture. Lower damping factors are related with a more viscous behaviour of the mixtures, while the opposite is related with an elastic behaviour. Figure 6 presents the variation of damping factors of OG and OGP mixtures with the temperature, and the corresponding first resonant frequency. The OGP mixtures present lower damping factors at each temperature, which is related to the lower phase angles and higher stiffness of these mixtures. This could indicate that PEX may have interacted somehow with the binders during the production of OGP mixtures. The damping and the resonant frequency variation with the temperature is similar for all mixtures, and

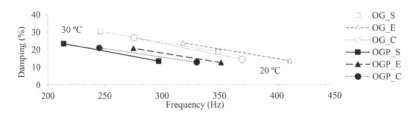

Figure 6. Damping factors for OG and OGP mixtures in the first resonant frequency.

thus the addition of PEX does not influence the temperature susceptibility. The less stiff mixture is OG_E, probably due to the lower viscosity of the EVA modified binder.

4 CONCLUSION

It can be concluded that the binders with plastic wastes developed in this study can promote the same challenging performance in open-graded asphalt mixtures as those produced with a commercial modified binder, having additional ecological and economic advantages. The use of PEX as partial substitute of the aggregates increased the particle loss, even though this is associated with higher air voids contents. Thus, the binder content of these mixtures should be increased to avoid this problem, probably without additional costs. However, the lower density of these mixtures with PEX and their higher rutting resistance could be beneficial. In terms of rutting resistance, all the new binders with plastic wastes promoted a good performance at high temperatures. These results were confirmed in mechanical impedance tests, where the mixtures with PEX as aggregates were stiffer than those only with mineral aggregates.

ACKNOWLEDGMENTS

The authors gratefully acknowledge the funding support of this work, namely the ERDF funds through the Operational Competitiveness Program – COMPETE and the National funds by FCT – Portuguese Foundation for Science and Technology in the scope of PLASTIROADS Project FCOMP-01-0124-FEDER-020335 (PTDC/ECM/119179/2010).

REFERENCES

Arrieta, V.S. & Maquilón, J.E.C. 2014. Resistance to Degradation or Cohesion Loss in Cantabro Test on Specimens of Porous Asphalt Friction Courses. *Procedia – Social and Behavioral Sciences,* 162, 290–299.
Becker, Y., Méndez, M.P. & Rodríguez, Y. 2001. Polymer Modified Asphalt. *Vision Tecnologia,* 9, 39–50.
Biligiri, K.P. 2013. Effect of pavement materials' damping properties on tyre/road noise characteristics. *Construction and Building Materials,* 49, 223–232.
Costa, L., Fernandes, S., Silva, H., Oliveira, J., Pereira, P. & Fonseca, P. 2013a. Valorização de materiais reciclados na produção de betumes modificados para pavimentos. *XVII Congreso Ibero-Latinoamericano del Asfalto, Guatemala.*
Costa, L.M.B., Silva, H.M.R.D., Oliveira, J.R.M. & Fernandes, S.R. M. 2013b. Incorporation of Waste Plastic in Asphalt Binders to Improve their Performance in the Pavement. *International journal of pavement research and technology,* Vol. 6, 457–464.
Kalantar, Z.N., Karim, M.R. & Mahrez, A. 2012. A review of using waste and virgin polymer in pavement. *Construction and Building Materials,* 33, 55–62.
Nils, R. 2009. Determining the asphalt mastercurve from free-free resonant testing on cylindrical samples. *Non-Destructive Testing in Civil Engineering* Nantes, France.
Robinson, H.L. 2004. *Polymers in asphalt.*
Sandberg, U. 1987. Road traffic noise—The influence of the road surface and its characterization. *Applied Acoustics,* 21, 97–118.

Wastes: Solutions, Treatments and Opportunities – Vilarinho, Castro & Russo (eds)
© 2015 Taylor & Francis Group, London, ISBN 978-1-138-02882-1

Physical analysis methods of municipal solid waste of Santo Andre

K.C.R. Drudi, A.M.P. Neto, G. Martins, G.C. Antonio, J.T.C.L. Toneli, R. Drudi,
C.H.S. Cenedese & L. Silva
Federal University of ABC, Santo Andre, São Paulo, Brazil

ABSTRACT: Due to the increase of the Municipal Solid Waste (MSW) production in Brazil, improvements in the treatment and disposal of MSW have been necessary. In this paper, a survey of the methods and standards commonly used for the collection and preparation of solid waste samples for gravimetric characterization are presented, in order to propose a methodology that allows the collecting of statistically representative samples of the Municipal Solid Waste of a city for the evaluation of their physical, chemical and thermal properties, identifying the most appropriate treatment method for both thermal and biological energy recovery. As a result, a methodology for collection and preparation of samples using as reference the MSW of Santo André was proposed, especially based in the American standards (ASTM) and the European standards (MODECOM™-used in France).

1 OVERVIEW MUNICIPAL SOLID WASTE

1.1 *Characterization of MSW for energy recovery*

The composition of MSW is influenced by various aspects. For instance, the MSW of developed countries has, as its main characteristic, a small amount of organic content, meanwhile, in developing countries this fraction is relatively greater. In addition, factors such as the season, geographical location, economic factors and the characteristic of neighborhood where it generated (commercial, industrial, residential), for example, also interfere on its composition.

The knowledge of the composition of MSW, and its variation along the year, and of the characteristics of the region where it is generated are critical to evaluate the potential of energy generation by different processes, either thermochemical or biochemical. Even the definition of the most appropriate energy recover process for a certain location depends on the amount and on the properties of the waste generated there.

Brazil has not a specific technical standard that describes a methodology for the collection and preparation of samples for characterization of MSW for energy recovery purposes. The Brazilian technical standard, the ABNT NBR 10007 (2004), is not as specific as in other countries.

Gravimetric analysis is the determination of constituents and their percentages in weight and volume. According to the ABNT NBR 10007 (2004), quarteringis the process of dividing a pre-homogenized sample into four equal parts, taking two opposing parts for a new sample and discardig the remaining parts. The two selected parts are homogenized again and the quartering process is repeated until the desired volume. This is the only method of preparing samples according to this Standard. The analyzed studies and the standards used in other countries are presented below.

2 METHODOLOGIES OF COLLECTION AND SAMPLE PREPARATION

2.1 *Methodologies survey*

At first, studies that were conducted in Brazil will be presented, highlighting the methodology applied for the characterization and sampling of the waste. Six studies that took place in Brazil and

Table 1. Analyzed papers.

Author	Region	Standards	Objectives	Quartered
Moura e Archanjo (2012)	Itaúna – MG (Brazil)	ABNT (2004)	Provides details of the gravimetric composition of municipal solid waste, showing the factors that determine their origin and formation.	Yes
Soares (2011)	Rio de Janeiro (Brazil)	ABNT (2004)	Studies the energetic potential of municipal solid waste generated in two cities in Rio de Janeiro.	Yes
Alcantara (2010)	Caceres – MT (Brazil)	Not mentioned	Characterizes the gravimetric composition of the waste produced in the city of Cáceres-MT, evaluates fertility and the levels of heavy metals in the soil of the landfill area.	No
Pessin et al. (2006)	Canela – RS (Brazil)	ABNT (2004)	Evaluates the factors that determine the origin and the formation of solid waste in touristic cities.	Yes
Tabalipa e Fiori (2006)	Pato Branco – PR (Brazil)	ABNT (2004)	Characterizes the waste that has been disposed in the city of Pato Branco.	Yes
Faria (2005)	Leopoldina MG (Brazil)	ABNT (2004)	Analyzes the practice of management of municipal solid waste collected in the city of Leopoldina-MG, as well as its mass characterization, proposing to the local administration to implement a sorting center, considering the amount of waste generated and its recycling potential.	Yes
Russo (2005)	Matosinhos (Portugal)	Modecom[TM]	Improves the knowledge of complex stabilization mechanisms of waste at landfills, to identify problems of the conception, operation and closure of the landfills, and settle methodologies for controlling the process, involving the operational procedures and the most appropriate analytical techniques to the effective control of operating parameters.	Yes

one that occurred in Portugal were analyzed. Table 1 presents information regarding the studies analyzed.

In many countries technical standards are used for managing and treating municipal solid waste. As previously mentioned, in Brazil the NBR 10007 of ABNT (2004) is commonly used. In the US, the standard used is ASTM – Method D 5231 (1992). Table 2 shows the analyzed standards.

It is noticeable that in the ARGUS, IBGE and EPA methodologies, the sampling is done at the residential waste collector, whereas in Modecom, the collection is done at the collector vehicle.

2.2 Considerations about the analyzed methodologies

Among the 7 papers analyzed (6 from Brazil and 1 from Portugal), just one of them (from Brazil) did not use the quartering method.

Considering the analyzed standards, ABNT (Brazilian) works with pre-homogenized samples and the quartering technique Among the international standards, only the US EPA, ARGUS, and MODECOM did not mention that the method used for the sample preparation was the quartering, while the others apply this method for sample preparation.

It is evident that none of the methodologies or standards analyzed are fully applicable to the situation of solid waste in Brazil, in particular in the city of Santo André. This result was expected, considering that each country or region has its own particularities. Accordingly, the custommization of some of these methodologies to the of Brazilian waste reality, both to determine the sample size, and for the preparation and collection, is necessary, since this kind of study is extremely important

Table 2. International standards.

Standards	Site	Wastes type	Collects	Statistic Base
ASTM (1992)	EUA	Not processed	The collection campaign took at least 5 days, considering the seasonal variation.	Yes
US EPA (1992)	EUA	Processed and not processed	The collection campaign took 7 days, considering the seasonal variation.	Yes
ERRA (1993)	European	Municipal waste	The collection campaign took 12 months, considering the seasonal variation and excluding festive periods.	Not mentioned
Argus (1998)	Germany (Remecom project)	Municipal waste	4 collections are made every campaign, 1 for each season, considering seasonal variation.	Not mentioned
IBGE (1998)	Belgium (Remecom project)	Municipal waste	2 collections are done every campaign, one in autumn and the other during the spring, considering seasonal variation.	Not mentioned
Modecom^TM (1998)	France (Remecom project)	Municipal waste	4 collections are done every campaign, 1 for each season, or 1 every 2 months, considering seasonal variation.	Not mentioned
EPA (1998)	Ireland (Remecom project)	Municipal waste	2 collections are done every collection campaign, in at least 6 months, repeating every trimester, considering the seasonal variation.	Yes

both for the definition of waste management system and for the most appropriate energy recovery technology.

The general characteristics of the MSW of Santo André city will be presented, as well as a proposal of methodology for collection and preparation of samples for waste characterization aiming the determination of the energy recovery potential of the municipality will be made.

3 METHODOLOGY PROPOSAL FOR COLLECTION AND PREPARATION OF SAMPLES FOR THE MUNICIPALITY OF SANTO ANDRE

3.1 *Municipality characteristics of Santo Andre*

The estimated population of Santo André in 2014, according to Brazilian Institute of Geography and Statistcs IBGE (2014) was 707,613. The Municipal Service of Environmental Sanitation of Santo Andre SEMASA collects about 700 tons of waste daily, which means that each inhabitant generates about 1 kg of waste daily.

In 2013, according to Geotech (2013), a gravimetric characterization of solid waste from the city of Santo André was performed, and the results are shown in Figure 1.

The methodology used for the characterization was ABNT NBR 10007 (2004). Some elements, such as Styrofoam, diapers and rubber were grouped in the category Miscellaneous, and metals, construction, demolition, technological and sanitary wastes were grouped in the category others.

Presently, SEMASA collects 100% of the waste generated in Santo André municipality. The districts are divided into 15 sectors, and each sector is subdivided into subsectors, totalizing the number of 116 subsectors.

To determine the size of a sample that is statistically representative, the standards mentioned above were analyzed, as well as papers published in Brazil and Portugal, besides the literature related to statistical approaches.

The Brazilian standard has no methodology with statistical basis for determining the sample size. Among the studies reviewed, only the ASTM (1992) and EPA (1992) deal with a statistical basis for a sample size validation.

Physical Component Analysis

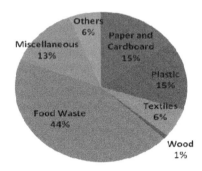

Figure 1.   Physical components analysis in Santo Andre. Drudi & Martins (2014).

Table 3.   Sample size.

Equation	Sample size
1	59
2	54

ASTM (1992) indicates that the confidence interval of parameters to be used may be between 90% and 95% and a margin of error of 10%. The EPA standard also considers the time of arrival of the trucks in the landfill, and from these data, determines the confidence interval and the error involved.

Due to the fact that there is no information about the logistics of garbage trucks in the municipal landfill of Santo Andre, the EPA standard methodology was discarded.

Based on this analysis, the statistical parameters of confidence interval and margin of error were simulated from the formula of ASTM (1) and from the standard distribution (2), with the correction by the amount of the population:

$$n = \left(\frac{z \times s}{e \times \bar{x}} \right)^2 \tag{1}$$

$$n = z^2 \frac{\left(\bar{x}/n \right)\left[1 - \left(\bar{x}/n \right) \right]}{e^2} \sqrt{\frac{N - n}{N - 1}} \tag{2}$$

where n = sample size; z = table value for confidence interval; e = error; $\bar{x}$ = average; and N = population size. With the 95% confidence interval and an error of 10%, the sample size for each one of the equations is shown in table 3.

When statistical parameters to be used, as margin of error and confidence interval, are unknown, it is perfectly valid to use parameters taken from related works.

After the analysis and the determination of the calculations, the proposal for collection and preparation of samples for solid wastes characterization in Santo André is to collect samples from 54 collection vehicles, in at least two distinct seasons, with the objective to consider the seasonality.

The procedure suggested for the collection and preparation of samples is the following.

After the collector vehicle dumps the waste in an appropriate place, 4 samples of 240 liters should be taken from the pile, with 3 samples from the bottom and 1 from the top of the pile, which should be placed over a plastic canvas, totaling 960 liters.

The waste should be removed from the bags with the help of an excavator, remaining over the plastic canvas. Then, the homogenization and quartering proceeds, obtaining approximately 400 liters of waste. The process should be repeated until a sample of 100 liters of waste is obtained, which would be used to perform the characterization. This methodology can be changed or adapted, if it is necessary.

As mentioned earlier, the municipality of Santo Andre is divided in 15 sectors for waste collection purposes, and the collection is done every other days, or every day, in the downtown sector. Moreover, in most sectors, the collection is done after 5 pm, and there are no collections on Sundays.

Other key information considered was the number of trucks per sector, which led to the calculation that the total sampling should be of 54 trucks per campaign. This amount of trucks to be sampled by sector was calculated based on the amount of waste collected in each sector.

According to this information, a preliminary schedule of sampling was prepared to run over 4 weeks. In addition, the days and the period of collection were considered, classified into day and night. During the first weeks, it will be necessary to collect samples from 3 to 4 trucks per day, and at the last week, just 1 to 2 trucks daily. The sampling will happen from Monday to Friday during business hours.

4 CONCLUSIONS

The objective of this study was to evaluate the methodologies and standards that deal with the collection and preparation of municipal solid waste samples, both national and international, and based on this survey, to propose a sampling method for the MSW of Santo André which will permit to evaluate its energy recovery potential.

About the Brazilian standard ABNT (2004), which deals with the waste issues, the standard is very superficial, when compared with the standards of other countries, especially about the amount of samples to be collected to preserve the representativeness of the entire universe. Of course, in such matters like solid waste, an heterogeneous content, which varies according to region, social class and seasonality, it is very difficult to create a standard that can be used all over the country, but some standardization is nevertheless necessary.

American and European standards are more comprehensive and better established, however, some changes need to be made to apply these standards in Brazil, either due to the characteristics of the Brazilian waste or due to the kind of collection.

The set of Remecom standards seems to be an interesting proposal, because, for each country involved, the original draft standard had a specificity to answer the country's expectations for to the collection and preparation of waste sample.

Brazil has recently started to improve its waste management and energy recovery. This process has been accelerated because of the National Solid Waste Policy, approved in 2010.

The most used method in studies made in Brazil for waste characterization was the quartering. It is believed to be an efficient method, as it was found in almost all literature on the subject, however other methods could be tested and compared.

As seen previously, municipal solid waste is composed by elements that can be used to recover energy. Therefore, a proper waste characterization is extremely important, because different kinds of waste have different physical, chemical and thermal properties, so they are more suitable for different types of energy conversion process. Waste with high heating value, for example, are good for thermochemical processes, whereas waste with higher moisture content and higher volatile solids content are best suited for the anaerobic digestion process.

This work is just the beginning of a long journey to effectively convert the content of the municipal solid waste from Santo Andre city, located in Sao Paulo state, Brazil, into useful energy and biogas.

REFERENCES

ABNT Associação Brasileira de Normas Técnicas 2004. *Resíduos Sólidos: Classificação – NBR 10.007/2004.*

Alcantara, A.J. de O. 2010. *Composição Gravimétrica dos Resíduos Sólidos Urbanos e Caracterização Química do Solo da Área de Disposição Final do Município de Cáceres – MT – Dissertação de mestrado.* Universidade do Estado do Mato Grosso.

Almeida, R.G. de 2012. Estudo da Geração de Resíduos Sólidos Domiciliares Urbanos do Município de Caçador SC, a Partir da Caracterização Física e Composição Gravimétrica. *Ignis| Caçador* 1(1).

ASTM Método D 5231–92 1992. *Standard Test Method for Determination of the Composition of Unprocessed Municipal Solid Waste.* American Society for Testing and Materials International. Reaprovado em 2003.

Carvalho, E.M.F.D.B. 2005. *Metodologias para a quantificação e caracterização física dos resíduos sólidos urbanos – 2005.*

CENSO 2010. Rio de Janeiro: IBGE, 2014. Available at: <http://www.ibge.gov.br/home/estatistica/populacao/condicaodevida/pnsb2008/PNSB_2008.pdf>. Accessed in: 25/08/2014.

Drudi, K.C.R. & Martins, G. 2014. *Evaluation of the Energy Potential of the Municipal Solid Waste of Santo Andre.* ISWA.

Faria, M.R.A. 2005. Caracterização do Resíduo Sólido Urbano da Cidade de Leopoldina – MG: Proposta de Implantação de um Centro de Triagem. *Revista APS* 8(2): 96–108, jul./dez.

Geotech 2013. *Geotecnia Ambiental. Caracterização gravimétrica dos resíduos sólidos urbanos domiciliares do município de Santo André. Relatório Técnico. Nov.*

Moura, A. A., Lima, W.S., Archanjo, C.R. 2012. *Análise da Composição Gravimétrica de Resíduos Sólidos Urbanos: Estudo de Caso – Município de Itaúna – MG.*

Pessin, N. Conto, S.M., Telh, M., Cadore, J., Rovatti, D., Boff, R.E. 2006. *Composição Gravimétrica de Resíduos Sólidos Urbanos: Estudo de Caso – Município de Canela – RS.*

Russo, M.A.T. 2005. *Avaliação dos Processos de Transformação de Resíduos Sólidos Urbanos em Aterro Sanitário – Tese doutorado.* Universidade do Minho.

Soares, E.L. de S.F. 2011. *Estudo da Caracterização Gravimétrica e Poder Calorífico dos Resíduos Sólidos Urbanos – Dissertação (mestrado).* Rio de Janeiro: UFRJ/COPPE.

Tabalipa, N.L. & Fiori, A.P. 2006. Caracterização e Classificação dos Resíduos Sólidos Urbanos do Município de Pato Branco, PR. *Revista Brasileira de Ciências Ambientais* 4.

Wastes: Solutions, Treatments and Opportunities – Vilarinho, Castro & Russo (eds)
© 2015 Taylor & Francis Group, London, ISBN 978-1-138-02882-1

Can oil, plastic and RAP wastes have a new life in novel asphalt mixtures?

S. Fernandes, L. Costa, H. Silva & J. Oliveira
CTAC – Territory, Environment and Construction Centre, University of Minho, Guimarães, Portugal

A. Machado & F. Duarte
IPC/I3N – Institute of Polymers and Composites, University of Minho, Guimarães, Portugal

ABSTRACT: The pavement recycling allows to reuse Reclaimed Asphalt Pavement (RAP) or other waste materials in new asphalt mixtures for road construction or rehabilitation, thus reducing the use of virgin materials (aggregates and bitumen). Thus, the main aim of this study is to minimize the use of natural resources through the reuse of three waste materials: HDPE, motor oil and RAP. Different amounts of waste motor oil and HDPE were added to an asphalt binder with 50% aged bitumen. The best solutions to produce the modified binders (4.5 to 5.0% HDPE and 10% waste motor oil) performed as well as a conventional bitumen although they only used 35% of virgin bitumen. Asphalt mixtures with 50% RAP were produced with the selected modified binders, improving some characteristics in comparison with conventional asphalt mixtures. In conclusion, these wastes can revive in new asphalt mixtures.

1 INTRODUCTION

The pavement recycling presents environmental and economic advantages for the paving Industry. From an environmental point of view, the pavement recycling enables to reduce the amount of new aggregates and bitumen used, the disposal area required for placing milled material and the energy consumption. Besides the environmental advantages, the pavement recycling can present economic benefits in comparison to the traditional pavement overlay and it can solve other problems, such as cracking, roughness and/or others pathologies (INIR, 2012).

According to data provided by EAPA (2011), 2000 tonnes of reclaimed asphalt were available in Portugal, in 2011, of which 60% were used in hot and warm recycling, 5% in cold recycling and 15% in unbound layers. However, the recycling rates in some European countries is very low although considerable amounts of reclaimed asphalt are available (EAPA, 2012). Thus, it is observed that asphalt recycling techniques are still poorly used in Europe.

Reclaimed asphalt pavement (RAP) used in asphalt recycling is a heterogeneous material (Karlsson & Isacsson, 2006), since it depends on the amount and type of materials used in the initial road construction, in the maintenance or conservation history of the pavement and in the RAP transportation and storing conditions (Vislavičius & Sivilevičius, 2013). RAP is a difficult material to characterize, which requires a high experience and knowledge when it is used in new asphalt mixtures comparatively to the conventional ones (Karlsson & Isacsson, 2006).

Several authors (Sengoz & Oylumluoglu, 2013, Reyes-Ortiz et al., 2012) studied and evaluated the use of different amounts of RAP in new recycled asphalt mixtures and their mechanical performance. The RAP already contains bitumen, but due to its short-term and long-term aging it is necessary to add a recycling agent (Karlsson & Isacsson, 2006). The recycling agent or rejuvenator should change the aged binder in order to give it typical mechanical proprieties of a conventional bitumen (Dony et al., 2013). Some examples of rejuvenators are soft grade bitumens, vegetal oil (Dony et al., 2013), extender oils (Karlsson & Isacsson, 2006) and waste motor oil (Silva et al., 2012).

The waste motor oil is a residue used for rejuvenating or preventing the bitumen aging, because it dilutes the asphaltenes from the aged binder by adding maltenes (Lesueur, 2009). The addition of waste motor oil reduces the mixing and compaction temperatures and the bitumen's viscosity (Silva et al., 2012) and its reuse can present benefits at an environmental and economic level. However, the use of high quantities of rejuvenator can cause rutting problems, which can be avoided by using plastic modification. Thus, waste plastic can be used in bitumen modification to improve the pavement performance, namely by decreasing rutting, cracking and thermal susceptibility (Costa et al., 2013). The high density polyethylene (HDPE) is available in large amounts at a low price (Casey et al., 2008), and is one of the waste plastics that can be used for bitumen modification. It decreases the penetration and increases the softening point temperature and the resistance to the variation of temperature (Al-Hadidy & Yi-Qiu, 2009).

Taking this into account, the main objective of this study is to reduce the amount of virgin materials used in new asphalt mixtures, by using 50% of RAP material and a new modified bitumen that maximizes the use of waste materials (waste motor oil and HDPE).

2 MATERIALS AND METHODS

2.1 *Materials*

The materials used in this study were a RAP, new aggregates, a conventional bitumen and two additives or modifiers (waste motor oil and waste HDPE).

The RAP material was obtained by milling a surface layer of a highway pavement, which was separated in two fractions (fine and coarse) by a classifier with a sieve opening of 8 mm. The aged bitumen recovered from the RAP material is a very hard binder with a penetration of 9 dmm and a softening temperature of 74°C.

The selected aggregates are crushed rock aggregates of granite origin, except the filler that is of limestone origin. The conventional bitumen from CEPSA has got a penetration of 35 dmm and a softening point of 54°C. This bitumen was modified with waste motor oil and waste HDPE. Then, it was mixed with aged bitumen from RAP and its characteristics were compared with the initial bitumen. The waste motor oil from heavy vehicles showed a viscosity between 0.1 and 0.005 Pa.s, over the temperature range studied (30 to 180°C). In turn, the HDPE was supplied with a maximum dimension of 4 mm and its melting temperature was 134°C.

2.2 *Methods*

The 35/50 bitumen was modified with the waste polymer (HDPE) in a high shear mixer at 7200 rpm, for 20 min, at temperature of 180°C, in order to obtain a homogeneous modified binder. Then, the necessary amount of waste motor oil was introduced into the modified bitumen by using a low shear mixer, during 10 min, at a temperature of 180°C. At the end, the bitumen modified with polymer and waste motor oil was added to the aged bitumen in a proportion of 50%, during 2 min, at a temperature of 180°C. It should be pointed out that some samples were only modified with waste motor oil for comparison reasons.

Initially, binders with different amounts of waste motor oil (10, 15 and 20%) were studied. These binders were modified with 5% of HDPE (maximum recommended percentage), and the most promising modified binder allowed to select the amount of waste motor oil to use in the next phase. Then, several percentages of HDPE (namely, 2.5, 3.5 and 4.5%) were studied with this amount of waste motor oil. The content of each material and their classification are shown in Table 1. The final modified binders, already mixed with the aged bitumen, were characterized by penetration at 25°C (EN 1426), softening point temperature (EN 1427) and dynamic viscosity (EN 13302) tests. Finally, the modified binders (with oil and HDPE) with characteristics similar to those of a conventional 35/50 pen bitumen (base bitumen used in this study) were selected to produce recycled asphalt mixtures in order to evaluate their mechanical performance.

Table 1. Percentage of each waste material used in the studied binders.

Modified binder	Aged bitumen	Bitumen 35/50	Polymer	Motor oil
B10/20	100.0	0.0	0.0	0.0
RBO10	50.0	40.0	0.0	10.0
RBO15	50.0	35.0	0.0	15.0
RBO20	50.0	30.0	0.0	20.0
RBP5O10	50.0	35.0	5.0	10.0
RBP5O15	50.0	30.0	5.0	15.0
RBP5O20	50.0	25.0	5.0	20.0
RBP2.5O10	50.0	37.5	2.5	10.0
RBP3.5O10	50.0	36.5	3.5	10.0
RBP4.5O10	50.0	35.5	4.5	10.0

Table 2. Penetration and R&B test results with different contents of waste motor oil and HDPE.

Binder	Penetration at 25°C dmm	Softening point temperature °C
B35/50	35	54
RBO10	77	47
RBO15	166	41
RBO20	304	34
RBP5O10	43	81
RBP5O15	98	52
RBP5O20	188	47

With regards to the production of recycled asphalt mixtures with RAP (AC 14 Surf mixtures with a binder content of 5%), the procedure starts with the addition of new aggregates at a temperature of 230°C together with 20% of the coarse RAP fraction. These materials should be mixing during 1 min, to ensure their homogeneity. After that, 30% of the fine RAP fraction is introduced at room temperature in the previous mixture, during 2 min, in order to homogenize the mixture's temperature. Then, the modified bitumen with waste materials (at 180°C) is added to the mixture, during 2 min, as recommended in EN 12697-35 standard. This production process is based in a previous study of Palha et al. (2013). All mixtures were tested for water sensitivity (EN 12697-23) and rutting resistance (EN 12697-22) in order to evaluate their performance.

3 RESULTS

3.1 *Selection of waste materials contents*

The basic properties of the studied binders are presented in Table 2. The increase of the waste motor oil content increases the penetration values and decreases the softening point temperatures. In turn, the addition of HDPE reduces the penetration values and raises the softening point temperatures. It is possible to verify that only the modified bitumen with 5% of waste HDPE and 10% of waste motor oil has penetration values similar to those of a conventional B35/50 bitumen, with the advantage of having a higher softening point temperature value. Thus, the study will proceed only with 10% of waste motor oil and with a range of HDPE contents in order to find best modified binders that are able to maximize the use of waste materials.

According to the Figure 1, for the same percentage of waste motor oil, the increase of HDPE content decreases the penetration values and increases the softening point values. It should be pointed out that the softening point values are very similar for HDPE amounts between 2.5 and

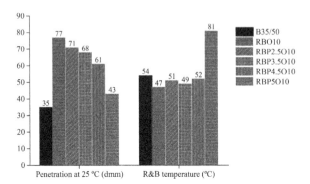

Figure 1. Penetration and softening point test results of binders with different contents of HDPE and 10% of waste motor oil.

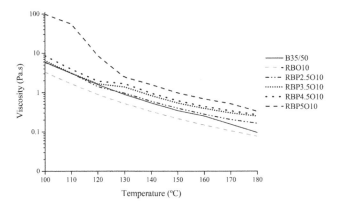

Figure 2. Dynamic viscosity results of binders with different contents of HDPE and 10% of waste motor oil.

4.5%, largely increasing for a HDPE amount of 5%. This sharp variation can be related with a phase change in the binder, when HDPE concentration is enough to create internal structures.

The dynamic viscosity results (Figure 2) confirm the previous observations. The viscosity of the binder modified only with waste motor oil is lowest one, while the other modified binders (with waste motor oil and HDPE) present a viscosity similar or higher than that of the conventional bitumen. Again, the modified bitumen with 5% of HDPE and 10% of waste motor oil (RBP5O10) exhibits the highest viscosity, followed by the RBP4.5O10 binder.

In view of the above results, only one of the modified bitumens, i.e. RBP5O10, showed similar or higher proprieties than those of a conventional B35/50 pen bitumen. However, the modified bitumen RBP4.5O10 presents characteristics similar to those of a conventional B50/70 pen bitumen (also used for road construction), while avoiding some possible problems related with the lower workability of binders with higher viscosity. Thus, these two modified bitumens (RBP5O10 and RBP4.5O10) were selected to produce recycled asphalt mixtures in order to evaluate their performance, which was compared with that of a conventional mixture produced with the base bitumen used in this study (B35/50).

3.2 *Performance of conventional and recycled HMA mixtures*

Figure 3 presents the indirect tensile strengths ratio (ITSR), related to water sensitivity, and the air voids contents results of the studied mixtures determined based on EN 12697-8 standard (method A). The different asphalt mixtures showed similar water sensitivity values that ranged between 66%

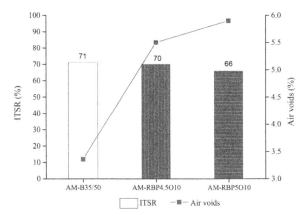

Figure 3. Water sensitivity and air voids results of the studied mixtures.

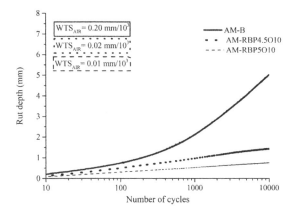

Figure 4. Variation of rut depth with the number of cycles for the studied mixtures.

and 71%, and among these values the conventional asphalt mixture presented the lowest water sensitivity. With respect to the air voids content, the novel asphalt mixtures with modified binder revealed air voids contents slightly higher than those specified for an AC 14 Surf mixture (3% to 5%). Nevertheless, the higher voids contents of these new mixtures should not pose a problem because their water sensitivity was very close to that of the conventional asphalt mixture. The higher air voids contents of these new recycled mixtures could have been caused by the use of RAP material or by the higher viscosity of the binders modified with HDPE at high working temperatures.

The resistance to permanent deformation (rutting) was determined through the measurement of the rut depth formed by successive passages of a loading wheel over the tested mixture. The evolution of the rut depth with the number of cycles was used to rank the permanent deformation performance of asphalt mixtures (Figure 4), namely by means of the wheel-tracking slope in air (WTS$_{AIR}$) parameter. The asphalt mixtures with 4.5% or 5.0% of HDPE and 10% of waste motor oil (AM-RBP4.5O10 and AM-RBP5O10) presented higher rutting resistance in comparison with the conventional mixture (AM-B). Moreover, both asphalt mixtures with the new modified binders presented very low WTS$_{AIR}$ values, clearly lower than those of the conventional mixture, which is an indicator of the good rutting performance of these new mixtures.

4 CONCLUSIONS

The present study presents the main results concerning the development of new asphalt mixtures with significant quantities of different waste materials (RAP, motor oil and HDPE), which is expected to have a performance similar or higher than that of a conventional mixture. During the initial basic characterization of the binders, the best performance was obtained by the binders modified with 4.5% to 5.0% of HDPE and 10% of waste motor oil. Then, recycled asphalt mixtures produced with 50% RAP and the selected modified binders presented higher rutting resistance and similar water sensitivity in comparison with a conventional mixture with 35/50 bitumen, despite their higher air voids contents. It was concluded that it is possible to give a new life to different waste materials (RAP, motor oil, HDPE) in new asphalt mixtures, greatly reducing in 65% the amount of virgin bitumen used. Thus, these new asphalt mixtures with waste modified binders showed an environmental and economic way to reuse waste materials.

ACKNOWLEDGMENTS

The authors gratefully acknowledge the funding support of this work, namely the ERDF funds through the Operational Competitiveness Program – COMPETE and the National funds by FCT – Portuguese Foundation for Science and Technology in the scope of PLASTIROADS Project FCOMP-01-0124-FEDER-020335 (PTDC/ECM/119179/2010) and the EU/FSE funds through the POPH/QREN Program in the scope of a FCT Ph.D. grant (SFRH/BD98379/2013).

REFERENCES

Al-Hadidy, A. I. & Yi-Qiu, T. 2009. Effect of polyethylene on life of flexible pavements. *Construction and Building Materials* 23: 1456–1464.

Casey, D., Mcnally, C., Gibney, A. & Gilchrist, M. D. 2008. Development of a recycled polymer modified binder for use in stone mastic asphalt. *Resources, Conservation and Recycling* 52: 1167–1174.

Costa, L. M. B., Silva, H. M. R. D., Oliveira, J. R. M. & Fernandes, S. R. M. 2013. Incorporation of waste plastic in asphalt binders to improve their performance in the pavement. *International Journal of Pavement Research and Technology* 6: 457–464.

Dony, A., Colin, J., Bruneau, D., Drouadaine, I. & Navaro, J. 2013. Reclaimed asphalt concretes with high recycling rates: Changes in reclaimed binder properties according to rejuvenating agent. *Construction and Building Materials* 41: 175–181.

Eapa 2011. Asphalt in Figures 2011. Brussels, Belgium, European Asphalt Pavement Association.

Eapa 2012. Asphalt in Figures 2012. Brussels, Belgium, European Asphalt Pavement Association.

Inir 2012. Construção e Reabilitação de pavimentos – Reciclagem de Pavimentos. Instituto de Infra-Estruturas Rodoviárias IP.

Karlsson, R. & Isacsson, U. 2006. Material-Related Aspects of Asphalt Recycling—State-of-the-Art. *Journal of Materials in Civil Engineering* 18: 81–92.

Lesueur, D. 2009. The colloidal structure of bitumen: Consequences on the rheology and on the mechanisms of bitumen modification. *Advances in Colloid and Interface Science* 145: 42–82.

Palha, D., Fonseca, P., Abreu, L., Silva, H. & Oliveira, J. 2013. Solutions to improve the recycling rate and quality of plant produced hot mix asphalt. *WASTES: Solutions, Treatments and Opportunities.* Braga, Portugal.

Reyes-Ortiz, O., Berardinelli, E., Alvarez, A. E., Carvajal-Muñoz, J. S. & Fuentes, L. G. 2012. Evaluation of Hot Mix Asphalt Mixtures with Replacement of Aggregates by Reclaimed Asphalt Pavement (RAP) Material. *Procedia – Social and Behavioral Sciences* 53: 379–388.

Sengoz, B. & Oylumluoglu, J. 2013. Utilization of recycled asphalt concrete with different warm mix asphalt additives prepared with different penetration grades bitumen. *Construction and Building Materials* 45: 173–183.

Silva, H. M. R. D., Oliveira, J. R. M. & Jesus, C. M. G. 2012. Are totally recycled hot mix asphalts a sustainable alternative for road paving? *Resources, Conservation and Recycling* 60: 38–48.

Vislavičius, K. & Sivilevičius, H. 2013. Effect of reclaimed asphalt pavement gradation variation on the homogeneity of recycled hot-mix asphalt. *Archives of Civil and Mechanical Engineering* 13: 345–353.

Wastes: Solutions, Treatments and Opportunities – Vilarinho, Castro & Russo (eds)
© 2015 Taylor & Francis Group, London, ISBN 978-1-138-02882-1

Organic load of substrata for wastewater treatment constructed wetlands

A.B. Ferreira, L.M. Oliveira & I.A.-P. Mina
Biology Department, University of Minho School of Sciences, Braga, Portugal

A.M. Almeida
Physics Department, University of Minho School of Sciences, Braga, Portugal

ABSTRACT: In Portugal the use of Constructed Wetlands (CW) for wastewater treatment has been increasing. However a number of these facilities need new strategies to achieve better efficiency. Keeping the culms of reeds on the CW beds not always results as desired, but the use of widely available agro-forest wastes, may be suitable as CW support matrix. This study was performed at lab-scale with dried culms of *Phragmites* and eucalyptus bark maintained in tap water, to assess them as CW substrata. With a 7 days residence time in water, *Phragmites* culms added a high organic load (about 400 mg L^{-1} BOD$_5$) to the medium, while the eucalyptus bark added only, about 60 mg L^{-1} BOD$_5$. However, by lixiviation, the organic load decreased to about 25 mg L^{-1} BOD$_5$ in 5 weeks. With the organic load reduction of the leachate water, its surface tension increase approaching the surface tension of tap water, corroborating the observed organic load decrease.

1 INTRODUCTION

Constructed Wetlands (CW) are artificial ecosystems for wastewater treatment that rely on the physicochemical and biological processes that occur in nature. These natural processes take place in plant's root zone (rhizosphere) and involve interactions between plants, microorganisms, CW matrix or substratum and pollutants. The different types of existing CW systems vary according to the specific main elements used, e.g. the plants, the substratum and the hydraulic design for water flow (Stottmeister *et al.* 2003; Dordio *et al.* 2007; Vymazal 2007; Faulwetter *et al.* 2009; Vymazal 2010).

The substratum presents itself as a component with the greatest influence in the process, as it is the main support for the development of both the plants and the microbiological communities and it helps in the isolation of pollutants present in the wastewater through the adsorption process (Stottmeister *et al.* 2003; Calheiros *et al.* 2008; Calheiros *et al.* 2009).

The nature of the material used as substratum influences the environmental conditions present in its pores, such as the redox potential (Saeed *et al.* 2012) and its physical characteristics impact the wastewater flow through the reed bed eventually affecting the efficiency of the system (Albuquerque *et al.* 2010; Stottmeister *et al.* 2003). A support matrix with a greater capacity to retain contaminants by sorption phenomena, ionic exchange or other physico-chemical processes, may significantly enhanced the efficiency of the CW systems in the removal of pollutants (Dordio *et al.* 2007).

The use of organic substrata, isolated or mixed with inorganic ones as CW support matrix, aims to provide an internal source of carbon to promote nitrogen removal, enabling that the denitrification process do not depend entirely on the carbon present in the wastewater. The mineral substrata most commonly used do not assure this supplying of carbon, thereby frequently limiting the denitrification process (Saeed *et al.* 2011; Saeed *et al.* 2012).

The materials of natural origin with good adsorption capacity have gain importance showing good affinity for some pollutants. Another advantage of these materials is their low cost, since they are by-products of other activities (Dordio *et al.* 2011). Agro-wastes, due to their high carbon content, have also been increasingly used as low cost precursors for the production of cheaper types of activated carbon (Dordio & Carvalho, 2013).

In Portugal the use of CW as a technology for the treatment of municipal and domestic wastewater has spread at the turn of the XXI century. This is easy to understand once, as low technology systems, the CW presents low costs, low energy requirements, low technical skills operators, providing environmentally friendly landscapes with good efficiencies in treating wastewater. These characteristics are unquestionably advantages for its application in Portuguese rural communities (Matos *et al.* 2009; Duarte *et al.* 2010). However in our country a considerable number of theses facilities need more knowledge and new strategies to achieved better efficiency. For example, one of the practices in some of these facilities consists in maintaining the dried culms of the reeds on the CW beds during the cold season. This is one of the practices that require the knowledge of the advantages and disadvantages. Besides, the use of widely available agro-forest wastes may be one of the main criteria for the choice of a support matrix (Dordio & Carvalho, 2013), mixed or not with inorganic materials (Brix *et al.* 2001).

This study was performed at lab-scale to evaluate the organic load added to a CW by dried culms of reeds and eucalyptus bark. Besides the Biochemical Oxygen Demand (BOD$_5$) to estimate the organic load of water samples, measurement of their surface tension was performed, to attempt a better understanding of the organic matter effect in water. This parameter may be a supplementary tool to evaluate potential CW support matrix materials.

2 MATERIALS AND METHODS

A full-scale CW can be mimicked in laboratory, by microcosms (Kangas, 2004). To study these low-tech systems the first characteristic of microcosms should be the reuse of materials for its construction. Thus absorption and organic load assays were performed on reused plastic carboys (5000 mL). Their bottoms were removed and its openings with the stoppers on, inner isolated with cotton gauze, were maintained upside down in supports. These "microcosms" were set up filling each one with a certain amount of the material to assay: *Phragmites australis* dried culms (collected on a CW from the central region of Portugal, in April 2013) and *Eucalyptus globulus* bark (collected in April 2014 in the Alto Alentejo region). After addition of a given volume of tap water (generally 2000 mL) they were left in the laboratory at room temperature during seven days (resident time). Weekly the water was gathered from the microcosms to further analyses and new tap water was added to each of them. All the microcosms were weighed before and after the water gathering and after the addition of new tap water, to evaluate the absorption capacity of the materials. The weights were determined on electronic scales Mettler PJ3000 and/or PJ600. The Biochemical Oxygen Demand (BOD$_5$, mgL^{-1}), and the water surface tension (mN^{-1}) were assessed respectively by the manometric method (using OxiTop®self-check measuring systems) and the *du Nouy* ring method (du Nouy, 1919) using a tensiometer (Surface Tension Analyser model DST 60).

The assays were done in three experimental periods: the first one, between May 29th and July 13th of 2013 (six weeks period) with only three microcosms with *Phragmites australis* dried culms – Pa1; the second assay, between September 25th to November 13th of 2013 (eight weeks) also with three microcosms with *Phragmites australis* dried culms – Pa2 and the third assay realized between April 23rd to June 23rd of 2014 (during eight weeks) included two microcosms with *Phragmites australis* dried culms (Pa3), two microcosms with *Eucalyptus globulus* barks (Eg) and two more microcosms with both *Phragmites australis* dried culms and *Eucalyptus globulus* barks (PE).

2.1 *Surface tension determination*

The surface tension is defined as the work required to increase the area of a liquid surface isothermally and reversibly by one unit. It is expressed as surface energy per unit of area and alternatively

Table 1. Absorption capacity (% of weight gain) of the substrata.

Microcosms	Substrata initial weight g	Substrata final weight g	Weight gain %
Pa2a	200.1	514.2	157
Pa2b	201.7	456.9	126
Pa2c	200.4	407.5	103
Pa3a	200.1	477.4	139
Pa3b	200.6	495.5	147
Ega	200.2	249.9	25
Egb	200.2	280.0	40
PEa	201.9	348.3	72
PEb	204.6	404.1	98

Legend: Microcosms with *Phragmites australis* dried culms (Pa) of the second (2) and third (3) assay with replicates (a,b,c); microcosms with *Eucalyptus globulus* barks (Eg) and microcosms with both *Phragmites australis* dried culms and *Eucalyptus globulus* barks (PE).

as force per unit length (Ebnesajjad, 2011). Surface tension is related to the surface energy and to the inter-molecular forces in water (Tariq *el al*. 2012). In theory, any particle dissolved in water may alter this balance and thereby alter the values of the surface tension force.

The water surface tension was determined in three samples of the recovered water from each microcosm using the *du Nouy* ring method (du Nouy 1919). About 8 to 10 mL of each sample were analysed on a tensiometer (Surface Tension Analyser model DST 60), with a ring that, according with the manufacturer, its circumference (R) and the ratio of the ring dimensions (R/R') were 6.015 cm and 53.883 cm respectively. The surface tension values were computed according to the mathematical model proposed by *Zuidema & Waters* (Zuidma & Waters 1914; Lee *et al.* 2012) using Equation 1.

$$\lambda/\lambda_{ap} = 0.7250 + \sqrt{[(0.03678/r^2)\times(\lambda_{ap}/\Delta\rho)+0.04534 - 1.679(R/R')]} \quad (1)$$

where λ stands for surface tension; λ_{ap} stands for apparent surface tension; r is the radius of the ring; $\Delta\rho$ is the density difference between fluid and air, R is the circumference of the ring and R' is the dimensions ratio.

3 RESULTS

3.1 *Substrata absorption capacity*

The water volumes weekly recovered from the microcosms, were always smaller on the first week – between 70–80% of the 2000 mL initially added.

Thus the amount of water not recovered from the microcosms varied according to the substratum capacity to absorb water and to the evaporation rate. The absorption capacities of the materials tested (Table 1) were estimated considering the substratum mass initially introduced in each microcosm and their final mass after a one week drying period.

3.2 *Organic load profiles*

The organic load profiles, assessed only by BOD_5 measurements showed a similar trend on all the assays – a progressive decrease over time (only shown for third assay on Fig. 1). On the first weeks BOD_5 values of the water recovered from the microcosms were always the highest. On the

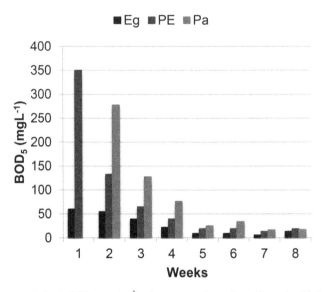

Figure 1. Average organic load (BOD$_5$, mg L^{-1}) of water samples gathered from the third assay microcosms.

third assay (Fig. 1) BOD$_5$ of water samples from eucalyptus bark microcosms (Eg) never exceeded 60 mg L^{-1}, but BOD$_5$ of water samples from microcosms with *Phragmithes* dried culms alone (Pa) or mixed with eucalyptus bark (PE), only got similar BOD$_5$ values from the third or forth week. To reach the BOD$_5$ values allowed by Portuguese legislation for discharges of wastewater from CW (Decreto-Lei n°152/97 de 19 de junho) i.e. 25 mgL^{-1} O$_2$, these substrata *per se* need about 7 to 8 weeks.

3.3 *Surface tension profiles*

As with the BOD$_5$ values, the surface tension estimated on water samples from microcosms showed a similar profile (Fig. 2): over time, these values approached the surface tension values of tap water, which corroborate a decrease on organic load content, reflected on the increase of the surface tension. Although the tap water surface tension values showed some fluctuations, the trend of leachate samples from microcosms was toward an increase of surface tension values.

4 DISCUSSION

According to our results *Phragmites australis* dried culms are able to retain more water than *Eucalyptus globulus* bark, which was evidenced by the continuous increase of its weight during the experimental period. The average percentage of weight gain by *Phragmites* culms used on second and third assay (Pa2 and Pa3) was 134 ± 21% (n = 5). On the other hand, *Eucalyptus globulus* bark mixed or not with *Phragmites* culms (PE and Eg) shown lower capacity to retain water and the differences between replicates are not negligible (Table 1).

Concerning the organic load, *Phragmites australis* dried culms added a high amount during the first weeks (higher than 400 mgL^{-1}), but this load was leached over time as no organic load was added by the tap water. The *Eucalyptus globulus* bark added less organic content (approximately 60 mgL^{-1} BOD$_5$) to the medium and as with *Phragmites* culms, this low organic load was also leached over time. As so, it takes about five weeks for these leached substrata, to reach the organic load contents allowed by Portuguese legislation, for discharges of wastewater from CW. As the dried

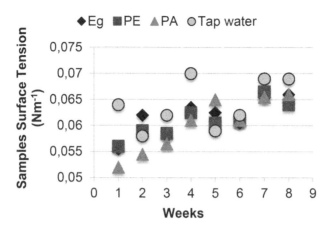

Figure 2. Water Surface Tension (Nm^{-1}) of tap water samples, and gathered samples from the third assay microcosms.

culms of *Phragmites australis* were collected in a CW they could be contaminated with wastewater from that system.

The surface tension values of the leached were occasionally conflicting with the tap water values, but were generally lower, as almost all the tap water surface tension values were very close to the surface tension values tabulated for pure water at $20°C–0.0728\ Nm^{-1}$ (Vazquez *et al.* 1995). Theoretically, the addition of foreign substances (dissolved or suspended) may change the value of the water surface tension, and may even weaken the cohesive forces between water molecules. Therefore the results obtained seem to support this principle, since the surface tension values of the leached waters from the microcosms increased over time while the organic load values decreased in all assays along the experimental period.

Although preliminary, these results suggested that *Phragmites* culms might be held in CW beds, if we considered the incoming organic load and the residence time of the effluent. Besides, the use a woody material like eucalyptus bark for a CW substratum (alone or mixed with other materials), may have advantages since it does not add organic load to the medium. However more studies are required to assess the toxicity of its chemical constituents (namely COD assays) on the biocenoses needed for wastewater treatment.

Lastly, the surface tension is a parameter with great potential to provide information on the presence of suspended or dissolved substances in water, with the advantage of its low cost.

REFERENCES

Albuquerque, A., Oliveira, J., Semitela, S. & Amaral, L. 2010. Evaluation of the effectiveness of horizontal subsurface flow constructed wetlands for different media. *Journal of Environmental Sciences* 22 (6): 820–825.

Brix, H., Arias, C.A. & del Bubba, M. 2001. Media selection for sustainable phosphorus removal in subsurface flow constructed wetlands. *Water Science and Technology?* 44 (11–12): 47–54.

Calheiros, C.S.C., Rangel, A.O.S.S. & Castro, P.M.L. 2008. Evaluation of different substrates to support the growth of *Typha latifolia* in constructed wetlands treating tannery wastewater over long-term operation. *Bioresource technology* 99(15): 6866–6877.

Calheiros, C.S.C., Duque, A.F., Moura, A., Henriques, I.S., Correia, A., Rangel, A.O.S.S. & Castro, P.M.L. 2009. Substrate effect on bacterial communities from constructed wetlands planted with *Typha latifolia* treating industrial wastewater. *Ecological Engineering* 35(5): 744–753.

Dordio, A.V., Teimão, J., Ramalho, I., Palace Carvalho, A.J. & Candeias, A.J.E. 2007. Selection of a support matrix for the removal of some phenoxyacetic compounds in constructed wetlands systems. *Science of the Total Environment* 380: 237–246.

Dordio, A.V., Gonçalves, P., Texeira D., Candeias A.J., Castanheiro J.E., Pinto, A.P. & Palace Carvalho A.J. 2011. Pharmaceuticals sorption behaviour in granulated cork for the selection of a support matrix for a constructed wetlands system. *International Journal of Environmental Analytical Chemistry* 91(7–8): 615–631.

Dordio, A.V. & Carvalho A.J.P. 2013. Organic xenobiotics removal in constructed wetlands, with emphasis on the importance of the support matrix. *Journal of Hazardous Materials* 252–253: 272–292.

Duarte, A.A.L.S., Canais-Seco, T., Peres, J.A., Bentes, I. & Pinto, J. 2010. Performance evaluation of Portuguese constructed wetlands for municipal wastewater treatment. In *Advances in Waste Management*. WSEAS Press: 124–129.

Du Nouy, P.L., 1919. A New Apparatus For Measuring Surface Tension. *The Journal of General Physiology*: 521–524.

Ebnesajjad, S., 2011. Surface tension and its measurement. In S. Ebnesajjad, ed. *Handbook of Adhesives and Surface Preparation*. William Andrew: 21–30.

Faulwetter, J.L., Gagnon, V., Sundberg, C., Chazarenc, F., Burr, M.D., Brisson, J., Camper, A.K. & Stein, O.R. 2009. Microbial processes influencing performance of treatment wetlands: A review. *Ecological Engineering* 35: 987–1004.

Kangas P. 2004. *Ecological Engineering: Principles and Practices*. CRC Press. Boca Raton, FL.

Lee, B.-B., Chan, E.-S., Ravindra, P. & Khan, T.A. 2012. Surface tension of viscous biopolymer solutions measured using the du Nouy ring method and the drop weight methods. *Polymer Bulletin* 69: 471–489.

Matos J., Galvão A., Carreira M. & Ferreira, F. 2009. Small wastewater systems in Portugal: Current situation and trends for the future. *Desalination and Water Treatment* 4: 6–11.

Saeed, T. & Sun, G. 2011. A comparative study on the removal of nutrients and organic matter in wetland reactors employing organic media. *Chemical Engineering Journal* 171(2): 439–447.

Saeed, T. & Sun, G., 2012. A review on nitrogen and organics removal mechanisms in subsurface flow constructed wetlands: dependency on environmental parameters, operating conditions and supporting media. *Journal of environmental management* 112: 429–48.

Stottmeister, U., Wießner, A., Kuschk, P., Kappelmeyer, U., Kästner, M., Bederski, O., Müller, R.A. & Moormann, H. 2003. Effects of plants and microorganisms in constructed wetlands for wastewater treatment. *Biotechnology Advances* 22 (1–2): 93–117.

Tariq, M., Freire, M.G., Saramago, B., Coutinho, J.A.P. & Canongia Lopes, J.N. 2012. Surface tension of ionic liquids and ionic liquid solutions. *Chemical Society reviews* 41(2): 829–68.

Vazquez, G., Alvarez, E. & Navaza, J.M. 1995. Surface tension of alcohol water + water from 20 to 50°C. *Journal of Chemical & Engineering Data* 40 (3): 611–614.

Vymazal, J., 2007. Removal of nutrients in various types of constructed wetlands. *The Science of the total environment* 380(1–3): 48–65.

Vymazal, J., 2010. Constructed Wetlands for Wastewater Treatment. *Water* 2: 530–549.

Zuidema, H.H. & Waters, G.W. 1941. Ring method for the determination of interfacial tension. *Industrial and Engineering Chemistry* 13: 312–313.

Wastes: Solutions, Treatments and Opportunities – Vilarinho, Castro & Russo (eds)
© 2015 Taylor & Francis Group, London, ISBN 978-1-138-02882-1

A decision support system for a waste collection vehicle routing problem

J.A. Ferreira, J.A. Oliveira & M. Figueiredo
Centro Algoritmi, University of Minho, Guimarães, Portugal

ABSTRACT: The selective collection of municipal solid waste for recycling is a very complex and expensive process, where a major issue is to perform cost-efficient waste collection routes. Despite the abundance of commercially available software for fleet management, they often lack the capability to deal properly with sequencing problems and dynamic revision of plans and schedules during process execution. Our approach to achieve better solutions for the waste collection process is to model it as a vehicle routing problem, more specifically as a team orienteering problem where capacity constraints on the vehicles are considered, as well as time windows for the waste collection points and for the vehicles. The final model is called Capacitated Team Orienteering Problem with double Time Windows (CTOPdTW). We developed a genetic algorithm to solve routing problems in waste collection modelled as a CTOPdTW. The results achieved suggest possible reductions of logistic costs in selective waste collection.

1 INTRODUCTION

Over the last few decades the European Union has published several directives and reports that are very relevant for waste management. Some of these documents are very rich in terms of data and statistics. The information provided by these documents should be well known by each citizen because it could alert them to the importance of the collection of waste and recycling. According to Eurostat in 2011, each European generates an average of 500 kg of waste per year (Eurostat 2015). In terms of recycling, only 122 kg is processed, representing around 25%. The amount of packaging waste generated in the EU-27 was, on average, 153 kg per capita in 2009 and 157 kg per capita in 2010 (Eurostat 2015).

In Portugal, the Ponto Verde System is coordinated by Sociedade Ponto Verde, S.A., (SPV), an organisation responsible for the collection and recycling of household, commercial, and industrial packaging waste of 308 municipalities. SPV promotes the selective collection, take-back, and recycling of packaging waste in order to guarantee the achievement of the recycling and recovery targets defined in the packaging Portuguese law.

In Portugal, household packing waste (HPW) is separated by citizens at the local recycling site, named Ecoponto ("ecological point"). Here, the waste is divided into three main containers, typically identified with different colours to help people to separate the waste: 1) glass (green) 2) paper/cardboard (blue), and 3) plastic/metal (yellow). These Ecoponto sites are provided by the municipalities for household waste only to be recycled in these containers. Currently in Portugal there are around 30,000 recycling banks and this number is continuously rising (Ponto Verde 2015). Given the goals Portugal has to fulfil for the recycling and recovery of HPW, there is a permanent need for increased efficiency in waste collection. The objective of this paper is to present a new model and a methodology that deals with the selective collection of HPW and helps improve the efficiency of the routes to visit Ecoponto sites. Section 2 presents the real problem, and discusses models and techniques of Operational Research to solve the Team Orienteering Problems with Time Windows (TOPTW), which we use to model the selective collection of HPW. Mathematical models developed for this problem are analyzed in the third section. The methodology based on Genetic Algorithms is presented in section 4. Computational experiments are discussed in the fifth Section, followed by the main conclusions.

2 PROBLEM DESCRIPTION

Braval is an inter-municipal waste management company in the Cávado sub region of northern Portugal and is established as an association of six municipalities: Braga (79%), Vieira do Minho (5.7%), Vila Verde (5%) Póvoa do Lanhoso (4.2%), Amares (3.7%) and Terras de Bouro (2.4%) (Braval 2015). Braval currently operates a network of 1,208 local recycling sites where residents can start the recycling process of their HPW. These sites are located across the municipalities in a variety of easily accessible areas. Braval covers a total area of 1,120 square kilometres and serves a population of around 290,000 inhabitants, which produces roughly 98,000 tons per year (on average 1 kg per capita/day) of municipal solid waste (MSW) (Braval 2015). In Braval's area of operation there is a mix of urban and rural areas, and so, different strategies for waste management must be employed.

In 2012 Braval achieved 14,121 tons of selective collection of HPW: 42% glass, 44% paper, and 14% plastic/metal. Most recycling banks are emptied on a regular basis. The banks are emptied by a vehicle equipped with an onboard crane. Each type of material (glass, paper, plastic/metal) is collected in different trucks; that is, each Ecoponto is visited by three different trucks. Braval uses a heterogeneous fleet of 10 trucks.

ERSUC reported that, on average, their trucks need to travel around 64 km to collect one ton of HPW (Ersuc 2015). If we use this index in the Braval case, and considering there are 340 working days per year, to collect 14,121 tons each truck of the Braval fleet would have to travel around 240 km per working day, which seems to be hard to achieve. Given the differences between urban and rural areas, it is necessary to produce adequate routes to collect the HPW, because one truck in an urban area could collect more than two tons in less than 60 km, while in rural areas it may travel more than 100 km to collect less than one ton.

Since it is obvious that the Braval fleet does not visit all the Ecoponto sites every workday, it is necessary to select a subset of Ecopontos to visit every workday. This situation is crucial to the model we chose. Given the differences between urban and rural areas, it is necessary to model properly to produce efficient routes to collect the HPW in both areas. Thus, given a one week or one month planning horizon, Braval must decide which Ecoponto sites must be visited, which Ecoponto sites can be visited, and which sites are not necessary to be visited, and then make effective routes to perform the selective collection of HPW. Taking in to account to its quantity of waste, an Ecoponto is either selected or not to be visited. The quantity of waste will define the prize to be considered. This situation configures a special type of routing problem, as described next. This is a novel approach, which we have been pursuing for the past few years and has proven to be amenable to mathematical modeling. This allows the use of optimization techniques, which promote the recycling and recovery of waste.

3 LITERATURE REVIEW

Routing is probably one of the subjects most studied in the operations research area. Routing research is applied widespread across real problems, and is also applied to the waste problems. Mainly, arc routing is applied for the collection of municipal solid waste, while vehicle routing problems (VRP) are applied for selective collection. The waste collection problem consists of routing vehicles to collect customer waste while minimizing travel costs.

Golden et al. (Golden et al. 2002) classified as one early classics in the VRP the work of (Beltrami & Bodin 1974), precisely a study about municipal waste collection. At that time Beltrami and Bodin already realized that waste collection encompasses a variety of problems. Golden et al.

described the waste collection problem studied in (Beltrami & Bodin 1974) as a very rich and challenging VRP attending the defined model. In (Golden et al. 2002) the authors performed a useful literature review about waste collection problems. The reader may consult the book by Toth and Vigo (Toth & Vigo 2002) for an interesting introduction to VRP. Due to the huge volume of VRP literature it is not possible to present the most important works in this field. However, an interesting point of view about the problem is included in the works of Laporte (Laporte 2007, Laporte 2009) and Cordeau (Cordeau 2006) which are mandatory reading.

3.1 *Waste collection and VRP variants*

Nuortio et al. claimed in their work that "The collection of waste is a highly visible and important municipal service" (Nuortio 2006). The importance of the subject is translated by the huge number of published works about the topic. These works focus on different subjects of real problems, like the distance to place the containers (Alvarez 2009). Collection frequency and time of collection are other topics studied. Several works analyze the effect in service effectiveness (Williams 2003, Williams 2013, El-Hamouz 2008, Kim 2006, Kaseva 2005). Chu et al. (Chu 2013) provide results of multinomial logistic regression and propose improvements to the collection service. Nabila et al. (Nabila 2009) and Fobil et al. (Fobil 2008) discuss waste management policies such as communal container collection and house-to-house waste collection. Several studies (Huang 2011, Gallardo 2010, García-Sanchez 2008) have looked at different systems for collecting household waste in order to evaluate the performance of municipal solid waste collection.

In this paper we address the optimization problem that arises with the waste collection performed with a fleet of vehicles. A significant number of recent studies about these issues is available in the literature. Cetinkaya et al. (Cetinkaya 2013) presented a new variant of VRP with Arc Time Windows. These Time Window restrictions may be encountered in urban waste collection. Buhrkal et al. (Buhrkal 2012) studied the Waste Collection Vehicle Routing Problem with Time Windows to find routes for garbage trucks. According to Buhrkal et al., their methodology produces solutions to collect waste in an efficient way. They define the Waste Collection Vehicle Routing Problem with Time Windows (WCVRPTW), which differs from the traditional VRPTW in three main points: 1) the waste collecting vehicles must empty their load at disposal sites; 2) multiple trips to disposal sites are allowed for the vehicles; and 3) the vehicles must be empty when returning to the depot.

In the case of (vehicle) routing problems with profits, two different decisions have to be simultaneously considered: which customers to serve and how to sequence them in one or several routes. This situation is different in the case of the most classical vehicle routing problems, because it is mandatory that all customers must be visited. Archetti et al. (Archetti 2012) explain that, in general, a profit is associated with each customer, which makes each customer more or less attractive to visit. There are many real life problems that could be modelled in this way, and one of them was presented by Aras et al. (Aras 2011): the "selective multi-depot vehicle routing problem with pricing", and denoted as SMDVRPP. They modelled a reverse logistics problem of a durable goods manufacturing firm that collects cores returned by the consumers that are then accumulated at its dealerships. The authors consider this problem very close to the Team Orienteering Problem (TOP) introduced by Chao et al. (Chao 1996), the multiple tour maximum collection problem (Butt & Cavalier 1994) that is a variant of the TOP (introducing heterogeneous vehicles), and selective VRP with time windows (SVRPTW) proposed by Gueguen (Gueguen 1999) that incorporates additional capacity and time window constraints over TOP. Besides the particular reverse logistics problem studied by Aras et al. (Aras 2011), to the best of our knowledge the selective collection of HPW has never been modelled as a variant of the Team Orienteering Problem.

4 MATHEMATICAL MODELS

Due to the importance of routing problems, one can say that for each problem or variant there is a mathematical model. Laporte (Laporte 2007, Laporte 2009) provides a very useful set of mathematical formulations for some variants of VRP. Baldacci et al. (Baldacci 2007, Baldacci 2010) review the most recent developments in the exact solution of the CVRP. They present the different mathematical formulations used in the literature and discuss the properties that were exploited in the most successful recent exact approaches.

In the case of TOP set of problems is available in the literature mathematical formulations that are close to VRP formulations. Mota et al. (Mota 2013) presented the first mathematical formulation for the Team Orienteering Problem with double Time Windows (TOPdTW), which is a TOPTW with availability time for the vehicles. Their model is an extension of the mathematical formulation presented for the TOPTW by Labadie et al. (Labadie 2012). The mathematical formulation for the Capacitated Team Orienteering Problem with double Time Windows (CTOPdTW) results from merging the Capacitated Team Orienteering Problem (CTOP) and the TOPdTW formulations.

5 A GENETIC ALGORITHM FOR THE CTOPdTW

As similar to other studies on routing problems, such as VRP and TOP, the CTOPdTW is an NP hard problem. For many instances it is not possible to achieve optimal solutions in a very short computing time. An alternative is the use of heuristics methods that can provide good solutions in a reasonable amount of time.

We extend the genetic algorithm (GA) developed by Mota et al. (Mota 2013) for the TOPdTW to solve the CTOPdTW. In brief, the GA uses a list of vertices for each vehicle; according to an insertion constructive algorithm the GA selects and schedules a tour for each vehicle.

We define the Braval Ecopontos sites as the set of vertices to visit, and for the fleet we define time windows of availability. The time windows of availability allow us to use the same facility (vehicle) more than once along the planning horizon, considering the existent shifts.

We include the GA in a decision support system (DSS) to aid in the task of designing routes for the fleet. Figure 1 shows a screenshot of the DSS.

In general, the GA finds solutions that are 95.86% of optimal or best known solutions for the TOPTW, in a reasonable amount of computation time. For the TOPdTW the GA obtained a reduction of objective function relatively TOPTW because the vehicles had constraints to be considered. In the CTOPdTW the results are affected by the capacity of vehicles and the performance was decreased by 25.63%.

6 COMPUTATIONAL EXPERIMENTS

Several experiments were performed in order to evaluate the GA performance. Firstly, we use the TOPTW set of public instances. Secondly we used the instances presented in (Mota 2013). Finally, we test the GA with randomly generated instances for the CTOPdTW. These instances are generated from TOPTW, adding a capacity and an availably time windows for each vehicle. In each vertex we consider the quantity equal to the prize defined in TOPTW instances.

7 CONCLUSIONS

This paper considers the selective collection of household packaging waste as a process that could be model as a combinatorial optimization problem. In particular we model it as a new variant of the well-known team orienteering problem that considers double time windows, one for the vertices to visit and another for the vehicles used in the process. We discuss some important works published either in the waste management field or in the routing optimization area. Some statistics of a real company that operates in the north of Portugal are presented and give an idea about the dimension and the complexity of the real life problem.

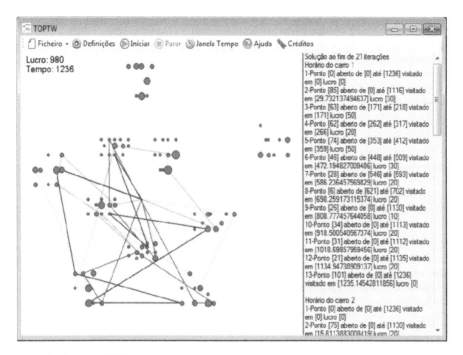

Figure 1.   The developed DSS.

An extended version of the GA developed for the TOPdTW was presented and the computation experiments show the performance of this technique. The GA was integrated in a DSS that could be used to aid the generation of routes in a real life process of selective collection of HPW.

As future work, we intend to apply the GA into real instances of Braval ecoponto sites to validate this methodology when applied to real life instances.

ACKNOWLEDGEMENTS

This work was funded by the "Programa Operacional Fatores de Competitividade – COMPETE" and by the FCT – Fundação para a Ciência e Tecnologia in the scope of the project: FCOMP-01-0124-FEDER-022674.

REFERENCES

Aras, N., Aksen, D., Tekin, M.T. 2011. Selective multi-depot vehicle routing problem with pricing. *Transportation Research Part C: Emerging Technologies* 19: 866–884.
Archetti, C., Speranza, M.G., Vigo, D. 2012. *Vehicle routing problems with profits*. Working paper.
Baldacci, R., Toth, P., Vigo, D. 2007. Recent advances in vehicle routing exact algorithms. *4OR: A Quarterly Journal of Operations Research* 5: 269–298.
Baldacci, R., Toth, P., Vigo, D. 2010. Exact algorithms for routing problems under vehicle capacity constraints. *Annals of Operations Research* 175: 213–245.
Beltrami, E.J. & Bodin, L.D. 1974. Networks and vehicle routing for municipal waste collection. *Networks* 4: 65–94.
Braval, 2015. Available from: <http://www.braval.pt/>, [February 2015].
Buhrkal, K., Larsen, A., Ropke, S. 2012. The waste collection vehicle routing problem with time windows in a city logistics context. *Procedia-Social and Behavioral Sciences* 39: 241–254.

Butt, S.E. & Cavalier, T.M. 1994. A heuristic for the multiple tour maximum collection problem. *Computers and Operations Research* 21: 101–111.

Çetinkaya, C., Karaoglan, I., Gokçen, H. 2013. Two-stage vehicle routing problem with arc time windows: a mixed integer programming formulation and a heuristic approach. *European Journal of Operational Research* 230: 539–550.

Chao, I., Golden, B.L., Wasil, E.A. 1996. The team orienteering problem. *European Journal of Operational Research* 88: 464–474.

Chu, Z., Xi, B., Song, Y., Crampton, E. 2013. Taking out the trash: Household preferences over municipal solid waste collection in Harbin, China. *Habitat International* 40: 194–200.

Cordeau, J.F., Laporte, G., Savelsbergh, M.W., Vigo, D. 2006. Vehicle routing. *Transportation* 14: 367–428.

El-Hamouz, A.M. 2008. Logistical management and private sector involvement in reducing the cost of municipal solid waste collection service in the Tubas area of the West Bank. *Waste Management* 28: 260–271.

Ersuc, 2015. Available from: <http://ersuc.pt/www/>, [February 2015].

Eurostat, 2015. Available from: <http://appsso.eurostat.ec.europa.eu/>, [February 2015].

Fobil, J.N., Armah, N.A., Hogarh, J.N., Carboo, D. 2008. The influence of institutions and organizations on urban waste collection systems: an analysis of waste collection system in Accra, Ghana (1985–2000). *Journal of Environmental Management* 86: 262–271.

Gallardo, A., Bovea, M.D., Colomer, F.J., Prades, M., Carlos, M. 2010. Comparison of different collection systems for sorted household waste in Spain. *Waste Management* 30: 2430–2439.

García-Sánchez, I.M. 2008. The performance of Spanish solid waste collection. *Waste Management & Research* 26: 327–336.

Gueguen, C. 1999. *Méthodes de résolution exacte pour les problèmes de tournées de véhicules*, Doctoral dissertation. Ecole Centrale Paris.

Huang, Y.T., Pan, T.C., Kao, J.J. 2011. Performance assessment for municipal solid waste collection in Taiwan. *Journal of Environmental Management* 92: 1277–1283.

Kaseva, M.E. & Mbuligwe, S.E. 2005. Appraisal of solid waste collection following private sector involvement in Dar es Salaam city. Tanzania. *Habitat International* 29: 353–366.

Kim, B.I., Kim, S., Sahoo, S. 2006. Waste collection vehicle routing problem with time windows. *Computers and Operations Research* 33: 3624–3642.

Labadie, N., Mansini, R., Melechovský, J., Calvo, R.W. 2012. The team orienteering problem with time windows: An lp-based granular variable neighborhood search. *European Journal of Operational Research* 220: 15–27.

Laporte, G. 2007. What you should know about the vehicle routing problem. *Naval Research Logistics* 5: 811–819.

Laporte, G. 2009. Fifty years of vehicle routing. *Transportation Science* 43: 408–416.

Mota, G., Abreu, M., Quintas, A., Ferreira, J., Dias, L.S., Pereira, G.A., Oliveira, J.A. 2013. A Genetic Algorithm for the TOPdTW at Operating Rooms. *Computational Science and Its Applications – ICCSA* 2013: 304–317. Springer Berlin Heidelberg.

Nabila, J.S., Yangyuoru, Y., Avle, S., Bosque-Hamilton, E.K., Amponsah, P.E., Alhassan, O., Satterthwaite, D. 2009. *Environmental Health Watch and Disaster Monitoring in the Greater Accra Metropolitan Area (GAMA)*. Accra: Ghana Universities Press.

Nuortio, T., Kytöjoki, J., Niska, H., Bräysy, O. 2006. Improved route planning and scheduling of waste collection and transport. *Expert Systems with Applications* 30: 223–232.

Ponto Verde, 2015. Available from: <http://www.pontoverde.pt/>, [February 2015].

Toth, P. & Vigo, D. 2002. The vehicle routing problem. SIAM Monographs on Discrete Mathematics. Philadelphia.

Williams, I.D. & Kelly, J. 2003. Green waste collection and the public's recycling behaviour in the Borough of Wyre, England. *Resources, Conservation and Recycling* 38: 139–159.

Williams, I.D. & Cole, C. 2013. The impact of alternate weekly collections on waste arisings. Science of the Total Environment 445: 29–40.

Wastes: Solutions, Treatments and Opportunities – Vilarinho, Castro & Russo (eds)
© 2015 Taylor & Francis Group, London, ISBN 978-1-138-02882-1

Quality requirements for the biomass in the SUDOE region

J. Ferreira, J.C.F. Teixeira & M.E.C. Ferreira
CT2M – Mechanical and Materials Technologies Centre, University of Minho, Guimarães, Portugal

J. Araújo
CVR – Centre for Waste Valorisation, Guimarães, Portugal

ABSTRACT: Contrary to standardized solid biofuels, such as wood pellets, wood chips and briquettes, Southern Western Europe (SUDOE) endogenous solid biomass were not covered by a quality referential, stablishing minimum quality criterion for its use for heat production. Furthermore, reliable information related to the SUDOE biomass characteristics wasn't available. The aim of this work was to identify the main physic and chemical characteristics of the endogenous SUDOE biomass, perform a comparative analysis with the existing quality standards for solid biofuels and establish reasonable thresholds for important parameters. SUDOE endogenous biomass shown to have very similar properties to the ones found in processed solid biofuels, like wood pellets. Therefore, it's reasonable to assume a great capability of the SUDOE biomass to be used for heat production and even cover a market share, now occupied by wood pellets.

1 INTRODUCTION

Economic and environmental factors, have increased the interest in solid biofuels over the last decade. However, the use of biomass for energy purposes is limited by both its availability and its quality.

Biomass is characterized by having relatively low calorific values and bulk density, when compared to fossil fuels. Withal, it has a higher ash content, whereas the presence of inorganic minerals such as alkali metals, calcium, phosphorous, chlorine and sulphur lead to ash related problems such as slag formation, fouling deposits as well as corrosion in combustion equipment's.

Biomass quality is a relevant issue nowadays, since state-of-the-art combustion equipment require standardized and high-end quality fuels. However, the variable quality of the raw materials may have an impact on the process efficiency, increasing maintenance requirements as well as costs.

The development of solid biofuels standards such as the EN 14961 series, established quality classes for wood pellets, briquettes and wood chips. Nonetheless, there are some endogenous biomass resources that are not covered by a quality referential. In the South Western Europe some specific endogenous resources, such as olive stones and nut shells, may represent a very high potential.

Addressing this potential market, the Biomasud project proposed, as main goal, to establish a quality label certification scheme for the SUDOE region of Europe endogenous biomass, by balancing its characteristics and the consumer's needs. This project was developed within the Interreg IV – SUDOE framework.

Biomass characterization, through laboratory analyses, allows the prediction of the combustion behaviour of equipment and pollutants emissions. This provides steady ground to establish reasonable thresholds for critic elements on biomass composition and set quality classes and specifications for different biomass types.

2 MATERIALS AND METHODS

2.1 *Biomass provenance and potential*

The types of biomass selected for this project are common in the SUDOE region and all represent subproducts of main agro industrial or forestry activities, namely: (1) olive stones (OS); (2) pine nut shells (PS) and (3) almond shells (AS); (4) pine cones (PC) and (5) wood chips (WC). Wood pellet samples were also taken into account in this study in order to establish a comparison benchmark for laboratory test results.

Around 100 biomass samples were taken from Portugal, Spain and South of France and sent to CEDER – CIEMAT laboratory in Soria, Spain, where they were analyzed.

In the scope of the task 2.3.1 – Inventory of the available biomass in SUDOE – it was identified a total potential of 6 million tonnes per year of biomass in the SUDOE region. WC represent more than 4 million tonnes, while a potential of 400 thousand tonnes was identified for OS. AS shells represent 182 thousand tonnes whereas the potential for PC and PS was 35 and 10 thousand tonnes, respectively.

2.2 *Sample testing*

The samples were tested for the following parameters: (1) moisture content; (2) net calorific value (NCV); (3) bulk density (4) ash content and ash melting behavior; (5) particle size distribution; (6) Chlorine and Sulphur; (7) Nitrogen and (8) other minor elements. Characterization of the biomass was performed according to the European standards defined by the Technical Committee CEN/TC 335, namely, EN 14774 (biomass moisture), EN 15104 (nitrogen content), EN 14775 (biomass ash), EN 15103 (bulk density), EN15289 (Sulphur and Chlorine), EN15297 (minor elements), EN15370 (ash melting behavior) and EN15149 (particle size distribution).

2.2.1 *NCV and moisture*

Moisture content has a large effect on the equipment performance, reducing thermal efficiency and sometimes, furnace temperature, which can lead to condensation and corrosion, as well as, changes in the emissions patterns (create unburned hydrocarbons). The moisture content has a deep effect on the biomass NCV, due to the amount of energy spent to evaporate the water molecules, and through that, not used to produce useful heat (Chen et al. 2010), (Koppmann, von Czapiewski, and Reid 2005).

The pre-flame smouldering of wet biomass mainly emits particles with mean diameters tipically over $0.55\,\mu$m. Particulate matter emissions associated with flaming combustion are relatively minor in quantity and smaller in size (e.g., 0.34–$0.45\,\mu$m) (Boman 2005).

2.2.2 *Ash and melting behavior*

There are two distinct types of ashes. Fly ash, being the lightest-weight component, rises with the flue gases and has higher concentrations of heavy metals and dioxins. Bottom ash is the material that falls to the bottom of the burner. (Barrows and Rawson 2011).

The major ash forming elements (Al, Ca, Fe, K, Mg, Na, P, Si) have relevance for the ash melting behaviour and deposit formation. Volatile ash forming elements such as Cl, S, Na, K, As, Cd, Hg, Pb, Zn play a major role regarding gaseous and especially aerosol emissions as well as deposit formation, corrosion. (Biedermann and Obernberger 2005). Ca and Mg tend to increase the ash melting point, while K and Na produce the opposite effect, as well as chlorides and low melting alkali-and aluminium silicates (Miles et al. 1995).

The key technical ash-related problems experienced by operators of biomass combustors and boilers are associated with (1) formation of fused or partly-fused agglomerates and slag deposits at high temperatures within furnaces and stoves; (2) formation of bonded ash deposits and accumulations of ash materials at lower temperatures on surfaces in the convective sections of boilers.

2.2.3 *Bulk density*

The bulk density influences the transport efficiency, storage area and combustion properties. The bulk density represents the weight contained in a determined volume and, for this reason, a low bulk density solid fuel requires a higher amount of storage space for the same energy content when compared with a higher bulk density one. This statement remains true for the transport operations, increasing the transport cost per energy content for lower bulk density biofuels and, ultimately, its carbon footprint.

2.2.4 *Fines*

Acording the EN14961-2 standard for wood pellets, fines are considered as particles with dimensions under 3.15 mm. Fines are produced by the mechanical wear of the solid biofuels or due to defects on sieving processes. Fine particles can influence several processes in the biofuel value chain, namely: (1) Obstruction or damage in the supply and transport equipment; (2) Particles behaviour during combustion; (3) Dust formation during transport and transfer; (4) Risk of dust explosions.

2.2.5 *Nitrogen, chlorine and sulphur*

Nitrogen is a common element in the atmosphere and it is commonly used as part of fertilizers in agriculture in the form of nitrous oxide (N_2O). Thereby, it is expected a higher quantity of N in agricultural biomass than in forestry based by-products. Regarding the combustion processes, fuel nitrogen and oxygen in the atmosphere combine, producing NOx, a pollutant gas that can lead to acid rain and smog formation. Biomass combustion systems operate at low temperatures, avoiding the occurrence of thermal NOx, produced by the combination of N and O in the atmosphere air. Thereby, the existing N in the fuel has a much more important role on emissions performance (Jenkins et al. 1998), (J.H. Seinfeld 1998).

The presence of sulphur in the flue gas can lead to sulphuric acid formation, and consequently corrosion at the heat exchangers surfaces. There is a higher potential for agriculture based biomass to have higher sulphur content, as this element is used in fertilizers (Hindiyarti 2007).

A major concern during the combustion of chlorinated fuels is the formation of chlorinated organic pollutants, which are potentially dangerous to human health and to the environment. During combustion, oxygen and hydrogen compete for the available hydrogen, forming H_2O and HCl, respectively (Procaccini 1999), (Biedermann and Obernberger 2005).

2.2.6 *Minor elements*

Heavy metals in solid fuels are mostly found at the fly and bottom ashes during combustion. All these elements, are highly volatile and at a certain concentration become toxic and represent a health hazard. Depending on the temperature, the heavy metals become part of the flue gas, aggregating with other chemical elements and molecules, taking part of the fly ashes. The volatilization of the heavy metals is mainly influenced by the presence of chlorine, combustion atmosphere (fuel rich or lean, temperature) and obviously by its concentration in the fuel composition. (Lu-shi et al. 2004).

Chlorine enhances the heavy metals volatilization through the increase of metal chlorides formation. However, when the combustion atmosphere is rich in oxygen, the chlorinating reaction is hindered, which leads to the delay of heavy metals release.

The heavy metals which represent the greatest environmental hazard are Lead, Cadmium and Zinc and they are, at the same time, the most volatile. Thereby, these elements appear predominantly in fly ashes and because of that are more prone to be emitted to the atmosphere.

3 RESULTS AND DISCUSSION

A comparative analysis of the laboratory test results was made, setting the wood pellets as a benchmark, from which SUDOE biomass was compared, in order to verify if its properties fulfill the requirements already established by quality certification schemes for wood pellets such as

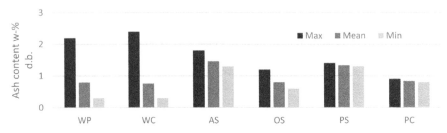

Figure 1.   Ash content (w-% d.b.).

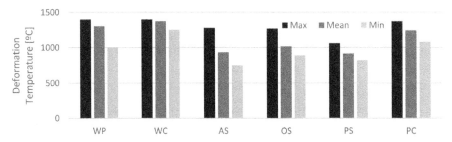

Figure 2.   Deformation temperature of the ash [°C].

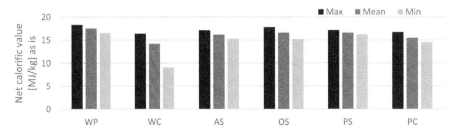

Figure 3.   Net calorific value [MJ/kg].

ENplus. For each type of biomass, the average, maximum and minimum value of the laboratorial tests data was found, for each parameter tested.

3.1 *Ash content and ash melting behavior*

Figure 1 shows a clear tendency of the nutshells to have a higher ash content than wood pellets, wood chips, olive stones and chopped pine cone. The ash content of almond and pine nut shells is approximately 30% higher than that found on the other types of biomass. The ENplus-A1 class wood pellets, set a threshold of 0.7% for the ash content, which is not consistently met by PS and AS.

The ash melting behavior is defined in the ENplus handbook by the deformation temperature, and specifies a minimum limit of 1200°C. This criteria is met by the general woody biomass and biofuels such as wood chips and wood pellets (Figure 2). On the other hand, nutshells showed an average deformation temperature lower than 1000°C. At the combustion bed of a biomass burning equipment reaches temperatures near 1000°C and there is a risk of slagging using this source of biomass.

3.2 *Net calorific value*

Wood pellets present a generally higher net calorific value than the SUDOE biomass, though, only wood chips show a significantly lower value, below 15 MJ/kg and would not comply with the minimum threshold set for wood pellets by the ENplus certification scheme of 16.5 MJ/kg.

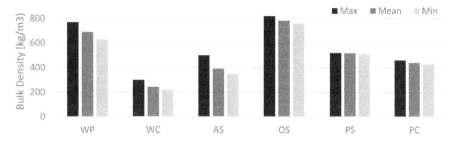

Figure 4.　Bulk density [kg/m^3].

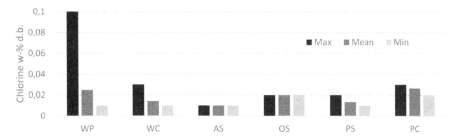

Figure 5.　Chlorine content [w-%] d.b.

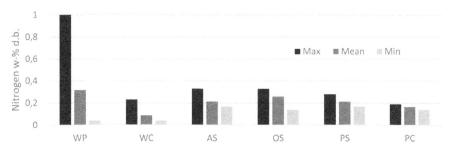

Figure 6.　Nitrogen content [w-%] d.b.

3.3　Bulk density

Because WP are a densified form of biomass, it is observed that nearly all the biomasses tested show a lower bulk density than WP, being OS the exception.

3.4　Chorine, Nitrogen and Sulphur

Through the analysis of the values taken from the biomass tests, it is possible to identify similarities between chlorine contents of all biomass types and wood pellets, despite the value found in one wood pellet sample, to be clearly above the average. The maximum threshold for chlorine defined for ENplus-A1 pellets is 0.02%. The majority of the tested biomass, with the exception of pine cone, would comply with this limit, even though the average value of the wood pellets exceeds it. Nevertheless, the wood pellet sample represented in the Figure 5 with a chlorine content above 0.1% might have had a significant influence on the average calculation.

The types of biomass studied comply consistently with the maximum nitrogen content threshold defined for the ENplus-A1 class wood pellets, which is 0.3%. There is, also, a similarity between wood pellets and the other types of biomass, suggesting that, regarding the nitrogen content there is no significant difference between them (Figure 6).

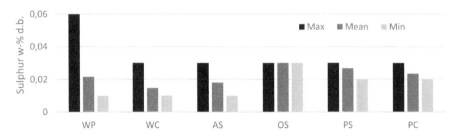

Figure 7. Sulphur content [w-%] d.b.

The biomass in this project comply consistently with the maximum Sulphur content threshold defined for the ENplus-A1 class wood pellets, which is 0.03% (Figure 7).

4 CONCLUSIONS

The main conclusions to take from this study are: (1) The properties of SUDOE solid biomass are very similar to standardized solid biofuels like wood pellets; (2) The SUDOE solid biomass will easily meet the quality requirements established in this report, which are very similar to the ones required in the EN14961-2 for wood pellets.

Establishing a set of quality requirements for SUDOE solid biomass will allow: (1) the development of equipment's technology supported by well know fuel characteristics in order to enhance their efficiency and performance; (2) to increase the consumers trust and enhance the market to introduce new renewable energy alternatives; (3) a better use of resources, adding value to otherwise waste materials, with normalized quality and characteristics.

ACKNOWLEDGEMENTS

The authors acknowledge the financial support from European Regional Development Fund through grant SOE2/P2/E414 – INTEREG IV B SUDOE for the Biomasud project.

REFERENCES

Barrows, B. & Rawson, S. 2011. "Management of Wood Ash Generated from Biomass Combustion Facilities".
Biedermann, F. & Obernberger, I. 2005. *Ash-Related Problems during Biomass Combustion and Possibilities for a Sustainable Ash Utilisation.*
Boman, C. 2005. *Particulate and Gaseous Emissions from Residential Biomass Combustion.*
Chen, L.W. et al. 2010. "Moisture Effects on Carbon and Nitrogen Emission from Burning of Wildland Biomass." *Atmospheric Chemistry and Physics* 10(14):6617–25. Retrieved July 9, 2012 (http://www.atmos-chem-phys.net/10/6617/2010/).
Hindiyarti, L. 2007. "Gas Phase Sulfur , Chlorine and Potassium Chemistry in Biomass Combustion."
J.H. Seinfeld. 1998. *Atmospheric Chemistry and Physics of Air Pollution.* Wiley, New York.
Jenkins, B.M., Baxter, L.L., Miles jr, T.R. & Miles, T. R.. 1998. "Combustion Properties of Biomass."
Koppmann, R., von Czapiewski, K. & Reid, J.S. 2005. "A Review of Biomass Burning Emissions, Part I: Gaseous Emissions of Carbon Monoxide, Methane, Volatile Organic Compounds, and Nitrogen Containing Compounds." *Atmospheric Chemistry and Physics Discussions* 5(5):10455–516. Retrieved (http://www.atmos-chem-phys-discuss.net/5/10455/2005/).
Lu-shi, S., Abanades, S., , Lu, J.D., Flamant, G. & Gauthier, D. 2004. "Volatilization of Heavy Metals during Incineration of Municipal Solid Wastes." *Journal of Environmental Sciences*, 5.
Miles, T.R., Baxter, L.L., Bryers, R.W., Jenkins, B.M., and Oden, L.L. 1995. *Alkali Deposits Found in Biomass Power Plants,* Vol 2. II:1–122.
Procaccini, C. 1999. *44602824_B-.pdf.* Cambridge: Massachusetts Institute of Technology.

Wastes: Solutions, Treatments and Opportunities – Vilarinho, Castro & Russo (eds)
© 2015 Taylor & Francis Group, London, ISBN 978-1-138-02882-1

Evaluation of briquettes made from amazonian biomass waste

R.F.B. Gonçalves, M.O. Silva, D.R.S. Guerra & M.F.M Nogueira
Faculdade de Engenharia Mecânica, Universidade Federal do Pará, Belém/PA, Brasil

ABSTRACT: The energy density of biomass is usually small and its moisture concentration is high, causing an increase in energy needed for its consumption. After torrefaction process, the biomass gets more interesting energy characteristics and could be used to replace partially or totally coal in industrial processes. This work contains the laboratory test results of briquettes produced with açai berry waste, largely found in Amazonia region. The material is suitable for combustion as biomass, with high content of volatiles (>70%) and low content of fixed carbon (<25%) and ashes (<1.5%). A procedure for torrefying the waste was developed, resulting in significant removal of moisture and low reabsorption of ambient moisture. The briquettes produced with the pre-treated waste were robust and resistant to mechanical impacts, while maintaining the high energetic values of the torrefied biomass.

1 INTRODUCTION

The great demand, greenhouse effect, price increase and reduction in the availability of fossil fuels have forced the world to seek new sources of energy. Parthasarathy & Narayanan (2013). One of the variations is biomass, a renewable carbon source and plentiful. Klass (2004); McKendry (2002); White *et al.* (2011).

Due to the low conversion rate in energy, biomass is still not widely used worldwide. However, there are treatment methods of this material aimed at increasing energy efficiency, such as pyrolysis, gasification, liquefaction or roasting, for example.

The biomass is composed of three major components: cellulose, hemicellulose and lignin, and the proportion of each vary with the biomass. Consequently, the energy behavior varies proportionally.

Compared to coal, the energy density of biomass is small and its moisture concentration is high, which causes an increase in energy needed for its consumption. However, when converted (roasting or pyrolysis) moisture is reduced dramatically and the biomass gets more interesting energy characteristics. In view of this, biomass could be used to replace partially or totally coal in industrial processes.

The national production of açaí was approximately 202,000 ton in 2013, according to the Brazilian Institute of Geography and Statistics. The State of Pará, the main producer located at Amazon region, participated with 54.9% of this production. In the Amazonian region there is great availability of biomass, which are in general disposed as waste. One of them is the açaí fruit which in the agro industry produces a large amount of waste, mainly seeds and fibers. The acaí berry is one of the major components of this waste; after retrieving the extracts, the core and its fibers are disposed like common waste, promoting a serious environmental and public health problem. With the proper treatment, this waste can become a very powerful fuel and significant energy may be obtained.

When a biomass is torrefied, the moisture is removed (which increases both the gross and net calorific values) and the lignin present is melted, then diffusing to the surface creating a hydrophobic barrier. This is known as mild torrefaction process that is carried out at maximum temperature of 300°C, in which hemicellulose, the most reactive fraction of wood, is decomposed. Mark *et al.* (2006); Felfli *et al.* (2005).

Table 1. Briquetting process evaluation.

Variables	Trials								
Pressure (MPa)	7.85	7.85	7.85	7.85	7.85	7.85	7.85	7.85	7.85
Temperature (°C)	120	150	180	120	150	180	120	150	180
Time (min)	5	5	5	7	7	7	9	9	9
Pressure (MPa)	9.80	9.80	9.80	9.80	9.80	9.80	9.80	9.80	9.80
Temperature (°C)	120	150	180	120	150	180	120	150	180
Time (min)	5	5	5	7	7	7	9	9	9

Table 2. Torrefaction trials of açai berry waste.

Temperature (°C)	Time (min)	Temperature (°C)	Time (min)
200	5	225	10
200	10	225	12.5
200	15	225	15
200	20	250	5
200	30	250	7.5
225	5	250	10
225	7.5		

For combustion the biomass is prepared to the suitable size and form. Wood fuels are usually supplied as logs, bark, chips, sawdust, shavings, briquettes and pellets. Brožek *et al.* (2012). In general, biomass are collected and transported, and then converted into energy by the burning process. Ngusale *et al.* (2014). However, in order to achieve higher energy per volume, a densification process is usually needed. The densification process involves drying, shredding, and pressing, which needs costly equipments and a few energy sources. Zhang & Guo (2014).

The aim of the present work is to present the development of briquettes made from açai berry waste, its thermal treatment and thermodynamic evaluations. Also, a comparison between non torrefied and torrefied briquettes was made.

2 METHODOLOGY

Quality of briquettes depends on both the material used and the process. For example, varying the structure of the material, its moisture content and granulometry, leads to briquettes with completely different characteristics (physical and mechanical). For instance, briquettes made from un-milled açai berry are extremely fragile due to the physical form of the material. The conditions applied for the production are also extremely significant to guarantee a robust briquette. Table 1 below presents all the developed trials for the process optimization. All briquettes had their hardness and density evaluated.

The samples of açaí seed were briquetted at ambient temperature and elevated temperatures in a calibrated briquetter machine for laboratories model LB 32, under pressures of 7.85–9.80 MPa. The machine is made of porcelain and stainless steel, engaging a hydraulic motor 2.21 kW and its resistance can reach up to 300°C. The briquettes produced were cylindrical in shape, have the average size of 60 mm with 32 mm of diameter.

In the briquetting press, the material (located in the press chamber) is pushed by pistons connected to a short screw conveyor which is placed in the hopper bottom. Then the briquette weight and length depends on the material amount which gets at the piston working stroke in the press chamber. For the present work, the mass used for each trial was 40 g.

A procedure for torrefying the biomass was developed in a muffle (Table 2). Different temperatures and residence times were evaluated, until optimization of the process. All samples obtained

Table 3. Açai berry analysis.

Proximate Analysis Property	Dry	Humid
Humidity (%)	0.00	19.85
Volatiles (%)	77.59	62.19
Ashes (%)	1.48	1.18
Fixed Carbon (%)	20.93	16.78
Ultimate Analysis	C (%)	43.26
	H (%)	5.69
	N (%)	1.10
	S (%)	1.46
Gross Calorific Value (MJ/kg)		17.382

had the moisture content analyzed, as well as their gross calorific value (GCV) and the water re-absorption after a week exposed to environment.

The biomass waste samples (in natura and torrefied) were submitted to proximate analysis, and calorific value determination. The proximate analysis was done according to the following standards: ASTM D 1102–84, D 3173-03, E 871–82, E 872–82, E 1755, E 1757-01. For the gross calorific value (GCV), the determination was made in a calorimeter Ika Werke C2000, according to ASTM E711-87.

3   RESULTS AND DISCUSSION

Before preparing the briquettes, it is important to evaluate the biomass via proximate and ultimate analyses, as well as the GCV. Table 3 presents the results of the biomass waste obtained.

The material contains about 20 percent of moisture, which becomes responsible for a significant decrease on the calorific value. However, it has a high content of volatiles and low content of fixed carbon and ashes, which are desirable characteristics for combustible materials. Nitrogen levels are normal ($<3.5\%$), indicating low influence on NOx production, but there's a high level of sulfur ($>0.45\%$), which may lead to SO_2 production during combustion.

The briquettes were produced with the açai berry waste, according to the previously stated conditions. Table 4 shows the characteristics of each sample obtained. All samples showed densities in a desirable range (1000–1400 kg/m^3), indicating a good cohesion among the particles. Surprisingly, hardness evaluations showed a poor correlation among the variables. It was expected at first that the hardness would increase with the temperature, pressure and residence time, but several samples produced with milder properties ended up with higher hardness.

The briquettes were very stable and robust at first, but some of them (especially the ones produced at lower temperatures) disintegrated after a week exposed to the environment. This phenomenon happened due to the re-absorption of moisture by the sample. Although the material was heated to mild temperatures, it was not enough to melt and diffuse the lignin for creating a hydrophobic surface. Therefore, it was interesting to pre-treat thermally (torrefy) the material to avoid further fragilization of the briquette.

Table 5 presents the results of the açai berry torrefied in a muffle in different conditions. The moisture content (wet basis) represents the residual water in the material after the treatment, and is proportional to the time and temperature, as expected. As observed, it is interesting to treat the samples for more than 5 minutes, thus guaranteeing moisture content below 2%. If the material is held too long in the muffle, there is loss of volatile matter and consequent reduction of the GCV. After a week exposed to the environment, the moisture was once more evaluated, showing the re-absorption of the material.

Table 4. Briquettes produced of milled in natura biomass waste.

Pressure (MPa)	7.85	7.85	7.85	7.85	7.85	7.85	7.85	7.85	7.85
Temperature (°C)	120	150	180	120	150	180	120	150	180
Time (min)	5	5	5	7	7	7	9	9	9
Mass (g)	39.6	39.5	39.0	39.4	39.4	38.7	39.9	39.7	39.4
Diameter (cm)	3.23	3.20	3.18	3.15	3.20	3.17	3.25	3.22	3.18
Height (cm)	3.90	3.90	3.70	3.88	3.73	3.61	4.02	3.92	3.68
Density (kg/m^3)	1246	1259	1330	1307	1313	1363	1201	1246	1350
Hardness (mN)	7120	7412	7341	7216	7591	7670	–	7070	7412
Pressure (MPa)	9.80	9.80	9.80	9.80	9.80	9.80	9.80	9.80	9.80
Temperature (°C)	120	150	180	120	150	180	120	150	180
Time (min)	5	5	5	7	7	7	9	9	9
Mass (g)	39.1	39.4	39.4	39.1	39.1	39.	39.2	39.5	39.3
Diameter (cm)	3.20	3.20	3.24	3.20	3.26	3.63	3.20	3.25	3.13
Height (cm)	4.23	3.92	3.73	4.26	4.10	3.14	4.23	3.96	3.52
Density (kg/m^3)	1149	1252	1282	1143	1144	1211	1156	1203	1457
Hardness (mN)	6741	6812	7516	6275	7125	7625	–	6716	7775

Table 5. Torrefication results.

Temperature (°C)	Time (min)	Moisture (%)	Moisture re-absorption (%)
200	5	3.49	6.91
200	10	1.13	6.81
200	15	1.24	6.57
200	20	1.56	5.41
200	30	1.51	6.54
225	5	4.74	6.23
225	7.5	1.18	5.88
225	10	1.14	5.71
225	12.5	1.70	4.69
225	15	0.22	4.49
250	5	2.81	5.59
250	7.5	1.09	5.44
250	10	0.22	4.72

Table 6. Energetic evaluation of chosen torrefied waste samples.

Temperature (°C)	Time (min)	GCV (MJ/kg)
200	20	19.039
200	30	19.079
225	12.5	18.193

Focusing on low moisture content, low re-absorption and avoiding volatile matter loss, three conditions were chosen for GCV evaluation as shown in Table 6. When comparing to the raw material, there is an increase of almost 10% in the GCV, due to the low moisture of the torrefied waste.

When producing briquettes with the pre-treated waste, the density and hardness are kept, but there's a significant increase in the calorific value and in the resistance, as the re-absorption of moisture is greatly reduced due to the presence and homogeneity of lignin in the surface.

4 CONCLUSION

This work contains the laboratory test results of briquettes produced with açai berry waste, found in Amazonia region. Proximate and ultimate analyses were done in the waste, followed by the evaluation of the gross calorific value to complete the characterization. The material is suitable for combustion as biomass, with high content of volatiles ($>70\%$) and low content of fixed carbon ($<25\%$) and ashes ($<1.5\%$).

Several briquetting parameters were tested, while varying temperature, pressure and residence time in the press chamber. All briquettes fulfilled the demands of density ($1000-1400\,kg/m^3$), but showed no linear correlation with the hardness. The samples were robust until the re-absorption of moisture, due to the lack of a protective lignin barrier.

A procedure for torrefying the waste was developed, resulting in significant removal of moisture and low reabsorption of ambient moisture. All briquettes obtained with the torrefied waste followed the properties previously obtained and also remained intact and resistant.

REFERENCES

Brožek M., Nováková A. & Kolářová M. 2012. Quality evaluation of briquettes made from wood waste, *Res. Agr. Eng.*, 58(1): 30–35.

Felfli, F.F., Luengo, C.A. & Rocha, J.D. 2005. Torrefied briquettes: technical and economic feasibility and perspectives in the Brazilian market, *Energy for Sustainable Development,* IX(3):23–29.

Ngusale, G.K., Yonghao, L. & Kiplagat, J.K. 2014. Briquette making in Kenya: Nairobi and peri-urbanareas, *Renewable and Sustainable Energy Reviews*, 40:749–759.

Zhang, J. & Guo, Y. 2014. Physical properties of solid fuel briquettes made from Caragan korshinskii Kom, *Powder Technology*, 256: 293–299.

Klass, D.L. 2004. Biomass for renewable energy and fuels. *Encyclopedia of Energy (Volume 1)*, Elsevier Inc. Elsevier Science Publishers, Amsterdam, The Netherlands.

Prins, M.J., Ptasinski, K.J. & Janssen, F.J.J.G. 2006. Torrefaction of wood Part 2. Analysis of products, *J. Anal. Appl. Pyrolysis*, 77: 35–40.

McKendry, P. 2002. Energy production from biomass (Part 1): Overview of biomass, *Bioresource Technology*, 83: 37–46.

Parthasarathy, P. & Narayanan, S. K. 2013. Determination of Kinetic Parameters of Biomass Samples Using Thermogravimetric Analysis, *Environmental Progress & Sustainable Energy,* 33(1): 256–266.

White, J.E., Catallo, W.J. & Legendre, B.L. 2011. Biomass pyrolysis kinetics: A comparative critical review with relevant agricultural residue case studies, *Journal of Analytical and Applied Pyrolysis*, 91: 1–33.

Wastes: Solutions, Treatments and Opportunities – Vilarinho, Castro & Russo (eds)
© 2015 Taylor & Francis Group, London, ISBN 978-1-138-02882-1

Evaluation the feasibility of AIMD waste treatment focused in metal recovery

R. Guimarães, V. Leal & A. Guerner Dias
Faculty of Science University of Porto, Porto, Portugal

J. Carvalho
Aveiro University, Aveiro, Portugal

ABSTRACT: This study analyses the feasibility of Active Implantable Medical Devices (AIMD) waste treatment focused in metal recovery. After contacts with some responsible for the management of these devices in Portugal, it is possible to verify that none have a treatment process for this devices after been used, so they are stored in the hospitals. The number of devices implanted increase every year, so is important propose an efficient treatment process for this waste with the goal of recovery precious metals and other materials. With the aim of development an efficient treatment process was realized a characterization of AIMD. The results indicate that printed circuit board from AIMD present quantities of precious metals, particularly gold (Au), superiors of printed circuit board of other electronic devices. Is possible to conclude that exist technology able to valorize the active implantable medical devices waste, since these are subject to a pretreatment after their explantation.

1 INTRODUCTION

Population growth is more pronounced every day who leads to the rise of consumption, generating many problems in relation to treatment and final destination for the waste produced. With the improvement of life conditions, more waste is generated, particularly hospital waste, turning to several legal and social issues, in relation to environmental effective treatment solutions.

The technology industry growth leads to inordinate use of metals and ensure their supply is seen as critical (UNEP, 2010). In addition, that fact leads to an increase of waste electric and electronic equipment (WEEE) which rounds 8.3 to 9.1 million metric tons with an annual increase of 3–5% in twenty seven countries of European Union (Huisman et al., 2008).

Despite the implementation of WEEE directive and the Restriction of Hazardous Substances (RoHS) directive in the European Union, to properly manage ever-growing stream of WEEE, other countries with large production of EEE (e.g. China, USA) have not implemented specific legislation on the issue (European Parliament, 2003).

Medical devices are contained on 8th category of EEE in the directive 2012/19/EU of the European Parliament and of the Council of 4 July 2012 on WEEE, which states that recycling goals shall not apply to *"medical devices and in vitro diagnostic medical devices, where such devices are expected to be infective prior to end of life, and active implantable medical devices"* (article 2, point 4, paragraph g), where pacemakers and implantable cardioverter defibrillator (ICD) are included. Due this fact, there's no actual solution to recycle AIMD.

The hospital waste is divided in four groups (Agência Portuguesa do Ambiente, 2011), in which, groups I and II obey in an efficient way the current waste management hierarchy, while groups III and IV, after a previously treatment, are sent for incineration or landfilled.

The AIMD, after explantation and disinfection, belong to group III. According to European Waste Catalogue (EWC) (Ministério da Economia, 2004), these devices should be classified in chapter 18 *"Waste from human or animal health care and/or related research"* however they do not represent a specific waste flow. Therefore, arises the necessity to identify their components, with particular relevance to high value components, with the aim of development an efficient treatment process.

Devices' functionality is increased by precious and special metals, and then is expectable that the use of gold and other technology metals grow further. Electronics represents 12% of total annual mine production of gold. Their efficient recovery from electronic scrap represents a potential recycling source. Gold's demand had an increasing during the last decade, not only due to the jewelry market, but also due to the increasing uses of gold in industry, as well as medical devices (Hagelüken and Corti, 2010).

Printed circuit boards (PCB) are the essential parts of electronic devices and hold the major fraction of metals present in WEEE. PCBs are particularly rich in copper and precious metals with high economic potential (80% of the total intrinsic value even though the amount is less than 1 wt.%, such as gold (Park and Fray, 2009). Every year 300 ton of gold are used in electronic components such as integrated circuits, contacts and bonding wires (Hagelüken and Corti, 2010).

The rigid PCB multilayers are fabricated from copper-clad dielectric materials. The dielectric consists of an organic resin reinforced with fibers. The organic media can be of a wide formulation and include flame-retardant phenolic, epoxy, polyfunctional epoxy, and polyimide resins (Harper and Sampson, 1994). Multilayer circuits are produced by building a structure of conductive layers and are filled with the same metallization in a separate screening step (Minges and Committee, 1989).

The gold is not deposited on the PCB surfaces directly but on an aluminum or titan pre-coat layer (Harper and Sampson, 1994). Conductors that use gold are easily bonded and when alloyed with small amount of platinum or palladium, it is easy to solder and form reliable joints (Minges and Committee, 1989).

In AIMD are included the pacemaker, the ICD and others. In this study only pacemakers and ICDs were included.

A pacemaker is an electronic biomedical device that can regulate the human heartbeat when its natural regulating mechanisms break down. The durability of these devices varies from 5 to 10 years due the degree of battery's use. In Portugal, the annually average number of implanted devices is about 7200 (Direção Geral de Saúde, 2013).

The pacemaker's motherboard contains all the electrical circuitry of the pacemaker and using hybridization, all components are combined to form a single complex circuit, allowing the use of materials which cannot be included in a monolithic integrated circuitry. The primordial conductor used in pacemakers is gold because of its stable and inert status, does not oxidize or migrate (Minges and Committee, 1989).

An ICD is a small device that is placed in the human body, specifically in chest or abdomen. Are used to detect dangerously fast heartbeats and give a lifesaving shock to correct the heart's rhythm. Like pacemakers that are explained previously, ICD contain a generator containing a computer, battery, and wires called "leads" that go through a vein into the heart. The leads stay in contact with the heart muscle on one end, while the other end is connected to the generator. The battery in the generator lasts in average 5–8 years and must be replaced when it runs out (National Institutes of Health). According to the Portuguese Directorate-General of Health in the year 2012 were implanted 888 ICD (Direção Geral de Saúde, 2013).

The aim of this work is the characterization of AIMD to study the technical/economic feasibility of their treatment with special attention on metal recovery.

2 MATERIAL AND METHODS

A sample of AIMD, previously disinfected by autoclaving, were collected at several hospitals in the north of Portugal, and transported to Faculty of Science of the University of Porto (FCUP) laboratory's where they were identified and weighed. The devices were opened with a handsaw and a lathe. After opening these devices were examined macroscopically to find out the best way

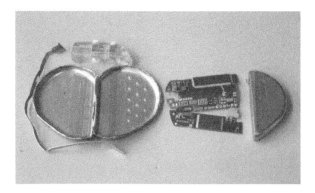

Figure 1. Exterior casing, electrodes, PCB and batteries from Pacemaker (Left to right).

Figure 2. Identification of the points analyzed in the Printed Circuit Board.

to analyze them with more detail. The separation into three major components appeared to be the best option. The separation in smallest parts was done physically with tweezers and identified the three main components are: i) Exterior casing and electrodes; ii) PCB's; iii) Batteries as showed in Figure 1.

All of the components were identified and weighed individually, and were prepared to be analyzed by Scanning Electron Microscopy (SEM).

The electrodes were manually detached from exterior casing in order to individually analyze both components. A chainsaw precision was used to make transverse and longitudinal sections in the polymer which coats the electrode and to cut a part of the exterior case. From the small pieces obtained two samples were prepared with Epoxy resin in order to be analyzed in Materials Centre of the University of Porto (CEMUP) by SEM technique. The PCB were also prepared with Epoxy resin to ensure that all conditions were satisfied to be analyzed by SEM technique. The spots analyzed with SEM were previously selected with careful observations in the magnifying glass and the spots analyzed (Z1, Z2, Z4, Z5, Z6, Z8) are showed in Figure 2.

The SEM technique does not allow the quantification of materials present in PCB, therefore, several samples were prepared in order to quantify the percentage of their metal content. One pacemakers' PCB (P1) and two samples of the metallic components (ICD6, ICD7) present in the surface of two PCB from ICD were selected and grinded in an agate mill, and sent for an international laboratory (ALS Scandinavia AB) where several elements were analyzed. For the sample P1 was quantified the presence Au, Ag and Pt, while for the samples ICD6 and ICD7 was quantified the presence of Ag, Au, Cu, Ga, Ni and Pt.

At this time the analysis of batteries was not performed due to lack of resources.

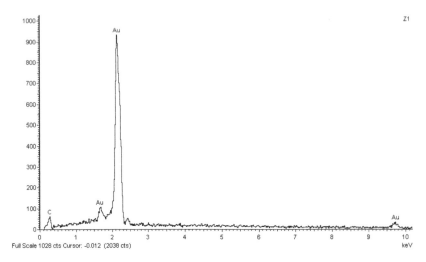

Figure 3. Spectrum of point Z1 (Z5, Z6) of PCB identify in figure 2 (SEM-EDS in CEMUP).

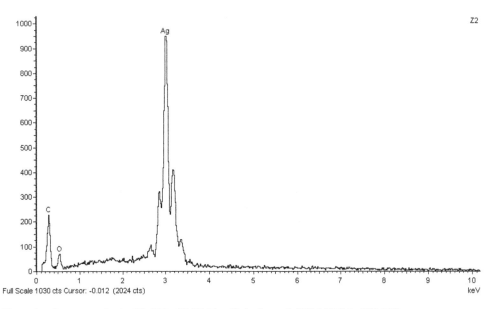

Figure 4. Spectrum of point Z2 (Z4) of PCB identify in figure 2 (SEM-EDS in CEMUP).

3 RESULTS AND DISCUSSION

This section presents the most relevant values obtained during the analysis. Focus is given to each component (exterior casing, electrodes and PCB), with particular emphasis on the presence of precious metals in their structure.

The exterior casing is constituted by titanium (Ti) with very small inclusions of copper (Cu). The results of SEM technique for the electrodes show that they are constituted by a variety of metals with higher incidence in iron (Fe) and manganese (Mn) and lowest incidence in nickel (Ni), chromium (Cr) and others. For PCB, the results show that they have a huge variety of metals in their composition The analysis of the point Z1 is presented in the figure 3 and represents as well the results for the points Z5 and Z6, being these points mainly composed by gold (Au). The point

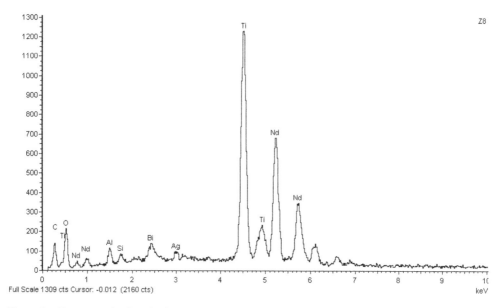

Figure 5.   Spectrum of point Z8 of PCB identify in figure 2 (SEM-EDS in CEMUP).

Table 1.   Composition of a typical PCB present in WEEE (Goosey and Kellner, 2002).

Element	% mass of each metallic component	% financial of each metallic component
Gold	0.025	65.40
Palladium	0.010	11.40
Silver	0.100	4.600
Copper	16.000	9.700
Aluminum	5.000	1.100
Iron	5.000	0.100
Tin	3.000	4.500
Lead	2.000	0.500
Nickel	1.000	2.400
Zinc	1.000	0.200

Z2, showed in figure 4, is mainly composed by Silver (Ag). This result was also verified for the point Z4. Point Z8 was found to be mainly composed by Titanium (Ti) as well as in others types of PCB (Table 1), (Goosey and Kellner, 2002).

Table 2 shows the quantitative analysis of two ICD samples and one pacemaker sample. For the ICD samples the analysis were realized for 6 elements (silver, gold, copper, gallium, nickel and platinum), while for the pacemaker sample analysis was performed for 3 elements (silver, gold and platinum). In the sample "ICD7", platinum was not possible to quantify because the concentration was under the detection limit.

These results reveal to be very promising as the quantity of precious metals present in this type of PCB is much higher than in PCB used in other technologies such as laptops and cellphones.

As previously mentioned, the batteries were not analyzed due to lack of time. Nevertheless, information about their internal composition is readily available on site manufactures. For example, by consulting factsheets from ICD and pacemakers manufacturers (Medtronic and St. Jude Medical) it is possible to verify that lithium iodine batteries are the most used in AIMD's. Since the market value of these metals is far below from gold, their interest for this study becomes minor. Also, a

Table 2. Results of quantitative analyses of ICD and pacemaker from ALS laboratory.

SAMPLE	Element (mg/kg)					
	Ag	Au	Cu	Ga	Ni	Pt
ICD6	12979	2396	15974	103	30950	113
ICD7	1302	1742	33186	55	53171	n.d.
P1	162	5776	n.a.	n.a.	n.a.	81

n.d. – not detected; n.a. – not analysed.

survey found that there are a great number of companies operating recycling process for lithium iodine batteries.

The results, even considering that the analytic methods had focused in particular components, showed that AIMD are composed by a relevant quantity of metals with high economic value, such as gold and silver, along with other metals with lowest economic value (titanium, neodymium, bismuth, copper, etc.).

4 CONCLUSIONS

The obtained results allow us to conclude that these devices have conditions to be treat, in compliance for the sustainable criteria. The recovery of the precious metals during the AIMD recycling process appears to be the most important step to achieve the economic feasibility, and, by consequence, the sustainability of the process since the technological availability, the social benefits and the environmental criteria are guaranteed. The implementation of an AIMD treatment stream can generate jobs (waste collecting and handling) and eliminates a management problems verified today in the European Hospital. For the environment perspective, the elimination of a residue, by transforming it in valuable products and by properly disposing the harmful components of these devices, represents a protective measure of the human health and the environment.

The study, which was realized in small scale, will continue with the purpose to determinate the feasibility of the development of this project in a larger scale.

REFERENCES

Agência Portuguesa do Ambiente, D.G.d.S.e.D.G.d.V., 2011. Plano Estratégico dos Resíduos Hospitalares.

Direção Geral de Saúde. 2013. Doenças Cérebro-Cardiovasculares em números.

European Parliament. 2003. Restriction of the use of certain hazardous substances (RoHS) in electrical and electronic equipment, 2002/95/EC.

Goosey, M. & Kellner, R. 2002. End-of Life Printed Circuit Boards, http://www.cfsd.org.uk/seeba/TD/reports/PCB.

Hagelüken, C. & Corti, C. 2010. Recycling of gold from electronics: Cost-effective use through 'Design for Recycling'. *Gold Bulletin* 43 (3): 209–220.

Harper, C.A. & Sampson, R.N. 1994. *Electronic materials and processes handbook*. McGraw-Hill.

Huisman, J., Magalini, F., Kuehr, R. & Maurer, C. 2008. 2008 Review of directive 2002/96 on waste electrical and electronic equipment (WEEE) final report. United Nations University.

Minges, M.L. & Committee, A.S.M.I.H. 1989. *Electronic Materials Handbook*: Packaging. Taylor & Francis.

Ministério da Economia, d.A., desenvolvimento Rural e Pescas, da Saúde e das Cidades, Ordenamento do Território e Ambiente, 2004. Portaria n.° 209/2004, de 3 de março. Diário da República – I Série – A. 19:1188–1206.

National Institutes of Health, What Is an Implantable Cardioverter Defibrillator.

Park, Y.J. & Fray, D.J., 2009. Recovery of high purity precious metals from printed circuit boards. *Journal of Hazardous Materials* 164 (2–3): 1152–1158.

UNEP, 2010. *Metal Stocks in Society-Scientific Synthesis*, Paris.

Wastes: Solutions, Treatments and Opportunities – Vilarinho, Castro & Russo (eds)
© 2015 Taylor & Francis Group, London, ISBN 978-1-138-02882-1

Wastes materials in geopolymers

I. Lancellotti, C. Ponzoni, L. Barbieri & C. Leonelli
Department of Engineering "Enzo Ferrari", University of Modena and Reggio Emilia, Modena, Italy

ABSTRACT: A promising process, often indicated as *geopolymerization*, describes the formation of an aluminosilicate amorphous matrix where Ca is absent or contained in low amount. Geopolymers are obtained from an aluminosilicate powder activated by alkaline solutions and the synthesis process involves three separate stages: dissolution, condensation and polymerization. Sodium silicate is used for the activation of many kinds of aluminosilicate precursors and hazardous wastes. In the present work, three case studies of solid and liquid, hazardous and non-hazardous wastes added in geopolymers will be presented. The results show the versatility of this technique and the possibility to obtain geopolymers from aluminosilicate-based waste and to immobilize their hazardous ions.

1 INTRODUCTION

A promising process, which is already being used for the immobilization of several kinds of waste is *geopolymerization*, a term used to describe the formation of an aluminosilicate amorphous matrix where Ca is absent or in low content (Duxson et al., 2005). Geopolymers are obtained from an aluminosilicate source, which is then organized in a three dimensional repeating structures of (Si-O-Al-O-)n units, similar to those of an aluminosilicate glass (Davidovits, J., 1991, Duxson et al., 2005, Schmücker & MacKenzie, 2005). However, unlike a glass, these materials are formed at room or low (40–90°C) temperature during the reaction between an alkaline solution, the chemical activator, and an aluminosilicate powder. The synthesis process involves three separate stages: dissolution, condensation and polymerization.

Different alkaline solutions based on alkali metal hydroxides, carbonates and silicates have been used in the last years. Among these solutions, sodium silicate has been extensively used in the activation of many kinds of aluminosilicate precursors such as metakaolin, fly ash, incinerator bottom ash or blast furnace slag, mainly thanks to its activation capacity which generally promotes materials with high mechanical strength (Wang et al., 1994, Fernández-Jiménez et al., 2003).

Furthermore, this technology has received an increasing attention due to its sustainability for room temperature hardening, flexibility and capability to immobilize and stabilize hazardous wastes such as incinerator fly ash, ferronickel slag, lead smelting slag, medical wastes etc. (Lancellotti, et al., 2010, Tzanakos, 2014, Komnitsas, 2013, Ogundiran, 2013) and ultimately nuclear waste (Aly, et al., 2008).

In the present work, three cases study of solid and liquid, hazardous and non-hazardous wastes in geopolymers will be presented:

– Case study 1: Non-hazardous solid waste as aluminosilicate source
– Case study 2: Hazardous solid waste inertized in geopolymers
– Case study 3: Hazardous liquid waste inertized in geopolymers

Considering the lack of landfill space and the contamination of the environment, municipal solid waste (MSW) is usually incinerated to reduce its volume, usually by 90% and recover energy. The not hazardous solid waste studied in this paper is incinerator bottom ash, while the hazardous solid wastes are fly ashes (electrofilters and fabric filters) collected from the flue gas by the air pollution control devices in incinerator plants. These ashes are classified in Italy as hazardous waste because both the volatile metals (Pb, Cd, Hg, ...) and metals which form volatile compounds (chlorides ...) released in the combustion chamber, where the solid waste is burnt, are successively condensed when the gases are cooled down to around 200°C, in this way, they are concentrated into the fly ashes.

Finally, the hazardous liquid waste investigated derives from the colouring process of ceramic tiles surfaces. This waste is in liquid homogeneous form composed prevalently of aqueous solutions of metal compounds which develop colours during the firing cycle. The colorant solution can contain Fe, Mo, Mn, Co, Cr, depending on the final desired color, together with mineralizers and complexes.

2 MATERIALS AND METHODS

Metakaolin with $SiO_2/Al_2O_3 = 1.5$ (wt%), containing 2.5 wt% K_2O, 1% Fe_2O_3 and 0.5% TiO_2, produced by calcination at 700°C for 4 h of kaolinite, was used as source of aluminosilicate for all geopolymers; in high percentages for hazardous wastes immobilization (90-60wt%), while lower percentages for aluminosilicate waste (30%).

Incinerator bottom ash (BA) shows the presence of many crystalline phases: α quartz (α-SiO_2, JCPDF file 33-1161), calcite ($CaCO_3$, ICCD #5-586), anhydrite ($CaSO_4$,ICCD #37-1496), gehlenite ($Ca_2Al(Al,Si)O_7$, ICCD #35-755) and plagioclase (($Na,Ca)(Si,Al)_4O_8$) and an amorphous fraction. The main components are: Si 33.26 wt%, Ca 21.27 wt%, Al 3.96 wt%, Na 3.21 wt%.

Incinerator fly ashes have a fine grain heterogeneous mix of heavy metals (e.g. Pb, Cd, Cr, Mn, Hg) and a variable part (10–60%) of soluble salts such as Na^+, K^+, Cl^-, HCO_3^-, SO_4^-.

The chemical characterization of liquid waste derived from the colouring process of ceramic tiles surfaces, elsewhere detailed (Ponzoni et al., 2015), shows that the water content is 67 wt% while the dry residue represents 33 wt%. In the dry residue the content of Cr is around 25 wt%. Additionally, high content of soluble salts and C determined by elemental analysis demonstrated the presence of a significant organic fraction and content of N compounds could indicate the presence of ammonium ions.

The procedure for the preparation of all sample was carried out according to the following steps:

- preparation of an alkaline solution by dissolving sodium hydroxide pellets in a sodium silicate solution (SiO_2/Na_2O molar ratio: 3.1);
- addition of hazardous waste to the alkaline solution; the addition is made separately from metakaolin in order to maximize the contact with the binding agent;
- addition of metakaolin and/or waste aluminium source and intensive/thorough stirring until a homogeneous and fluid paste is formed; the paste is poured into greased moulds;
- setting stage maintaining the cast at room temperature;
- curing stage at room temperature for different times.

For hazardous wastes the stability of the matrix and its capability to immobilize heavy metals has been followed by leaching tests accordingly to EN 12457 regulation.

For geopolymers containing non-hazardous waste mineralogical and microstructural character-izations were performed.

3 RESULTS AND DISCUSSION

3.1 Case study 1: Non-hazardous solid waste as aluminosilicate source

With the aim of study the possibility to use an inorganic industrial waste as aluminosilicate precursor for geopolymers, bottom ash from urban waste incineration has been activated via alkali solutions.

(a)   (b)

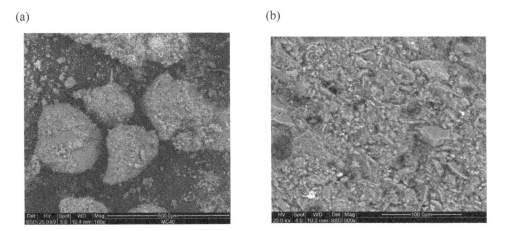

Figure 1.   SEM micrographs of incinerator ash (a) and geopolymers containing 70 wt% of incinerator ash (b).

This ash presents a chemical composition with insufficient Al content with respect to silicon, therefore the ash has been admixed with metakaolin (70wt% ash-30wt% metakaolin) to ensure Si/Al ratio in the range 2–3 for geopolymerization of structural materials (Davidovits, 1999).

Comparing the morphologies of bottom ash and corresponding geopolymer reported in Figure 1, it can be seen that, after 30 days of curing, the geopolymeric matrix appears denser for the formation of gel, but particles of 30–50 μm are still evident. Geopolymer morphology appears as partially reacted particles embedded in a geopolymeric gel rich in Si, Al and Na.

This observation demonstrates that incinerator ash is reactive in alkaline environment and can be transformed in geopolymers, but an optimization of the formulation could help to obtain higher reactivity of the aluminosilicate powder as elsewhere reported (Lancellotti et al., 2013). This is due to the complexity of the wastes, rich in both crystalline phases and amorphous fraction which are characterized by a different reactivity in the alkaline environment.

The crystalline phases present in the starting ash are still present in the consolidated product (except for calcium sulfate, due to its solubility) together with muscovite, present in metakaolin. Beside crystalline phases, a wide amorphous hump, typical of gel formed in geopolymeric materials is formed (Lancellotti et al., 2013).

3.2   *Case study 2: Hazardous solid waste inertized in geopolymers*

Incinerator fly ashes, hereafter indicated as EF and FF, are classified as hazardous waste in Italy. These ashes have been added to the metakaolin geopolymer matrix in percentage of 20 and 40 wt%, indicated GP20EF, GP20FF and GP40EF, GP40FF, respectively.

The chemical efficiency of the process adopted has been evaluated by comparing the results of release tests applied to the as-received waste and to the geopolymerized materials in order to verify the effectiveness of the matrix to fix the different heavy metals (Table 1) (Lancellotti et al., 2010). For all the fly ash-geopolymeric compositions, leachable metals in the eluates fall within limit values set by regulation for non-dangerous waste landfill disposal (DM 03/08/2005, Italy), differently with respect to the fly ashes which show high releases. In particular, the release of cadmium is negligible in the ash based geopolymer, probably due to the formation of cadmium hydroxide, insoluble at high pH, which is partially responsible for the immobilization of Cd within the geopolymer. Cr shows higher release value, related to the content in the ash, with respect to Pb and Cu. Comparing the released amount of chlorides to the total content introduced by the ash (values in brackets, Table 1), it is evident that the geopolymer containing the EF ash is more able to retain chlorides with respect to the one containing the FF ash.

Table 1. Release values for the fly ashes-geopolymers and as-received fly ashes, EF and FF, according to EN 12457 (mg/l).

Element/ Law Limits (for not dangerous dump)	Cr 1 ppm	Cd 0.02 ppm	Cu 5 ppm	Pb 1 ppm	Cl⁻ (ppm)
EF	2.41	3.55	0.01	0.12	–
FM	2.31	3.22	2.63	3.94	–
GPEF20	0.02	0.00	0.04	0.00	240 (800)
GPFF20	0.53	0.00	0.03	0.10	1750 (2160)
GPEF40	0.64	0.01	0.07	0.59	745 (1600)
GPFF40	0.01	0.00	0.01	0.87	2021 (4320)

(a)   (b)

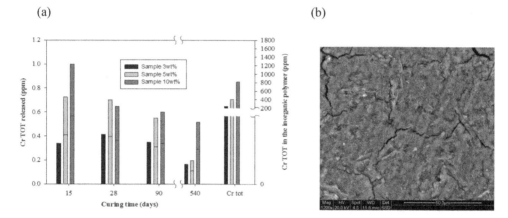

Figure 2. (a) Cr leaching as a function of curing time and compared to chromium in the waste and (b) SEM micrograph of geopolymers after 15 days of curing.

3.3 Case study 3: Hazardous liquid waste inertized in geopolymers

In this case study, the waste derived from the colouring process of ceramic tiles surfaces, was immobilized in a geopolymeric inorganic matrix. One of the innovative aspects of this research is the exploitation of water content of the waste and the absence of the drying step, the latter being a common step in the management of liquid hazardous wastes.

Four different formulations of MK-based geopolymers were prepared, containing 3, 5, 10 or 20wt% of liquid waste. All the samples were then tested by determining the chromium content in the eluate solutions after different curing times (15, 28, 90, and 540 days).

For compositions containing up to 10 wt% of liquid waste, Cr release falls within limit values set by Italian regulation (DM 30/08/2005) for non-dangerous waste landfill disposal, confirming the effectiveness of the matrix to fix this metal (Fig. 2(a)). The chemical efficiency in terms of immobilisation of heavy metals has been evaluated by comparing such values with the chromium content in the original waste, calculated taking into account the percentage of waste introduced (right part of the graph in Fig. 2(a)).

All the samples, independently from the waste amount introduced and from the curing times, show an homogeneous porosity characterized by pore size ranged from 2 to 3 nm; no macroporosity was present, as below reported in SEM images (Fig. 2(b)). This suggests that a homogeneous geopolymerization process occurs after only 15 day of curing time. As a consequence, all the geopolymers are dense and hardly pulverized or granulated. All these physical properties, confirmed by the decreasing of leaching values at long curing times (540 days), allow to assert the long term stability of these materials in the conditions required by the norm EN 12457 for the disposal in landfill.

REFERENCES

Aly, Z., Vance, E.R., Perera, D.S., Hanna, J.V., Griffith, C.S., Davis, J. & Durce, D. 2008. Aqueous leach-ability of metakaolin-based geopolymers with molar ratios of Si/Al=1.5–4, *Journal of Nuclear Materials*, 378(2), 172–179.

Davidovits, J. 1991. Geopolymers: inorganic polymeric new materials, *Journal of thermal analysis* 37, 1633–1656.

Davidovits, J. 1999. Chemistry of geopolymeric systems. In: J. Davidovits, R. Davidovits, C. James (Eds.), Terminology, Geopolymere '99, Geopolymer, International Conference, Proceedings, 30 June–2 July, 1999. Institute Geopolymere, Saint Quentin, France, 9–39.

Duxson, P., Provis, J.L., Lukey, G.C., Mallicoat, S.W., Kriven, W. M. & van Deventer, J.S.J. 2005. Understanding the relationship between geopolymer composition, microstructure and mechanical properties. *Coll. and Surf. A: Physiochem. Eng. Aspects* 269, 47–58.

Fernández-Jiménez, A. & Puertas, F., 2003. Effect of activator mix on the hydration and strength behaviour of alkali-activated slag cements. *Adv. Cem. Res.* 15(3), 129–136.

Komnitsas K., Zaharaki D. & Bartzas G. 2013. Effect of sulphate and nitrate anions on heavy metal immobilisation in ferronickel slag geopolymers, *Applied Clay Science,* 73, 103–109.

Lancellotti I., Kamseu E., Michelazzi M., Barbieri L., Corradi A. & Leonelli C. 2010. Chemical stability of geo-polymers containing municipal solid waste incinerator fly ash, *Waste Management*, 30, 673–679.

Lancellotti I., Ponzoni C., Barbieri L. & Leonelli C. 2013. Alkali activation processes for incinerator residues management. *Waste Management*, 33(8), 1740–1749.

Ogundiran M.B., Nugteren H.W. & Witkamp G.J. 2013. Immobilisation of lead smelting slag within spent aluminate—fly ash based geopolymers, *Journal of Hazardous Materials*, 248–249, 29–36.

Ponzoni C., Lancellotti I., Barbieri L., Spinella A., Saladino M. L., Chillura Martino D., Caponetti E., Armetta F. & Leonelli C. 2015. Chromium liquid waste inertization in an inorganic alkali activated matrix: Leaching and NMR multinuclear approach, *Journal of Hazardous Materials*, 286, 474–483.

Schmücker, M. & MacKenzie, K.J.D. 2005. Microstructure of sodium polysialate siloxo geopolymer. *Ceramics International* 31, 433–437.

Tzanakos K., Mimilidou A., Anastasiadou K., Stratakis A. & Gidarakos E. 2014. Solidification/stabilization of ash from medical waste incineration into geopolymers, *Waste Management,* 34(10), 1823–1828.

Wang S.D., Scrivener K.L. & Pratt P.L. 1994. Factors affecting the strength of alkali-activated slag. *Cement and Concrete Research,* 24(6), 1033–1043.

Wastes: Solutions, Treatments and Opportunities – Vilarinho, Castro & Russo (eds)
© 2015 Taylor & Francis Group, London, ISBN 978-1-138-02882-1

The collection and transport of differentiated waste and CO_2 emission

A.R. Leitão, S. Paixão, N. Sá & A. Ferreira
Department of Environmental Health from ESTeSC, Coimbra, Portugal

J.P. Figueiredo
Department of Complementary Sciences from ESTeSC, Coimbra, Portugal

ABSTRACT: This research has been an observational study, retrospective in nature, level II (descriptive-correlational). This study aims at relating the amount of waste produced by the inhabitants of a town, in the central area of Portugal, with emissions of carbon dioxide (CO_2) produced by vehicles which collect and transport. This estimate of CO_2 emissions was observed with the program COPERT 4 (Computer Programme to Calculate Emissions from Road Transport). In relation to the CO_2 emissions produced during the collection and transport of waste, the differences were statistically significant for the Cardboard-Waste-Bank (CWB – blue container where waste paper/card was placed) between 2010 and 2012 and between 2011 and 2012. We have come to the conclusion that the optimization of collection routes must be a priority commitment for the Portuguese towns, so that vehicles move with a full load, maximizing the processes and consequently minimizing of CO_2 emissions.

1 INTRODUCTION

Municipal waste (MW) is a major environmental problem for cities in the 21 st century. Waste production has continued to assert itself increasingly, with great importance on the economical and technological sectors. Thus, proper waste management contributes to the preservation of natural resources, through prevention or through recycling and recovery, to ensure the optimization of environmental objectives. In this way, the fulfilment of these goals requires an orientation of the behavior of economic operators and final consumers in order to reduce the waste production and its more efficient treatment. Therefore, it is critical that the identification and characterization of wastes and the use of appropriate logistical means for their collection and transport, according to their specific characteristics (Duarte et al. 2008).

Municipal residues may be classified according to their deposition in differentiated and non-differentiated waste. In this study we look at differentiated waste, more specifically at the packaging wastes which are properly separated by categories (paper and card; plastic and metal; glass) and deposited in the recycling bins. The different residues can be separated into three categories Cardboard – Waste-Bank (CWB – blue container where we place paper and cardboard); Glass Waste Bank (GWB – green container where we place glass) and Metal/Plastic Waste Bank (M/PWB – yellow container where we place plastic and metal).

Another activity that assumes high importance in terms of urban waste management, regarding gaseous emissions, is the collection and transportation of waste. The assessment of these emissions is essential to set priorities on the optimization of collection/transport circuits in order to reduce fuel consumption and hence GHG emissions (Larsen et al. 2009, Fontaras et al. 2012).

After their deposition in containers, the MW is collected by the responsible entities, which forwards them to the various final destinations. The collection is performed using a vehicle-based service that, as a general rule, has diesel engines and emits polluting gases. Urban transport accounts for a very significant percentage of the transport sector as a whole, influencing both air quality parameters (CO, HC, NOx, PM) in local pollution, and climate change through emissions of CO_2 (Armstrong & Khan 2004).

An investigation carried out by Oliveira (2009) found that when compared to diesel and biodiesel fuels, diesel was the one which caused higher CO_2 emissions in the collection and transport of MW. Also noted is that its replacement by biodiesel contributed significantly to the reduction of GHG emissions. It was also found that the use of natural gas vehicles decreased CO_2 emissions by about 16% per year.

The collection, transfer and transport of waste are basic activities of waste management systems worldwide. These activities' use of fossil fuels, according to the study developed by the authors, has a strong contribution to the production of greenhouse gases, contributing in this way thus for the global warming potential (GWP). In this study, (Eisted et al. 2009) concluded that the optimization of long-distance transport of wastes, for example by using the railways, may be a more suitable option in terms of environment .

Armstrong and Khan (2004) described an integrated system for the calculation of these emissions, assisted by GIS-T (geographic information Systems for transport), taking into account the activity and the fleet of automobiles, as well as the climate and characteristics of fuels in the region under study. These authors concluded that the emission of harmful gases to the environment is closely linked to the quality of road infrastructures. It is essential to note that, regardless of recognized environmental impact resulting from the use of motor vehicles for waste transport, energy and environmental benefits of recycling programs, compared with the use of materials that are not recycled are not, generally, compromised even when they involve long distance transport (Salhofer et al. 2007).

2 MATERIAL AND METHODS

For this study, we have made a retrospective observational, level II (descriptive-correlational) (Fortin 1999). This study follows a distinctive design, not an experimental or post-fact study, which aims at measuring the volume of waste produced in the town centre area, and the impact that the collecting and transporting of such waste have produced in the environment, from the point of view of CO_2 emissions generated by vehicles.

The data were analyzed for the waste produced in the municipality of the central region of the country, in a given time window, from January 2010 to December 2012. In this sense, we had access to the total quantity of waste of the municipality and not a sample.

Data were obtained from the electronic platform of the company ERSUC (Resíduos Sólidos do Centro, SA), located in the district of Coimbra. This company is responsible for the management of differentiated and non-differentiated waste in the central region of Portugal.

Regarding the use of technical resources, we resorted to the program COPERT4 (Computer Programme to Calculate Emissions from Road Transport), computer software that estimates CO_2 emissions from road transport (Agência Portuguesa do Ambiente 2013, Gkatzoflias et al. 2007). To obtain the estimated values of CO_2, this program uses as main variables the miles driven, characteristics of vehicles (registration, fuel type), type of route (rural/urban/highway), speed, weather conditions including temperature and relative humidity, *Reid Vapor Pressure* (RVP) and the population. Relation to temperature and relative humidity were used in their average monthly values for Continental Portugal. As for the vapor pressure of diesel vehicles, we used the values stipulated in Decree-Law no. 142/2010 (Ministério da Economia, da Inovação e do Desenvolvimento e do Trabalho e da Solidariedade Social 2010) for the months in study. The route carried by the vehicle was the same for all three types of waste. Since the vehicles were used for the GWB a vehicle dated in 2002 and the CWB and M/PWB a vehicle dated from 2010, thus the collection of waste from CWB and M/PWB was performed by the same vehicle in different moments. The route was determined through Google Earth software with 40% in urban and 60% in rural areas. The speed was an average of 56 km/h for rural area and 25 km/h in urban one.

For treatment of the data, we use the IBM SPSS Statistics Statistical program 22 version for Windows. Descriptive statistics were used as measures of location (the mean) and standard deviation of dispersion. There was also the resort to a statistical parametric ANOVA one factor test, using the Bonferroni post hoc test for comparisons between groups. The interpretation of statistical tests was performed based on a significance level of p-value ≤ 0.05 with 95% confidence interval.

3 RESULTS

The data considered for the selective collection of waste in the years 2010, 2011 and 2012, in the central region of Portugal in the district of Coimbra. This town had in 2010 a total of 17544 inhabitants, in 2011 there were 17584 inhabitants and in 2012 the population consisted of 17494 inhabitants. Considering the three years, the population average is 17541 inhabitants (Pordata 2009).

The concern with environmental issues is a priority for the town in this region, because it has a significant network of differentiated waste collection points. Specifically, this town has a network of 57 points of differentiated waste collection, of which 30 are in the centre and the other in adjacent locations.

The data obtained from the ERSUC platform for the months January 2010 to December 2012, allowed us a detailed analysis for differentiated waste, GWB, CWB and M/PWB, which are presented in the form of tables and figures of CO_2 emissions in the collection and transportation.

Regarding CO_2 emissions, we noted the absence of average change in the type of waste GWB and M/PWB during the three years under study. However, with regards to CO_2 emissions of waste CWB, there were mean differences statistically significant on the amount of emissions between the three years under review. These differences, according to the Bonferroni test, demonstrated a reduction of CO_2 emissions from 2010 until 2012 significantly.

Figures 1 to 3 refer to the CO_2 emissions produced by vehicles for the collection and transport of differentiated waste from different recycling bins.

Regarding CO_2 emissions in the transport of waste from GWB (Fig. 1), we found that the largest emissions were in August and October. As we can see, in September, from 2010 to 2012, there has been a significant average reduction of CO_2 emissions. The lowest emissions were recorded during November for three years, and in March, July and September for the years 2011 and 2012, with particular relevance for September 2012, in which was found the lowest value of CO_2 emissions.

Table 1. Mean difference in CO_2 emissions produced by vehicles of collection and transport of waste in the years 2010 to 2012.

CO_2 emissions (kg)		N	$\bar{x} \pm s$	F; df; sig
Glass Waste Bank	2010	12	3709.42 ± 1059.92	1.052; 2.33; 0.361
	2011	12	3117.83 ± 1054.02	
	2012	12	3201.58 ± 1127.63	
Carboard Waste Bank	2010	12	13706.33 ± 1080.40	32.167; 2.33; 0.000*
	2011	12	10773.33 ± 1415.73	
	2012	12	10014.17 ± 1040.83	
Metal/Plastic Waste Bank	2010	12	7590.42 ± 1326.13	1.465; 2.33; 0.246
	2011	12	7473.00 ± 689.56	
	2012	12	8236.75 ± 1386.79	

Test: ANOVA one factor.
*Mean difference statistically significant (post hoc Bonferroni test, between 2010 and 2012, and 2011 and 2012).

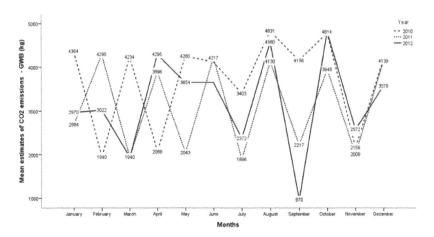

Figure 1. Graph of the mean estimates of CO_2 emissions produced during transport of waste of GWB according to the months in a calendar year.

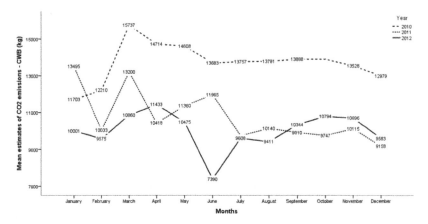

Figure 2. Graph of the mean of CO_2 emissions produced during transport of waste of CWB according to the months in a calendar year.

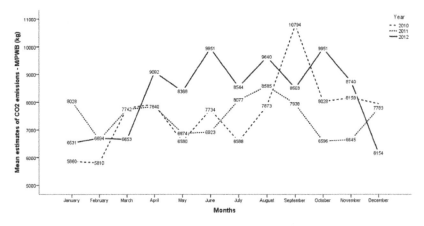

Figure 3. Graph of the mean estimates of CO_2 emissions produced during transport of waste of M/PWB according to the months in a calendar year.

In relation to CO_2 emissions in the transport and collection of waste CWB (Fig. 2) it was found that in 2010 emissions were higher compared to the other years. In the second half of 2012, a tendency to increase emissions was confirmed, noting a slight increase in 2012 from September until December, but fairly similar to the previous year.

The average estimates of CO_2 emissions related to transport of wastes from M/PWB (Fig. 3) show some monthly variability during the course of the three years, but with a tendency to higher production in the months April to November, especially in 2012. Overall, the year 2012 was characterised by an increase in emissions in comparison with the previous years.

4 DISCUSSION

We concluded that there was been a statistically significant mean difference for the production of CWB waste, and these differences between the years 2010 and 2012, and 2011 and 2012, with the lowest value in the year 2012. These results differ from what was the previous behavior of the general Portuguese population during the years 2004–2009 relatively to selective waste collection. In fact, according to data from the National Statistics Institute (INE) (Instituto Nacional de Estatística 2010), in this three-year period there was an increase in waste generation, which did not happen between 2010–2012. We believe that this decrease can be attributed to several factors, including the financial problems felt in 2011 at the level of living conditions (Dennison et al. 1996). We have found that in the case of waste from CWB, all months of 2012 were less productive than the other years under study. Due to this diminishing of business trade we have reached the conclusion that there has been an enormous decrease in the input of either gift wrapping, magazines, newspapers or other kind of disposals. Although this can be tricky, when people have a lower level their garbage waste is lower too, which is indeed beneficial to the environment.

The mean differences of CO_2 emissions in the collection and transportation of waste were statistically significant only for the waste of CWB. These differences were between 2010 and 2012, and between 2011 and 2012, and the lowest value was in year 2012.

However, we have noticed that months with higher CO_2 emissions do not correspond to an effective increase of the quantities of the waste collected, which led us to realize that in these situations there was no optimization of routes (Oliveira 2009). This is the case of the collection route CWB with higher values of CO_2 emissions in the year 2012, when the amount of this type of waste that year was lower.

Although in this study, the age of the vehicles used has not been one of the variables analyzed, we have concluded that the older vehicle (2002), produced smaller amounts of CO_2 emissions than the latest vehicle (2010). So the age of the vehicle is not always the most important factor when it comes to emissions of pollutants (Larsen et al. 2009). Indeed factors which contribute to different results range from optimizing routes, speed of the vehicles and weather conditions (Oliveira 2009).

5 CONCLUSION

Through the analysis of the results, we are aware of the environmental impact caused by the vehicles for waste transport, the energy and also the benefits from recycling programs. If we compare the use of materials that are not recycled the benefits of recycling out waste the disadvantages of the long travelled distances to the recycling centres (Salhofer et al. 2007). It is mandatory that campaigns aimed at the population to recycle must go on. Governments should invest in the education for a better environment as being essential in the building of a truly citizenship.

Much of the research in this area is manly worried in minimizing operational costs of collection processes (Armstrong & Khan 2004). However we must not discard the environmental area and the respective costs associated (Armstrong & Khan 2004, Sonesson 2000, Komilis 2008).

REFERENCES

Agência Portuguesa do Ambiente. 2013. Portuguese Informative Inventory. Amadora.

Armstrong, J. M. & Khan, A. M. 2004. Modelling urban transportation emissions: role of GIS. *Computers, Environment and Urban Systems* vol. 28: 421–433.

Dennison, G.J. et al. 1996. A socio-economic based survey of household waste characteristics in the city of Dublin, Ireland – I Waste composition. *Resources, Conservation and Recycling* vol. 17: 227–244.

Duarte, et al. 2008. Gestão e Tratamento de Resíduos. Coimbra: Almedina.

Eisted, R. et al. 2009. Collection, transfer and transport of waste: accounting of greenhouse gases and global warming contribution. *Waste Management & Research*, vol. 27: 738–745.

Fontaras, G. et al. 2012. Assessment of on-road emissions of four Euro V diesel and CNG Waste collection trucks for supporting air-quality improvement initiatives in the city of Milan. *Science of the Total Environment* vol. 426: 65–72.

Fortin, M. 1999. O processo de investigação: da concepção à realização. Loures: Lusociência.

Instituto Nacional de Estatística, 2010. Gestão de Resíduos em Portugal de 2004–2009. Destaque – informação á comunidade social agosto 2010.

Gkatzoflias, D. et al. 2007. *COPERT 4* Computer programme to calculate emissions from road transport. Thessaloniki.

Komilis, D.P. 2008. Conceptual modeling to optimize the haul and transfer of municipal solid waste. *Waste Management* vol 28: 2355–2365.

Larsen, A.W. et al. 2009. Diesel consumption in waste collection and transport and its environmental significance. *Waste Management & Research*, vol. 27: 652–659.

Oliveira N. 2009. Avaliação de sistemas de recolha e transporte de Resíduos Sólidos urbanos: Eficiência energética e emissões poluentes. Coimbra. Faculdade de Ciências e Tecnologias da Universidade de Coimbra.

Pordata.base de dados Portugal Contemporâneo (webpage) Lisboa: FFMS 2009 avaiable from: http://www.pordata.pt.

Portugal. Ministério da Economia, da Inovação e do Desenvolvimento e do Trabalho e da Solidariedade Social Dec-Lei n.º 142 de 31 de Dezembro de 2010. Diário da República. Lisboa, nº 253 – Série I: 6098– 6119.

Salhofer, S. et al. 2007. The ecological relevance of transport in waste disposal systems in Western Europe. *Waste Management* vol. 27: 47–57.

Sonesson, U. 2000. Modelling of waste collection – a general approach to calculate fuel consumption and time. *Waste management and Research* vol 18: 115–123.

Wastes: Solutions, Treatments and Opportunities – Vilarinho, Castro & Russo (eds)
© *2015 Taylor & Francis Group, London, ISBN 978-1-138-02882-1*

Sonication of olive pomace to improve xylanases production by SSF

P. Leite, J.M. Salgado, L. Abrunhosa, A. Venâncio & I. Belo
Center of Biological Engineering, University of Minho, Braga, Portugal

ABSTRACT: Olive mill wastes are an important environmental problem in the areas where they are generated. Portugal is an important producer of olive oil in the world, thus it is interesting to find the most suitable valorization strategy to exploit their wastes which are generated in huge quantities in short periods of time. In this study, it was used the olive pomace as solid substrate to produce xylanases and cellulases by filamentous fungi. To improve the enzymes production ultrasounds pretreatment of olive pomace was evaluated. The results showed that the sonication led to a 3-fold increase of xylanase activity and a decrease of cellulase activity, indicating that ultrasounds treatment attacked the integrity of cell walls and increased the accessibility of hemicelluloses inducing the xylanases production by fungi.

1 INTRODUCTION

Olive oil is an important component of the Mediterranean diet and its extraction is one of the dominant economic activities in the southern regions Europe. There are three processes of extracting oil, traditional pressed, three-phase system and two-phase system. The two-phase system, a recent process, allows the production of olive oil with economic and environmental benefits because it dramatically reduces the water consumption during the process and the generation of wastewater. This system produces a semi-solid waste, termed two-phase olive mill waste (TPOMW) or olive pomace (OP). In Portugal, about 0.5 million of tons of OP was produced in 2013 from the two-phase olive mills (INE, 2013). As a result of the processes of extraction, the oil industry generates large amounts of waste in a short period of time, and these become an increasing problem of environmental pollution.

Different studies have demonstrated that olive industry by-products are harmful to the environment and that cause negative effects on soil microbial populations (Paredes et al., 1987), on aquatic ecosystems (Dellagreca et al., 2001) and even in air through phenol and sulfur dioxide emissions (Rana et al., 2007). The toxicity and antimicrobial activity of the olive phenols are major contributors to this pollution and hinder the biological treatment of wastes, needed to reduce their pollutant load. Therefore there is an urgent need to find ways of treating these liquid and solid residues from the olive oil industry (Demerche et al., 2013). The biological treatment by fungi can be an efficient use of these wastes to obtain enzymes with industrial interest as cellulases and xylanases.

Cellulases and xylanases are enzymes used to degrade lignocellulosic materials hydrolyzing cellulose and hemicellulose. Their main application is in the saccharification of lignocellulosic materials to obtain fermentable sugars to produce bioethanol. In addition, they have a wide range of applications, including detergents and textile industry, pulp and paper industry, animal feeding, extraction of fruit and vegetable juices, and starch processing (Dogaris et al., 2009). The solid-state fermentation is a technology that allows to produce these enzymes in low cost conditions, mainly due to the use of cheaper substrate such as olive pomace. Numerous fungi have been identified as hydrolyzers of cellulose and hemicellulose, however the filamentous fungi considered as strong cellulases and xylanases secreting strains, perform better using SSF (Ang et al., 2013). On an industrial scale, cellulases and xylanases are secreted mainly by *Aspergillus* and *Trichoderma* spp. (Kulkarni et al., 1999).

The physical pre-treatments can increase size of pores and accessible surface area, and decrease degrees of polymerization of cellulose and crystallinity, which can be used to improve the biodegradability or enzymatic hydrolysis of these residues (Taherzadeh et al., 2008). Ultrasound pre-treatment causes a cavitation bubbles formation in the liquid phase (Tiehm et al., 2001), the bubbles grow and then violently collapse when they reach a critical size. Cavitational collapse produces turbulence, intense local heating and high pressure at the liquid-gas interface, high shearing phenomena in the liquid phase and formation of radicals (Atchley et al., 1988; Gonze et al., 1998). It was also proven that the degradation of excess sludge is more efficient when using low frequencies: mechanical effects facilitate particles solubilization (Tiehm et al., 2001).

The aim of the present work was to prepare olive pomace for solid-state fermentation by *A. niger* using ultrasounds pretreatment to improve the accessibility of fungi to hemicellulose and cellulose. Thus, the ultrasounds pretreatment was optimized by full factorial experimental design (3^2) to maximize the production of xylanases and cellulases varying the time of ultrasounds treatment and the liquid and solid ratio.

2 MATERIALS AND METHODS

2.1 *Microorganisms*

The fungus used in this study was selected after previously screening, the fungus with higher capacity to degrade the lignocellulosic materials was *Aspergillus niger* CECT 2915 from the Spanish Type Culture Collection. The fungi was grown on MEA slants at 25°C for several days and stored at 4°C until used.

2.2 *Characterization of solid substrate*

Two types of olive wastes (crude olive pomace and exhausted olive pomace) were provide by a Portuguese olive oil industrial plant. After olive oil extraction, crude olive pomace (COP) was recovered and stored at −20°C. The exhausted olive pomace (EOP) was obtained after recovery residual olive oil and dried, then it was stored at room temperature in dry conditions.

Both wastes were analysed in terms of physical and chemical characteristics, including cellulose, hemicellulose and lignin by quantitative acid hydrolysis (Salgado et al., 2014) . The moisture content and total solids content of wastes was obtained by drying in an oven at 105°C during 24 h to a constant weight. Solid wastes were ovendried to constant weight at 550°C to analyse the ashes content. Total nitrogen and organic carbon were determined by a Thermo Finningan Flash Elemental Analyzer 1112 series, San Jose, CA (USA) meanwhile Ca, Mg, Zn, Cu, Fe, Mn, Cr, Ni, Pb, Na, K were analyzed in ashes using Flame Atomic Absorption and Atomic Emission Spectrometry (FFAS/FAES) FAAS/FAES (Rodríguez-Solana et al., 2013). Reduction sugars, total phenols and protein was analysed in the liquid obtained after extraction process with water (ratio solid:liquid, 1:5 w/v). Reduction sugars was analysed by dinitrosalicylic acid method, protein was determined by Bradford method, total phenols were determined by the Folin–Ciocalteau method using caffeic acid as a standard. Lipids were extracted with diethyl eter and extracts were extracted with ethanol in a Soxtec System HT2 1045 extraction unit.

2.3 *Solid state fermentation*

SSF were carried out in Erlenmeyer flasks with 10 g of dried solid substrate, moisture was adjusted with water and nutrient solution (5 g/L Peptone, 5 g/L yeast extract and 0.2 g/L KH_2PO_4). The solid substrates were COP and EOP for previous fermentation, then EOP was used in fermentations for the optimization of ultrasound treatments. In all experiments, the initial moisture, temperature and inoculated spores were 75% (wet basis), 30°C and 10^7 spores/mL, respectively. Erlenmeyers flasks with substrates were sterilized at 121°C and 15 min, cooled and inoculated with 1 mL of spore solution. The experiments were incubated 6 days, and then enzymes were extracted. All experiments were carried out in duplicate and a control without inoculum was also performed. Enzyme were extracted at fermentation end (6 days) as described by Salgado et al. (2014).

Table 1. Levels of independent variables and dimensionless coded variables definition (x_i) i.

Independent variables	Units	Levels			x_i
		-1	0	1	
Time (X_1)	min	5	10	15	($X_1 - 10/5$)
Ratio L/S (X_2)	g/g	3	7	11	($X_2 - 7/4$)
Dependent variables	Units				
Xylanase activity	U/g solid substrate				
Cellulase activity	U/g solid substrate				

2.4 Ultrasounds treatment

The sonication of EOP was carried out with an high intensity ultrasonic processor Cole-Parmer 750 model (Illinois, USA) operating at 750 W and 20 kHz. The solid waste was added to vessel and mixed with water. Different solid and liquid ratios and time of treatment were studied in an experimental design. The vessel was placed in a protective box and the tip (diameter 1/2″) allocated into the vessel. After treatment, the solid was recovered by vacuum filtration and used as solid substrate in SSF.

2.5 Experimental design

For evaluation of sonication effect of EOP in xylanases and cellulases production by SSF a full factorial design 3^2 was carried out. The two studied variables were the time of treatment and the ratio liquid:solid (v/w), the dependent variables studied were xylanase and cellulase activity. The independent variables considered and their variations ranges are shown in Table 1. The correspondence between coded and uncoded variables was established by linear equations deduced from their respective variation limits. This design allowed the estimation of the significance of the parameters and their interaction using Student's t-test. A second order polynomial model of the form shown in Eq. 1 was used to fit the data:

$$y = b_0 + b_1 \cdot x_1 + b_{11} \cdot x_1^2 + b_2 \cdot x_2 + b_{22} \cdot x_2^2 + b_{12} \cdot x_1 \cdot x_2 + b_{112} \cdot x_1^2 \cdot x_2 + b_{122} \cdot x_1 \cdot x_2^2 + b_{1122} \cdot x_1^2 \cdot x_2^2 \tag{1}$$

where y represents the dependent variable, b denotes the regression coefficients (calculated from experimental data by multiple regression using the least-squares method), and x denotes the independent variables. All experiments were carried out in triplicate and in randomized run order.

The experimental data were evaluated by response surface methodology using Statistica 5.0 software. Dependent variables were optimized using an application of commercial software (Solver, Microsoft Excel 2007, Redmon, WA, USA).

2.6 Enzymatic assays

Cellulase (endo-1,4-β-glucanase) activity was analysed using an enzymatic kit Azo-CM-Cellulase S-ACMC 04/07 (Megazyme International, Ireland). One unit of enzyme activity was defined as the amount of enzyme required to release 1 μmol of glucose reducing sugars equivalents from CM-cellulose in 1 min and pH 4.5.

Xylanase (endo-1,4-β-xylanase) activity was analysed using an enzymatic kit Azo wheat arabinoxylan AWX 10/2002 (Megazyme International, Ireland). One unit of enzyme activity was defined as the amount of enzyme required to release 1 μmol of xylose reducing sugar equivalents from wheat arabinoxylan in 1 min and pH 4.5.

Table 2. Characterization of olive mill wastes.

Parameter	Crude olive pomace	Exhausted olive pomace
Humidity (% fresh weight)	73.5 ± 0.4	9.86 ± 0.12
Total solids (%)	26.5 ± 0.4	90.12 ± 0.12
Extracts (%)	2 ± 0.7	4.19 ± 0.86
Lignin (%)	34.6 ± 1.1	37.16 ± 0.32
Hemicellulose (%)	39.1 ± 4.5	36.98 ± 4.07
Cellulose (%)	33.5 ± 1.1	30.03 ± 3.28
Lipids (%)	16.6 ± 0.1	3.93 ± 1.94
Reducing sugars (mg/g)	97.5 ± 7.1	41.49 ± 2.12
Proteins (mg/g)	3.6 ± 1.4	2.55 ± 0.26
Phenols (mg/g)	8.3 ± 0.3	8.77 ± 0.23
Ash (%)	6.6 ± 0.5	3.36 ± 0.18
N (%)	0.60 ± 0.10	1.27 ± 0.07
C (%)	49.72 ± 0.68	46.07 ± 1.29
Ca (g/kg)	1.16 ± 0.04	1.75 ± 0.18
K (g/kg)	16.86 ± 1.00	14.17 ± 0.72
Mg (mg/kg)	473.50 ± 21.92	473.00 ± 56.57
Zn (mg/kg)	12.00 ± 0.00	10.50 ± 0.71
Cu (mg/kg)	11.50 ± 0.71	11.00 ± 1.41
Fe (mg/kg)	41.50 ± 2.12	146.50 ± 33.23
Mn (mg/kg)	8.60 ± 0.14	10.20 ± 0.42
Cr (mg/kg)	<22	<22
Ni (mg/kg)	<22	<22
Pb (mg/kg)	<22	<22
Na (mg/kg)	373.00 ± 35.36	91.50 ± 4.95

3 RESULTS

3.1 Characterization of EOP and COP

The solids wastes from olive mill wastes were characterized to evaluate their potential as solid substrate in SSF. The crude olive pomace (COP) was directly collected after olive oil extraction obtaining a wet solid waste, however the exhausted olive pomace (EOP) was recovered after extraction of residual olive oil and dried, these processes were carried out to use the solid waste in combustion processes. As can be observed in Table 2 the moisture content and total solids were very different due to the dried processes. The free reduction sugars, protein were higher in COP, these compounds could have been extracted in the recovered of residual olive oil. This effect is clear in reduction of lipids content which were reduced from 16.65% to 3.93%.

The content in cellulose, hemicellulose and lignin were similar in both residues and other residues studied in literature (Roig et al., 2006), the higher content of hemicelluloses and lignin indicated that these wastes have potential to be used as solid substrate in solid-state fermentation.

The analysis of C and N showed an increase of N content and decrease of C after extraction of residual oil, in the mineral content the values was not different, except to the Fe content that was higher in EOP and the content of Na that was higher in COP.

3.2 Solid-state fermentation of EOP and COP

The solid wastes EOP and COP were evaluated as solid substrate for cellulases and xylanases production in SSF by A. niger. Table 3 shows the cellulase and xylanase activities achieved. A. niger showed higher cellulase and xylanase activity using exhausted olive pomade. This solid was selected to evaluate the effect of ultrasounds treatment on cellulase and xylanase production by A. niger in SSF.

Table 3. Enzymatic activity of extracts from solid-state fermentation of olive mill wastes.

	Cellulases		Xylanases	
	COP	EOP	COP	EOP
A. uvarum	0.47 ± 0.03	16.70 ± 0.97	0.13 ± 0.06	7.80 ± 0.37
A. ibericus	2.12 ± 0.66	18.87 ± 4.74	0.42 ± 0.19	7.22 ± 1.35
A. niger	1.22 ± 0.26	47.55 ± 1.79	0.81 ± 0.53	25.89 ± 2.72

Table 4. Design matrix and response values.

Run	Time (min)	Ratio L/S (mL/g)	Cellulases (U/g)	Xylanases (U/g)
1	5	3	28.94	27.351
2	5	7	18.31	28.839
3	5	11	37.28	7.533
4	10	3	16.59	11.463
5	10	7	7.49	69.107
6	10	11	8.40	26.881
7	15	3	20.96	55.518
8	15	7	18.13	69.863
9	15	11	17.30	48.639
10	10	7	8.58	70.862
11	10	7	7.91	70.166

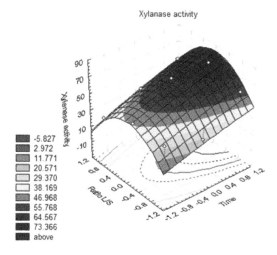

Figure 1. Estimated response surface from the full factorial design 3^2.

3.3 *Effect of ultrasounds pretreatment on enzymes production by SSF*

EOP was selected to study the effect of its sonication. Thus, two variables of ultrasound treatment were evaluated in a full factorial design $3^2 \cdot r$. Table 4 depicts the corresponding experimental matrix and the results obtained. The two responses studied were cellulase and xylanase activity. As can be observed, the ultrasounds treatment had a positive effect in xylanase production, however the cellulase activity decreased after ultrasound treatment. The positive effect of ultrasounds pretreatment on xylanase production by SSF of rice hull was also observed by Yang et al. (2011).

The optimal conditions that led to maximum xylanase activity (75.32 U/g) were calculated with Solver tool showing 12.41 min and 7.27 of liquid and solid ratio as optimal parameters of ultrasound treatment. Figure 1 displays the estimated response surface obtained from the data analysis of the matrix.

An optimal clear zone can be seen at intermediate liquid and solid ratios and maximum time studied. The validation of the model was performed in the optimal conditions, the xylanase activity achieved was close to the maximum value predicted by the model.

4 CONCLUSIONS

Olive mill wastes showed to be a suitable solid substrate for cellulase and xylanase production. The maximum cellulase and xylanase activities were achieved using exhausted olive pomace as solid substrate.

The ultrasounds treatment of exhausted olive pomace improved xylanase production by SSF using A. niger. The optimal conditions of treatment were 12.41 min and a liquid and solid ratio of 7.27. Ultrasounds treatment showed a negative effect on cellulase production. Thus the sonication can be an effective treatment to induce the production of xylanases by SSF. In future works, it will be studied other treatments in combination with sonication to improve cellulase production.

REFERENCES

Ang, S.K., Shaza, E.M., Adibah, Y., Suraini, A.A. & Madiha, M.S. 2013. Production of cellulases and xylanase by *Aspergillus fumigatus* SK1 using untreated oil palm trunk through solid state fermentation. Process Biochemistry, 48(9):1293–1302.

Atchley, A.A. & Crum, L.A. 1988. Acoustic cavitation and bubble dynamics. In K.S. Suslick (eds), *Ultrasound – its chemical, physical, and biological effects*: 1–64, VCH Publishers, New York.

DellaGreca, M., Monaco, P., Pinto, G., Pollio, A., Previtera, L. & Temussi, F. 2001. Phytotoxicity of Low-Molecular-Weight Phenols from Olive Mill Waste Waters. *Bulletin of environmental contamination and toxicology*, 67:352–359.

Dermeche, S., Nadour, M., Larroche, C., Moulti-Mati, F. & Michaud, P. 2013. Olive mill wastes: Biochemical characterizations and valorization strategies. *Process Biochem*. 48, 1532–1552.

Dogaris, I., Vakontios, G., Kalogeris, E., Mamma, D. & Kekos, D. 2009. Induction of cellulases and hemi-cellulases from *Neurospora crassa* under solid-state cultivation for bioconversion of sorghum bagasse into ethanol, 9:404–411.

Gonze, E., Gonthier, Y., Boldo, P. & Bernis, A. 1998. Standing waves in a high frequency sonoreactor: Visualization and effects. *Chemical Engineering Science* 53:523–532.

Kulkarni, N., Shendye, A. & Rao, M. 1999. Molecular and biotechnological aspects of xylanases. *FEMS Microbiology Reviews*, 23:411–456.

Paredes, M.J., Moreno, E., Ramos-Cormenzana, A. & Martinez, J. 1987. Characteristics of soil after pollution with waste waters from olive oil extraction plants. *Chemosphere* 16:1557–1564.

Rana, G., Rinaldi, M. & Introna, M. 2003. Volatilisation of substances after spreading olive oil waste water on the soil in a Mediterranean environment. *Agriculture, Ecosystems and Environment*, 96:49–58.

Roig, A., Cayuela, M. L. & Sánchez-Monedero, M. A. 2006. An overview on olive mill wastes and their valorisation methods. *Waste Management*, 26:960–969.

Salgado, J.M., Abrunhosa, L., Venâncio, A., Domínguez, J.M. & Belo, I. 2013. Screening of winery and olive mill wastes for lignocellulolytic enzyme production from Aspergillus species by solid-state fermentation. *Biomass Conversion and Biorefinery*, 4:201–209.

Taherzadeh, M.J. & Karimi, K. 2008. Pretreatment of lignocellulosic wastes to improve ethanol and biogas production: a review. *International Journal of Molecular Science*, 9:1621–1651.

Tiehm, A., Nickel, K., Zellhorn, M. & Neis, U. 2001. Ultrasonic waste activated sludge disintegration for improving anaerobic stabilization. *Water Research*, 35:2003–2009.

Yang, C., Sheih, I. & Fang, T. J. 2012. Fermentation of rice hull by aspergillus japonicus under ultrasonic pretreatment. *Ultrasonics Sonochemistry*, 19:687–691.

Wastes: Solutions, Treatments and Opportunities – Vilarinho, Castro & Russo (eds)
© 2015 Taylor & Francis Group, London, ISBN 978-1-138-02882-1

Estimation of residue biomass in the Oka river basin (Spain)

E. Mateos & J.M. Edeso
University of the Basque Country, (UPV/EHU), Bilbao, Spain

ABSTRACT: The aim of this paper is to present a methodology for the evaluation of residual forest biomass for the production of energy and the resources cartography in the Oka river basin. The Urdaibai estuary (Bizkaia) lies at the mouth of the River Oka and covers a surface area of 22,040 ha of great ecological value. It was declared a "Biosphere reserve" by UNESCO in 1984. The aim is to determine the potential residual biomass available, usable as an energy resource from the residue of the main forest species in the area under study, generated by forestry treatments. El total forest residues in Oka river basin are estimated as 13,093 Mg year^{-1}. The availability of such biomass potential for energy production is strongly conditioned to the inherent difficulties during the extraction process. The available forest resources estimated after the application of restrictions reach 7776 Mg year^{-1}.

1 INTRODUCTION

The overall energy potential of the biomass that the biosphere would be capable of generating on an annual basis is estimated at 68,080 Gtoe. This represents an enormous biomass production potential worldwide. In Spain, the target set out in the new 2011–2020 Renewable Energies Plan (PANER) is to use renewable energies to provide 20% of the final energy consumption in Spain. The use of forest biomass is a major feature of such energy sources. In the Autonomous Community of the Basque Country (ACBC), biomass is currently the most widely used source of renewable energy. For example, in 2009, the contribution of renewable energies to final energy consumption in the Basque Country was 6.8%. Of this amount, 85% was biomass (Hormaeche, 2011).

The use of forest biomass as an energy resource, instead of fossil fuels, involves a number of major environmental advantages: minimum emissions of pollutants and particles and the reduction of forest fire risk as well as plagues of insects. Moreover, the energy exploitation of forest residues does not contribute to an increase in the greenhouse effect, since the balance of CO_2 emissions to the atmosphere is neutral. This fact is an aid in reaching international agreements on subjects such as polluting emissions, the fight against climate change and promotion of sustainable development. The use of biomass for energy purposes is closely linked to carbon sinks. The carbon stock in the aboveground forest biomass in the Basque Community is 14.7 Tg of C and, including roots, this figure rises to 18.43 Tg of C (Balboa et al., 2006).

One of the main hindrances to biomass energy management is the difficulty of ensuring a steady supply for heat or electricity generating plants. Nevertheless, the job of identifying and assessing the potential of different forest species for energy purposes is a priority task undertaken in most developed regions worldwide in order to replace fossil fuels and hence contribute to biosphere sustainability. As a result, the amount of biomass is an essential consideration since its supply to the energy plant must be guaranteed (Tolosana, 2009).

Oka river basin.

Figure 1.   Oka river basin's landscape.

2   METHODOLOGY

2.1   *Study area*

The study area is located within a farming system in Oka river basin (N. Spain), between the coordinates 43°12′ and 43°28′ latitude north and 2°33′W and 2°46′W longitude. This area is situated on the banks of the Cantabria Sea, in the north of the Iberian Peninsula (Figure 1). In order to conserve its high ecological value, it was declared Biosphere Reserve by UNESCO. The Biosphere Reserve presents the same environmental problems as in other areas of the Basque-Cantabria mountains: erosional processes, landscape homogeneity, and so on, and about the 90% of the land here is private (Álvarez et al., 2008).

The biogeographical situation of the area favours intensive forestry activities. According to data collected during the Third National Forestry Inventory, IFN3 (IFN3, 2006), most of the Oka river basin (66%) is occupied by forest (15,000 ha). It involves, therefore, a highly suitable area for forestry activities. The Pine Radiata and Eucalyptus plantations occupy 82.37% of the wooded surface area, and, except for the Cantabria holm oak woods, few vestiges of native vegetation remain. The regulations to control forestry activities in the area are set out in the Use and Management Master Plan (UMMP), passed in 2003. The Master Plan defines forestry areas with higher risks of erosion, such as those with slopes of more than 50%, as well as those with slopes of more than 30% and less than 50% with soils with erosion-sensitive lithology. In our case, this involves restricting those exploitation techniques that accentuate soil erosion.

2.2   *Preliminary studies*

The study of forest biomass is based on the application of updated national databases such as the Third National Forest Inventory (IFN3, 2006) and the Spanish Forest Map (SFM). An appropriate method for estimating biomass stocks is the use of the "stratum" concept, which is defined in the 3NFI (López-Rodríguez et al., 2009). Every stratum is formed by grouping the forest surfaces of similar features. Once the forest masses have been classified per strata, the most adequate forest treatments that can be carried out in each stratum are identified. The forest biomass residue quantities that could be obtained in each stratum are estimated in accordance with those treatments.

In view of the special characteristics of the study area, which is of great ecological value, it is appropriate in this paper to assess the erosion status of the soil by means of the USLE/RUSLE model (Edeso et al., 1998). According to a number of different researchers, this model is particularly appropriate for calculating the average soil removed by laminar erosion or in the irrigation channels of a river basin (Merino et al., 2004). The procedure utilized for approaching such estimation is shown schematically in Figure 2.

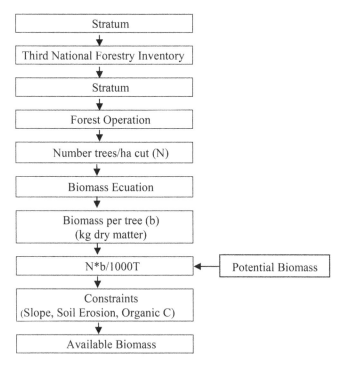

Figure 2. Schematic diagram of the method followed to determine the available Biomass.

Table 1. Description of the applied allometric equations for calculating above-ground tree biomass (kg).

Specie	Equation	CF	R^2	SEE
P. radiata	$0.0749D_n^{2.48774}$	1.018852	0.977	0.193270
E. globulus	$0.2678D_n^{2.19404}$	1.012536	0.980	0.157850

Source: Montero et al., 2005.

2.3 *Estimate of the production of potential forest residues*

The methodology used to determine the annual quantity of forest biomass in the area of study consists primarily of determining two factors: a) The forest residue per unit of surface and time derived from a forest mass ($Mg\,ha^{-1}\,year^{-1}$) according to the species and forest treatment to which each mass has been subjected to, and b) the surface (ha) occupied by the forest mass that is going to generate such residue (Gil et al., 2011). For estimating the amount of annual biomass that might be generated by current forest masses in Oka river basin, the methodology applied uses a Geographic Information System (GIS) with the help of the Arcview GIS 10 program. In this way, it has been possible to process an important amount of data and to manage the results obtained (Mateos et al., 2012).

After reviewing different estimating methods for forest biomass residue (Ter-Mikaeliann et al., 1997; Zianis et al., 2004), an indirect methodology based on the use of linealized alometric equations as logarithmic model has been selected (Table 1). The aim of these equations is to link the total dry biomass of the tree, or some of its components, with the average diameter:

$$W = CF\ e^a D_n^{\ b} \qquad (1)$$

135

Table 2. Percentage of available biomass in different conditions of slope, erosion risk and organic carbon in top soil.

Slope		≤35%	35–60%	≥60%
K	0–0.15	100	80	60
	0.15–0.30	90	70	50
	030–0.45	0	0	0
OC	>2	100	80	60
	1–2	80	50	0
	0–1	0	0	0

where W is the aboveground biomass (kg) (oven-dry weight), D_n is the diameter at breast height (DBH) or trunk diameter at 1.30 m, a and b are two specific regression parameters and CF is a correction factor which is calculated from the standard error of estimate (SEE) of the regression:

$$CF = \exp(SEE^2/2) \tag{2}$$

2.4 Available residual biomass

The available biomass is the biomass resulting from the application of certain restrictions to the potential biomass as a consequence of limitations that make its extraction impossible. For both technical and economical reasons, the collection of residual biomass should not be carried out in areas of steep slopes, poor soil organic carbon and erosion risk. The information utilised for the abovementioned restrictions was gathered from the following sources:

– Organic carbon (OC) content in soils was obtained by analysing the soil sample. For this purpose, field work was done in order to collect samples and subsequently analyse these in the laboratory using the dichromate method.
– Erosion risk was obtained from the soil erodibility (Factor K) of the USLE/RUSLE model.
– Slope was derived from the digital elevations model. The digital elevation model was made using a pixel size of 2 metres.

The slope effect is implemented with a collection or availability factor of this resource, as steep slopes make transporting the residue to the transformation plant very difficult, increasing the cost of the process. To do this, three ranges of slopes have been delimited: under 35%; 35–60% and slopes of over 60% (Table 2).

The annual available quantities of dry biomass (expressed in t) Q will be obtained as:

$$Q = A_n W_n \tag{3}$$

where A_n represents the suitable area for collection in tile n (ha) and W_n the annual residue estimator (Mg ha^{-1} year^{-1}).

3 RESULTS

Table 3 lists the annual quantities of forestry biomass for the Oka river basin in Urdaibai. The quantity of forest biomass residue estimated in Oka river basin was 7775 Mg y^{-1} from which 6334 Mg y^{-1} correspond to *Pinus radiata* residue and 1441 Mg y^{-1} to *Eucalyptus globulos*. Table 3 also presents the availability of the biomass exploitable for energy purposes under the constraints of the difficulties in the extraction and the environmental restrictions (the slope, the organic carbon content in top soil and the erosion risk.

Table 3. Potential of residues production (ton/year) and annual residue estimators (t haL1).

Residues source	Covered area [ha]	Potential biomass [Mg y^{-1}]	Available biomass [Mg y^{-1}]	Estimator of residue [Mg ha^{-1} y^{-1}]
P. radiata	8764	10,473	6334	0.72
E. globulus	1321	2620	1442	1.09
All	10,085	13,093	7776	

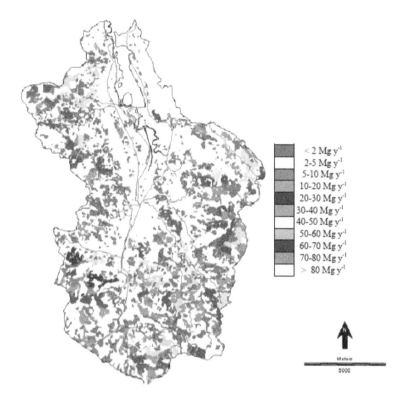

Figure 3. Final maps of annual quantities of forest residual biomass in the Oka river basin (Urdaibai).

The spatial distribution of forestry residual biomass (Mg year^{-1}) can be seen in Figure 3.

The use of residual forest biomass for energy purposes would permit the exploitation of residues, increasing regional socio-economic development through additional local employment and infrastructures. The major challenges faced by these energy conversion technologies include how to collect, prepare (chipping, briquetting, etc.) and transport biomass to electric power plants with the lowest possible financial and environmental costs, and how to assure supplies over time. Steep slopes play an extremely important role in biomass extraction, and reduce the effective area. This factor must be taken into consideration when estimating the amount of biomass generated.

GIS is a powerful tool of great use for evaluating forest biomass resources since it efficiently combines both cartographic data and information from different databases that facilitate the work of mapping the results. Through the use of forest biomass, an important reduction is expected in forest fires, which at this time are a cause for concern in the study area. Using residual forest biomass as an energy source provides employment and creates new business opportunities in rural communities. Biomass fuels reduce the dependence on fossil fuels and increase levels of energy self-sufficiency.

4 CONCLUSIONS

The forest residues considered in the present study could come to replace, at least in part, the fuels that are currently being used.

We believe that in the Basque Country, the most promising applications are cogeneration in medium-sized power stations and combustion in small and medium-sized plants for heat and steam generation.

The specific features of the economy of scale restrict the design of energy plants to those with a minimum production capacity of at least 1.5 MW. This corresponds to a need for 7000 t y^{-1} of raw material (dry mass).

An increase in the use of residual forest biomass as a source of energy instead of conventional fossil fuels will represent important environmental advantages and will help to follow the aims of the EU policies in terms of energy rationalization.

ACKNOWLEDGEMENT

The authors express their gratitude to the Gobierno Vasco and Vicerrectorado de Investigación de la Universidad del País Vasco-Euskal Herriko Unibertsitatea for the financial support through projects SAI10/147-SPE10UN90, and NUPV10/10, respectively.

REFERENCES

Álvarez, K. & Rosas, M. 2008. Seguimiento de la actividad forestal en la Reserva de la Biosfera de Urdaibai. XV Jornadas forestales de Gran Canaria.

Balboa, M., Álvarez-González J. G., Rodríguez-Soalleiro R. & Merino A. 2006. Temporal variations and distribution of carbon stocks in aboveground biomass of radiata pine and maritime pine pure stands under different silvicultural regimes. For. Ecol. *Manage* 237: 29–28.

Edeso, J. M., Merino, A., González, M. J. & Marauri, P. 1998. Manejo de explotaciones forestales y pérdida de suelo en zonas de elevada pendiente del País Vasco. *Cuaternario y Geomorfología* 12: 105–116.

Gil, M.V., Blanco, D., Carballo, M.T. & Calvo, L.F. 2011. Carbon stock estimates for forests in the Castilla y León región, Spain. A GIS based method for evaluating spatial distribution of residual biomass for bio-energy. *Biomass and Bioenergy* 35: 243–252.

Hormaeche, J.I. 2011. La energía de la biomasa convertida en calefacción. Jornadas sobre la energía renovable de la biomasa. EVE. Gobierno Vasco.

López-Rodríguez, F., Pérez Atanet, C. F., Cuadros-Blázquez, F. & Ruiz-Celma, A. 2009. Spatial assessment of the bioenergy potential of forest residues in the western province of Spain, Caceres. *Biomass and Bioenergy* 33(10): 1358–1366.

Mateos, E., Fito, J. M. E., Izaguirre, A. B. & Tojal, L. T. 2012. Estimación de la biomasa residual procedente de la gestión forestal en Bizkaia. *Lurralde: Investigación y espacio*, (35): 13–30.

Merino, A., Fernández-López, A., Solla-Gullón, F. & Edeso J.M. 2004. Soil changes and tree growth in intensively managed Pinus radiata in northern Spain. *Forest Ecology and Management* 196: 393–404.

Montero, G., Ruiz-Peinado, R. & Muñoz, M. 2005. Producción de biomasa y fijación de CO2 en los montes españoles. Monografías INIA.

Tercer Inventario Forestal Nacional en versión digital para la provincia de Vizcaya (IFN3) Ministerio de Medio Ambiente, (2006). http://www.mma.es/portal/secciones/ Accessed 15 January 2014.

Ter-Mikaeliann, M.T. & Korzukhin, M.D. 1997. Biomass equations for sixty-five North American tree species. *Forest Ecology and Management* 97(1): 1–24.

Tolosana, E. 2009. *Manual técnico para el aprovechamiento y elaboración de biomasa forestal*. Fucovasa y Mundi-Prensa Madrid.

Zianis, D., Mencuccini, M. 2004. On simplifying algometric analyses of forest biomass. *Forest Ecology and Management* 187: 311–332.

Wastes: Solutions, Treatments and Opportunities – Vilarinho, Castro & Russo (eds)
© *2015 Taylor & Francis Group, London, ISBN 978-1-138-02882-1*

Valorization of cooked quail egg residues

C.E. Mathieu, A. Lung, L. Candy & C. Raynaud
Université de Toulouse, INP-ENSIACET, Laboratoire de Chimie Agro-industrielle (LCA),
Toulouse, France
INRA, UMR 1010 CAI, Toulouse, France

ABSTRACT: This work describes a study of valorization of quail eggs waste and the potential uses in fields such as food, cosmetic industry, and agronomy. Fractionation techniques have been applied to generate fractions rich in proteins, lipids and minerals. Global characterization of cooked quail eggs has been determined. Upon fractionation three extracts revealed firstly a potential use in food/cosmetic industry for its content in lecithin, secondly a product rich in proteins for pet-food industry, and finally a solid organo-mineral residue for pet-food or fertilizers.

1 INTRODUCTION

Less developed than chicken industry, quail industry still represents a valuable pool of by-products. The production of quails-meat and eggs for food industry generates residues of various nature. With a concern of sustainable economy, the waste of quail industry, especially the eggs, comprises a range of relevant compounds. Eggs shell for their mineral content is the major subject treated in the field (Oliveira et al., 2013). But the organic part has not been extensively regarded. Usually the wastes of eggs, mainly hen eggs, arise from quality processes and represent the largest production. The investigation of their possible re-use of discarded products relies mainly on the mineral part of the shell. These minerals presents adsorption properties for metals and therefore offered perspectives for water treatment solutions (Carvalho et al., 2013). Valorization of the egg shell is already industrially possible. On another hand, researchers found interesting features in albumen proteins which have been characterized and studied for a possible use in food industry or pharmaceuticals (Awadé et al., 1995).

The case of quail eggs has not extensively regarded but still represents a valuable pool of compound. Compared to hen eggs, chemical composition of mineral and organic parts present macroscopic differences, and their nutrient properties vary. Applications of quails eggs has found features in nutraceutical and pharmaceutical industries in relation to their potentialities in health promoters and anti-allergic nutrients (Sahin et al., 2008, Tashiro, 2003). Further the lipids of quail eggs can also be a source of functional ingredient for food, cosmetic and nutraceutical industries. We propose here to investigate the re-use, the recycle and the recovery of a real industry waste to access to new extracts bearing high value products.

2 MATERIAL AND METHODS

2.1 *Material*

The waste is composed of eggs from *Coturnix coturnix japonica* provided by Caillor company based in Sabazam (Landes, France). These are discarded during the cooking step occurring at 80°C in a water bath. After cooking, they were stored at −24°C until use. The shell was in some case separated manually, and the matters were dried and grinded. This sample was treated seperatly with and without shell taking to account the removal of the shell is at this stage not operational in food industry. The material is kept under food conditions. Mineral content was determined by weighting after 5 hours at 550°C.

2.2 Proteins content

The protein content was determined by the Kjeldahl method. These analyses were carried out in triplicate. The conversion factor of N total in proteins is 6.25.

2.3 Amino-acids characterization

For the hydrolysis procedure, dried and defatted samples were weighed in the screw-capped tubes (50–100 mg) and 5 mL of HCl (6.0 N) was added to each tube. The tubes were attached to a system; which allows the connection of nitrogen and vacuum lines without disturbing the sample. The tubes were placed in an oven at 110°C for 24 hours [20]. The pH was corrected to 6 and the volume of the solution completed to 10 mL with 0.2 M sodium citrate buffer (Loading buffer at pH 2.2) and the solution filtered on a 0.2 μm membrane filter. Hydrolyzed sample solution was introduced to the column (20 μL).

Amino acids were separated using buffers with different pH and molarities. Three citrate buffers were used to elute 16 amino acids, buffer 1(0.2 M, pH 3.20) and buffer 2 (0.2 M, pH 4.25) elute the acidic and neutral amino acids while buffer 3 (0.2 M, pH 6.45) elute the basic amino acids. In addition, a loading dilution citrate buffer (0.2 M, pH 2.2) and a column-regeneration solution (0.4 M NaOH) was used. All buffers and NaOH solution were purchased in Serlabo compagny. Ninhydrin detection reagent was used which consisted of Ultrasolve (2.0 L) and ninhydrin (20 g). Amino acid analyzer Biochrom 30 equipped with stainless steel column (200 × 4.6 mm) packed with Ultropac 8 (8 μm ± 0.5 μm) cation exchange resin. The following program was used for the separation and detection of the amino acids: Buffer 1 was pumped for 9 minutes followed by buffer 2 for 12 minutes and buffer 3 for 17 minutes. The column was regenerated using 0.4 M NaOH for 4 minutes followed by equilibration in buffer, for 16 min. The column was initially heated at 53°C for 9 minutes. The temperature was changed to 58°C for 13 min then changed to 95°C for 24 minutes, finally cooled down to 53°C for the remainder of the cycle (12 minutes) The cycle time from injection to injection was 58 minutes. The flow rate was 25 mL/hr for ninhydrin reagent and 35 mL/hr for the buffers. The reaction between the amino acids and ninhydrin occurred at 135°C in a 10 mL PTFE reaction coil (0.3 mm I.D) immersed in silicon oil. Detection was performed at two wavelengths (570 and 440 nm). The data of each chromatogram was analyzed by EZ Chromatography Version 6.7.

2.4 Lipid content

Lipid content was determined by sohxlet extraction with cyclohexane for 5 hours. For FAs analyses, 20 mg of oil was stabilized with 1 mL TBME. Then, FAs were derivatized into their corresponding methyl esters (FAMEs) by adding of 50 μL TMSH (Triméthyl Sulphonium Hydroxide) with 0.5 M in methanol. The analyses were performed by gas chromatography (Varian, France). The injection was split 1:100 at 250°C. The capillary column used was a Select CB for FAME fused silica WCOT (50 m × 0.25 mm; film thickness 0.25 mm). The temperature gradient was 185°C for 40 min, then at 15°C/min to 250°C, and 250°C for 10 min. The analysis time was 55 min. The detector FID was set up at 250°C. The helium was the carrier gas at a constant flow of 1.2 mL/min.

2.5 Phospholipids characterization

Phosophoslipids was quantified as the insoluble acetone (Collins and Shotlan, 1960). Phosphatidyl choline (PC) and phosphatidyl ethanolamine (PE) was quantified by HPTLC. Commercial standards were purchased at Aldrich (Germany).

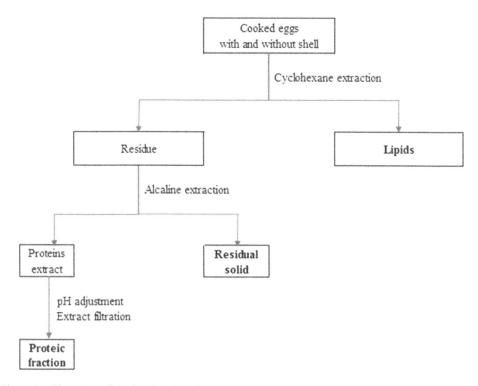

Figure 1.   Flow chart of the fractionation of cooked quail eggs.

Oil was dissolved in chloroforme at a concentration of 20 mg/mL, and deposits of 300 μg were spotted on silica gel plates (Silica gel Merck F254) using a ATS3-01081 (CAMAG, France). Elution uses a system of eluents composed of X: Chloroforme/Methanol/water 10 : 5 : 1 and Y: Cyclohexane/Diethylether/acetic acid 40 : 20 : 1, respectively). Two following migrations with X up to half plate, and one to the top with Y were carried on. Each migration occurs after saturation of the developing chamber for 2 hours, and drying between them. Standards of phosphatidyl choline (PC) and phosphatidyl ethanolamine (PE) were also analyzed at 254 nm with a TLC Scanner 3 (CAMAG, France) and Rf of each compound were compared to lipid extract.

2.6   *Fractionation*

Dried cooked eggs (60 g) with and without shell were extracted at reflux with cyclohexane using a soxhlet apparatus for 6 hours. The lipid concentrated extract was dried under nitrogen prior to analyses. The solid residue (51 g) was extracted with 0.5 L of a solution of sodium hydroxide (0.37 M) at 40°C during 70 min. After centrifugation, pH of basic extract is lowered at 7 by adding HCl solution and the extract is freeze-dried. The solid residue is freeze-dried. The flow chart of the fractionation is shown in Figure 1.

3   RESULTS AND DISCUSSION

3.1   *Global chemical composition*

The results obtained for the chemical composition of the eggs samples after drying and grinding are presented in Table 1. The main components are the minerals from the egg shell representing

Table 1. Chemical composition of cooked quail eggs.

(%) without shell	Fresh eggs	Grinded dried	Grinded dried
Dry matter	40.0	98.4	97.6
Mineral matter	–	49.2	5.2
Total nitrogen	–	4.4	8.4
Proteins	–	27.5	52.4
Lipid extractives	–	23.6	39.4

Table 2. Fatty acid composition of the lipid fraction.

Fatty Acid	Unsaturation position	%
Myristic acid	C14:0	0.51
Palmitic acid	C16:0	25.50
Pamitoleic acid	C16:1	3.68
Stearic acid	C18:0	10.35
Oleic acid ($\omega - 9$)	C18:1n9c	38.93
Linoleic ($\omega - 6$)	C18:2n6c	13.57
Linolenic acid ($\omega - 3$)	C18:3n3	0.34
gamma	C18:3n6	0.27
Gadoleic acid	C20:1	0.08
Σpoly-unsaturated (PUFA)		14.18
Σmonounsaturated (MUFA)		42.69
Σsaturated (SFA)		36.36
UFA/SFA		1.17

49% of the dry matter, and organic matter is mainly composed of proteins 27% and lipids 23%. According to the recently published value by Gomathi et al., the level of proteins of quail eggs is around 13% in fresh weight (Gomathi et al., 2014). Similar results up to 48 % of dry weight of the organic part only were also obtained by Song et al. (Song et al., 2000). Lipids part is 39%, close to those found by Ricklefs (Ricklefs, 1977).

3.2 Composition of lipid fraction

Fatty acids profile showed the major are oleic, palmitic and linoleic acids respectively with 38.9, 25.5 and 13.6% (Table 2). The ratio UFA/SFA is of 1.17 (1.17 and 0.39 respectively for MUFA/SFA and for PUFA/SFA) similar to literature (Sinanoglou et al., 2011). The phospholipids analyzed by HPTLC in the total lipids count for 11% PC and 3% PE.

3.3 Amino-acids profile

The major amino acids found in the cooked eggs were glutamate, asparagine, serine and leucine, with 12.7, 9.7, 9.5 and 8% respectively of the analyzed amino acids. The essential amino acids leucine, valine and threonine appear in same order as described by Tunsaringkarn et al. (Tunsaringkarn et al., 2012). The analyses of the different extracts obtained through fractionation exhibit a similar profile with no variation of level of amino acids.

3.4 Fractionation/obtaining extracts

Characterization of the three obtained fractions allows the identification of different ways to re-use of such a complex waste. Firstly lipid extracts are recovered at 13.6 % and 39.4% for the dried cooked eggs with and without shell respectively. Lipids only come from the egg yolk. The *insoluble in acetone* represents 24% of the lipid fraction, comprising mainly phospholipids. Polar lipids in quail egg yolk have been evaluated up to 38% of the total lipids (Sinanoglou et al., 2011). Our value appeared lower than this value obtained on fresh egg yolk. The differences could be explained by the cooking step affecting either the extractability of polar lipids or their stability. The lipid fraction contains an important part of lecithins which could be isolated with acetone to gain purity. Lecithins are commonly found in a wide range of food products, and almost in every transformed food, and in cosmetics (Pardun, 1989). There chemical properties offered an enhanced stability to formulation of dairy, fruits, meat or fish products. The main supplier of lecithins is the oil industry, especially the soy industry (well known as E322) which is also a by-product of soy oil, and sunflower industry. Finding a new source of lecithins appears interesting to vary the provenance of the product which these days could be an alternative to the soy coming from genetically modified seeds. Lecithins are already produced from hen eggs but these eggs are fresh and no coagulated. This study showed phospholipids, PC and PE, from quail eggs even cooked eggs are still accessible and can recovered at 14% in the lipid fraction, e.g. 1.8% of the starting dried eggs with shell. The fractionation of the lipids extract on TLC showed the phosphatidyl choline (PC) and phosphatidylethanolamine (PE) count for 11 and 4%. These two compounds represent 95% of the polar lipids in literature (Palacios and Wang, 2005, Sinanoglou et al., 2011).

Secondly the proteic fraction contains 61.4% and 32.5% of proteins for the dried cooked eggs without and with shell respectively. It corresponds to the proteins able to be solubilized in basic treatment. The main protein of quail egg are lysozyme, ovalbumin, ovomucoids and ovotranferrin (Segura-Campos et al., 2014). These proteins were denatured under the cooking treatment in food industry, therefore their solubility and other properties can be affected. A part of them were solvated and extracted, but the remaining residue formed a gel still containing an important level of proteins 64% in the sample without shell. The content of minerals is of 23% in both samples with and without shell, because of the addition of base and acid salts during the extraction phase. In literature, the proteic part of quail eggs has been precisely investigated with a therapeutic prospect. Indeed, proteins of quail albumen varies compared to hen albumen proteins and present anti-allergic properties (Betend Dit Bon, 2010). Ovomucoids have been studied to understand their anti-trypsin activities (Takahashi et al., 1994). Such application for quail albumen is viable when the eggs are not cooked. In our case, cooked eggs even kept under food conditions are not a suitable source of biologically active proteins: partial denaturation of proteins occurs during the cooking step, affecting also their extractability (Santos et al., 2015). Nevertheless, the applied extraction procedure generates a protein rich extract, containing essential amino-acids (leucine, threonine and lysine) yielded 30% and 57% of the starting material treated with and without shell respectively. Its potential use for nutraceutical specialties in pet-food or as emulsifier is a valuable alternative of valorization of this waste.

On last part, the solid residue is rich in minerals (and in non-extractible proteins), therefore the pet-food or fertilizers industries could capitalize on this fraction. This fraction count for 59% and 12% of the starting material treated with and without shell respectively. The solid residue still contains 18% and 20% of non-solvated proteins, and 73% and 20% of minerals with and without shell respectively.

4 CONCLUSION

The main goal of this study was to evaluate how to add value to specific poultry wastes. From the cooked eggs, it has been demonstrated that three new fractions with complementary features were isolated: lipid fraction rich in lecithins, proteic fraction and proteo-mineral fraction. In two complex industrial samples, a realist study was carried out to treat and add value to wastes. The raw waste comprising cooked eggs and their shells was treated to generate three valuable fractions. The second sample without the shell exhibits best fractionation yields and facilities of treatment of the organic part. Nevertheless, the shell removal was manually achieved. Further technological development is needed to operate this step. Furthermore, necessary conditions of treating coagulated eggs have been set up for the extraction of proteins from quail eggs. Future research should therefore be concentrated on the investigation of the incorporation of the extracts in cosmetic, food, or pet-food formulations.

REFERENCES

Awadé, A. C., Guérin-Dubiard, C., Nau, F. & Thapon, J. L. 1995. Utilisations actuelles et potentielles des protéines de blanc d'oeuf de poule. *Comptes rendus Journées de la recherche avicole, Angers, FRA (1995-03-28–1995-03-30).* 295–300.

Betend Dit Bon, M. 2010. New formulations based on lyophilizates of ass's milk and inhibitors of human proteases, indicated in milk intolerance, and allergic and infectious inflammation.

Carvalho, J., Ribeiro, A., Araujo, J. & Castro, F. 2013. Technical aspects of adsorption process onto an innovative eggshell-derived low-cost adsorbent. *Mater. Sci. Forum*, 730–732, 648–652.

Collins, F. D. & Shotlan, V. L. 1960. Low temperature chromatography of lipids on cellulose. *Journal of Lipid research*, 1, 352–353.

Gomathi, S., Vijitha, M. & Rathnasamy, S. 2014. Affinity separation of lysozyme from quail egg (Coturnix ypsilophora) and its antimicrobial characterization. *Int. J. PharmTech Res.*, 6, 1286–1291, 6 pp.

Oliveira, D. A., Benelli, P. & Amante, E. R. 2013. A literature review on adding value to solid residues: egg shells. *Journal of Cleaner Production*, 46, 42–47.

Palacios, L. E. & Wang, T. 2005. Egg-yolk lipid fractionation and lecithin characterization. *J. Am. Oil Chem. Soc.*, 82, 571–578.

Pardun, H. 1989. Pflanzenlecithine-wertvolle Hilfs und Wirkstoffe. *Fat Sci. Technol.*, 91, 45–57.

Ricklefs, R. E. 1977. Composition of eggs of several bird species. *The Auk*, 94, 350–356.

Sahin, N., Akdemir, F., Orhan, C., Kucuk, O., Hayirli, A. & Sahin, K. 2008. Lycopene-enriched quail egg as functional food for humans. *Food Res. Int.*, 41, 295–300.

Santos, D. O., Coimbra, J. S. R., Teixeira, C. R., Barreto, S. L. T., Silva, M. C. H. & Giraldo-Zuniga, A. D. 2015. Solubility of Proteins from Quail (Coturnix coturnix japonica) Egg White as Affected by Agitation Time, pH, and Salt Concentration. *Int. J. Food Prop.*, 18, 250–258.

Segura-Campos, M., Perez-Hernandez, R., Chel-Guerrero, L., Castellanos-Ruelas, A., Gallegos-Tintore, S. & Betancur-Ancona, D. 2014. Physicochemical and functional properties of dehydrated Japanese quail (Coturnix japonica) egg white. *Food Nutr. Sci.*, 4, 289–298, 10 pp.

Sinanoglou, V. J., Strati, I. F. & Miniadis-Meimaroglou, S. 2011. Lipid, fatty acid and carotenoid content of edible egg yolks from avian species: A comparative study. *Food Chem.*, 124, 971–977.

Song, K. T., Choi, S. H. & Oh, H. R. 2000. A Comparison of Egg Quality of Pheasant, Chukar, Quail and Guinea Fowl. *Asian Australas. J. Anim. Sci*, 13, 986–990.

Takahashi, K., Kitao, S., Tashiro, M., Asao, T. & Kanamori, M. 1994. Inhibitory specificity against various trypsins and stability of ovomucoid from Japanese quail egg white. *J. Nutr. Sci. Vitaminol.*, 40, 593–601.

Tashiro, M. 2003. Anti-nutrients and their health functionality. *New Food Ind.*, 45, 40–46.

Tunsaringkarn, T., Tungjaroenchai, W. & Siriwong, W. 2012. Nutrient benefits of quail (Coturnix coturnix japonica) eggs. *Ann.: Food Sci. Technol.*, 13, 122–131.

Wastes: Solutions, Treatments and Opportunities – Vilarinho, Castro & Russo (eds)
© 2015 Taylor & Francis Group, London, ISBN 978-1-138-02882-1

Waste prevention and education in five European countries

K. Matthes, G. Schelstraete, B. Sturm, J. Janela, M. Bours, A. Gutschi & A.P. Martinho
Virtual European Seminar, Open Universitait, The Netherlands,
Universidade Aberta, Portugal

ABSTRACT: This research assessed the different goals of the EU Waste Framework Directive (2008) concerning waste prevention in five countries of the EU. The countries under research are the home countries of the EVS group members: Austria, Belgium, Germany, Netherlands and Portugal. The research about the status quo of waste prevention revealed that waste prevention has not reached yet. All the countries have high recycling rates and low dumping rate, except Portugal. Although decoupling trends start to appear and the countries are climbing up the waste hierarchy, waste prevention is still a mountain too far. The EVS group also checked if education was included in the waste prevention programmes and education plays a very important role. The group looked for good examples of waste prevention and analyzed them with SWOT. The examples are a success in the home countries and can be implemented in other EU or even worldwide countries.

1 INTRODUCTION

The matter of waste is connected tightly to Sustainable Development. This means a development that meets the needs of the present generation, without compromising the ability of future generations to meet their own needs (Brundtland Report, 1987). Thus, it offers a vision of progress that integrates immediate and longer-term objectives, local and global action, and regards social, economic and environmental issues as inseparable and interdependent components of human progress (European Commission, 2014). Waste is a matter which contains challenges in all mentioned dimensions of Sustainable Development. It evokes various environmental, social and economic challenges, both local and global, which can have immediate as well as long-term consequences. Waste prevention bases on a simple concept: If you create less waste, you consume fewer resources and you will have to spend less effort like money or energy to recycle or dispose the waste. To become truly sustainable, there shouldn't be any waste production. This works either through re-use of materials, as they were designed, or waste prevention (Bartl, 2014).

Waste prevention is difficult to measure and to monitor. How to measure a thing that never existed? Recycling can be measured and quantified, but waste prevention is something that is probably eliminated from the waste stream (Bortoleto, 2014). Statements about the success of a particular waste manage activity therefore always require assumptions about the kind and amount of waste would have been generated without the measure (Dehoust, 2010).

Working in a research group with people from different countries implies the possibility to show variations of the application of the WFD in designated countries. Besides, waste prevention practices vary and there are plenty of different waste prevention programmes on international, national and regional level which are more or less successful. Statistics like the amount of waste per household or the recycling rate vary a lot between the different countries. They may reveal the status quo of waste prevention in the different countries and show in which areas improvements are required.

As the implementation of waste prevention and education varies a lot between the countries, it is important to see, in which way concrete actions take place. Therefore, this study presents interesting examples of waste prevention in the different countries and describes the advantages and disadvantages. This is an approach to show how the matter of waste prevention is realized and to learn from each other. Sharing good ideas is crucial for successful waste prevention.

2 METHODOLOGY

The countries under research are Austria, Belgium, Germany, Netherlands and Portugal. These countries are chosen because they are the home countries of the members of this research group. This research group worked in an e-learning of the course in Environmental Seminar of Sustainable Development with the support of a tutor and expert to develop this research. The methodology consists of literature research, the definition of indicators and the collection and analyzing of statistics.

To be able to compare the status quo of waste prevention in the different countries, the group defined nine indicators. These indicators are chosen on basis of what statistics are available in each country. The chosen indicators are meant to reveal the status quo of waste prevention in the countries. The group selected a data time frame from 2006 till 2012. The following indicators were selected:

1. Total amount of waste produced by households
2. Amount of waste per capita/year
3. Waste intensity – relation between GDP and total waste generation
4. Household waste destination
5. Number of separated waste streams
6. Measures in which money is an incentive to reduce the amount/production of waste
7. Re-use
8. GDP in relation with national direct material input
9. Environmental awareness

3 RESULTS

Waste prevention should be top priority in every EU country aiming at a sustainable waste management, in the table 1 we present the indicators selected and presented in methodology and for these we make use of the national or regional statistics to determine if the impacts of the waste prevention goals are visible by means of numbers.

3.1 Comparison of nine different indicators related with waste prevention in five different countries (Austria, Belgium; Germany; Netherlands; Portugal)

Although waste prevention is more then only numbers (the definition of waste prevention relates also to the hazardousness of substances), trends can become visible. Eventually, a comparison between the different countries will give insight in the differences between these countries, as can be seen in the Table 1.

3.1.1 Indicator 2: Household waste per capita
Less waste is produced per capita in 2012 than in 2006, but with the exception of Germany. In Germany the number keeps rising until 2011 and in 2012 there is a small decline of the number of waste produced per capita. Some countries (Belgium, Netherlands and Portugal) have a year where there is a maximum of produced waste, but altogether they have less waste produced in 2012 than in 2006.

3.1.2 Indicator 3: Relation between GDP and total waste
There is a general trend when the GDP raises, the waste per household declines. A trend is as if the GDP rises, most of the time the amount of waste declines. As seen in indicator 2, we can see

Table 1. Results from the different country related with waste prevention indicators.

Indicators	Austria	Belgium	Germany	Netherlands	Portugal
Total amount of waste produced by households	$4020\,(10^6)$	$5035\,(10^6)$	$44.188\,(10^6)$	$8800\,(10^6)$	$4782\,(10^6)$
Amount of waste per capita/year	477 kg /2012	459 kg /2012	549 kg/2012	530 kg/2012	454 kg/2012
Waste intensity – relation between GDP and total waste generation	Decoupling	Decoupling	dDecoupling	Decoupling	not been reached the goal of a decoupling waste generation of economic growth
Household waste destination (2012)	C&D – 33.9% MR – 28.03% TI – 34.66% L – 3.41%	C&D – 20.97% MR – 36.3% TI – 42.08% L –1.16 %	C&D – 97.5% MR – 66% TI – 31.6% L –< 0.5 %	MR – 47% TI – 51.2% L – 0.8%	C&D – 17% MR – 34% TI – 19% L – 55 %
Number of separated waste streams	13	6	9	9	5
Measures in which money is an incentive to reduce the amount/ production of waste	Waste fee 2012 – 669 mio increasing	Diftar system	Waste fee (garbage)	Diftar system	Waste fee
Re-use	decreasing trend of the re-use of glass visible	recycle center (is rising, but very slowly)	decreasing trend of system of returnable bottles	0.5% of house-holds waste was destined for re-use (2010)	
GDP in relation with national direct material input	Material intensity decreased by a quarter (1995–2010) and GDP increase 36%	Decoupling (but no systematic improvement of the material productivity of the economy)	long-term absolute decline in use of resources by 20%)	Decoupling (material intensity, has been reduced	No Decoupling
Environmental awareness	–	Several surveys in Flanders (survey of eco-friendly consumption)	–	–	–

C&D – Composting and Digesting; MR – Material Recycling; TI – Total Incineration; L – Landfill.

the amount of waste decline over the years. Most of the time the GDP keeps rising after 2009 until the end of the used data, with the exception of Portugal where the GDP keeps declining after 2009 together with the waste produced. So in all countries, except in Portugal, we encounter a decoupling trend of waste production and economic development.

3.1.3 *Indicator 4: Household waste destination*

While this is the bottom of the waste hierarchy, all the countries have a decreasing trend and therefore they all did a great effort to reduce this number. This is necessary if the countries are paying attention to the long term goals of the WFD.

Portugal is the only one who is above the European Union average, while the rest is beneath it. Austria did well in preventing this deposit of waste. They reduced their number by almost 70% in our data. It seems that Belgium has done a lot of effort, because their number of deposit decreased for almost 90% in the used data. Germany did it good from the start of our data and even decreased their number a little bit. The Netherlands has a decrease of 40% and, while their number was already low at the start of the used data.

Total incineration (including energy recovery): kg/capita.

The trend of the European Union is an increase in amount of waste incinerated over the years. There is in also an increase of incineration per capita in every studied country. Some countries have a rising trend over the entire period of time. This rise of incineration is due to more strict rules about the dumping and land filling. This way the countries are going up the waste hierarchy. Other countries, like Germany, Netherlands and Portugal have less waste incinerated in the last year than in previous years. They follow the global trend of the European Union. A reason for this will be the result of an european guideline for going even more up the waste hierarchy.

Material recycling: kg/capita.

There is a general positive trend in the European Union when it comes to recycling. More waste gets recycled per capita, while the global trend of waste per capita keeps declining. Not every country has the same definition concerning recycling, this makes the data difficult to compare. The best recycling country is Germany and the least is Portugal. Germany recycles 5 times more per capita than they do in Portugal.

Composting and digestion: kg/capita.

Not all compared countries follow the trend of the EU (table 4.27). The countries who have a similar positive trend are Germany and Portugal. The household waste that gets composted or digested in the other four countries decline throughout the years. In the Netherlands this number stays relative stable. However, countries do not follow the EU trend. They are all above the average, except Portugal. Portugal is doing a good job in increasing the number, but the number of composted and digested household waste stays under the average. Austria is doing the best in this topic, but their number is degrading at a fast rate.

3.1.4 *Indicator 5: Number of separated waste streams of household waste*

Belgium and Germany have numbers in percentages, so they are easier to compare. Glass is 100% recycled in most countries, so we can conclude that glass is totally recycled in the compared countries which is a good sign. Paper is in most countries a waste that is recycled well. Most of the time the number of recycling is near 90%. Again Germany does better with 99 to 100%. Plastic is more and more recycled in the countries. This is because more methods are found for the recycling of plastics and the efficiency of the methods keep rising. The recycled percentage keeps under 50% in every country because you cannot recycle the same plastic many times. When we recycle plastics, most of the time it is a downcycle, so the same plastic has a less attractive function and can never go to its previous function anymore. There are big differences of the recycling of biodegradable wastes. Germany has a large recycling rate (around 99%) of this kind of waste, while the other countries do not get this number above 50%. The recycling of different metals goes down in every country, except in Belgium, because of the use of more kinds of metals and smaller parts of them in electronic divices. The recycling of metals goes better and there are more efficient ways to recycle different kinds of metal. In Belgium some companies like Umicore earn a lot of money in this business.

3.1.5 *Indicator 6: Measures in which money is an incentive to reduce the waste*

Most of the countries have one or more incentives to reduce waste. Austria works with household waste fees, Belgium and Netherlands use the diftar system and Germany uses garbage fees. All these incentives are based on the polluter pays principle. Mostly the incentives are not the only measure, but strengthened in a broad range of measures. Portugal seems to have no real incentives yet for households, although is mentioned in the PERSU 2020. Normally more measures lead to a better recycling rates, as can be seen with Austria, Belgium, Netherlands and Germany.

3.1.6 *Indicator 7: Re-use of glass*
Austria works with fees and it works good. They have data about the re-use of glass, but this number declines from 2008 (from 32,5%) until 2013 (to 29,4%). In Belgium and Germany there is also a very high number of glass re-use. The number of glass recycling is not that high in Belgium because the same company has to refill the bottles to re-use it for the same products. The highest number of re-use of glass is in Germany with around 45% but it is declining.

3.1.7 *Indicator 8: GDP in relation with national direct material input*
In all countries the material input is declining while there is economic growth. We can talk about a decoupling trend. The more waste does not mean there is more economic growth. The decoupling starts around 2007. Only Portugal does not have this decoupling trend. Their GDP does not rise while the material input is, like in the rest of the countries, declining.

A decrease in total waste generation could indicate that there is a trend of waste prevention. So, the waste quantity reduction does not automatically indicate a lesser impact on the environment, since a lighter material can generate a greater impact during its life-cycle than a denser one (Prewaste, 2010). For example plastic versus glass. The amount of packaging can decline by due to a shift from glass to plastic. But the lifecycle energy used required and the GHG emissions to create the packaging increases. This results in a quantitative, but not a qualitative waste reduction (Bortoleto, 2014).

Waste prevention can be reached through changing the behaviour of people (Bortoleto, 2014). People can be made aware by making them pay for their waste, for example through the diftar system (see indicator 6, in table 1), but also through education and good examples.

The study on managing municipal waste (EEA, 2013) indicates that countries using a broad range of instruments have a higher municipal waste recycling rate than countries using very few or no instruments.

3.2 *Examples for waste prevention*

One project of each of the countries Austria, Belgium, Germany, Portugal and the Netherlands were analysed. This was conducted with a SWOT analysis, also to see the implementation of the EU WFD in the projects. These examples should give people all over Europe ideas for finding new ways to prevent waste.

3.2.1 *Austria – Award winning project "Iss mich" 2014*
Since 2012 the waste-cooking-show wastecooking has been raising awareness for the topic of food waste. Founding the food brand iss mich! is an important move in stepping up against food waste. The slogan is "eat it, don't waste it". That's why iss mich! (eat me!) prepares delicious vegetarian dishes from perfectly healthy veggies that did not meet retail standards – not due to quality but due to aesthetics. The food, filled in glas jars is offered in Vienna and can be orderd for office lunch or home for up to 25 persons and for any events with more than 25 persons. The dishes are organic, healthy, tasty, reducing food waste, served in re-useable jars and delivered by bicycle. Thus, the food is prepared by people, who do hard getting a job. Women, who are living in mother-child houses of the organisation Caritas (Frauenhäuser Caritas) prepare the meals and have a chance to earn some money with it.

3.2.2 *Belgium – Robuust*
Robuust is a market shop in the city of Antwerp that calls itself 'a zero waste shop'. It is the first packaging free shop in Belgium. It is an idea of prerecycling, where they do not use packaging

in all ways. The method for this zero waste shop is that the customers have to bring their own refill box to the shop, define the empty weight and refill it with a product and weight it again. This way, they only pay for the product that is inside the refill box. This is a specialized biological shop where you can buy seasonal vegetables and fruits from local farmers. Besides, you can get a lot of dry foods in bulk; like coffee and tea, dairy products, sweets, nuts, grains, oils and fruit juices. The idea is that you bring your own packaging with you and you can refill it in the shop. The sizes of your buys are not a problem, because you can buy small or big quantities. Another way to narrow their ecological footprint, the company tries to work with local distributors.

3.2.3   *Germany – Collaborative consumption*

Using instead of owning, borrowing instead of buying – those are the principles of collaborative consumption. The idea is that not everybody needs to own everything but that it is rather useful to share and exchange objects. This concept is not new, though combined with the possibilities of the internet and the current debate on sustainability it brought a renaissance of collaborative consumption in Germany and other countries. The core advantages are the conservation of resources and the money saving for people. A much known form of collaborative consumption is car-sharing or bike-sharing, which meanwhile exists in many German cities. But there is more: for example you can share your parking space (www.unserparkplatz.de), your garden (www.meine-ernte.de) and even your dinner (www.foodsharing.de). There is a platform which concentrates on objects for the daily life (www.frents.com), which will be analyzed in detail. The goal of this platform is to become a social network for people who want to live the idea of collaborative consumption (or shareconomy how it is also called) as often as possible. This is how it works: Frents is the platform for sharing your stuff with your neighbours and friends. It works similar to other community platforms. You can create a profile, have contact with your friends and send them messages. The special feature of Frents is that you can open your collections in all topics you are interested in.

3.2.4   *Netherlands – Wecycle*

Wecycle is an organisation which promotes the recycling and re-use of electronics and light saving bulbs. It is an organisation that has been with a minimal governmental influence for several years in the Netherlands. Only for separate collection of electronic waste, agreements have to be made local governments. They work for importers and producers. It is financed mostly from the trade in electronic waste and it's parts (Wecycle.nl, 2015). There is a lot of attention for education in Wecycle, from elementary school, secondary school, to companies and households. Especially in electronic devices the re-use is promoted by disassembling electronics after collection and re-using electronic parts, like capacitors. Apart from the goal to recycle and re-use as much as possible, another goal is to reduce electronic waste to 4 kg/person/year, which is now about 8 kg/person/year. This can be achieved by ensuring a longer life-time for electronics (Wecycle.nl, 2015). For now most of the materials are being recycled instead of re-used.

3.2.5   *Portugal – Green Cork*

Implemented since 2008, Green Cork is a Quercus (National Association for Nature Conservation) project for collecting cork stoppers for recycling. It is developed in partnership with Amorim group, Continente supermarkets, the Dolce Vita Commercial Centers, schools, scouts, municipalities, waste collection companies, wineries, wine producers and other entities that locally make this project a success. Its main objectives are to collect cork stoppers and finance the planting of indigenous trees through the Joint "Floresta Comum" project. The Green Cork is a project that works in cycle, from the tree comes cork, recycling gives new uses for cork that was previously the stopper, and even allows to plant new trees. Things that came from nature return to nature. Through amounts which Quercus received for delivery to recycling about 235 tonnes of cork stoppers were already planted about 200,000 trees (2014 data).

4 CONCLUSIONS

All countries chosen for the research have implemented the long term goals of the EU WFD. The waste prevention programs are well developed in all countries and include targets similar to the EU WFD, per example the recycling quota of 50% by 2015. Some countries like Austria, Belgium and Germany and the Netherlands can be seen as role models for recycling because in those countries the recycling quote is already about 60%.

Looking at the status quo of the countries, one main difference can be seen. The countries Austria, Belgium and the Netherlands do a good job reducing waste. Portugal does a good job as well, but still has a long way to go to reach the environmental level of the other countries. While in all other researched countries a decoupling trend of GDP and waste production can be found, in Portugal the GDP declines with the amount of waste. Furthermore, Portugal is the only country which is above the EU average concerning landfill as a waste destination. Thus, Portugal hasn't reached the EU average in recycling and composting.

There is an increasing trend for incineration in all countries (incineration with energy recuperation is subsidized by the government) and this indicate contradictions which exist between the different levels in the waste hierarchy, and may be a reason why waste prevention is difficult to reach. Very positive is fast declining landfill rate. Austria has a landfill ban since 2009 which led to almost zero landfill. Also in Germany the landfill rate is about zero.

Even though most of the countries researched are role models for waste management there are still things to be improved. Germany, for example, does a good job in treating waste, but it is the only country where the amount of waste increases. Also, all countries could work on re-using of products like glass, since the rate is declining in general.

All countries see education as very important for waste prevention. Therefore, parts of the national waste prevention programs describe the possibilities for education. People need to get aware of waste management issues and sustainable consumption. But there are also already lot of innovative ways for waste prevention. This shows the examples given by each country. Projects like collaborative consumption, using groceries which don't meet the retail standards, zero waste shops, recycling and re-use of electronics and recycling of cork stoppers are great innovative projects which could be developed in all EU countries.

REFERENCES

APA 2013. Resíduos Urbanos Relatório Anual 2012. Amadora: Agência Portuguesa do Ambiente, I.P.
Bartl, A. 2014. Moving from recycling to waste prevention: A review of barriers and enables. *Waste Management & Research* 32 (9).
Bortoleto, A. P. 2014. Waste Prevention Policy and Behaviour: *New Approaches to Reducing Waste Generation and Its Environmental Impacts*. Routledge.
Dehoust, G. & Küppers, P. 2010. For the Federal Environment Agency – Umweltbundesamt. *Development of scientific and technical foundations for a national waste prevention programme*. Dessau.
EEA 2013. Managing municipal solid waste. European Environment Agency.
European Environment Agency 2012. Material Resources and waste – 2012 update, Copenhagen.
European Commission 2014. Sustainable Development.
Prewaste 2010. Pre-waste is a European project (2010–2012) whose aim is to help cities and regions to improve their waste prevention policies.

Wastes: Solutions, Treatments and Opportunities – Vilarinho, Castro & Russo (eds)
© *2015 Taylor & Francis Group, London, ISBN 978-1-138-02882-1*

Toxicity measurement techniques for building materials with wastes

P. Mendes
Department of Civil Engineering, University of Minho, Guimarães, Portugal

H. Silva & J. Aguiar
CTAC, Department of Civil Engineering, University of Minho, Guimarães, Portugal

ABSTRACT: The innovative Horizon 2020 program sponsored by the European Union (EU) aims to promote and develop processes of waste integration in construction materials. However, several potential health hazards caused by building materials have been identified and, therefore, there is an ongoing need to develop new recycling methods for hazardous wastes and efficient barriers in order to prevent toxic releases from the new construction solutions with wastes. This paper presents an overview that focus on two main aspects: the identification of the health risks related to radioactivity and heavy metals present in building materials and identification of these toxic substances in new construction solutions that contain recycled wastes. Different waste materials were selected and distinct methodologies of toxicity evaluation are presented to analyse the potential hazardous, the feasibility of using those wastes and the achievement of optimal construction solutions involving wastes.

1 INTRODUCTION

The importance of evaluating the potential hazards of toxicity of building materials with wastes is the result of the rising awareness of several human living conditions, environmental consequences and massive damaging of industrial track (Hopwood et al., 2005; McDougall et al., 2001; Sullivan et al., 2014). In fact, construction materials generate about 31% of all waste in Europe. Moreover, recent investigation (Torgal et al., 2012; Fucic et al., 2010; Fucic, 2012, Kovler, 2009; Orisakwe, 2012; Prazakova et al., 2013) showed that several health problems may be caused by the chronic exposure to hazardous substances present in the building materials, such as materials that contain asbestos, heavy metals, radioactive substances or release volatile organic compounds and nanoparticles. For instances, asbestos exposure has been pointed out as a cause of health problems such as mesothelioma, lung cancer and asbestosis. Therefore, the Directive 2003/18/EC prohibited the production of asbestos-based products (Torgal et al., 2012). For all these reasons, the European Union (EU) seeks to create frameworks of policies and methodologies that will promote a substantial reduction of the waste generation in the future, as well as increase the control of the exposure to hazardous wastes.

The integration of recycled waste materials in new construction materials, as well as in engineering structures, is foreseen as a promising sustainable approach to close the material loop towards circular materials' economy, reducing hazardous emissions to air, land, water and human living conditions. Several waste materials have been integrated in construction solutions, such as: fly ash, phosphorous slag, tin slag, copper slag, residues from aluminium production, residues from steel production, textile, rubber tires, plastic, sewage sludge and foundry sands (Torgal et al., 2012). Nevertheless, the success of this approach implies the characterization of the material properties (e.g. chemical, structural and energy content) and also requires the evaluation of its toxicity level (Tiruta-Barna et al., 2012; Perrodin et al., 2000). To evaluate the toxic effects of pollutants for human and environment, standard methodologies based on classical human Health Risk Assessment (HRA), Ecological Risk Assessment (ERA) and laboratorial tools exist as measures to quantify

Table 1. List of building materials and health effects associated to the presence of radioactive substances.

Isotope:	^{226}Ra, ^{232}Th, ^{222}Rn, ^{40}K
Building materials:	Granite, gneiss, porphyries, syenite, basalt, tuff, pozzolana, lava, phosphor-gypsum, alum-shale, coal fly ash, oil shale ash, tin slag, phosphate fertilizer, oil industry, red mud from aluminium production residues from steel production, from steel production residues from steel production, concrete, clay, bricks, sand-lime, tiles, cement.
Health effects:	Cancer, cardiovascular diseases, immunological disturbances, chromosomal abnormalities and gene mutations.
References:	(Torgal et al., 2012; Nisnevich et al., 2008; Yao et al., 2015, Kovler, 2012a; Kovler, 2009; Dikstra et al., 2004; Marocchi et al., 2011; El-Thaer, 2012; Uosif, 2014; Kinsara et al., 2014)

the potential hazardous pollution in specific disposal, elimination or reuse scenarios (Tiruta-Barna et al., 2012).

There are two main tasks that are carried out in parallel in this paper. The first one is to describe two sources of toxicity (radioactivity and heavy metals), which is the core of section 2. The second task is to describe some techniques used to determine toxic substances in construction solutions with wastes materials. Finally the conclusion points out the state of the current practice in the field of toxicity evaluation of construction materials and directions for future research are considered.

2 TOXIC SOURCES AND MAIN HEALTH HAZARDS OF BUILDING MATERIALS

Radioactivity is a process by which certain naturally occurring or artificial nuclides undergo spontaneous decay or loss of energy releasing a new energy, therefore resulting in an atom "the parent nuclide" transforming to an atom "the daughter nuclide".

Radiation sources can be natural or artificial. In case of natural, it is estimated that 85% can be cosmic (14%) and telluric radiation (71%) (Kovler, 2012b). The telluric radiation is caused by Naturally Occurring Radioactive Materials (NORM), ^{235}U and ^{238}U – uranium; ^{40}K – potassium; ^{226}Ra – Radium and ^{232}Th – thorium, which represents an external gamma radiation, both inside and outside buildings (Kovler, 2011; Kovler, 2012a; Kovler, 2012b). The radionuclides ^{226}Ra, ^{232}Th and ^{40}K, which are present in the Earth's crust, are also in building materials or in additives of building materials (Table 1). Radon, ^{222}Rn, is a radioactive, colourless, odourless, tasteless noble gas, occurring naturally as the decay product of radium ^{226}Ra (external radiation). The internal radiation in the human body is caused by natural radionuclides in food, water and air. Moreover, the internal radiation exposure is due to radon gas ^{222}Rn, which belongs to the ^{238}U decay chain, and marginally to its isotope thoron ^{220}Rn, which belongs to the ^{235}U decay chain, and their short lived decay products, exhaled from building materials into the room air (Kovler, 2011). Being an inert gas, radon can move rather freely through the materials. For instance, Kovler (2012b) mentioned that radon emanates naturally from the ground and from mineral building materials, wherever traces of uranium or thorium can be found (e.g. granite or shale). Radon is also present in high concentration in granitic residual soils, rocks and fly ash (Kovler, 2012b; ECRP-112, 1999; ECRP-96, 1997). Table 1 lists examples of building materials where the radioisotopes are present and the corresponding health problems associated to radioactivity exposure.

The new Council Directive 2013/59/EURATOM, which sets out basic safety standards for protection against the hazards arising from exposure to ionising radiation, recognized the importance of protection against natural radiation sources, in particular from industries processing materials containing naturally-occurring radionuclides. Establishing reference levels for indoor radon concentrations and for indoor gamma radiation emitted from building materials and introducing requirements on the recycling of residues from industries processing naturally-occurring radioactive

Table 2. Heavy metals in building materials and health effects associated to chronic exposure.

Heavy metal	Health effects	Building materials	References
Cadmium	Lung cancer, osteomalacia, proteinuria.	Pigments, chemical stabilizers, nickel-cadmium batteries.	(Orisakwe, 2012)
Mercury	Stomatitis, nausea, nephrotic syndrome, neurasthenia tremor.	Fluorescent lamps.	(Orisakwe, 2012)
Lead	Nephropathy, anaemia, encephalopathy.	Lead connection pipe.	(Hayes, 2012)
Chromium	Pulmonary fibrosis, lung cancer.	Dyes and paints, stainless steel, metallurgy, chrome platin.	(Orisakwe, 2012)
Arsenic	Diabetes, hypopigmentation, cancer, liver disease.	Chromated copper arsenic in pressure treated wood.	(Hall, 2002)

Table 3. Estimation of heavy metals present in wastes of case studies.

Wastes	Heavy metals	Case studies	References
Coal fly ash	Cr, Cu, Ni, Pb, V, Zn, As, Hg, Cd, Sn.	Incorporation of coal fly ash in Portland cement concrete.	(Torgal et al., 2012) (Yao et al., 2015) (Azevedo et al., 2012) (EPA, 2015)
Rubber tires	Zn, Se, Cd, Pb.	Incorporation of crumb rubber in asphalt concrete.	(Peralta et al., 2012) (EHHI, 2015)

materials into building materials. Other main hazards of interior building materials are the heavy metals, which are present in pigments, electric lamps, leads pipelines and various other household wares. Heavy metals are also present in various wastes used in concrete. Table 2 lists examples of building materials that contain heavy metals (cadmium (Cd),mercury (Mg), lead (Pb), chromium (Cr), arsenic and the corresponding health effects associated to heavy metals chronic exposure (Orisakwe, 2012; Hayes, 2012).

3 INTEGRATION OF WASTE MATERIALS IN CONSTRUCTION SOLUTIONS

3.1 *Case studies of waste incorporation in building materials*

Integrating waste materials in building materials requires a good knowledge of the properties and characteristics of those waste materials. Basic characterization constitutes a full characterization of the waste materials by gathering all the necessary information for a safe management of the waste in the short and long term, such as: type and origin, composition, consistency, leachability, among other (EN 14735). The Technical Committee CEN/TC 351 'Construction products: Assessment of release of dangerous substances' and CEN/TR 16098:2010 (Concept of horizontal testing procedures in support of requirements under the CPD) work underpinned of standards in the field of the assessment of building materials for their reuse.

The final rule of Environmental Protection Agency (EPA), EPA "Hazardous and Solid Waste Management System; Disposal of Coal Combustion Residuals from Electric Utilities." address serious human health and environmental risks from unsafe coal ash disposal (EPA, 2015). Table 3 shows two case studies of incorporation of wastes (coal fly ash and rubber tires) into concrete (both for Portland cement and/or asphalt concrete) and shows the heavy metals associated to each waste. In accordance with research (Torgal et al., 2012; Yao et al., 2015; EPA, 2015), the coal fly ashes contain: chromium (Cr), copper (Cu), nickel (Ni), lead (Pb), zinc (Zi), arsenic (As), mercury (Hg), cadmium (Cd), tin (Sn) and selenium (Se). A study on the properties of recycled tire rubber (Cao, 2007) shows that its chemical composition is acetone extract (15.5%), ash content (6.0%), carbon black (29.5%) and rubber hydrocarbon (49.0%).

In the Waste Framework Directive 2008/98/EC, (European Commission, 2008), waste should be classified in accordance with the European List of Waste (Commission Decision 2000/532/EC), where 405 waste types are defined as hazardous waste material. Other 200 waste types are also listed as "mirror entries", what means "Waste with potential to be either hazardous or non-hazardous depending on their composition and the concentration of dangerous substances" (Römbke, 2009). Therefore, fly ash can be classified as 19 01 13 and considered with hazardous properties if containing dangerous substances (H13 and H14 "Ecotoxic"). On the other hand, it can be classified as 19 01 14 fly ash is non-hazardous. The classification to end-of-life tyres is 16 01 03 ("end-of-life tyres"), non-hazardous.

3.2 *Toxicity measurement techniques of buildings materials: radioactivity and heavy metals*

The release of soluble constituents upon contact with water is regarded as a main mechanism of release which results in a potential risk to the environment during the disposal of waste materials (Stiernström et al., 2014). Therefore, the heavy metals can be determined through an assessment of leaching potential of materials with laboratory prepared samples and on site samples. These tests aim to identify the leaching properties of waste materials. Therefore, surfacing materials are measured under controlled laboratory conditions in accordance with technical specification CEN/TS 15862: 2012 – "Characterization of waste. Compliance leaching test. One stage batch leaching test for monoliths at fixed liquid to surface area ratio (L/A) for test portions with fixed minimum dimension". The main factors to control leachability are: pH, redox potential and physical parameters.

Laboratorial determination of radioactive properties can be through gamma spectrometry and exhalation with representative site samples. The first determination of ^{232}Th, ^{226}Ra and ^{40}K on material is made in order to estimate the potential hazard of it from a radiotoxicity perspective. The radionuclide measurements can be carried out by high resolution gamma-ray spectrometry. The HPGe detectors can be lead shielded with cooper and tin lining. The Genie 2000 software can be used for data acquisition and analysis. The detection efficiency is determined using NIST-traceable radioactive standards, to reproduce the exact geometry of samples, in a water-equivalent epoxy resin matrix. The measurement of radon aims to determine the material potential to generate high indoor radon concentrations and establish a classification based on its potential risk. The composition of samples can be analysed when placed inside cylindrical sealed containers. The radon exhalation rate is measured by using an active continuous radon monitor, in order to follow the ^{222}Rn activity growth as a function of time and determining the average equilibrium of the gas concentration during the exposure period (Madruga et al., 2011).

Some simple actions such as sealing around loft-hatches, sealing large openings in floors and extra ventilation do not reduce radon levels on their own. To lower radon levels, homeowners can use passive or active mitigation methods. The first are associated to radon levels above the 4 pCi/L limit (Henschel, 1993). Such methods include sealing foundation openings and around areas where pipes enter the home, reducing 15% of radon level (UkRadon, 2014). Other methods include applying coatings or membranes on walls to reduce permeability and installing a passive soil ventilation system and also active systems with mechanical and electrical devices to reverse air flow and ventilate soil beneath the foundation (Koff et al., 2007; UkRadon, 2014).

4 CONCLUSIONS

Two main sources of toxicity have been identified in building materials as causes of health problems: radioactivity and heavy metals. Two examples of incorporation of waste solutions were shown to estimate the potential hazards. Fly ash and tire rubbers contain different chemicals which need to be tested accurately in laboratory to determine how much substances contaminate the indoor air and the environment.

The lack of safety procedures related to the process of integration of waste materials in new construction solutions constitutes a current problem in terms of toxic compounds. Moreover, when evaluating the influence of the radioactivity and heavy metals release from construction materials with wastes it is fundamental to understand which are the main factors to consider. Such studies should include both laboratory tests of those material specimens of the same geometry and also boundary conditions under well-controlled ventilation (radioactivity). Moreover, they should include assessment of the toxicity potential of the solutions with wastes in their utilization scenarios, especially via emissions in water and dispersion in the surrounding natural environment (e.g., heavy metals).

REFERENCES

Azevedo, F., Torgal, F, Jesus, C., Barroso de Aguiar, J. & Camões, A. 2012. Properties and durability of HPC with tyre rubber wastes. *Construction and Building Materials*, 34, 186–191.

Cao, W. 2007. Study on properties of recycled tire rubber modified asphalt mixtures using dry process. *Construction and Building Materials*, 21, 1011–1015.

Commission Decision 2000/532/EC of 3 May 2000 establishing a list of wastes pursuant to Article 1(a) of Council Directive 75/442/EEC on waste and Council Decision 94/904/EC establishing a list of hazardous waste pursuant to Article 1(4) of Council Directive 91/689/EEC on hazardous waste.

Council Directive 2013/59/Euratom, 2013. The council of the European Union of 5 December 2013 on Laying down basic safety standards for protection against the dangers arising from exposure to ionising radiation. European Commission.

Dikstra, J., Meeussen, J. & Comans, A. 2004. Leaching of Heavy Metals from Contaminated Soils: An Experimental and Modelling Study. *Energy Research Centre of the Netherlands (ECN)*. ZG Petten, The Netherlands.

EHHI, 2015. *Report – Exposures to ground up rubber tires*. Environment and Human Health, Inc. Available at: www.ehhi.org (accessed in 20 April 2015).

El-Thaer, A. 2012. Assessment of natural radioactivity levels and radiation hazards for building materials used in Qassim area, Saudi Arabia. *Rom. Journal Phys.*, 57 (3–4), 726–735.

ECRP-96, 1997.European Commission, Radiation Protection 96. Enhanced radioactivity of building materials. Luxembourg: European Commission.

ECRP-112, 1999, European Commission Radiation Protection 112. Radiological Protection Principles concerning the Natural Radioactivity of Building Materials. Luxembourg: European Commission.

EPA, 2015. Final Rule: Hazardous and Solid Waste Management System; Disposal of Coal Combustion Residuals from Electric Utilities. Environmental Protection Agency.

European Commission, 2003. Directive 2003/18/EC of the European Parliament and of the Council of 27 March 2003 on the protection of workers from the risks related to exposure to asbestos at work, Brussels, Belgium.

European Commission, 2008. *Directive 2008/98/EC of the European parliament and of the Council of 19 November 2008 on waste*. Brussels, Belgium.

European Parliament and the Council, 2003. Directive 2003/18/EC of 27 March 2003 amending Council Directive 83/477/EEC on the protection of workers from the risks related to exposure to asbestos at work, Official Journal L 97/48, 15.4.2003, Brussels

Fucic, A., Gamulin, M., Ferencic, Z, Rokotov, D.S., Katic, J., Bartonova, A., Lovasic, I.B. & Merlo, D.F. 2010. Lung cancer and environment chemical exposure: a review of our current state of knowledge with reference to the role of hormones and hormone receptors as an increased risk factor for developing lung cancer in man. *Toxicologic Pathology*, 000, 1–7

Fucic, A., Fucic, L., Katic, J., Stojkovic, R., Gamulin, M. & Seferovic, E. 2011. Radiochemical indoor environment and possible health risks in current building technology. *Building and Environment*, 48, 2609–2614.

Fucic, A. 2012. The main health hazards from building materials. *Toxicity of building materials*, Editors: F. Pacheco Torgal, S. Jalali, A. Fucic. Chapter 1, 283–296. Cambridge: Woodhead Publishing.

Hall, A. 2002. Chronic arsenic poisoning. *Toxicology Letters*, 128 (1–3), 69–72.

Hayes, C. 2012. Heavy metals: lead. *Toxicity of building materials*, Editors: F. Pacheco Torgal, S. Jalali, A. Fucic. Chapter 10, Cambridge: Woodhead Publishing.

Henschel, D.B. 1993. Radon reduction techniques for existing detached houses: technical guidance for active soil depressurization systems. EPA 625-R-93-011. 3rd ed. EPA, Washington, DC.

Hopwood, B., Mellor, M. & O'Brien, G. 2005. *UK Sustainable Development*. Sustainable Cities Research Institute, University of Northumbria, Newcastle on Tyne. Available at: www.interscience.wiley.com (accessed in 15 March 2015).

Kinsara, A. A., Shabana, E. I. & Qutub, M. M. T. 2014. Natural radioactivity in some building materials originating from a high background radiation area. *International Journal for Innovation Education and Research*, 70–78.

Koff, J. P., Lee, B. D. & Ziemer, P. L. 2007. *Radon: how to assess the risks and protect your home*. Purdue University Cooperative Extension Service.

Kovler, K. 2009. Radiological constraints of using building materials and industrial by-products in construction. *Construction and Building Materials*, 23, 1–253.

Kovler, K. 2011. Legislative aspects of radiation hazards from both gamma emitters and radon exhalation of concrete containing coal fly ash. *Construction and Building Materials*, 25, 3404–3409.

Kovler, K. 2012a. Does the utilization of coal fly ash in concrete construction present a radiation hazard? *Construction and Building Materials*, 29, 158–166.

Kovler, K. 2012b. Radioactive materials. *Toxicity of building materials*, Editors: F. Pacheco Torgal, S. Jalali, A. Fucic. Chapter 8, 196–240. Cambridge: Woodhead Publishing.

Madruga, M., Carvalho, F., Reis, M., Alves, J. & Curisco, J. 2011. *Programas de Monitorização Radiológica Ambiental*. Report *LPSR-A* n°39/13. Laboratório de Proteção e Segurança Radiológica. IST.

Marocchi, M., Righi, S., Bargossi, G. M. & Gasparotto, G. 2011. Natural radionuclides content and radiological hazard of commercial ornamental stones: an integrated radiometric and mineralogical-petrographic study. *Radiat. Meas.*, 46, 538–545.

McDougall, F., White, P. R., Franke, M. & Hindle, P. 2001. *Integrated Solid Waste Management: A Life Cycle Inventory*. Oxford: Blackwell Publishing.

Nisnevich, M., Sirotin, G., Schlesinger, T., Eshel, Y. (2008). Radiological safety aspects of utilizing coal ashes for production of lightweight concrete. *Fuel*. 87, 1610–1616.

Orisakwe, O. 2012. Other heavy metals: antimony, cadmium, chromium and mercury. *Toxicity of building materials*, Editors: F. Pacheco Torgal, S. Jalali, A. Fucic. Chapter 11, 297–333. Cambridge: Woodhead Publishing.

Perrodin, Y., Grelier-Volatier, L., Barna, R., & Gobbey, A. 2000. Assessment of the Ecocompatibility of waste disposal or waste use scenarios: towards the elaboration and implementation of a comprehensive methodology.Waste Materials in Construction. 504–512. Oxford: Elsevier Science Ltd.

Peralta, J., Silva, H., Hilliou, L., Machado, A., Pais, J. & Williams, C. 2012. Mutual changes in bitumen and rubber related to the production of asphalt rubber binders. *Construction and Building Materials*, 36, 557–565.

Prazakova, S., Thomas, P. S., Sandrini, A. & Yates, D. H. 2013. Asbestos and the lung in the 21st century: an update. *The Clinical Respiratory Journal*.

Römbke, J., Moser, T. & Moser, H. 2009. Ecotoxicological Characterization of Waste: Results and Experiences of an International Ring Test. Springer Science & Business Media.

Stiernström, S., Enell, A., Wik, O., Hemström, K. & Breitholtz, M. 2014 Influence of leaching conditions for ecotoxicological classification of ash. *Waste Management* 3, 421–429.

Sofilic, T., Barisik, D., Sofilic, U. & Durokovic, M. 2011. Radioactivity of some building and raw materials used in Croatia. *Polish Journal of Chemical Technology*, 13(3), 23–27.

Tiruta-Barna, L. & Barna, R. 2012 Potential hazards from waste based/recycled building materials, Toxicity of building materials. Editors: F. Pacheco Torgal, S. Jalali, A. Fucic. Chapter 14, 391–426. Cambridge: Woodhead Publishing.

Torgal, F., Jalali, S. & Fucic, A. 2012. *Toxicity of Building Materials*; Woodhead Publishing Limited, Cambridge.

Uosif, M. A. M. 2014. Estimation of radiological hazards of some Egyptian building materials due to natural radioactivity. *International Journal of u- and e- Service, Science and Technology*, 7(2), 63–76.

UkRadon. 2014. *Measuring Radon*. Available at: http://www.ukradon.org/information/measuringradon (accessed in 28 December 2014).

Yao, Z.T., Ji, X.S., Sarker, P.K., Tang, J.H., Ge, L.Q., Xia, M.S. & Xi, Y.Q. 2015. A comprehensive review on the applications of coal fly ash. *Earth-Science Reviews*, 141, 105–121.

Wastes: Solutions, Treatments and Opportunities – Vilarinho, Castro & Russo (eds)
© 2015 Taylor & Francis Group, London, ISBN 978-1-138-02882-1

Shared responsibility and reverse logistics: Study of a sectorial agreement in Brazil

C.B. Milano, A. Santi, C.D. Santiago, E. Pugliesi & F.L. Lizarelli
Federal University of São Carlos, São Carlos, São Paulo, Brazil

ABSTRACT: The waste matter in Brazil has been broadly discussed by many people since the approval of the National Waste Policy (Law number 12,305/2010), due to legal aspects in execution of preventive, corrective and punitive activities linked to waste generation. Among the tools presented to mitigate the impacts of waste accumulation, we find two instruments that might be applied together: Reverse logistics and shared responsibility. Based on this, both instruments are brought out by sectorial agreements and reference terms to guide joint efforts of the government and private enterprise. In Brazil, the opening of public consultation for general packing sectorial agreement was the first step of this joint work. The present paper describes the history of this agreement in order to identify potential and fragile aspects of the process until March 2015.

1 INTRODUCTION

Increase of human consumption is a reality all across the world; in a rampant and scary scale, this situation promotes direct increase of product-discard, since their lifespan and real usage are now less important when manufactured than *fashion* and *outer beauty*, for instance.

Brazilian government, looking for alternatives to restrain this immediate and overwhelming reality approved on August 2010, the National Waste Policy (Law 12,305/2010), leading off the adequacy process of the society, government and private enterprise to improve waste management in Brazil. After more than 20 years in discussion on the senate, this law expands the operating-horizon of managers and fulfillment of legal and environmental adequacy regarding waste management in Brazil.

Reverse Logistics (RL) is one amongst many other tools brought by the National Waste Policy. RL aims at analyzing, organizing and managing environmentally adequate destination of post-sale and post-consumption waste, in order to add value to materials, which can still be reintroduced on the supply chain. RL is news for the industry in Brazil, helping on the search for more sustainable alternatives for products manufactured and traded on a daily basis.

Shared responsibility (SR) is another tool, strongly connected to RL. SR matches interests of economic, social and environmental agents, developing sustainable strategies with involved stakeholders, besides promoting waste reuse, returning them to the supply chain (Brasil, 2010). With this initiative, this principle seeks decrease of waste generation by manufacturers, importers, distributors, traders and consumers, along with a better use of the materials as a whole. Through SR, RL joins forces between government and private sector, covering involved stakeholders.

In order to coordinate this tool, Brazilian environment ministry, *MMA*, responsible to evaluate and manage environmental developing actions, brought together a guiding committee for RL establishment (CORI). This committee will be responsible for the sectorial agreements, regulating terms among government and private enterprise. The latter can choose to: 1) be part of sectorial agreements on which the government can publish a calling notice through MMA for the interested waste type or sector (after approval by CORI of the technical and economic viability); 2) have a

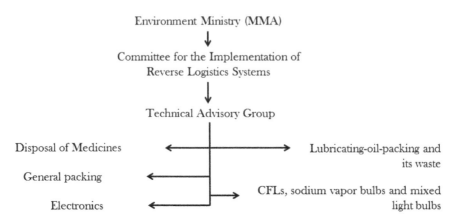

Figure 1.　Flowchart of the reverse logistics structure inside MMA (Milano & Pugliesi, 2014).

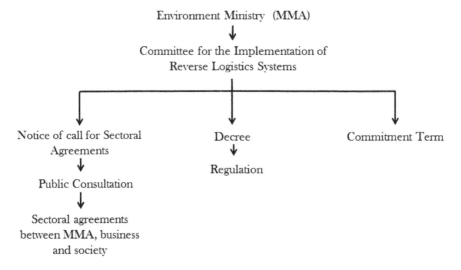

Figure 2.　Flowchart of the acting methods to apply Reverse Logistics – Government initiative possibilities (Milano & Pugliesi, 2014).

private enterprise initiative, presenting a formal proposal of reverse logistics, called *Commitment Term*, to MMA.

The environment-responsible bureau, in order to guarantee the best operation of the committee, divided the first products to be studied in 5 technical-work groups (GTT). The five technical-work groups created are: 1) Disposal of medicines, 2) General packing, 3) Electronics, 4) Lubricating-oil-packing and its waste, 5) CFLs, sodium vapor bulbs and mixed light bulbs. These groups must then study the viability and possible structure of those sectorial agreements, like the following flowchart displays (Figure 1).

Brazilian decree number 7,404/2010 regulates the National Waste Policy. This decree describes shared responsibilities between government and private enterprise, in order to develop and apply acting methods towards the implementation of reverse logistics to several supply chains in Brazil.

From the acting methods described on the previous decree, we detach sectorial agreements, commitment terms and regulations. These three can happen by the initiative of the government along with involved companies and other civil departments (Figure 2).

As a result of this format, environment ministry, MMA, began working with two work groups, the sectorial agreement proposal for reverse logistics system of CFLs, sodium vapor bulbs and mixed light bulbs and the sectorial agreement proposal for a reverse logistics system of general packing. Both groups aimed at forming partnerships and beginning efforts on shared responsibility of these products.

2 OBJECTIVE

The objective of this paper is to follow the process of the sectorial agreement proposal for application of the general packing reverse logistics system. This paper also verifies potential and fragile aspects along this process until March 2015.

3 RESULTS AND DISCUSSION: SECTORIAL AGREEMENT PROCESS

In July 2012, the discussions about the general packing sectorial agreement began because of a public calling notice from Brazilian environment minister to elaborate proposals for this agreement, with a 180-day deadline (MMA, 2012). The environment ministry received four sectorial agreement proposals, from December 2012 to January 2013, and CORI validated three of these proposals for negotiation (SINIR, 2015).

After studies, read of the proposals and texts elaboration, MMA released on the website *www.consultas.governoeletronico.gov.br* in September 2014 the general packing sectorial agreement for public consultation, so that the population could contribute on the process of structuring this RL system. The document was online for one month but, due to the expressive amount of contributions, MMA decided to let it online until November 2014, with a total sixty-six-day-period of contribution.

The agreement that was available to public consultation presents as possible signatories MMA (as a stakeholder) and 20 groups (also stakeholders) among associations (18), unions (1) and organizations (1), besides 7 consenting parties organizations, in order to agree as a coalition for the implementation of the RL system.

The document compiles the transcription of technical terms meanings, so that there will not be misunderstanding or miscomprehension, making the text coherent and of easy understanding. All the associations and companies actions will compose this coalition, and annual reports will be presented, in order to follow and verify goals accomplishment. This information will be fed to the National System of Waste Information, *SINIR* (MMA, 2014).

The general packing sectorial agreement foresees the dry-waste-material gathering, disregarding potentially polluting products, such as lubricating-oil-packing and its waste, and also potentially contaminated packing. Besides that, it is a national-range agreement, prevailing over regional or municipal sectorial agreements.

In order to have an adequate operation, the agreement will follow the agreement will keep on with the subsequent flowchart (Figure 3).

Through these steps, the agreement previses a 2-step system. The first step covers the application of the system in some of the cities that are part of the metropolitan regions of the 2014 World Cup host cities, and in other cities classified as conurbations. The second step covers several analysis of step one's results, identifying potential and fragile aspects, in order to promote continuous improvement and expansion of the system, with national range.

Concerning the management of this process, the coalition will form as follows (Figure 4).

Along with MMA, associations and other members' commitment, beyond the coalition we perceive a wide shared accountability from all signatories and civil society. The latter must be informed and encouraged to be part of this process, achieving the goal of the National Waste Policy. Another thing is the probation and support of waste-picker's associations, regarding the insertion and encouragement of these workers on the supply chain. Therefore, the agreement describes the

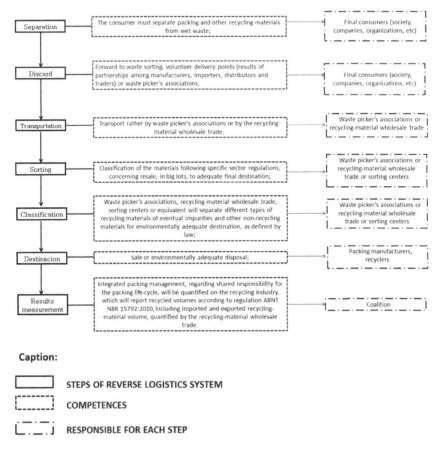

Caption:

☐ STEPS OF REVERSE LOGISTICS SYSTEM

┌┈┈┈┐
┊ ┊ COMPETENCES
└┈┈┈┘

┌ ⋅ ─ ⋅ ┐
└ ⋅ ─ ⋅ ┘ RESPONSIBLE FOR EACH STEP

Figure 3.   Reverse Logistic System Steps, as proposed by the agreement (Adapted from MMA, 2014).

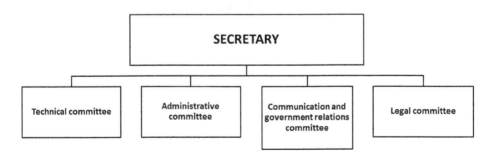

Figure 4.   Coalition Organization Chart (Adapted from MMA, 2014).

responsibilities of each link on the chain, along with goals and penalties, if any member does not carry out with its duties. This description aims at a continuous system's well-functioning (MMA, 2014).

Another document that comes along in this process and is presented as part of the agreement is an Economic Viability and Social and Environmental Impacts Study, with a view to present the current scenario of waste disposal amounts, along with future projections with the appliance of the agreement, contributing for the system well-functioning.

The sectorial agreement is in the process of analysis until now (March 2015). It shall be ruling by the first term of 2015.

4 CLOSING REMARKS

This paper aimed at analyzing potential and fragile aspects of the Sectorial Agreement Proposal for General Packing Reverse Logistics. One of the potential aspects identified concerns the accountability clauses of every part involved, signatories or not, like the civil society for instance. The civil society must carry out their responsibility as citizens to accomplish the National Waste Policy. This action shows that the system is concerned about signing and accomplishing the shared responsibility, in order to distribute bonus and onus among all those involved with the general packing supply chain.

Another potential aspect of the deal is the division in two steps to implement the system, addressing scenarios of action, contemplated by the analysis of the Economic Viability and Social and Environmental Impacts Study, so that the activities will really evolve and reach the proposed goals. With the coalition, the agreement will help a better management and accompaniment of the execution of the system on each step.

Regarding the public consultation, it was efficient, since each item had more than 10 contributions. However, disclosure was indirectly restricted to immediate interested personnel, the signatories, leaving civil society on the edge of this process. We consider this condition a fragile aspect.

Another fragile aspect is the fact that the penalty matter is not detailed on the agreement, and to improve this situation the agreement could incorporate laws concerning penalties, such as law 9,605/98 (Environmental crimes law), or other applicable regulations.

Under such situation, the general-packing reverse logistics system is broadly promising and built on a solid basis, what can guarantee effectiveness and activeness. With the agreement's signature, companies, associations, MMA and civil society should be active and work on a joint effort, in order to achieve the progress of this system and improve the environmentally adequate destination of generated waste.

REFERENCES

Brasil. Lei nº 12.305 de 02 de agosto de 2010. Institui a Política Nacional de Resíduos Sólidos; altera a Lei no 9.605, de 12 de fevereiro de 1998; e dá outras providências.

Brasil. Decreto nº 7.404 de 23 de dezembro de 2010. Regulamenta a Lei no 12.305, de 2 de agosto de 2010, que institui a Política Nacional de Residuos Solidos, cria o Comitê Interministerial da Política Nacional de Residuos Solidos e o Comitê Orientador para a Implantação dos Sistemas de Logística Reversa, e dá outras providências.

Milano, C. B. & Pugliesi, E. 2014. Reverse logistics: studies and complementation. In: *Fifth International Symposium on Energy from Biomass and Waste*, 2014, Venice. Proceedings Venice.

Ministério do Meio Ambiente (MMA) 2012. Edital nº02/2012. Chamamento para a elaboração de Acordo Setorial para a Implementação de Sistema de Logística Reversa de Embalagens em geral.

Ministério do Meio Ambiente (MMA) 2014. 140 – Proposta de Acordo Setorial para a implantação de Sistema de Logística Reversa de Embalagens em Geral.

Sistema Nacional de informações sobre a gestão dos resíduos sólidos (SINIR) 2015. Portal de Logística Reversa. Disponível em: http://sinir.gov.br/web/guest/inicio.

Wastes: Solutions, Treatments and Opportunities – Vilarinho, Castro & Russo (eds)
© 2015 Taylor & Francis Group, London, ISBN 978-1-138-02882-1

Decolorization of metal complex dye reactive blue 221 by *pseudomonas aeruginosa* BDS23

S. Mumtaz, N.H. Malik & S. Ahmed
Department of Microbiology, Quaid-i-Azam University, Islamabad, Pakistan

ABSTRACT: The present study was conducted to evaluate the potential of native bacterial species for the degradation of a reactive textile dye, i.e. Reactive Blue 221 (RB 221). Bacterial strain was isolated for the degradation of RB 221 from a textile industry soil sample by enrichment technique and then various parameters were optimized for maximum dye decolorization. The strain was identified as *Pseudomonas aeruginosa* and showed better decolorization under static as compared to shaking condition. Isolate displayed 96.82% decolorization in glucose and yeast extract (0.1%) supplemented Minimal Salt Medium at pH 7, temperature 30°C and an inoculum size of 10%. The strain was found to be tolerant to higher concentrations of dyes, different metal compounds and up to 6% NaCl concentration. UV-Visible spectroscopy indicated that decolorization was achieved by biodegradation. The non-toxic nature of degradation metabolites was revealed by phytotoxicity and cytotoxicity assays.

1 INTRODUCTION

Throughout the world wastewater generated by textile industries poses a great environmental threat. Discharge of colored effluents is objectionable not only because of esthetically unacceptable color but also due to breakdown products i.e. aromatic amines are toxic, mutagenic and carcinogenic (Aravind *et al.* 2010). Almost 45% of annually manufactured textile dyes are reactive dyes. Due to water soluble nature, these dyes are most problematic and challenging agents in textile effluents (Tunc *et al.* 2009).

Over the decades treatment of textile wastewater has emerged as one of the greatest challenges. Nowadays, biological methods for treatment of textile wastewater have provided a better alternative to physicochemical technologies (Khan *et al.* 2012). Bacteria are among the most frequently used microorganisms for the bioremediation of textile wastewater as they can rapidly grow and are usually easy to cultivate, adapt to extreme environmental conditions, and produce different types of oxidoreductases (Ola *et al.* 2010). The aim of present study was to evaluate the biodegradation of a textile dye RB 221 by native bacterial species and to assess the eco-toxicity of metabolites of degradation.

2 MATERIAL & METHODS

2.1 *Isolation of pure culture and bacterial identification*

The textile dye Reactive Blue 221 (RB 221) subjected to investigation was courteously provided by a local textile industry. The bacterial strain designated as BDS23 was isolated from the soil sample collected from waste disposal site of textile industry by enrichment technique in minimal salt medium (MSM) (Khehra *et al.* 2005) amended with 50 mg/L of RB 221 and identified according to Bergey's Manual (Holt *et al.* 1994).

2.2 Determination of decolorization potential of the isolate

Decolorization was checked by inoculating (10% v/v) overnight culture of bacterial isolate BDS 23 in liquid medium (MSM) supplemented with 50 ppm of RB 221 dye and incubated under shaking (150 rpm) as well as static conditions at 37°C for 8 days. Aliquots (3 mL) of culture media were withdrawn at regular intervals of 24 hrs. Decolorization was monitored by measuring absorbance of cell free supernatant at λ_{max} 615 nm by UV-Visible spectrophotometer (Agilent 8453) against media blank.

2.3 Effect of different environmental and culture conditions on dye decolorization

RB 221 decolorization was studied at different pH (5–9), temperature (30–45°C), NaCl concentration (0–10% w/v), inoculum size (2–10% v/v), carbon sources (glucose, lactose, sucrose, starch and mannitol), nitrogen sources including organic and inorganic sources (yeast extract, peptone, potassium nitrate, ammonium sulphate, sodium nitrate), concentrations of glucose (0–4 g/L), yeast extract (0–2 g/L), different initial RB 221 dye concentration i.e. 50–300 mg/L, NaCl (0–10% w/v), metal compounds including $CuCl_2$, $CoCl_2$, $CdCl_2$, $AgNO_3$ and $HgCl_2$, i.e. 10 mg/L.

2.4 Biodegradation analysis

UV-Visible absorption spectrum of the control and treated dye samples was studied in the range of 200–800 nm by using UV-Vis spectrophotometer (Agilent 8453).

2.5 Phytotoxicity & cytotoxicity assay

Phytotoxicity of the treated and untreated samples was determined by using *Raphanus sativus* (radish) seeds (Turker and Camper 2002) and Brine Shrimp (*Artemia salina*) Lethality Assay was performed for cytotoxicity analysis (Meyer *et al.* 1982).

3 RESULTS AND DISCUSSION

3.1 Isolation of pure cultures and bacterial identification

Over the years it has been observed that microorganisms indigenous to the contaminated environment are efficient in degrading recalcitrant xenobiotic compounds, as such places provide a major likelihood of presence of such type of microbes (Jothimani and Prabakaran, 2003). Bacterial strain BDS23 was found competent in decolorizing Reactive Blue 221 (RB 221) and exhibited more than 80% decolorization along with consistent growth under static conditions. Isolated bacterial strain was gram negative, short scattered rods and was identified as *Pseudomonas aeruginosa* BDS 23 on the basis of morphological and biochemical tests results. In several other studies, *Pseudomonas* strains are well documented for their dye decolorizing potential (Joe *et al.* 2011; Surwase *et al.* 2013). Under static (microaerophilic environment), *P. aeruginosa* BDS 23 exhibited better decolorization activity as compared to shaking, which indicate less oxygen demand for decolorization activity. Reason for lower decolorization under shaking conditions could be competition of oxygen and the azo compounds for the reduced electron carriers (NADH) and oxygen may dominate utilization of NADH; thus impeding the electron transfer from NADH to azo bonds (Kalme *et al.* 2006).

3.2 Effect of different environmental and culture conditions on dye decolorization

Dyes being deficient in easily utilizable carbon and nitrogen sources, their biodegradation without supplementation of carbon and nitrogen source are very difficult and slow. Therefore, the effect of various carbon sources (0.1 g/L) on decolorization of RB 221 was studied. When sucrose and

166

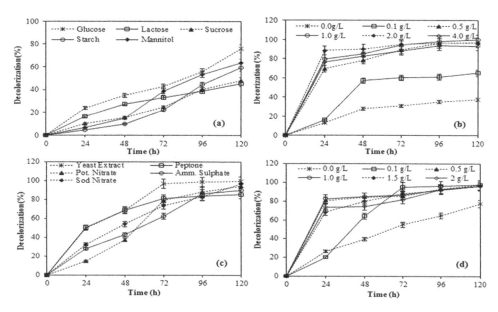

Figure 1. Optimization of nutritional parameters (a) different carbon sources; (b) glucose concentrations; (c) Nitrogen sources; (d) Yeast extract concentration.

lactose were used as carbon source decolorization was 48% and 45.1%, respectively (Figure 1a). Decolorization in presence of mannitol and starch was 59.3% and 63.4%, while glucose being the simplest carbohydrate was a preferable carbon source in decolorization by BDS23 and conferred better decolorization (76.1%) of RB 221 among all carbon sources. Likewise, for decolorization of Remazol Orange 3R by *P. aeruginosa* BCH (Surwase *et al.* 2013) and Direct Red 22 by *Bacillus cohnii* (Prasad and Rao 2013), glucose was found as the best carbon source. Results of effective glucose concentration evaluation depicted that with progressive increase in glucose concentration from 0.1 to 1 g/L, rate of decolorization also increased. Maximum decolorization (99.6%) was observed in the presence of 1 g/L of glucose concentration (Figure 1b). But, surprisingly upon further increase in glucose concentration upto 4 g/L, decolorization potential of the strain BDS 23 was reduced to 93.2% (Figure 1b). Parshetti *et al.* (2006) observed that *Kocuria rosea* MTCC 1532 failed to decolorize Malachite Green in the presence of 10 g/L glucose and at 30 g/L initial glucose concentration, only 31% decolorization was observed. Explaining such a phenomena of negative effect of glucose on decolorization, Chang *et al.* (2001) stated that it might be related to metabolic regulation known as glucose/catabolic repression.

Among different nitrogen sources including organic and inorganic sources tested, peptone exhibited 85.8% decolorization of the dye, whereas sodium and potassium nitrate decolorized 90.4% and 95.4% respectively. Likewise, in presence of $(NH_4)_2SO_4$, 96.9% of RB 221 was decolorized. In accordance with another studies carried out by Prasad and Rao (2013), yeast extract confer better decolorization efficiency (99.7%) in comparison to other nitrogen sources (Figure 1c). Organic nitrogen sources such as yeast extract are considered essential for regeneration of NADH which acts as electron donor in azo bond reduction, therefore involved in enhancing the rate of decolorization (Khehra *et al.* 2005; Asad *et al.* 2007; Moosvi *et al.* 2007). Besides the type, concentration of the nitrogen source is also crucial for the decolorization process. Increase in concentration of yeast extract (0.1–1 g/L) has no significant effect on decolorization rate (Figure 2d). Slight decrease in decolorization percent was observed with further increase in yeast extract concentration upto 2 g/L.

Incubation temperature is another very critical parameter and it was found that change in the temperature significantly affected the decolorization rate (Figure 2a). The optimum decolorization was

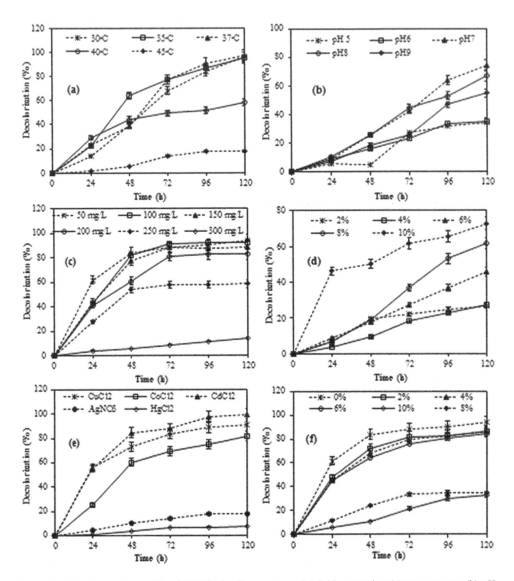

Figure 2. Decolorization profile of RB 221 by *P. aeruginosa* BDS 23 at varying (a) temperature; (b) pH; (c) dye concentration; (d) inoculum size; (e) metal compounds; (f) NaCl concentration.

recorded at 37°C (97.2%). It was noted that further increase or decrease in temperature adversely affects the rate of decolorization. These results are in accordance with previous report in which optimum temperature for decolorization of Reactive Brilliant Blue K-GR by *Shewanella decolorationis* S12 was 30–37°C (Asad *et al.* 2007). The pH plays a critical role for the optimal physiological performance of microbial cells. At pH values of 5.0 and 6.0 decolorization rate was less than 40%. Highest decolorization of 74.7% was observed at about pH 7. With further increase in pH, extent of decolorization was decreased gradually (Figure 2b). The variation in pH may affect the stability of enzyme systems involved in dye degradation, resulting in decreased performance in decolorization activity and also affect the viability of strain (Jadhav *et al.* 2011).

Among various concentrations of dyes tested, optimum decolorization was observed to be 93.6% % for 50 mg/L of dye concentration. As the concentration of dye RB 221 was increased from

100 to 300 mg/L, decolorization activity progressively decreased from 92.5 to 14.0% (Figure 2c). The results indicated that increase in dye concentration might be affecting overall growth and enzyme systems involved in decolorization of RB 221, ultimately resulting into the reduction of decolorization rate. Similar inverse relation in decolorization rate and dye concentration has been previously reported (Moosvi *et al.* 2007; Ayed *et al.* 2010; Jadhav *et al.* 2011). Decline in decolorization rate might be attributed to the poisonous effect of the dye on bacteria, inactivation of transport system of the dye, obstruction of active sites of azoreductase enzyme by complex dye structure (Stolz 2001). At an inoculum size of 2% only 26.5% decolorization was achieved. Upon further increase in the inoculum sizes (4–8%), a progressive increase in the decolorization ability (27.3–61.7%) was observed and maximum decolorization (72.8%) occurred at an inoculum size of 10% (Figure 2d). Textile effluents frequently contain different metal compounds along with different salts, which can interfere with microbial mediated decolorization process. Results of medium supplemented with different metal compounds showed that *P. aueroginosa* BDS23 displayed a variable response as it exhibit different decolorization percent with $CuCl_2$ (90.9%), $CdCl_2$, (99%), $CoCl_2$ (81.5%), $AgNO_3$ (17.8%) and $HgCl_2$ (7.3%) (Figure 2e). Along with unused dyes, textile effluent generally contains chloride salts of sodium and potassium. Therefore, salt tolerance in dye degrading bacteria is very critical. Results indicated that *P. aeruginosa* BDS 23 displayed more than 80% decolorization upto 6% NaCl concentration (Figure 2f). However, at 8 and 10% salt concentration, color removal percentage was reduced sharply below 35%. The decline in the decolorization performance at higher salt concentrations can be result of inhibition of microbial growth and activity at these concentrations due to plasmolysis (Peyton *et al.* 2002).

3.3 *Biodegradation analysis*

UV-visible spectral analysis (200–800 nm) of RB 221 showed single peak in visible region at 615 nm corresponding to its (λ_{max}) and two intense peaks in UV region near 250 and 300 nm respectively. Decolorized dye samples produced by *P. aeruginosa* BDS 23 displayed spectra in which the peak at 615 nm decreased without showing any shift in λ_{max} till complete decolorization, but the peak near 250 and 300 nm broadened up and became more intense and a less intense peak was also observed near 400 nm. This may have resulted due to accumulation of low molecular weight aromatic or aliphatic compounds that show absorbance in the these UV regions (Jain *et al.* 2012).

3.4 *Phytotoxicity & cytotoxicity assay*

The relative sensitivity towards the RB 221 and its degradation products in relation to *Raphanus sativus* seeds were studied and the 50% inhibition in seed germination of *R. sativus* in dye as compared to MSM (control) confirmed the toxicity of textile dye. But inhibition in germination of *R. sativus* was reduced to 30% in *P. aeruginosa* BDS 23 treated samples, indicating the reduction in toxicity of RB 221. Reduction in shoot and root length of dye sample was observed as compared to significant improvement of the same in the treated sample. The significant reduction in toxicity of dyes after microbial treatment is in accordance with previous reports (Ayed *et al.* 2010, Jadhav *et al.* 2011, Prasad and Rao 2013). In case of cytotoxicity assay, with control dye sample the percentage mortality of brine shrimp larvae (*Artemia salina*) was 40% but no cell death was observed with *P. aeruginosa* BDS 23 treated samples. Ayed *et al.* (2010) performed cytotoxicity test with *Artemia salina* nauplii and reported decrease in the toxicity of Methyl Red following treatment with *Sphingomonas paucimobilis*.

4 CONCLUSION

In the present scenario of environmental pollution, its restoration is a huge challenge. The present study results showed that *P. aeruginosa* BDS 23 has the potential to remediate the dye RB 221 effectively into nontoxic metabolites with minimal nutritional requirements. The results suggest that due to salt and metal tolerant ability, *P. aeruginosa* BDS 23 has potential for future application towards treatment of real textile wastewaters by using appropriate bioreactor.

REFERENCES

Aravind, U., George B., Baburaj, M., Thomas, S., Thomas, A. & Aravindakumar, C. 2010. Treatment of industrial effluents using polyelectrolyte membranes. *Desalination* 252: 27–32.

Asad, S., Amoozegar, M.A., Pourbabaee, A.A., Sarbolouki, M.N. & Dastgheib, S.M.M. 2007. Decolorization of textile azo dyes by newly isolated halophilic and halotoerant bacteria. *Bioresource Technology* 98: 2082–2088.

Ayed, L., Achour, S., Khelifi, E., Cheref, A. & Bakhrouf, A. 2010. Use of active consortia of constructed ternary bacterial cultures via mixture design for congo red decolorization enhancement. *Journal of Chemical Engineering* 162: 495–502.

Chang, J.S., Chou, C., Lin, Y.C., Lin, P.J., Ho, J.Y., Ho, T.L. & Lee Hu, T. 2001. Kinetic characteristic of bacterial azo-dye decolorization by *Pseudomonas luteola. Water Research.* 35: 2841–2850.

Holt, G. J., Sneath, P. H. & Krieg, N. R. 1994. Bergey's Manual of Determinative Bacteriology 9th edn. Baltimore, USA: Lippincott Williams and Wilkins.

Jadhav, U.U., Dawkar, V.V., Jadhav, M.U. & Govindwar, S.P. 2011. Decolorization of textile dyes using purified banana pulp polyphenol oxidase. *International Journal of Phytoremediation.* 13: 357–372.

Jain, K., Shah, V., Chapla, D. & Madamwar, D. 2012. Decolorization and degradation of azo dye-Reactive Violet 5R by an acclimatized indigenous bacterial mixed cultures-SB4 isolated from anthropogenic dye contaminated soil. *Journal of Hazardous Materials.* 213: 378–386.

Joe, J., Kothari, R.K., Raval, C.M., Kothari, C.R., Akbari, V.G. & Singh, S.P. 2011. Decolorization of Textile Dye Remazol Black B by *Pseudomonas aeruginosa* CR-25 Isolated from the Common Effluent Treatment Plant. *Journal of Bioremediation and Biodegradation.* 2: DOI: 10.4172/2155 6199.1000118.

Jothimani, P. & Prabakaran, J. 2003. Dye factory effluent decolorization by fungal cultures under static condition. *Journal of Ecobiology* 15: 255–260.

Kalme, S.D., Parshetti, G.K., Jadhav, S.U. & Govindwar, S.P. 2006. Biodegradation of benzidine based dye Direct Blue-6 by Pseudomonas desmolyticum NCIM 2112. *Bioresource Technology* 98: 1405–1410.

Khan, T.A., Dahiya, S. & Ali, I. 2012. Use of kaolinite as adsorbent: Equilibrium, dynamics and thermodynamic studies on the adsorption of Rhodamine B from aqueous solution. *Applied Clay Science* 69: 58–66.

Khehra, A.S., Saini, H.S., Sharma, D.K., Chadha, A.S. & Chimni, S.S. 2005. Decolorization of various azo dyes by bacterial consortium, Dyes and Pigments. 67: 55–61.

Meyer, B.N., Ferrigni, N.R., Putnam, J.E., Jacobsen, L.B., Nichols, D.J. & McLaughlin, J.L. 1982. Brine shrimp: a convenient general bioassay for active plant constituents. *Planta Medica* 45: 31–34.

Moosvi, S., Khera, X. & Madamwar, D. 2007. Isolation, characterization and decolorization of textile dyes by a mixed bacterial consortium JW-2. *Dyes and Pigments* 74: 723–729.

Ola, I., Akintokun, A., Akpan, I., Omomowo, I. & Areo, V. 2010. Aerobic decolourization of two reactive azo dyes under varying carbon and nitrogen source by Bacillus cereus. *African Journal of Biotechnology* 9.

Parshetti, G., Kalme, S., Saratale, G. & Govindwar, S. 2006. Biodegradation of Malachite Green by *Kocuria rosea* MTCC 1532. *Acta Chimica Slovenica* 53: 492–498.

Peyton, B.M. Wilson, T. & Yonge, D.R. 2002. Kinetics of phenol biodegradation in high salt s,olutions. *Water Research* 36: 4811–4820.

Prasad, A.A.S. & Rao, V.B.K. 2013. Aerobic biodegradation of Azo dye by *Bacillus cohnii* MTCC 3616; an obligately alkaliphilic bacterium and toxicity evaluation of metabolites by different bioassay systems. *Applied Microbiology and Biotechnology* 97: 7469–7481.

Stolz, A. 2001. Basic and applied aspects in the microbial degradation of azo dyes. *Applied Microbiology and Biotechnology.* 56: 69–80.

Surwase, V.S., Deshpande, K.K., Phugare, S.S. & Jadhav, P.J. 2013. Biotransformation studies of textile dye Remazol Orange 3R. *3 Biotechnology* 3: 267–275.

Tunc, Ö., Tanac, H. & Aksu, Z. 2009. Potential use of cotton plant wastes for the removal of Remazol Black B reactive dye. *Journal of hazardous materials* 163: 187–198.

Turker, A.U. & Camper, N.D. 2002. Biological activity of common mullein, a medicinal plant. *J. Ethnopharmacology*, 82: 117–125.

Torrefaction effects on composition and quality of biomass wastes pellets

C. Nobre, M. Gonçalves & B. Mendes
Mechanical Engineering and Resources Sustainability Center, Department of Sciences and Technology of Biomass, Faculty of Sciences and Technology, New University of Lisbon, Caparica, Portugal

C. Vilarinho & J. Teixeira
Mechanical Engineering and Resources Sustainability Center, Mechanical Engineering Department, University of Minho, Guimarães, Portugal

ABSTRACT: The torrefaction of pellets produced from different biomass wastes was performed at different temperatures and residence times to evaluate the influence of those parameters on the pellet quality. The fixed carbon, ash content and high heating value increased with the torrefaction temperature and time while the volatile matter content, apparent density and mechanical durability decreased. Using torrefaction temperatures lower than 250°C or adding a binder may allow to obtain pellets with good fuel properties and still preserve adequate mechanical properties.

1 INTRODUCTION

The conversion of biomass into solid biofuels such as pellets is a strategy developed to overcome questions such as low energy density, high moisture content and decentralized location of the raw biomass resources.

The consistent growth of the global market for wood pellets and the increasing awareness of the benefits brought by the use of biofuels, have triggered the need to diversify the raw materials used as feedstock. Pelletization is a flexible way to process biomass, compatible with the use of dedicated wood materials or biomass wastes such as fruit tree prunings (Arranz et al., 2015), food industry wastes (Ruiz Celma et al., 2012), wood industry wastes (Rabaçal et al., 2013), agricultural wastes (Miranda et al., 2011, Niedziółka et al., 2015) or forestry wastes (Acda M., 2014).

Diversification of raw materials for pellet production could enable the energetic valorisation of various biomass wastes available in significant amounts and with good fuel characteristics that are currently landfilled or burned in open air.

The material and energetic densification that results from pelletization, can be further improved by torrefaction, a thermochemical treatment also so known as mild pyrolysis, that uses low temperatures (between 200°C and 300°C), in an inert or oxygen-deficient atmosphere.

Torrefaction converts raw biomass in a solid homogeneous fuel, with low moisture content and a higher calorific value (Medic et al., 2012, Grigiante and Antolini, 2015, Wang et al., 2013).

During this process biomass loses water and part of its volatile matter, becoming dark, dry, friable and with a heating value closer to that of coal (Ghiasi et al., 2014).

The increased hydrophobicity after torrefaction is justified by the elimination of superficial hydroxyl groups that causes a decrease in the ability to establish hydrogen bonds and therefore interact with water molecules (Bergman and Kiel, 2005). Also, this process promotes the thermo-chemical formation of non-polar unsaturated structures, a property that can also contribute to a reduction of biodegradability (Patel et al., 2011).

Usually torrefaction is carried out as a pre-treatment directly on raw material and then the raw material is subjected to a densification process, such as pelletizing. The energy density of torrefied biomass pellets reaches $18\,GJ/m^3$, a value 20% higher than commercial pellets, although lower than the energy density of coal, typically higher than $20\,GJ/m^3$ (Uslu et al., 2008).

On the other hand, recent studies indicate that it is harder to compress torrefied biomass into pellets using regular densification conditions because the friction in the pellet mill increases, leading to higher temperatures in the die (Wang et al., 2013, Li et al., 2012, Peng et al., 2013). This situation can be controlled by pre-conditioning the torrefied biomass to increase its moisture and by increasing the mechanical force applied (Peng et al., 2015, Shang et al., 2012). Therefore the use of torrefied biomass for pellet production can increase the energy requirements of the process and reduce the lifetime of the pelletizing equipment while yielding pellets with poor mechanical properties that can be extensively degraded during transportation and handling.

An alternative to this methodology is to produce pellets and then subject them to the torrefaction process. Although a negative correlation was still observed between the torrefaction conditions and the pellet mechanical properties (Peng et al., 2015, Shang et al., 2012) the process of torrefaction after pelletization requires less energy than the pelletization of torrefied biomass and provides a similar final product (Ghiasi et al., 2014).

The aim of this work was to study the torrefaction of pellets produced from different biomass wastes and to determine the conditions that produce an improvement in the pellet fuel quality without causing a significant degradation of their mechanical properties.

2 MATERIAL AND METHODS

2.1 Pellet formulation and production

The industrial wood wastes pellets were produced with a mixture of end-of-life wood materials, furniture wastes and lignocellulosic materials from waste management units.

Pellets produced from orchard wastes contained prunings from *Malus domestica*, *Pyrus communis*, *Prunus persica* and *Prunus domestica*, provided by a fruit producer (COOPERFRUTAS).

Urban lignocellulosic wastes included aerial parts and branches from fifteen different species: *Arundo donax L., Sophora japonica, Tipuana speciosa, Prunus cerasifera var pissardii, Melia azedarach, Celtis australis, Pinus pinea, Fraxinus angustifolia, Grevillia robusta, Betula celtiberica, Platanus x hybrida, Olea europaea, Eucalyptus globulus, Causarina equisetifolia, Acer negundo*. These wastes were collected in cleaning operations of urban green spaces in Almada, Setúbal, Portugal.

The three types of pellet were produced in an industrial pellet production unit (Casal e Carreira Biomassa Lda, Alcobaça, Portugal) with a production capacity of 2.5 t/h. The raw materials were milled to a diameter lower than 4 mm and admitted to the press for densification. During pelletization the biomass was heated up to 90°C and extruded through die orifices with a diameter of 6 mm for the industrial wood wastes pellets and a diameter of 8 mm for the orchard and urban lignocellulosic wastes pellets.

2.2 Torrefaction experiments

The biomass pellets were subject to torrefaction at temperatures from 200°C to 250°C and residence times of 30, 60 and 120 minutes, using a muffle furnace (Nabertherm). The torrefied samples were placed in a desiccator and allowed to cool to room temperature.

2.3 Proximate analysis and higher heating value

Total moisture content and ash content were determined gravimetrically according to standards BS EN 14774-1:2009 and BS EN 14775:2009, respectively. The volatile matter content of each sample was determined according to the standard BS EN 15148: 2009 and the fixed carbon was calculated by difference. The higher heating value was determined according to the equation established by Parikh et al. (2005) (Equation 1).

$$HHV\ (MJ/kg) = 0.3536FC + 0.1559VM - 0.0078A \qquad (1)$$

where FC, VM and A correspond to fixed carbon, volatile matter and ash contents, respectively.

2.4 *Quality parameters*

The same pellets were subjected to torrefaction at 250°C for 60 minutes, in a larger scale (3.2 kg to 8.7 kg), using an industrial rotary pyrolysis furnace (MJ Amaral, model FR 100). The mechanical durability and content of fines of the torrefied pellets were determined following the standard CEN/TS 15210-1:2009 and their bulk density was determined through an adaptation of standard EN 15103:2009.

3 RESULTS AND DISCUSSION

The torrefaction conditions (temperature and time) had a clear effect in the pellet appearance, conferring a darker colour, a more uniform appearance and slightly lower dimensions (Figure 1).

It was observed that the pellet becomes darker for increasing residence time this process is faster with increasing temperatures, as stated by other authors (Unsal and Ayrilmis, 2005). These colour changes are related with the vaporization of the lighter biomass components and the decomposition and rearrangement of the heavier ones, namely cellulose and lignin (González-Peña and Hale, 2009).

The pellets also developed a rougher surface and became more brittle with increasing temperature and residence time. These features were more noticeable for torrefaction tests with longer residence times (120 min).

Volatile matter, fixed carbon and ash content and high heating value of the reference pellets and the corresponding torrefied pellets are presented in Figure 2.

The reference pellets contained 5–8% of residual moisture that was eliminated during torrefaction; since the torrefied pellets were always kept in a desiccator their moisture content was considered to be approximately zero. Even after re-equilibration with the atmospheric humidity, torrefied pellets typically regain only 1–6% of the initial moisture content (Bergman et al., 2005).

The pellets produced from urban biomass wastes presented a higher content in volatile matter and lower contents of ash and fixed carbon than the other two pellet types, but a similar high heating value.

Regardless of the raw materials composing each pellet, the torrefaction process resulted in a decrease of the volatile matter and increase of the fixed carbon and ash contents. These changes

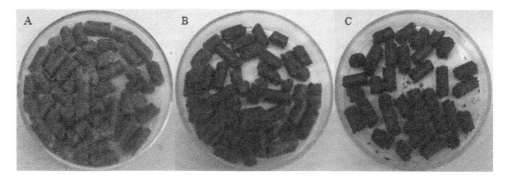

Figure 1.   Orchard wastes pellets: A – Reference; B – 200°C, 120 min; C – 250°C, 120 min.

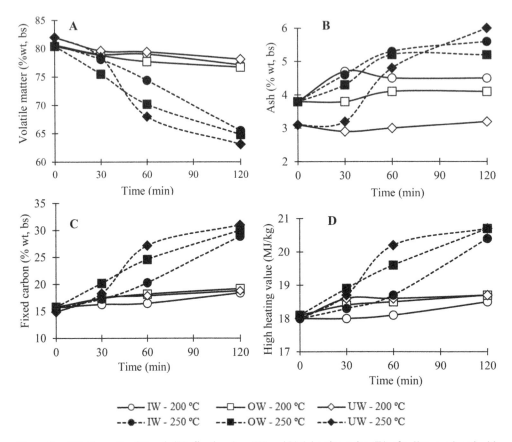

Figure 2. Volatile matter (A), ash (B), fixed carbon (C) and high heating value (D) of pellets produced with industrial wood wastes (IW), orchard wastes (OW) and urban lignocellulosic wastes (UW), after torrefaction at 200°C and 250°C and different residence times.

were proportional to the temperature and residence time and contributed to an increase in the pellet high heating value, as evaluated from the proximate analysis. The increase in the fixed carbon and the ash contents correspond to a concentration of those components in the torrefied pellets as more volatile components are eliminated and have opposite effects in the pellet heating value.

The volatile matter content decreased less than 5% (wt.) for torrefaction at 200°C, and around 25% (wt.), for torrefaction at 250°C (Fig. 2-A), indicating that at this condition, not only moisture and volatile organic compounds are lost but also an important fraction of structurally relevant components such as hemicellulose (Medic et al., 2012). The enhancement in fixed carbon is also more pronounced at 250°C, reaching values that are 15% (wt.) higher than the ones of non torrefied pellets (Fig. 2-C). The ash content of the torrefied pellets suffered almost no change at 200°C and increased of 1%–2.5% (wt.) at 250°C (Fig. 2-B).

In order to evaluate the effect of the different torrefaction conditions on the high heating value (HHV) of the studied pellets, the correlation proposed by Parikh et al. (2005) was used. The HHV of the reference pellets was improved at every torrefaction condition and the increase was directly proportional to the torrefaction temperature and residence time.

Urban lignocellulosic wastes pellets were the most affected by the torrefaction process. These pellets show a greater decrease in volatile matter (from 82.0% to 63.1%) and a more significant increase in fixed carbon (from 14.9% to 31.0%) and ash content (from 3.1% to 6.0%) at the most severe torrefaction conditions (Fig. 2).

Table 1. Quality parameters of the reference pellets and respective torrefied pellets.

Torrefaction conditions	Quality parameters	Industrial wood wastes pellets	Orchard wastes pellets	Urban lignocellulosic wastes pellets
No torrefaction (reference)	Mass, kg	7.8	8.7	3.7
	Bulk density, kg/m^3	673	638	605
	Mechanical durability, %	98.7	94.0	88.1
	Fine content, %	0.6	1.1	2.9
250°C, 60 min	Mass, kg	7.2	8.0	3.2
	Bulk density, kg/m^3	580	568	508
	Mechanical durability, %	93.1	92.6	82.7
	Fine content, %	10.1	10.1	21.9

The effect of the torrefaction process on the pellet mechanical properties was then evaluated at 250°C and 60 min, a condition in which the three types of pellets presented different proximate compositions. This test was also performed in a larger scale in order to obtain representative samples for the subsequent measurements (Table 1).

The registered mass loss of the pellets after torrefaction is between 7–14%. This can correspond to the loss of residual moisture and volatiles and the larger loss corresponds to the urban lignocellulosic pellets which presented the higher volatile content. At higher temperatures, the mass loss during torrefaction can be positively correlated with a loss of energy content (Shang et al., 2012). The bulk density also decrease after torrefaction, leaving the three pellet types below the normative value ($\geq$600 kg/m^3) and therefore these pellets will have a lower energy density, influencing costs for transportation and storage (Obernberger and Thek, 2004).

The torrefaction process also had a negative impact in the mechanical durability and fines content of the pellets, especially for the urban lignocellulosic wastes pellets that reached a fines content of 21.9%. The mechanical durability of the three types of torrefied pellets was bellow values the ENplus normative value for mechanical durability for domestic or industrial pellets ($\geq$97.5% and $\geq$96.5%, respectively). This indicates that some of the lost mass had a binding effect essential for the pellet integrity and milder torrefaction conditions should be used to preserve it. Another option to avoid degradation of the pellet mechanical properties is to add an appropriate binder before or after the torrefaction process.

4 CONCLUSIONS

The use of the torrefaction process after pelletization promotes changes in the chemical and quality properties of the pellets namely a uniform appearance, a significant increase in fixed carbon content and a consequent increase in their calorific value. These changes were more extensive at higher torrefaction temperatures and longer residence times. The variations in proximate composition and quality parameters between the different types of pellets are also influenced by the feedstock from which they were produced.

However, this treatment is negatively correlated with the mechanical properties of the pellets, decreasing their mechanical durability and their bulk density which can affect transport, storage and performance of the pellets in combustion systems. The use of torrefaction temperatures lower than 250°C and the addition of binders should be further investigated in order to retain the positive aspects of this thermal treatment regarding proximate composition and minimize the negative impact in the mechanical properties.

ACKNOWLEDGEMENTS

The authors gratefully acknowledge the support from project PROPELLET (Vale Inovação, Projeto n°37769) and the project promotor CMC Biomassas, Lda.

REFERENCES

Acda, M.N. & Devera, E.E. 2014. Physico-chemical properties of wood pellets from forest residues. *Journal of Tropical Forest Science*, 26, 589–595.

Arranz, J. I., Miranda, M. T., Montero, I., Sepúlveda, F. J. & Rojas, C. V. 2015. Characterization and combustion behaviour of commercial and experimental wood pellets in South West Europe. *Fuel*, 142, 199–207.

Bergman, P. & Kiel, J. 2005. Torrefaction for biomass upgrading. *14th European Biomass Conference & Exhibition*. Paris.

Ghiasi, B., Kumar, L., Furubayashi, T., Lim, C. J., Bi, X., Kim, C. S. & Sokhansanj, S. 2014. Densified biocoal from woodchips: Is it better to do torrefaction before or after densification? *Applied Energy*, 134, 133–142.

González-Peña, M. M. & Hale, M. D. C. 2009. Colour in thermally modified wood of beech, Norway spruce and Scots pine. Part 1: Colour evolution and colour changes. *Holzforschung*, 63.

Grigiante, M. & Antolini, D. 2015. Mass yield as guide parameter of the torrefaction process. An experimental study of the solid fuel properties referred to two types of biomass. *Fuel*, 153, 499–509.

Li, H., Liu, X., Legros, R., Bi, X. T., Jim Lim, C. & Sokhansanj, S. 2012. Pelletization of torrefied sawdust and properties of torrefied pellets. *Applied Energy*, 93, 680–685.

Medic, D., Darr, M., Shah, A., Potter, B. & Zimmerman, J. 2012. Effects of torrefaction process parameters on biomass feedstock upgrading. *Fuel*, 91, 147–154.

Miranda, M. T., Arranz, J. I., Román, S., Rojas, S., Montero, I., López, M. & Cruz, J. A. 2011. Characterization of grape pomace and pyrenean oak pellets. *Fuel Processing Technology*, 92, 278–283.

Niedziółka, I., Szpryngiel, M., Kachel-Jakubowska, M., Kraszkiewicz, A., Zawiślak, K., Sobczak, P. & Nadulski, R. 2015. Assessment of the energetic and mechanical properties of pellets produced from agricultural biomass. *Renewable Energy*, 76, 312–317.

Obernberger, I. & Thek, G. 2004. Physical characterisation and chemical composition of densified biomass fuels with regard to their combustion behaviour. *Biomass and Bioenergy*, 27, 653–669.

Parikh, J., Channiwala, S. & Ghosal, G. 2005. A correlation for calculating HHV from proximate analysis of solid fuels. *Fuel*, 84, 487–494.

Patel, B., Gami, B. & Bhimani, H. 2011. Improved fuel characteristics of cotton stalk, prosopis and sugarcane bagasse through torrefaction. *Energy for Sustainable Development*, 15, 372–375.

Peng, J., Wang, J., Bi, X. T., Lim, C. J., Sokhansanj, S., Peng, H. & Jia, D. 2015. Effects of thermal treatment on energy density and hardness of torrefied wood pellets. *Fuel Processing Technology*, 129, 168–173.

Peng, J. H., Bi, X. T., Sokhansanj, S. & Lim, C. J. 2013. Torrefaction and densification of different species of softwood residues. *Fuel*, 111, 411–421.

Rabaçal, M., Fernandes, U. & Costa, M. 2013. Combustion and emission characteristics of a domestic boiler fired with pellets of pine, industrial wood wastes and peach stones. *Renewable Energy*, 51, 220–226.

Ruiz Celma, A., Cuadros, F. & López-Rodríguez, F. 2012. Characterization of pellets from industrial tomato residues. *Food and Bioproducts Processing*, 90, 700–706.

Shang, L., Nielsen, N. P. K., Dahl, J., Stelte, W., Ahrenfeldt, J., Holm, J. K., Thomsen, T. & Henriksen, U. B. 2012. Quality effects caused by torrefaction of pellets made from Scots pine. *Fuel Processing Technology*, 101, 23–28.

Unsal, O. & Ayrilmis, N. 2005. Variations in compression strength and surface roughness of heat-treated Turkish river red gum (Eucalyptus camaldulensis) wood. *Journal of Wood Science*, 51, 405–409.

Uslu, A., Faaij, A. P. C. & Bergman, P. C. A. 2008. Pre-treatment technologies, and their effect on international bioenergy supply chain logistics. Techno-economic evaluation of torrefaction, fast pyrolysis and pelletisation. *Energy*, 33, 1206–1223.

Wang, C., Peng, J., Li, H., Bi, X. T., Legros, R., Lim, C. J. & Sokhansanj, S. 2013. Oxidative torrefaction of biomass residues and densification of torrefied sawdust to pellets. *Bioresour Technol*, 127, 318–325.

Wastes: Solutions, Treatments and Opportunities – Vilarinho, Castro & Russo (eds)
© 2015 Taylor & Francis Group, London, ISBN 978-1-138-02882-1

The role of automated sorting in the recovery of aluminium alloys waste

C.A. Nogueira, M.A. Trancoso, F. Pedrosa, A.T. Crujeira,
P.C. Oliveira & A.M. Gonçalves
Laboratório Nacional de Energia e Geologia – LNEG, Lisboa, Portugal

F. Margarido & R. Novais Santos
Instituto Superior Técnico, Universidade de Lisboa, IN+, Lisboa, Portugal

F. Durão & C. Guimarães
Instituto Superior Técnico, Universidade de Lisboa, CERENA, Lisboa, Portugal

ABSTRACT: A large number of aluminium alloys with varying alloying elements are present in vehicle structures and components, as well as in other household equipment. The recycling of these alloys is nowadays processed to low quality metal products due to high level of contamination, hindering the upgrading of recycling rates. The development and application of automated sorting technologies capable to detect, select and separate different alloy types could be of crucial importance in the progression of the recycling loop. This paper addresses the importance of sorting based on a study on the characterization of Al alloys in non-ferrous fraction of shredder plants.

1 INTRODUCTION

Light alloys, namely based of aluminium and, more recently, magnesium, have been progressively utilized in many equipments such as vehicles, aiming at reducing weight and fuel consumption. The driving forces for substituting classic alloys (namely those steel-based) for such new ones have been obvious economic but also environmental, trying to save resources and decreasing emissions. To answer to several specifications for a wide range of applications, numerous alloy types have been developed and applied in products. Such alloys have different properties (mechanical, chemical, electrical, etc.) conferred by the addition of alloying elements (e.g. Si, Mg, Mn, Cu, Zn) with different combinations and contents.

Due to the high variety of alloys, recycling of aluminium fractions produced in shredder plants is only suitable for manufacture of low purity Al alloys. Regarding most common wrought type alloys, typically with low Si content, the presence of cast alloy fragments, with high Si concentration, would be deleterious. So recycling of aluminium for producing wrought alloys becomes difficult. This is just one example of multiple cases where the production of specific Al alloys can be seriously affected by using recycled materials. Thus, the option is the use of substantial quantities of pure aluminium coming from primary resources or as much use recycled aluminium collected from very specific sources. Aluminium from recycling metal-bearing residues such as end-of-life vehicles, electric and electronic wastes or other household appliances is mostly destined to downcycling schemes.

To overcome these bottlenecks the development of efficient technologies for separating some relevant Al alloy groups would be highly welcome. In this domain, automated sorting devices that can detect alloying elements can play a significant role in boosting Al recycling industry. This paper discusses some aspects regarding the separation of the most important alloy series based

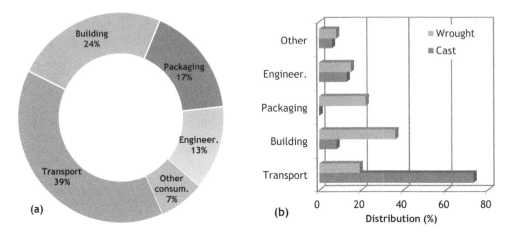

Figure 1. Aluminium end-use market distribution: (a) global; (b) by type of alloy (cast or wrought).

on analysis of samples taken from shredder plants, within activities of the project ShredderSort, aiming to contribute to the development is this important field.

2   THE ALUMINIUM RECYCLING LOOP

Aluminium is a metal that combines a series of properties like high electrical conductivity, strength, flexibility, corrosion resistance, lightweight and excellent recyclability, making it a unique material for a wide range of applications (European Aluminium Association 2013a). Figure 1a shows the distribution of aluminium markets by end-use application, illustrating as major uses the transport and building sectors, representing together more than 60% of the share. In average each European citizen uses about 20 kg Al yearly. Looking for the main applications of cast and wrought alloys (Fig. 1b) it is clear that cast alloys use in the transport market is predominantly, while wrought alloys use is much more distributed by different markets.

The automotive industry is clearly the predominant application since the aluminium alloys allow substantial reduction of weight, contributing to the decrease of fuel consumption and CO_2 emissions. The potential for weight savings depends strongly from the car component (European Aluminium Association, 2013b): for engine, suspension and hang-on parts, savings can be in the range 30–60%, while chassis, suspension and wheel rims can allow 10–50% savings and bumper devices in the range 30–50%. Today the average content of aluminium in European vehicles is about 140 kg. Annual fuel savings are estimated as 65 litres per vehicle as result of application of light alloys.

Aluminium production in Europe represents about 15–20% of world's total production, being near 50% the contribution of secondary sources. Although recycling of aluminium in Europe is continuously growing, Europe is a net exporter of scrap due to several reasons namely the high energy prices. In terms of energy efficiency, aluminium recycling is strongly advisable since provides savings in energy near 95% when compared with extraction from primary sources.

The recycling loop of aluminium is illustrated in Figure 2, showing that aluminium is a circular economy. The efficiency of the loop depends on many key players, starting from the end-of-life scrap collectors and dismantlers. They usually make a preliminary separation of aluminium types, contributing to generate specific scrap that can be adequately transformed to new products. For complex residues, shredder companies make their role by separating non-ferrous fractions, rich in

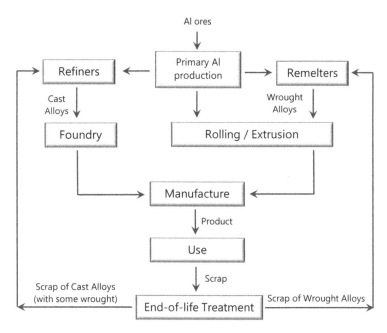

Figure 2. Simplified flowchart of the aluminium recycling loop.

aluminium, using technologies such as magnetic separation, eddy-current separation and gravity separation. Improving quality of aluminium fractions can be attained by hand sorting or, sometimes, by color sorting (Green 2007; Gaustad et al. 2012). However, separation of aluminium by alloy groups is not achievable by these processes.

Two important key players follow in the loop of the recycling industry: the refiners and the remelters. Both treat, compose and melt several types of scrap: the first ones supply the foundries with cast alloys while the second ones supply the rolling mills and extrusion plants with wrought alloys. The raw materials used by refiners and remelters are wide; besides Al fractions generated in end-of-life treatment (called "old scrap"), several wastes and scrap of manufacturing processes (designated as "home scrap") are also common feed materials. Examples are: skimmings and dross (from primary melters); billet ends of extrusion, edge trimmings from rolling; millings, turnings and borings from machining; off-cuts from stamping; and rejected goods and materials ("new scrap") from all process phases. Since remelters produce more pure wrought alloys (thus more sensitive to contaminants) they use clean scrap from a known origin or even primary aluminium. Refiners are less exigent in raw-materials. Both refiners and remelters comprise nearly similar operations in their processes, including an initial treatment, the alloy compiling, the melting and the casting. After melting the refiners have usually a refining step while remelters finish their processes with a heat treatment operation (European Aluminium Association 2013a).

The aluminium can be considered as "used" and not "consumed", meaning that can be recycled almost infinitely maintaining its quality. If aluminium alloys are properly sorted, e.g. by separating wrought and cast alloys, and eventually some alloy groups within each one, the cycle could be completed and the scrap can be used for almost all applications. It can be also noticed that by recycling aluminium, the other elements constituting the alloys are also conserved. For about 4 million tons of aluminium recycled yearly in Europe, about 200 thousand tons of alloying metals are also maintained in the consumption circuit. Development of high-performance sorting technologies is therefore essential.

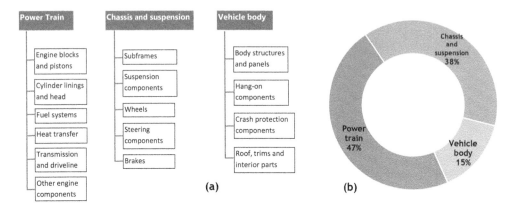

Figure 3. Aluminium application in vehicle components: (a) list of the main parts of automotive construction: (b) distribution (wt%) of application of Al alloys by the main parts of vehicles.

3 METHODOLOGY AND EXPERIMENTAL PROCEDURES

The research work here reported was focused in the non-ferrous fraction of auto-shredder plants, the main providers of aluminium-rich fractions produced in the treatment of end-of-life products. The work involved firstly a study on the utilization of aluminium alloys in vehicles based on available literature and other information gathered from specialists. In order to evaluate the composition of real fractions of non-ferrous wastes, samples (about 30 kg each) from five companies were collected and characterized in the laboratory in order to identify the main Al alloy series present. The experimental methodology involved the use of several techniques, namely visual inspection, determination of density, semi-quantitative analysis by X-ray fluorescence and quantitative chemical analysis by atomic absorption spectrometry (the later one after appropriate digestions of samples using standardized methods).

4 RESULTS AND DISCUSSION

4.1 *Aluminium alloys in vehicles*

Although being less than 10% (on average) in weight, aluminium application in vehicles is progressively increasing. Both casting and wrought alloys are used in car components, but generally it can be said that castings have still nowadays higher utilization when considering the volume/weight, but not necessarily when considering the area of utilization. Castings are essentially used in engine blocks, cylinder heads and some chassis components while wrought alloys (stampings, extrudates and forgings) are mainly used in car body components (European Aluminium Association 2013b,c; Musfirah & Jaharah 2012). The gradual substitution of iron-made engine parts by aluminium is still in progress. Figure 3 shows the main parts of a car where light alloys can be used. As can be seen, virtually, almost all important vehicle metallic components can be made in Al alloys. That is the case of power train parts, the chassis and suspension systems and the vehicle body.

Aluminium alloys for engine blocks must meet a series of requirements such as excellent castability, good machinability, moderate strength at elevated temperatures and low cost. Examples are the cast alloys of 3xx series (e.g. A380 and A319), some of them also used in cylinder heads. Heat exchangers have been for many years a successful application of Al wrought alloys in cars.

Regarding chassis, casting technologies are gaining importance in construction of structural components such as axles and subframes. Such components are made by several casting technologies

180

Table 1. Main Al alloys identified in samples, respective distribution and composition range of alloying elements.

Alloy series		Distribution (%)	Composition range (%)
Wrought	1xxx	13	(Al > 99%)
	3xxx	8	0.4–1% Mn
	4xxx	5	3–6% Si
	5xxx	12	1–6% Mg; 0–1.1% Mn
	6xxx	19	0.2–2% Mg; 0.3–2% Si
	7xxx	3	5–6% Zn
Cast	1xx	2	(Al > 99%)
	2xx	3	2–4% Cu
	3xx	30	9–15% Si; 0.5–3% Cu
	4xx	5	10–15% Si

(e.g. 3xx series, mainly A356), but also using wrought alloys (such as sheet stampings and extrusions made by 5xxx and 6xxx series). Suspension parts are also important components for Al alloys application, examples being forgings and extrusions made by 6xxx series. Knuckles are made by casting due to its complex geometry. Wheels made in light alloys are getting important, being produced by casting (3xx series, e.g. A356) or by forging (6xxx series, e.g. 6062). The steering system, connecting the driver to the vehicles, can be also produced by casting or extrusion.

The vehicle body has high possibilities of progress regarding the application of light alloys, being nowadays still marginal. Panel components such as doors, roof, boot lids and hood are possible applications of sheet stampings (5xxx and 6xxx Al series) while the skeleton space frames are excellent examples for use of extrusions of 6xxx series.

4.2 *Characterization of aluminium alloys in non-ferrous fraction of shredders*

The samples collected in five auto-shredder plants were analysed in order to identify the most common Al alloys present. Firstly the other non-ferrous alloys (copper, zinc, stainless steel) were identified and separated by visual inspection and using results of density determination and/or XRF analysis. Aluminium represents nearly 85–95% of samples collected. The subsequent chemical analysis allowed identifying and quantifying the main alloying elements and consequently the respective alloy series. Table 1 summarizes the results attained. Wrought alloys are predominantly but when considering each series *per si*, the main alloys detected were those of the 3xx series (a cast series) with a contribution of about 30% of the samples analysed. 6xxx and 5xxx series were also relevant (with 19 and 12% respectively). All these series are important in automotive construction. Regarding other alloy types found, series 1xxx and 3xxx have also contributions to be considered, being these wrought alloys mainly utilized in several general purpose utensils and beverage cans. This result was expected since auto-shredders also process other wastes than end-of-life vehicles.

4.3 *Emerging separation technologies for sorting aluminium alloys*

The results obtained in the analysis of Al alloys in non-ferrous auto-shredder fractions demonstrated that the discrimination of alloys by chemical content can allow obtaining different streams. Silicon is a clear example of this possibility, since many alloys have different Si content and so the application of a sorting device can eventually separate them. Most cast alloys have high Si content (e.g. series 3xx and 4xx) and therefore separation of alloys with a Si threshold of 5–10% can produce a high Si stream with 35–40% of feed material, the remaining stream containing 60–65% of material having low Si content.

Figure 4 illustrates the application of LIBS technology in a sorting device, as pursued in the Shreddersort project. The LIBS spectra presented in the Figure shows good sensitivity for most

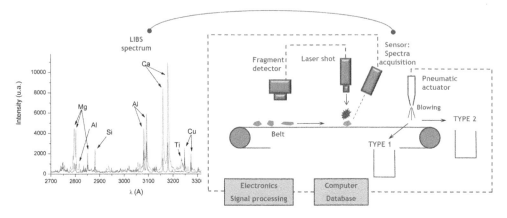

Figure 4. Schematic view of a sensor-detector sorting technology with LIBS.

peaks, including silicon, proving that a good discrimination can be attained for this element. It is also clear that peaks from other elements like magnesium can eventually be used for separating other alloy groups, such as 5xxx and 6xxx series from the other wrought alloys. Therefore the development of automated devices based on LIBS seems to be a very promising approach to separate Al alloys.

5 CONCLUSIONS

Sorting aluminium alloys contained in auto-shredder non-ferrous fractions is essential to improve the efficiency and the rates of recycling of this important metal. The separation of fragments from different alloy groups based on chemical composition of alloying elements seems to be the best approach. Silicon concentration can be an important target since this element content allows differentiating most important cast and wrought alloys, namely those of 3xx and 4xx series from those of 5xxx, 6xxx, 3xxx and 1xxx. The application of automated sorting devices based on LIBS technology seems to be adequate to attain such objective.

ACKNOWLEDGMENTS

This work was financed by the European Commission through FP7 (collaborative projects), grant agreement Nr. 603676, project ShredderSort.

REFERENCES

Gaustad, G., Olivetti, E. & Kirchain, R. 2012. Improving aluminum recycling: a survey of sorting and impurity removal technologies. *Resources, Conservation and Recycling* 58: 79–87.
Green, J.A.S. 2007. *Aluminium recycling and processing for energy conservation and sustainability*. Ohio: ASM International.
Musfirah, A.H. & Jaharah, A.G. 2012. Magnesium and aluminum alloys in automotive industry. *J. Applied Sciences Research* 8: 4865–4875.
European Aluminium Association 2013a. *Sustainable Development Indicators for the Aluminium Industry in Europe – 2012 Key facts and Figures*. Brussels: EAA.
European Aluminium Association 2013b. *Aluminium in Cars – Unlocking the Light-Weighting Potential*. Brussels: EAA.
European Aluminium Association 2013c. *The Aluminium Automotive Manual*. Brussels: EAA.

Wastes: Solutions, Treatments and Opportunities – Vilarinho, Castro & Russo (eds)
© 2015 Taylor & Francis Group, London, ISBN 978-1-138-02882-1

Mining tailing reuse in sulfobelitic clinker formulations

J. Nouairi & M. Medhioub
Department of Geology, Faculty of Sciences of Sfax, Sfax-Tunisia

L. Senff
Center of Mobility Engineering, Federal University of Santa Catarina (UFSC), Joinville, SC, Brazil

W. Hajjaji & F. Rocha
Geobiotec, Geosciences Department, University of Aveiro, Aveiro, Portugal

R. Novais, L. Beruberri & J.A. Labrincha
Ceramics and Glass Engineering Department & CICECO, University of Aveiro, Aveiro, Portugal

ABSTRACT: To neutralize the negative impact of contaminated mining discharges in northern Tunisia, sulfobelitic cements were produced using Zn-Pb mining tailings from Fej Lahdoum (FH) and Lakhouat (KH). As ore deposits for zinc and lead, these byproducts are mainly composed of calcite, phyllosilicates, dolomite, quartz, galena and gypsum. After sintering in presence of commercial $CaCO_3$ and alumina, the obtained clinkers showed compositions ($C_4A_3\check{S}$, C_2S and C_4AF) close to those predicted by the phase diagram. Concerning hydration temperatures, FH and KH based cements displayed similar behavior. Slight differences were observed in the reaction speed due to the variations in phase composition and probably in the particle size distribution. However, better workability is observed when compared to previously studied sulfobelitic products. With high amounts of $C_4A_3\check{S}$, the sulfobelitic cements show suitable mechanical behavior for earlier ages.

1 INTRODUCTION

Thousands of tons of waste containing important levels of Pb, Zn, Cd occupying large dumping areas without taking any special precaution regarding their harmful effects (Concas et al. 2006; Kouadio et al., 2007; Rodriguez et al., 2009). In a sedimentary carbonated context, as happens in Tunisia, the main risk of mining activities is the dispersion of tailings particles by prevalent winds and tailing erosion and leaching of hazardous elements during heavy raining seasons (Boussen et al., 2010; Chakroun et al., 2010; Destrigneville et al., 2010). In northern Tunisia, the main deposits of Zn-Pb ores are located in this triassic zones: Fej Lahdoum (Pb-Zn; 81.000 ton Pb and 150.000 ton Zn) and Lakhouat (Pb) (ONM, 2010; Babbou-Abdelmalek et al., 2011). Both have well established infrastructures and have been operating until quite recently (Mssedi, 2007; ONM, 2010). Nowadays, their mine tailings are problematic due to the high concentration of lead and zinc sulfides. Recent studies (Chakroun et al., 2010; Babbou-Abdelmalek et al., 2011) detected high levels of Pb and Zn and other heavy metals in these abandoned mine sites. For example, the concentration of Fej Lahdoum tailings varies from 5900 to 32 300 mg·kg^{-1} total Zn, while total Pb contents change between 800 and 6200 mg·kg^{-1} (Babbou-Abdelmalek et al., 2011).

Immobilization of these heavy metals is primordial to prevent health contamination effects, but studies on the reuse of such by-products are scarce. Some of the recent papers published by the authors of this investigation had been focused on the recycling of distinct mining by-products, such as red mud, tionite and phosphate sludge. Their incorporation in geopolymers (Hajjaji et al., 2013), ceramic pigments (Hajjaji et al., 2012) and lightweight aggregates (FakhFakh et al. 2007; Loutou et al., 2013), was tested. In particular, the preparation of clinker and cement is a promising

Figure 1. Mine tailings of Fej Lahdoum (FH, left) and Lakhouat (KH, right) regions.

way to recycle high volumes of wastes. The synthesis of sulfobelitic clinker/cement using red mud is a good example (Senff et al., 2011). Compared to OPC, the sulfobelitic clinkers include smaller amount of carbonate, and then CO_2 emissions upon firing of the raw mill are obviously lower. Furthermore, the maximum temperature and the energy required for their synthesis are also lower than in OPC (1400 vs. 1450°C) (Senff et al., 2011). Following that idea, the actual contribution describes the synthesis of sulfobelite clinkers incorporating the Zn/Pb mine tailings. While targeting the desirable composition and properties for the clinker/cement, the firing process should also assure the inertization of hazardous species, namely the heavy metals.

2 EXPERIMENTAL DETAILS

Two representative samples were collected from the mines of Fej Lahdoum (FH) and Lakhouat (KH) (Siliana district, located just 120 km southern of the capital Tunis) (Fig. 1).

The wastes were mixed with commercial raw materials, such as calcite (Calcitec M1) and alumina (Alcoa CT3000), in proportions adjusted to generate the desired phases of the clinker (C_2S, C_3A and C_4AF) as shown in Table 1. The chosen formulations were conditioned by the chemical composition of the by-products (especially, the proportion of SO3). In a previous work Sneff et al. (2011), the best results were obtained from this theoretical formulation; 60% $C_4A_3\check{S}$, 20% C_2S and 10% C_4AF. The mixtures were milled for 1 h in a dry mixer and then calcined in an electric furnace. The clinker cycle used was as follows: (a) heating to 1000°C at 15°C/min, (b) 1000°C for 45 min. to ensure complete decarbonation, (c) heating to 1350°C at 5°C/min, (d) 1 hour at 1350°C, (e) rapid cooling by air quenching.

The microstructure was studied by scanning electronic microscopy SEM and energy dispersive spectroscopy (Hitachi SU 70 coupled with an EDAX Bruker AXS detector).

Clinkers were characterized through X-ray powder diffraction (XRD), conducted in a Rigaku Geigerflex diffractometer with a Cu Ka radiation source in 10–80°2θ, 2 h scan, and scan rate of 0.02, 2 h, 4 s per step. The quantitative phase analysis was performed using GSAS-EXPGUI software following a Reference Intensity Ratio (RIR) and the Rietveld refinement techniques. Up to 40 independent variables were refined: scale-factors, zero-point, 15 coefficients of the shifted Chebyschev function to fit the background, unit cell dimensions, profile coefficients (1 Gaussian, GW, and 2 Lorentzian terms, LX and LY). The agreement indices, as defined in GSAS, for the final least-squares cycles of all refinements are represented by Rp (%), Rwp (%), v2 and R(F2) (%). In addition, SO_3 amounts in the clinkers were estimated based on oxides conversion from Rietveld quantification. Also, the particle size distribution was assessed by a Laser Coulter LS230.

Derived cements were obtained by adding 5 wt % calcium sulfate (gypsum) to the clinker. The heat upon curing was evaluated as a tool to estimate the speed of the hydration process. The temperature was monitored immediately after mixing in a quasi-adiabatic calorimeter (Langavant) using two test samples.

Table 1. Formulations and compositions of studied clinkers ($C_2S = 2CaO \cdot SiO2$; $C_4A_3\check{S} = Ca_4Al_6(SO_4)O_{12}$; $C_4AF = 4CaO \cdot Al_2O_3 \cdot Fe_2O_3$).

	CKH	CFH
Theoretical composition		
C_2S	30.0	30.0
$C_4A_3\check{S}$	60.0	60.0
C_4AF	10.0	10.0
Batch formulations		
$CaCO_3$	40	40
Al_2O_3	29	28.5
KH	70	0
FH	0	50
Rietveld quantification		
C_2S	34	32.9
$C_4A_3\check{S}$	64.4	63.2
C_4AF	1.6	3.9
Rafinement indices		
Rwp (%)	0.2269	0.2297
Rp (%)	0.1511	0.1531
χ^2	7.137	8.185
R(F2)	0.2212	0.2868

Finally, mortars were produced by combining the obtained cements with commercial sand (average particle of 0.6 mm). Mortars were prepared through dry mixing of raw materials in a bag for 1 min, followed by mixing with water during 1 min. The binder/aggregate and water/binder ratio (in weight) were 1:2 and 0.5, respectively. After demoulding, mortars were cured in a climate chamber (66% relative humidity and 22°C) up to 7, 14 and 21 days. For the compressive strength tests, cylindrical samples of mortars with 30×50 mm (diameter and height) were produced. The speed rate applied in the compressive strength test was 1 mm/min.

3 RESULTS AND DISCUSSION

3.1 *Mining tailing characterization*

The XRD of FH wastes showed the existence of calcite (30%), phyllosilicates (30%), dolomite (24%), quartz (5%), galena and gypsum. Concerning KH discharges, the mineralogical composition is broadly similar to FH samples; the presence of calcite, dolomite, quartz and gypsum is also noticed in the SEM analysis (Fig. 2). In fact, Northern Tunisia and especially the watershed Medjerda was the most important mining region of the country for the exploitation of base metals located in the old deposits (Sainfeld, 1952). The XRF measurements showed the following compositions; for FH: 15.48% SiO_2, 5.91% Fe_2O_3, 6.65% Al_2O_3, 21.92% SO_3 and 19.87% CaO; for KH: 23.44% SiO_2, 4.1% Fe_2O_3, 10.11% Al_2O_3, 9.12% SO_3 and 27.64% CaO. The accurate examination of these tailings showed high PTE (Potentially Toxic Elements) values; up to 42700, 15700, 24300 and 23700 mg/kg for Zn, Pb, Ba and Sr, for FH samples; while KH samples show 15400, 4600, 24300 and 23700 mg/kg for Zn, Pb, Ba and Sr, respectively. This difference in concentration between the two mines is a function of the mineralogical composition of each district, abundance of one element over another, the metal value during the period of exploitation and different processing techniques.

3.2 *Clinker/Cement phases*

From the diffractograms (Fig. 3), it is possible to identify the presence of the targeted crystalline phases: trialuminate sulfate ($C_4A_3\check{S}$), belite (C_2S), and ferrite (C_4AF). Such phases are present in

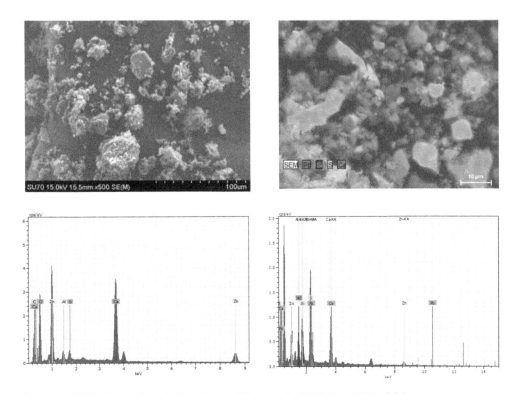

Figure 2.   SEM micrographs and elements quantifications for FH (left) and KH (right) wastes.

distinct relative amounts, depending on the formulations. Their quantification was conducted by Rietveld refinement. The obtained compositions are close to those predicted.

The SEM micrographs (also in Fig. 3) show the surface topography of samples cured for 7 days. As observed, the formation of needle-shape (A) and columnar (B) ettringite was among the hydration products of sulfobelitic cement. Moreover, a dense vitreous-like phase, forming a matrix, is associated to calcium silicate hydrates. A solid framework is formed by these linked hydration products.

3.3   *Temperature of hydration*

Figure 4 shows the evolution of the temperature during the hydration of cements. As noted, both cements showed a rapid reaction period, 4 hours for CKH and 6 hours for CFH after initial mixing. In general, and especially in the case of sulfobelitic cements, a rapid evolution of the temperature in the first moments of the process is usually observed (Senff et al., 2011), due to the reaction of water with gypsum and also to the hydration of aluminates. In contact with water, a strong chemical reaction is developed with the release of ions (Ca^{2+}, SO_4^{2-}, Na^+, K^+ and OH^-) to the solution. The first peak is attributed to gypsum and it is practically the same for both products. Further, the formation of hydrated phases is slower in case of CFH based cements, denoting a better workability. This could be due to the slight difference in phase composition between both clinkers. Moreover, these reactions are strongly influenced by the fineness of clinker/cement particles (CKH and CFH mean sizes are 20 and 16 μm, respectively) and might here assume a crucial and distinctive role between the tested samples (Cohen and Richards, 1982).

3.4   *Compressive strength*

Differences in the mechanical behavior were exhibited after 14 and 28 days of curing as is noted in Table 2. The differences are noticeable and are, in general, related to compositional variations

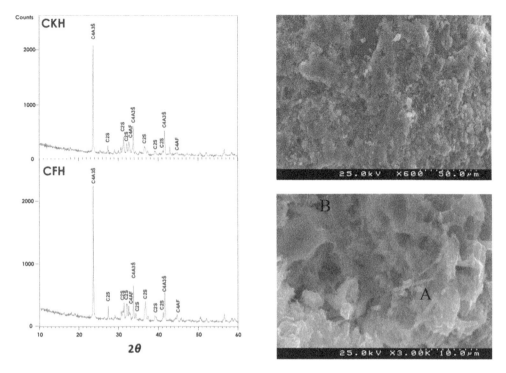

Figure 3. XRD patterns and SEM micrographs for the obtained clinker and cements, respectively. needle-shape (A) and columnar (B) ettringite.

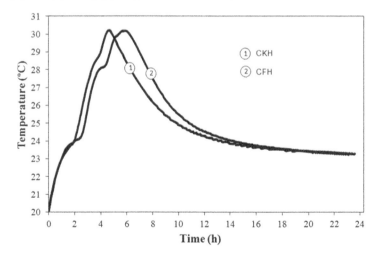

Figure 4. Hydration temperatures of CFH and CKH.

of the processed clinkers, in particular concerning the concentration of the major cement phases. The compressive strength of CFH decreases due to the excess of sulfur oxide. Moreover, as for Portland cements, the higher amount of C_4AF (3.9%) induced slower hydration and could lead to an increase in the expansion of sulfobelitic cements.

After 28 days of curing, the best performance of the compressive strength was obtained for samples with higher amounts of $C_4A_3\check{S}$ (CKH). Nevertheless, the slight difference in the amounts of this phase between both cements could suggests that superior CaO and less sulfur content are affecting positively the compressive strength of CKH.

Table 2. The compressive strength of mortars with 14 and 28 days of curing.

	14 days max stress (MPa)	28 days max stress (MPa)
CFH	12.27	11.51
CKH	8.46	13.56

4 CONCLUSIONS

The results proved that the reuse mining wastes in sulfobelitic clinkers is feasible. The obtained clinkers generate cements with distinct but tailored properties. X-Ray diffraction confirms the formation of expected crystalline phases and their relative amounts, determined by Rietveld refinement estimations, are close to the theoretical predictions from equilibria phase diagram.

The studied samples showed a slight difference in hydratation temperatures due to the particle size and the composition variations. The same factor affected the compressive strength of mortars cured at 14 and 28 days. Also, excess of sulfur oxides could lead to negative evolution of mechanical strength of the CFH studied clinker. With higher amounts of CaO in initial formulation and a slight increase of $C_4A_3\check{S}$, the CKH cement showed a better mechanical behavior.

REFERENCES

Boussen, S., Sebei, A., Soubrand-Colin, M., Bril, H.F., Chaabani, A. & Abdeljaouad, S. 2010. Mobilization of lead-zinc rich particles from mine tailings in northern Tunisia by aeolian and run-off processes. *Bull. Soc. Géol. Fr.*, 181(5): 371–379.

Chakroun, H.K., Souissi, F., Bouchardon, J.L., Souissi, R., Moutte, J., Faure, O., Remon, E. & Abdeljaoued, S. 2010. Transfer and accumulation of lead, zinc, cadmium and copper in plants growing in abandoned miningdistrict area. *Afr. J. Environ. Sci. Technol.*, 4(10): 651–659.

Cohen, M.D. & Richards, C.W. 1982. Effects of the particle sizes of expansive clinker on strength-expansion characteristics of type K expansive cements. *Cement and Concrete Research*, 12(6): 717–725.

Concas, A., Ardau, C., Cristini, A., Zuddas, P. & Cao, G. 2006. Mobility of heavy metals from tailings to stream waters in a mining activity contaminated site. *Chemosphere*, 63(2): 244–253.

Fakhfakh, E., Hajjaji, W., Medhioub, M., Rocha, F., López-Galindo, A., Setti, M., Kooli, F., Zargouni, F. & Jamoussi, F. 2007. Effects of sand addition on production of Lightweight Aggregates from Tunisian smectite-rich clayey rocks. *Clay Science*, Volume 35, Issues 3–4, pp. 228–237.

Hajjaji, W., Andrejkovicová, S., Zanelli, C., Alshaaer, M., Dondi, M., Labrincha, J.A. & Rocha, F. 2013. Composition and technological properties of geopolymers based on metakaolin and red mud. Materials and Design. *Materials and Design*, Volume 52 , pp. 648–654.

Hajjaji, W., Costa, G., Zanelli, C., Ribeiro, M.J., Seabra, M.P., Dondi, M. & Labrincha, J.A. 2012. An overview of using solid wastes for pigment industry. *Journal of the European Ceramic Society*, Volume 32, pp. 753–764.

Loutou, M., Hajjaji, M., Mansori, M., Favotto, C. & Hakkou, R. 2013. Phosphate sludge: thermal transformation and use as lightweight aggregate material. *Journal of Environmental Management*, (130) 354–360.

Office National des Mines.2010. Internet address: http://www.onm.nat.tn

Rodriguez, L., Ruiz, E., Alonso-Azcarate, J. & Rincon, J. 2009. Heavy metal distribution and chemical speciation in tailings and soils around a Pb–Zn mine in Spain. *J. Environ. Manage.*, 90(2): 1106–1111.

Sainfeld, P. 1952. Annales des mines et de la géologie numéro 9 les Gîtes plombozincifères de la Tunisie. *Imprimerie S.E.F.A.N.* Tunis. 252 p.

Senff, L., Castela, A., Hajjaji, W., Hotza, D. & Labrincha, J.A.2011. Formulations of sulfobelite cement through design of experiments. *Construction and Building Materials* 25, pp. 3410–3416.

Wastes: Solutions, Treatments and Opportunities – Vilarinho, Castro & Russo (eds)
© 2015 Taylor & Francis Group, London, ISBN 978-1-138-02882-1

Fractionation of oily sludges produced in the treatment of hydrocarbon wastes

A.P. Oliveira, M. Gonçalves, C. Nobre & B. Mendes
Mechanical Engineering and Resources Sustainability Center, Department of Science and Technology of the Biomass, Faculty of Science and Technology, New University of Lisbon, Caparica, Portugal

M. Vilarinho & F. Castro
Mechanical Engineering and Resources Sustainability Center, Mechanical Engineering Department, University of Minho, Guimarães, Portugal

ABSTRACT: Oily sludges are hazardous wastes that contain large amounts of hydrocarbons, water and dissolved solids. A simple distillation procedure was used to fractionate this sludge and isolate a gas phase (10%), and an organic liquid phase (30%) that can be used as fuels. This procedure also separates the water phase from the dissolved solids enabling the independent management of those phases with more appropriated methods.

1 INTRODUCTION

The cleaning operations of oil tanks and ships or the recycling of used mineral oils produce large amounts of oily sludges that can be classified as hazardous wastes. The oily sludge is a complex emulsion of hydrocarbons, water and mineral components, classified with code 05 01 (Sludge and Solid Residues with hydrocarbon components) in the List of Waste (LoW) (EU 2000) and Annex III to Directive 2008/98/ EC on waste (EU 2008), and take place in the list of hazardous waste, which have special and specific manipulation many of which are toxic, so there is an increasing interest in the development of new technologies for remediation or stabilization of this sludge (Mazlova 1999), (Hu et al. 2013).

In Portugal these sludges have been, so far, co-incinerated or landfilled. Nevertheless, their high contents in polyaromatic hydrocarbons and in some cases heavy metals result in high environmental pressures associated with these disposal methods. The aim of this work is to explore the oil sludge fractionation by distillation as a methodology for the isolation of homogeneous fractions that can be further remediated or valorized by different technologies. The components of this sludge are toxic, and when landfilled, the soil will be contaminated and it is a dangerous situation for the humanity and the environment.

The contaminated soils are a huge preoccupation for the governments worldwide. This concern is justified because of the polycyclic aromatics hydrocarbons (PAH), an oil contaminant (Reddy et al. 2011).

According to INE (National Statistical Institute) and APA (Environmental Portuguese Agency), Portugal in 2012, have produced ca 29 809 ton of used mineral oil residues, from which 23 110 ton were recovered. These numbers are according to the yields of oil recover from sludge that are around 80%. Nevertheless to recover the sludge are also part of the recover process (sub product) (Jean et al. 1999), (API 2010).

The Portuguese law allows that some enterprises can treat this kind of sludge, but there are limits for the incorporation in cement industry and also for the amount to landfill. So there is always the question: what to do with the remaining wastes?

The physical and chemical properties of the sludge may vary with the nature of the oils, the operating parameters of the recycling process and the time and conditions of storage (Jean et al. 1999).

Nevertheless there is still a fraction of the hydrocarbon components that are retained in this sludge and that have a potential for energetic valorization.

In this work oil sludges from pre-treatment units were fractionated by distillation to yield homogeneous fractions that were characterized in order to evaluate possible valorization processes.

2 EXPERIMENTAL

2.1 *Samples*

The raw oily sludge used is this work was supplied by a local waste management industry (Carmona-Sociedade de Limpeza e Tratamento de Combustíveis, S.A.). The sludge was stored in large plastic bottles, and kept in the dark, at room temperature, until analysis.

2.2 *Sludge distillation*

Sludge subsamples (∼200g) were subject to simple distillation in an apparatus equipped with a vacuum receiver adapter that enabled the collection of gas products in a Tedlar bag. The distillate was collected from 98°C to 130°C and was composed by an aqueous phase and an organic immiscible phase although some degree of emulsification occurs between these phases.

The product yield was evaluated for all liquid products and for the solid residue and expressed in wt. %. The gas yield was evaluated by the weight difference between the initial crude sludge and the liquid and solid products from distillation.

2.3 *Product characterization*

The oily sludge, the distilled products and the solid residue were characterized by determination of their proximate analysis and elemental analysis.

The proximate analysis was performed according to standards BS EN 14474-2: 2009, BS EN 15148: 2009 and BS EN 14775: 2009. Fixed carbon was evaluated by difference.

The elemental analysis (N, C, H and S) was performed in duplicate, using an Elemental Analyzer (Thermo Finnigan-EC Instruments, Flash template and CHNS 112 series). Oxygen was evaluated by difference.

The gas phase composition (H_2, N_2, O_2, CH_4, CO and CO_2) was determined using GC-TCD (Thermo Electron Corporation) equipped with a Supelco-25467, Carboxen®-1010 PLOT capillary column.

The molecular weight distribution of the liquid products from the oily phase was evaluated by GC-FID (Trace GC, Thermounicam) and their qualitative composition was studied using GC-MS (Focus GC-Polaris Q, Thermounicam). The main hydrocarbon components were identified by co-injection of the corresponding standards. The aqueous phase organic components were sequentially extracted with petroleum ether and ethyl acetate; both extracts were dried and analyzed by GC-MS.

3 RESULTS AND DISCUSSION

3.1 *Characterization of the oily sludges*

The characteristics of the crude oily sludge (proximate analysis) were evaluated in three subsamples collected after homogenization of the sludge and are presented in Table 1.

The sludge included in this study presents high contents of moisture and volatile matter when compared with similar wastes studied by other authors (Table 1), but a separate aqueous phase is not visible.

Table 1. Proximate analysis of oily sludge.

Moisture (%, a.r.)	Volatile matter (%, d.b.)	Fixed carbon (%, d.b.)	Ash (%, d.b.)	Reference
84.8 ± 1.8	86.8 ± 0.1	0.1 ± 0.8	13.1 ± 0.2	This work
27.7 ± 6.3	–	–	11.7 ± 1.0	(Al-Futaisi et al. 2007)
16.95	44.73	3.28	51.99	(Zhou et al. 2009)
29.5	–	–	50.1	(Li et al. 1995)
32.64	73.98	9.35	16.67	(Qin et al. 2015)
29.26	–	–	–	(Jing et al. 2011)
59.86	88.09	3.30	8.61	(Xu et al. 2014)
26.3	31.7	–	31.6	(Karayildirim et al. 2006)
24.0	–	–	–	(Jasmine & Mukherji 2015)
–	–	–	35.0	(Conesa et al. 2014)

Table 2. Ultimate analysis of dried oily sludge.

Carbon (%, d.b.)	Hydrogen (%, d.b.)	Nitrogen (%, d.b.)	Sulfur (%, d.b.)	Oxygen (%, d.b.)	Reference
44.9 ± 1	6.4 ± 0.0	1.04 ± 0.0	1.7 ± 0.0	45,9	This work
58.0 ± 2.4	9.8 ± 2.4	0.1 ± 0.0	2.1 ± 0.1	30.0 ± 2.6	(Al-Futaisi et al. 2007)
20.85	2.7	1.4	0.11	6.0	(Zhou et al. 2009)
64.1	10.6	–	–	4.0	(Li et al. 1995)
72.72	5.20	4.05	2.07	14.96	(Qin et al. 2015)
33.16	7.18	0.45	0.68		(Xu et al. 2014)
16.0	–	0.3	2.3	13.0	(Jasmine & Mukherji 2015)
51.2	7.54	0.52	1.69	4.05	(Conesa et al. 2014)

The ash content although moderate is also a factor that limits some management/remediation solutions and contributes for the hazardous nature of those sludges.

The ultimate composition of the dried sludge (C, N, H, S, O) was evaluated and results are presented in Table 2.

The dried sludge has a high carbon content (44.9 wt. %) but also a high oxygen content that indicates the presence of a high concentration of oxygenated organic compounds. The fractionation of the crude sludge in different fractions with different composition and properties is therefore an interesting approach to valorize some of its components and reduce the volume of landfilled waste.

3.2 Distillation Yields

The fractionation of the oily sludge by simple distillation yielded the following products (n = 4): gas components (8.5 ± 5.4 wt %), an oily phase (28.0 ± 3.1 wt %) and an aqueous phase (26.4 ± 4.2 wt %). An additional liquid fraction was collected in the form of an oil:water emulsion (18.3 ± 8.7 wt %), that could not be separated. The solid residue left in the distillation flask corresponded to around 18.8 (wt %) of the original sludge (Fig. 1). The gas and liquid products were collected at temperatures from 40°C up to 130°C, and correspond to around 81.2 wt. % of the initial mass.

The distillation yields showed a high variability due to the heterogeneous nature of this sludge especially in what concerns the water content of the sample.

The aqueous phase which is the effluent to be treated corresponds to 37.4% wt of the original mass thus allowing a significant reduction of the waste volume. A considerable amount of solid components (~20% w/w) is isolated in this process reducing the soluble solids and ash content of the liquid phases.

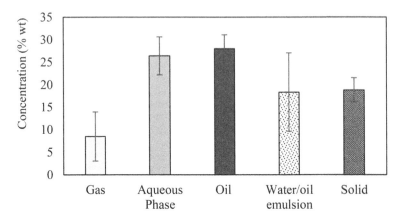

Figure 1.   Product yield (% w/w) for the distillation of the oily sludge.

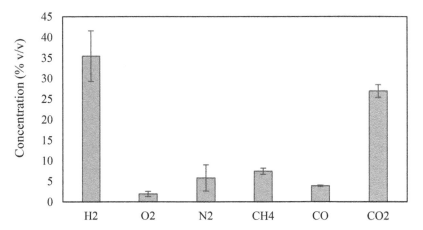

Figure 2.   Gas products collected during the distillation of the oily sludge.

3.3   *Product characterization*

The gas phase shows high concentrations of hydrogen and some methane that together account for 34.5% vol; the permanent gases included in this analysis correspond to around 50% vol of the gas phase and the remaining components are probably low molecular weight hydrocarbons that will certainly contribute to the gas heating value (Fig. 2).

Further analysis of hydrocarbon gases from the gas phase can eventually reinforce the conclusion that the distillation gas products can be used as a gaseous fuel to ensure the energetic requirements of the process. The oil phase was analyzed by GC-FID to evaluate its carbon number distribution and the boiling point range of their components (Fig. 3).

The bio-oil chromatogram was compared to the chromatogram of a hydrocarbon standard mix from C8 to C20 to determine the carbon number distribution. The main components of the oily phase are volatile or semi-volatile organic compounds (~89.8 wt%) with carbon numbers from C5 to C20 to which correspond boiling points from 36°C to 343°C. The lighter components of this phase with boiling points lower than 250°C account for 67.2% of the oil phase.

The analysis of the oil phase by gas chromatography and mass spectrometry provided some structural information concerning its components and confirmed that the oily phase is rich in saturated hydrocarbons, from C8 to C25 (Fig. 4), and also contains alkenes, cyclic hydrocarbons and aromatic hydrocarbons. The analysis of the organic components present in the water phase

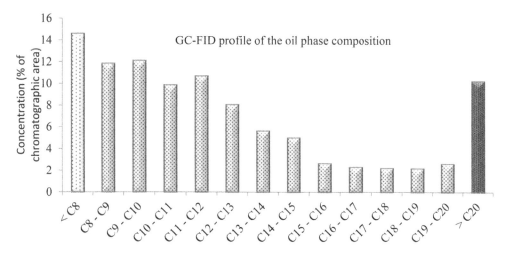

Figure 3.   GC-FID of the oil phase composition.

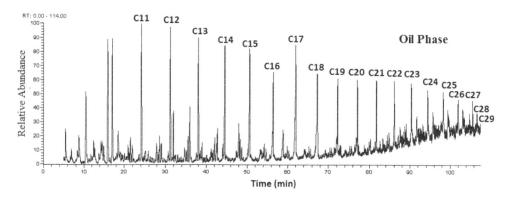

Figure 4.   Hydrocarbon distribution of the oily products collected during distillation.

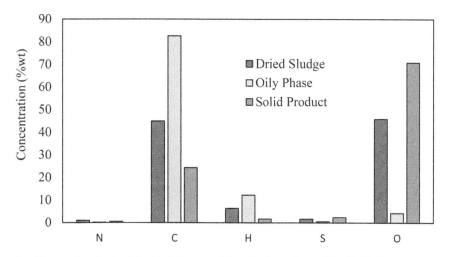

Figure 5.   Elemental analysis of the dried sludge and the oily phase obtained by distillation.

revealed high concentrations of aromatic hydrocarbons and oxygenated compounds both aliphatic and aromatic, confirming that this phase constitutes a highly contaminated effluent that must be subject to a remediation process adequate to its hazardous nature.

The oily phase and the solid residue were also subject to elemental analysis in order to evaluate the fractionation of the different elements between dried sludges and distillation products.

The oil product is a carbon-rich liquid with a low content of oxygen, nitrogen or sulfur that could be reintegrated in the refinery raw materials or used as raw fuel for industrial boilers.

The dried sludge and the solid product have high oxygen content indicating the presence of polar functional groups. The solid product still represents a significant fraction from the total sludge so its material valorization as adsorbent or raw material should be exploited (Fig. 5).

REFERENCES

Al-Futaisi, A. et al., 2007. Assessment of alternative management techniques of tank bottom petroleum sludge in Oman. *Journal of hazardous materials*, 141(3), pp. 557–564.

API, 2010. *Category Assessment Document for Reclaimed Petroleum Hydrocarbons: Residual Hydrocarbon Wastes from Petroleum Refinning*, Washington DC.

Conesa, J.A. et al., 2014. Study of the thermal decomposition of petrochemical sludge in a pilot plant reactor. *Journal of Analytical and Applied Pyrolysis*, 107, pp. 101–106.

Hu, G., Li, J. & Zeng, G., 2013. Recent development in the treatment of oily sludge from petroleum industry: a review. *Journal of hazardous materials*, 261, pp. 470–90.

Jasmine, J. & Mukherji, S., 2015. Characterization of oily sludge from a refinery and biodegradability assessment using various hydrocarbon degrading strains and reconstituted consortia. *Journal of Environmental Management*, 149, pp. 118–125.

Jean, D.S., Lee, D.J. & Wu, J.C.S., 1999. Separation of oil from oily sludge by freezing and thawing. *Water Research*, 33(7), pp. 1756–1759.

Jing, G., Chen, T. & Luan, M., 2011. Studying oily sludge treatment by thermo chemistry. *Arabian Journal of Chemistry*, pp. 7–10.

Karayildirim, T. et al., 2006. Characterisation of products from pyrolysis of waste sludges. *Fuel*, 85(10–11), pp. 1498–1508.

Li, C.-T. et al., 1995. PAH emission from the incineration of waste oily sludge and PE plastic mixtures. *Science of The Total Environment*, 170(3), pp. 171–183.

Mazlova, E.A.M., 1999. Ecological characteristics of oil sluges., pp. 49–53.

Qin, L. et al., 2015. Recovery of energy and iron from oily sludge pyrolysis in a fluidized bed reactor. *Journal of Environmental Management*, 154, pp. 177–182.

Reddy, M.V. et al., 2011. Aerobic remediation of petroleum sludge through soil supplementation: microbial community analysis. *Journal of hazardous materials*, 197, pp. 80–7.

Xu, M. et al., 2014. The resource utilization of oily sludge by co-gasification with coal. *Fuel*, 126, pp. 55–61.

Zhou, L., Jiang, X. & Liu, J., 2009. Characteristics of oily sludge combustion in circulating fluidized beds. *Journal of Hazardous Materials*, 170(1), pp. 175–179.

Wastes: Solutions, Treatments and Opportunities – Vilarinho, Castro & Russo (eds)
© *2015 Taylor & Francis Group, London, ISBN 978-1-138-02882-1*

Aspergillus ibericus lipase production by solid-state fermentation of olive pomace

F. Oliveira
CEB-Centre of Biological Engineering, University of Minho, Braga, Portugal
Department of Chemical Engineering, Faculty of Sciences, University of Vigo, Ourense, Spain

L. Abrunhosa, A. Venâncio & I. Belo
CEB-Centre of Biological Engineering, University of Minho, Braga, Portugal

N. Pérez-Rodríguez & J.M. Domínguez
Department of Chemical Engineering, Faculty of Sciences, University of Vigo, Ourense, Spain

ABSTRACT: Lipases are versatile catalysts with many applications, such as in food and detergents industries. Lipases can be produced by solid-state fermentation (SSF) using agro-industrial wastes. In addition, *A. ibericus* has been pointed as an interesting fungus to produce lipase through SSF of olive pomace (OP). The aim of this work was to optimize the production of lipase by *A. ibericus* under SSF using OP and wheat bran (WB). Additionally, extraction conditions of lipase were optimized. At optimum conditions lipase production and recovery improved 2-fold, yielding 223 ± 5 U/g. Optimum SSF conditions were using 30 g in a ratio of 1:1 (dry basis) of OP and WB supplemented with 0.4 g $(NH_4)_2SO_4$ at 60% of moisture content and incubated at 30°C during 7 days. The extraction of lipase was improved using 7.5 mL/g solid residues of 1% Triton X-100 homogenized for 0.5 hour at 250 rpm and 24°C.

1 INTRODUCTION

Solid-state fermentation (SSF) has been used for the production of value-added products using agro-industrial wastes (Salihu et al. 2012). SSF is a fermentation technique which involves the culture of microorganism on moist solid supports (Pandey 2003). It is the preferred choice for growing filamentous fungi since it simulates their natural habitat.

Olive pomace (OP) is a sludgy waste generated by the olive oil two-phase extraction system. It is an acidic and very humid material, rich in organic matter, potassium, nitrogen and fats (Alburquerque et al. 2006), which can be valorized for the production of lipases while being treated.

Lipases (triacylglycerol acylhydrolases EC 3.1.1.3) are a class of hydrolases which catalyze the hydrolysis of triglycerides to glycerol and free fatty acids over oil water interface (Treichel et al. 2009). The interest in the production of lipases is associated with their applications as additives in food, fine chemicals, detergents, waste water treatment, cosmetics, pharmaceuticals, leather processing and biomedical assays (Salihu et al. 2012). In addition, lipases also have important application in the field of bioenergy, especially in biodiesel production. Microorganisms including bacteria, fungi and yeast that produce extracellular lipases are recognized as the preferred sources of these enzymes because their recovery from the culture broth is facilitated (Salihu et al. 2012).

Aspergillus species within the section *Nigri* are important in many biotechnological processes. Species such as *A. niger* have a GRAS status from the FDA and have been widely used by the bioindustry. *Aspergillus ibericus* is a new black *Aspergillus* species that was isolated from wine grapes (Serra et al. 2006), and is able to produce lipase on OP and wheat bran (WB) through SSF (Oliveira et al. 2013).

Thus, the aim of this work was to optimize the production of lipase by *Aspergillus ibericus* MUM 03.49 under SSF. The variables investigated were the source and concentration of phosphorus, nitrogen and carbon (mixtures of OP and WB). Finally, at optimum conditions, a profile of lipase production over time and productivity was performed. Additionally, lipase extraction conditions were optimized.

2 MATERIALS AND METHODS

2.1 *Microorganism and residues*

Aspergillus ibericus MUM 03.49 (MUM culture collection, Braga, Portugal) was used. The fungus was grown on malt extract agar (MEA) plates (2% (w/v) malt extract, 2% (w/v) glucose, 0.1% (w/v) peptone and 2% (w/v) agar) at 30°C for 7 days and stored at 4°C. Spore suspensions of the inoculum were prepared by adding peptone solution (0.1% (w/v) peptone and 0.001% (w/v) Tween 80) to plates cultures, and after agitation were transferred to a falcon. The spore concentration of the suspension was adjusted to 10^7 spores/mL. OP samples were collected from a two-phase olive mill plant in Vila Real, Portugal, during the 2012/2013 campaign, and stored at −20°C. OP presented 69% ± 1% of moisture content (wet basis). WB was purchased in a local supermarket.

2.2 *Optimization of residues composition*

SSFs were performed in cotton-plugged 500 mL Erlenmeyer flasks containing 30 g dry solid residues in a ratio of 1:1 (w/w, dry basis) of wet OP and WB. $NaNO_3$ (0.6 g/30 g) was added to supplement the residues (Oliveira et al. 2013). The mixture of OP with WB resulted in optimum moisture content between 57% and 60%, without the need for its adjustment.

Flasks were prepared, autoclaved at 121°C for 20 min, cooled, inoculated with 1 mL of inoculum suspension and incubated at 30°C during 7 days. After the incubation period the fermented residues were extracted has described below to obtain the enzymatic extracts.

Different sets of experiments were conducted to evaluate the effect of phosphorus, nitrogen and residues composition in the production of lipase by *A. ibericus* under SSF.

The phosphorus source studied was KH_2PO_4 and concentrations tested were 0, 0.15, 0.3 and 0.6 g/30 g. The nitrogen sources tested were $NaNO_3$, urea, NH_4Cl and $(NH_4)_2SO_4$ and concentration tested were 0, 0.3 and 0.6 g/30 g. $(NH_4)_2SO_4$ was further evaluated at concentrations that ranged from 0 to 1 g/30 g because it was the nitrogen source with the most pronounced effect.

To evaluate de effect of carbon sources, different ratios of dried and ground OP:WB were tested (1:0, 4:1, 3:2, 1:1, 2:3, 1:4 and 0:1, w/w). Experiments containing only WB supplemented with different percentages of olive oil (0, 1, 2.5, 5 and 10%) were also performed. Moisture content was adjusted to 60%.

After the optimization of SSF conditions, a set of experiments was performed to obtain a profile of lipase production and productivity over fermentation time. Flasks were prepared as described before and destructively sampled each 2 days over a period of 20 days.

2.3 *Extraction and lipase determination*

The fermented residues were homogenized with 5 mL of 1% Triton X-100 per g of dried residues at 170 rpm and 24°C for 2 h, using a shaker. Homogenates were then centrifuged (3000 rpm and 10 min at 4°C) and filtered. Lipase activity was determined by a spectrophotometric method, using *p*-nitrophenyl butyrate in potassium phosphate 50 mM at pH 7.0 and 37°C for 15 min. The absorbance was measured at 405 nm. One unit of lipase activity (U) was expressed as the amount of enzyme which produces 1 μmol of *p*-nitrophenol per minute, under the assay conditions. Lipase activity obtained was expressed as units per gram of dry solid residue (U/g).

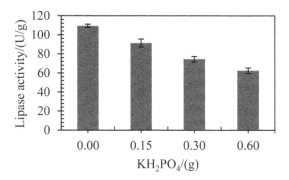

Figure 1. Effect of phosphorus source on lipase activity. Depicted values are the mean of triplicate analysis ± standard deviation.

2.4 *Optimization of lipase extraction conditions*

SSF were performed using the optimum conditions achieved. Initial conditions of extraction were 150 mL of extracting solvent, 2 h of shaking at 170 rpm and 24°C. Starting from this point, different variables were studied in order to improve lipase extraction. They were: type of solvent, volume of solvent, time of extraction, stirring rate and extraction temperature. Each of those variables was studied separately and following to the previous order. The best condition determined for each variable was used to optimize the following one by changing initial conditions. At the end of the lipase extraction optimization, consecutive extractions were performed in a same fermented residue in order to determine the extraction efficiency. The procedure was repeated 5 times, obtaining 5 extracts from the same fermented residue.

2.5 *Analysis of experimental data*

The data obtained were statistically analysed using SPSS (IBM SPSS Statistics, Version 22.0. Armonk, NY: IBM Corp.) in order to study the effect of variables on lipase production. Data were tested for homogeneity, submitted to one-way analysis of variance (ANOVA) and a pair-wise multiple comparison procedure (Tukey test) at a confidence level of 95%.

3 RESULTS AND DISCUSSION

3.1 *Optimization of residues composition*

In many cases, an additional source of phosphorus is important to induce an overproduction of some metabolite (Papagianni 2004). For example, Pérez-Rodríguez et al. (2014) found a positive effect of KH_2PO_4 on xylanase production by *A. niger* through SSF of corncob. In this work, the phosphorus source used (KH_2PO_4) produced a significant negative effect ($p < 0.0001$) on lipase production (Fig. 1). So, the mixture OP:WB seemed to supply enough phosphorus being unnecessary its supplementation.

Contrariwise, the nitrogen sources had a significant positive effect ($p < 0.001$) on lipase production (Table 1). All of them increased significantly the production of lipase when added to the OP:WB mixture, but the highest lipase activity was obtained with $(NH_4)_2SO_4$ at a concentration of 0.6 g/30 g of residues. Therefore, $(NH_4)_2SO_4$ was chosen to proceed with the following studies.

Figure 2 presents lipase yields of experiments conducted with different concentrations of $(NH_4)_2SO_4$. A significant effect ($p < 0.0001$) on lipase production was observed with increasing amounts of $(NH_4)_2SO_4$ until the maximum lipase activity of 138 ± 3 U/g was reached with a concentration of 0.4 g/30 g. This concentration was used in further SSFs.

Table 1. Lipase activity affected by the amount of different nitrogen sources. Values are the mean of triplicate analysis ± standard deviation. Means with the same letter do not differ significantly at $p > 0.05$.

Amount/(g)	Lipase activity ± standard deviation/(U/g)			
	$NaNO_3$	Urea	NH_4Cl	$(NH_4)_2SO_4$
0	89 ± 5			
0.3	111 ± 4[a]	128 ± 4	137 ± 3[b]	134 ± 5
0.6	115 ± 6[a]	106 ± 7	144 ± 5[b]	151 ± 7

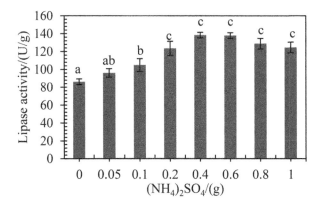

Figure 2. Profile of lipase activity as a function of the amount of $(NH_4)_2SO_4$. Depicted values are the mean of triplicate analysis ± standard deviation. Means with the same letter do not differ significantly at $p > 0.05$.

Lipase activity obtained for each tested OP:WB mixture is depicted in Figure 3a. The most favourable mixture for the production of lipases was the 1:1 ratio, yielding 144 ± 5 U/g of lipase activity. This yield corresponds to a 1.7-fold increase if compared to the SSF that contained only WB (0:1 ratio). It was also observed that *A. ibericus* could grow in all runs. However, it did not produce lipases in runs containing only OP or with low amount of WB (4:1 ratio). Mixed solid residues have been considered attractive for the growth of microorganisms on SSF, since they may act differently as support matrix, nutrient source and as inducers for the production of enzymes (Edwinoliver et al. 2010).

The effect of olive oil supplementation on WB was also studied to compare its effect with OP. It was observed that olive oil had a significant positive effect ($p < 0.0001$) on the production of lipases. An addition of 10% improved the production, reaching 152 ± 3 U/g (Fig. 3b). This activity was similar to one obtained with the OP:WB ratio of 1:1. Thus, the lipid content of OP can be a factor that is influencing most the production of lipases in experiments conducted with OP:WB mixtures. OP used in this work contained 10% of lipids, as determined by Oliveira et al. (2013). In this work, an OP:WB ratio of 1:1 was used in following studies.

Figure 4 presents results of lipase activity and its productivity over fermentation time. An increase of lipase production over time was observed, reaching 166 ± 5 U/g and 209 ± 10 U/g after 10 and 20 days of fermentation, respectively. However, the maximum productivity was obtained on the 6th day (1.8 ± 0.1 U/gh) with a lipase production of 127 ± 6 U/g.

3.2 *Optimization of lipase extraction conditions*

The effect of type of solvent, volume of solvent, time of extraction, stirring rate and extraction temperature on lipase recovery from fermented residues are presented in Table 2. It was found a

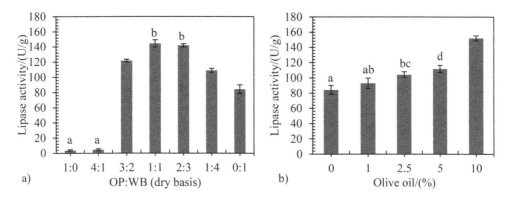

a)

b)

Figure 3.   Results of lipase activity a) from SSF using different ratios of olive pomace (OP) with wheat bran (WB) and b) from SSF using WB with different concentrations of olive oil. Depicted values of lipase activity are the mean of triplicate analysis ± standard deviation. Means with the same letter do not differ significantly at $p > 0.05$.

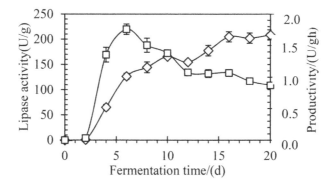

Figure 4.   Profiles of lipase activity (—◇—) and its productivity (—□—) over fermentation time. Depicted values are the mean of triplicate analysis ± standard deviation.

Table 2.   Conditions of extraction and respective lipase activity (LA). Values are the mean of triplicate analysis ± standard deviation (SD). Means with the same letter do not differ significantly at $p > 0.05$.

Solvent/ (5 mL/g)	LA ± SD/ (U/g)	Volume/ (mL/g)	LA ± SD/ (U/g)	Time/ (h)	LA ± SD/ (U/g)	Stirring/ rate/(rpm)	LA ± SD/ (U/g)	Temp./ (°C)	LA ± SD / (U/g)
Distilled water	17 ± 1^a	2.5	77 ± 3	0	127 ± 8	0	155 ± 5	24	218 ± 9^a
Phosphate buffer	23 ± 1^{ab}	5	156 ± 3	0.25	203 ± 6^{ab}	100	185 ± 8	26	217 ± 3^a
1% NaCl	5 ± 2	7.5	230 ± 6	0.75	217 ± 4^a	150	208 ± 7^a	30	133 ± 7^b
1%Tween 80	30 ± 3^b	10	198 ± 6^a	1	196 ± 4^b	200	205 ± 4^a	35	131 ± 4^b
1% Triton +		15	194 ± 7^a	2	189 ± 5^b	250	222 ± 7^a	40	176 ± 6
1% NaCl	153 ± 6^c	–	–	–	–	–	–	–	–
0.5% Triton	147 ± 9^c	–	–	–	–	–	–	–	–
1% Triton	175 ± 5^d	–	–	–	–	–	–	–	–
2% Triton	161 ± 4^d	–	–	–	–	–	–	–	–

Table 3. Results of lipase activity (LA) and respective extraction recovery (ER) from consecutive extractions of the fermented residue. Values are the mean of triplicate analysis ± standard deviation (SD).

	Number of consecutive extractions				
	1st	2nd	3rd	4th	5th
LA ± SD/(U/g)	223 ± 5	65 ± 2	3.5 ± 0.2	0.9 ± 0	0.5 ± 0
ER ± SD/(%)	76 ± 0	22 ± 0	1.2 ± 0.1	0.3 ± 0	0.2 ± 0

significant effect ($p < 0.0001$) of all variables on lipase extraction. The higher yields of lipase were recovered when the residue was extracted with 7.5 mL of 1% Triton X-100 per g of solid residue and the homogenization was done during 0.5 hour at 250 rpm and 24°C.

Consecutive extractions of fermented residues were also performed in order to determine the recovery of lipase at optimum extraction conditions. Table 3 presents lipase activity and the respective recovery percentages of those experiments. The first extraction yielded a lipase activity of 223 ± 5 U/g, which corresponds to an extraction recovery of 76 ± 0%. With a second extraction it was possible to extract almost all the lipase contained in the residue. The results agree with Rodriguez et al. (2006) who obtained a lipase recovery of 70% in the 1st extraction, using 1% Triton X-100.

This study shows that OP and WB are suitable residues to produce *A. ibericus* lipase under SSF. Mixtures of these agro-residues can be used without being necessary any processing step such as drying or grind. They can be applied directly without moisture adjustment, requiring only the addition of 0.4 g of $(NH_4)_2SO_4$ per 30 g of solid residues. With this study, an increase in lipase production of 2-fold was achieved.

4 CONCLUSIONS

In conclusion, it was observed that the addition of phosphorus did not improve the production of lipase by *A. ibericus*. On the contrary, the nitrogen source influenced positively lipase production. The best nitrogen source was found to be ammonium sulphate. Additionally, the carbon source and its composition influenced substantially its production. The mixture of OP with WB in a 1:1 ratio was found to be the most favourable composition. It was also observed that the recovery of lipase from residue is improved when the extraction is performed using 1% Triton X-100 (7.5 mL/g) at 250 rpm and 24°C during 30 min. Therefore, the production of lipase through SSF of OP with WB using *A. ibericus* is an interesting strategy for OP valorization.

ACKNOWLEDGEMENTS

Felisbela Oliveira acknowledges the financial support from Fundação para a Ciência e Tecnologia (FCT) of Portugal through grant SFRH/BD/87953/2012. Luís Abrunhosa was supported by the grant Incentivo/EQB/LA0023/2014 from ON.2 – O Novo Norte. The authors thank the Project "BioInd - Biotechnology and Bioengineering for improved Industrial and Agro-Food processes, REF. NORTE-07-0124-FEDER-000028" Co-funded by the Programa Operacional Regional do Norte (ON.2 – O Novo Norte), QREN, FEDER. The authors thank the FCT Strategic Project PEst-OE/EQB/LA0023/2013. Noelia Pérez-Rodríguez acknowledges the financial support of FPU from Spanish Ministry of Education, Culture and Sports. The authors thank to the Spanish Ministry of Science and Innovation for the financial support of this work (project CTQ2011-28967), which has partial financial support from the FEDER funds of the European Union.

REFERENCES

Alburquerque, J.A., Gonzálvez, J., García, D. & Cegarra, J. 2006. Effects of bulking agent on the composting of "alperujo", the solid by-product of the two-phase centrifugation method for olive oil extraction. *Process Biochemistry* 41(1): 127–132.

Edwinoliver, N.G., Thirunavukarasu, K., Naidu, R.B., Gowthaman, M.K., Kambe, T.N. & Kamini, N.R. 2010. Scale up of a novel tri-substrate fermentation for enhanced production of Aspergillus niger lipase for tallow hydrolysis. *Bioresource Technology* 101(17): 6791–6796.

Oliveira, F., Moreira, C., Salgado, J.M., Abrunhosa, L., Venâncio, A. & Belo, I. 2013. Valorisation of olive pomace by solid-state fermentation with Aspergillus species for lipase production. In Proceedings of *2nd International Conference Wastes: Solutions, Treatment and Opportunities*, 283–288, 11–13 September 2013, Braga.

Pandey, A. 2003. Solid-state fermentation. *Biochemical Engineering Journal* 13: 81–84.

Papagianni, M. 2004. Fungal morphology and metabolite production in submerged mycelial processes. *Biotechnology Advances* 22(3): 189–259.

Pérez-Rodríguez, N., Oliveira, F., Pérez-Bibbins, B., Belo, I., Torrado Agrasar, A. & Domínguez, J.M. 2014. Optimization of xylanase production by filamentous fungi in solid-state fermentation and scale-up to horizontal tube bioreactor. *Applied Biochemistry and Biotechnology* 173 (3): 803–825.

Rodriguez, J.A., Mateos, J.C., Nungaray, J., González, V., Bhagnagar, T., Roussos, S. & Baratti, J. 2006. Improving lipase production by nutrient source modification using *Rhizopus homothallicus* cultured in solid state fermentation. *Process Biochemistry* 41: 2264–2269.

Salihu, A., Alam, M.Z., AbdulKarim, M.I. & Salleh, H.M. 2012. Lipase production: An insight in the utilization of renewable agricultural residues. *Resources, Conservation and Recycling* 58: 36–44.

Serra, R., Cabañes, F.J., Perrone, G., Castellá, G., Venâncio, A., Mulè, G. & Kozakiewicz, Z. 2006. *Aspergillus ibericus*: a new species of section *Nigri* isolated from grapes. *Mycologia* 98(2): 295–306.

Treichel, H., Oliveira, D., Mazutti, M.A., Luccio, M. & Oliveira, J.V. 2009. A review on microbial lipases production. *Food and Bioprocess Technology* 3(2): 182–196.

Wastes: Solutions, Treatments and Opportunities – Vilarinho, Castro & Russo (eds)
© 2015 Taylor & Francis Group, London, ISBN 978-1-138-02882-1

Scale-up of *aspergillus ibericus* lipase production by solid-state fermentation

F. Oliveira
CEB-Centre of Biological Engineering, University of Minho, Braga, Portugal
Department of Chemical Engineering, Faculty of Sciences, University of Vigo, Ourense, Spain

A. Venâncio & I. Belo
CEB-Centre of Biological Engineering, University of Minho, Braga, Portugal

N. Pérez-Rodríguez & J.M. Domínguez
Department of Chemical Engineering, Faculty of Sciences, University of Vigo, Ourense, Spain

ABSTRACT: This work deals with the production of lipase by *Aspergillus ibericus* under Solid State Fermentation (SSF) of Olive Pomace (OP) with Wheat Bran (WB) at packed-bed, tray-type and pressurized bioreactors. Moreover, the effect of the sterilization step of the residues on lipase activity was studied. The sterilization at 121°C of residues benefits lipase production. SSF scale-up from flasks to different bioreactors was successfully performed with a slight lipase production decrease from 27% to 40% of the total lipase units produced per mass of residue in Erlenmeyer's flasks using autoclaved residues (183 ± 21 U/g). Contrariwise, results of specific activity obtained in flasks (51 ± 4 U/mg) were similar to the obtained in the packed-bed bioreactor (using autoclaved residues) and in the pressurized bioreactor (using no autoclaved residues).

1 INTRODUCTION

Solid-state fermentation (SSF) is defined as a fermentation process involving a moistened solid substrate as a support for growth and metabolism of microorganisms (Pandey 2003, Singhania et al. 2009). Due to low water activity in SSF, lower demand on sterility is needed (Singhania et al. 2009). Also, SSF offers the opportunity of the utilization of low cost agro-industrial residues for metabolites production while treating them (Pandey 2003).

Olive pomace (OP) is a sludgy waste generated by the olive oil two-phase extraction system. It is an acidic and very humid material, rich in organic matter, potassium and nitrogen, which also contains water-soluble carbohydrates, phenols and fats (Alburquerque et al. 2006), offering excellent properties to produce enzymes, particularly lipase, since it has residual content of olive oil. Previous work has demonstrated the ability of *Aspergillus ibericus* MUM 03.49 for the production of lipase (Oliveira et al. 2013) under SSF of OP with wheat bran (WB), and also under submerged fermentation of olive mill wastewaters (Abrunhosa et al. 2013).

In SSF processes, microorganism's growth may be limited by heat transfer and/or mass transfer of oxygen or nutrients, depending on the location in the substrate bed, the stage of the fermentation, and the design and operation of the bioreactor (Mitchell et al. 2000). Different bioreactor types have been used in SSF, including tray-type, packed-bed and horizontal rotary-drum, presenting their own advantages and disadvantages (Singhania et al. 2009).

The aim of this work was to scale-up the production of lipase from *Aspergillus ibericus* MUM 03.49 by SSF of OP with WB, studied in previous works on Erlenmeyer's flasks. (Oliveira et al. 2013). Packed-bed, tray-type and pressurized bioreactors were used and SSF conditions were optimized. Also, studies of the effect of the sterilization of residues on lipase production were performed.

2 MATERIALS AND METHODS

2.1 *Microorganism and residues*

Aspergillus ibericus MUM 03.49 (MUM culture collection, Braga, Portugal) was used. *A. ibericus* is a black *Aspergillus*, isolated from wine grapes in Portugal and in Spain (Serra et al. 2006). The fungus was grown on malt extract agar (MEA) plates (2% (w/v) malt extract, 2% (w/v) glucose, 0.1% (w/v) peptone and 2% (w/v) agar) at 30°C for 7 days and stored at 4°C. Spore suspensions of the inoculum were prepared by adding peptone solution (0.1% (w/v) peptone and 0.001% (w/v) Tween 80) to plates cultures, and after agitation were transferred to a falcon. The spore concentration of the suspension was adjusted to 10^7 spores/mL. OP samples were collected from a two-phase olive mill plant in Vila Real, Portugal, during the 2012/2013 campaign, and stored at $-20°C$. OP presented 69% $\pm$ 1% of moisture content (wet basis). WB was purchased in a local supermarket.

2.2 *SSF of OP with WB*

SSFs experiments were performed using OP mixed with WB. SSF conditions of lipase production were as follows: ratio of 1:1 (w/w, dry basis) of OP with WB, 0.0133 g/g of $(NH_4)_2SO_4$, 33.33 μL/g of 10^7 spores/mL inoculum suspension. The mixture of OP with WB resulted in optimum moisture content between 57% and 60%, without the need for its adjustment. Fermentations were carried out at 30°C during 7 days without agitation. SSF was performed with autoclaved (121°C, 200 kPa for 20 min) and also with non-autoclaved residues.

2.2.1 *Lipase production in 500 mL Erlenmeyer's flasks*
SSFs were carried out in cotton-plugged 500 mL Erlenmeyer's using 30 g of residues (OP + WB). SSF were performed in triplicate, where three flasks containing residues were autoclaved and another three were not. They were cooled, inoculated and incubated at conditions previous described. Also, controls were performed using the residues without autoclaving and without inoculating the fungus.

2.2.2 *Lipase production in packed-bed bioreactor*
The packed-bed bioreactor consisted in a double jacketed glass column (34 cm length and 3 cm internal diameter) connected to a filtered-air supply. The air was passed through a 0.45 μm cellulose filter and bubbled in distilled water before to enter in the column. The air flow was measured and controlled by a flowmeter (Aalborg Instruments & Controls, Inc., USA). The air outlet was bubbled in 1 M NaOH. The bioreactor and residues were previously autoclaved. The column was completely filled with 25 g of residue inoculated. SSF were performed at different aeration rates of 0.05 L/min, 0.1 L/min and 0.2 L/min. Experiments were performed in triplicate. Also, an additional SSF without aeration was performed.

2.2.3 *Lipase production tray-type bioreactor without forced aeration*
The tray-type bioreactor used consisted in a vertical incubator (112 × 48 × 45 cm) containing four stainless steel trays (38 × 26 × 5 cm). SSF was carried out in the 4 trays of the bioreactor at the same conditions. The residue was previously autoclaved. Each tray was filled with 300 g of residue, cooled and inoculated, resulting in a bed height of 2.5 cm. The incubator was opened once a day for monitoring and to allow aeration. SSF monitoring included temperature measurement and weight loss in each tray.

2.2.4 *Lipase production in pressurized bioreactor*
SSF was carried out in 19 dm^3 (42 cm height and 24 cm diameter) stainless steel stirred tank bioreactor, a pressurized bioreactor (4555, Parr Instrument Company, USA). SSF was performed using 500 g of residue no autoclaved. The bed height formed was around 8 cm. Air pressure of 200 kPa, 400 kPa and 700 kPa, was selected by the inlet air pressure setting and by controlling the regulatory outlet air valves. Also, different aeration rates were tested, 1 L/min and 2 L/min of outlet gas, at different pressures. SSFs were performed in duplicate.

2.3 Lipase and protein extraction and determination

Before extraction, a representative amount of the fermented residue was taken for moisture content determination. The fermented residues were homogenized with 5 mL of 1% Triton X-100 per g of dried residue at 170 rpm and 25°C for 2 h, using a shaker. Homogenates were then centrifuged (3000 rpm and 10 min at 4°C) and filtered. The enzymatic extracts were preserved at 4°C.

Lipase activity was determined by a spectrophotometric method, using p-nitrophenyl butyrate in potassium phosphate 50 mM at pH 7.0 and 37°C for 15 min. The absorbance was measured at 405 nm. One unit of lipase activity (U) was expressed as the amount of enzyme which produces 1 μmol of p-nitrophenol per minute, under the assay conditions. Lipase activity obtained was expressed as units per gram of dry solid residue (U/g).

The protein content was determined by Bradford's method, using BSA as the standard (Bradford 1976). Protein concentration was expressed as mg of protein per gram of dry solid residue (mg/g). Specific activity was obtained by the ratio between lipase activity and protein concentration. It was expressed as units of lipase activity per mg of total protein (U/mg). All analyses were performed in triplicate.

2.4 Analysis of experimental data

The data obtained were statistically analysed using SPSS (IBM SPSS Statistics, Version 22.0. Armonk, NY: IBM Corp.). Data were tested for homogeneity, submitted to one-way analysis of variance (ANOVA) and a pair-wise multiple comparison procedure (Tukey test), at a confidence level of 95%.

3 RESULTS AND DISCUSSION

3.1 Erlenmeyer's flasks

SSF were performed in Erlenmeyer's flaks with and without autoclaving the residues. The growth of A. ibericus was observed in both residues, where the prevalence of A. ibericus in the no autoclaved residues was observed, presenting black spores, while in the control experiment different green and brown spores were observed.

In what concern moisture content, there were no significant differences between the results of SFFs. With respect to the pH of the fermented residues, lower pH was obtained autoclaving the residues. As expected, autoclaving the residues eliminates the wild microbial population, contributing to a better colonization with the *Aspergillus* strain used as inoculum. Since black *Aspergillus* species are known producers of many organic acids, a significant decrease of the final pH was observed (from 4.8 ± 0.1 to 4.3 ± 0).

Similarly, a significantly higher lipase production was obtained when autoclaving the residue. Autoclaving the residue presented significant positive effect ($p < 0.01$) on lipase produced per mass of residue and on specific activity (units per mass of total protein), reaching 183 ± 21 U/g and 51 ± 4 U/mg, respectively. For another hand, residues sterilization led to the decrease of protein concentration (3.6 ± 0.3 mg/g) and, consequently, the specific activity was significantly higher. In the flasks using no autoclaving residues a lipase production and specific activity of 121 ± 11 U/g and 31 ± 2 U/mg, respectively, was obtained.

3.2 Packed-bed bioreactor

Different aeration rates were evaluated in a packed-bed bioreactor and results are presented in Figure 1. It was observed the dehydration of the residue at the beginning of the column over fermentation time, even using saturated air. Aeration rate presented significant effect ($p < 0.0001$) on lipase activity. Maximum lipase production and specific activity was obtained at 0.05 L/min, reaching 134 ± 2 U/g and 50 ± 13 U/mg, respectively.

The aeration rate favours the transport of oxygen to the solid residue, however, above the optimal aeration, lipase activity may decrease due to the fungal metabolism changes (Díaz et al. 2013).

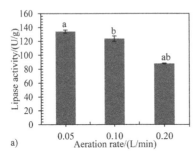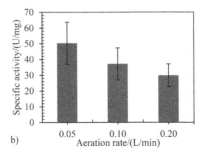

Figure 1. Effect of aeration rate on a) lipase production and b) specific activity of SSF in packed-bed bioreactor. Depicted values are the mean of three independent fermentation experiments ± standard deviation. Means with the same letter differ significantly at $p < 0.05$.

Pérez-Rodríguez et al. (2014) found an optimum aeration rate of 0.1 L/min in a packed-bed bioreactor using 20 g corncob for optimization of xylanase production under SSF with *A. niger*. A SSF without aeration was performed, where no fungal growth was observed.

Results of lipase obtained in packed-bed suffered a slight decrease of 27% of the lipase produced in flasks. However, the specific activity was similar (50 ± 13 U/mg) to the obtained in flasks. The slight reduction of lipase could be explained by the dehydration along the column derived from the aeration used.

3.3 *Tray-type bioreactor*

SSF in the tray-type bioreactor was performed using autoclaved residues without weight loss control. The weight loss of the trays was decreasing with time reaching 53% at the end of fermentation time. As a consequence, the moisture content of the fermented residues decreased to minimum values of 27% ± 3%, as shows Table 1. Lipase suffered a drastic decrease comparing to the values obtained in flasks, yielding 74 ± 10 U/g and 22 ± 2 U/mg, of lipase per g of residues and of specific activity, respectively.

Since the final moisture content was too low, another SSF was performed at similar conditions, but with control of weight loss by adding distilled water once a day, in the same amount of lost weight. Also, a SSF with no autoclaved residue with control of weight loss was performed. In all the fermentations, it was observed the increase of residues temperature till 34.3°C on the first and/or second day, due to the fungus's growth, after that the temperature was decreasing with fermentation time till 30°C. In both SSF with weight loss control, the weight loss at the end of fermentation was only 9%.

It was observed significant differences ($p < 0.005$) on lipase obtained. Lipase production was higher in SSF with weigh loss control, reaching 111 ± 27 U/g and 131 ± 16 U/g, using autoclaved and no autoclaved residues, respectively (Table 1). With respect to the specific activity, conditions used presented significant effect ($p < 0.005$) on specific activity. Similarly to the SSF in flasks, the specific activity was higher on SSF with autoclaved residue with weight loss control (36 ± 6 U/mg) than in SSF without autoclaving the residue. These results demonstrated the importance of moisture control to improve lipase production and its specific activity, and again the sterilization of the residues to improve the specific activity.

At a tray-type bioreactor, with moisture control, a decrease from 28% to 40% of the total production in flasks (183 ± 21 U/g) was observed. And also, a reduction in specific activity between 29% and 52% of that in flasks was obtained. This reduction could be attributed to the dehydration of the residue over fermentation time, high bed height (2.5 cm) used and the residues temperature rise during the culture growth. For example, Vaseghi et al. (2013) determined an optimum bed height of 0.5 cm for SSF on tray-type bioreactor.

Table 1. Results of moisture content (MC), pH, lipase activity (LA), specific activity (SA) and protein concentration (PC) in tray-type bioreactor at different conditions. Values are the mean of four trays of the incubator ± standard deviation (SD). Means with the same letter differ significantly at p < 0.05.

Condition	MC ± SD/(%)	pH ± SD	LA ± SD/(U/g)	SA ± SD/(U/mg)	PC ± SD/(mg/g)
Autoclaved without weight loss control	27 ± 3^{ab}	4.5 ± 0^{a}	74 ± 10^{a}	22 ± 2^{a}	3.3 ± 0.2^{a}
Autoclaved with weight loss control	66 ± 3^{a}	4.4 ± 0^{b}	111 ± 27	36 ± 6^{ab}	3.0 ± 0.3^{b}
No autoclaved with weight loss control	67 ± 3^{b}	5.0 ± 0.1^{ab}	131 ± 16^{a}	25 ± 5^{b}	5.4 ± 0.3^{ab}

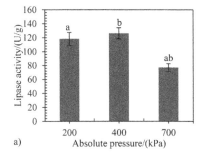
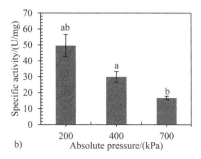

Figure 2. Results of a) lipase activity and b) specific activity of SSF in a pressurized bioreactor at different absolute air pressure and at aeration rate of 1 L/min. Depicted values are the mean of two independent fermentation experiments ± standard deviation. Means with the same letter differ significantly at $p < 0.05$.

3.4 Pressurized bioreactor

Pressurized air can be successfully applied to cultivations, as a way of improving the oxygen transfer rate to aerobic cultures (Belo et al. 2003). However, above certain limits the increased air pressure and the consequent increase in oxygen partial pressure may have detrimental effects on cells activity and product formation (Belo et al. 2005). Lopes et al. (2009, 2008) found significant improvements on lipase productivity from a nonconventional yeast under submerged fermentation, using air pressure up to 600 kPa.

Figure 2 presents results of SSF in a pressurized bioreactor, at aeration rate of 1 L/min and at different air pressures. Air pressure significantly affected ($p < 0.01$) lipase production. Maximum lipase activity was found at 400 kPa (126 ± 8 U/g). However, the specific activity was higher at 200 kPa (50 ± 7 U/mg). SSF at 2 L/min of aeration rate and at different air pressure were performed (data not shown). Similar to the results of SSF at 1 L/min, maximum lipase production and specific activity was obtained at 200 kPa (125 ± 5 U/g and 42 ± 4 U/mg, respectively). Results of SSF did not present significant differences using aeration rate of 1 L/min or 2 L/min.

In comparison to results from flasks, a decrease in 31% of lipase produced in pressurized bioreactor at 200 kPa and 1 L/min was observed. Contrariwise, the specific activity was similar to the flasks, even using no autoclaved residue (50 ± 7 U/g). The pressure of 200 kPa led to lower production of protein and consequently to maximum specific activities, presenting similar effects to the autoclaving residue.

4 CONCLUSIONS

A. ibericus presented good performance of growth and lipase production using no sterilized residue. For another hand, the effect of autoclaving residues for SSF led to higher yields of lipase and specific activity. SSF was successfully 10-fold and 17-fold scaled-up to tray-type and pressurized bioreactor, respectively, with a slight decrease (from 23% to 40%) in total lipase production obtained in flasks. However, no significant differences in specific activity of lipase were found in flasks and in the bioreactors with forced aeration (packed-bed and pressurized). In bioreactors as packed-bed and tray-type, further studies to control moisture content, temperature of the residue and aeration rate should be developed in order to optimize the these variables to improve lipase production.

ACKNOWLEDGEMENTS

Felisbela Oliveira acknowledges the financial support from Fundação para a Ciência e Tecnologia (FCT) of Portugal through grant SFRH/BD/87953/2012. The authors thank the Project "BioInd – Biotechnology and Bioengineering for improved Industrial and Agro-Food processes, REF. NORTE-07-0124-FEDER-000028" Co-funded by the Programa Operacional Regional do Norte (ON.2 – O Novo Norte), QREN, FEDER. The authors thank the FCT Strategic Project PEst-OE/EQB/LA0023/2013. Noelia Pérez-Rodríguez acknowledges the financial support of FPU from Spanish Ministry of Education, Culture and Sports. The authors thank to the Spanish Ministry of Science and Innovation for the financial support of this work (project CTQ2011-28967), which has partial financial support from the FEDER funds of the European Union.

REFERENCES

Abrunhosa, L., Oliveira, F., Dantas, D., Gonçalves, C. & Belo, I. 2013. Lipase production by Aspergillus ibericus using olive mill wastewater. *Bioprocess Biosystem Engineering* 36: 285–291.

Alburquerque, J.A., Gonzálvez, J., García, D. & Cegarra, J. 2006. Effects of bulking agent on the composting of "alperujo", the solid by-product of the two-phase centrifugation method for olive oil extraction. *Process Biochemistry* 41(1): 127–132.

Belo, I., Pinheiro, R. and Mota, M. 2003. Fed-batch cultivation of *Saccharomyces cerevisiae* in a hyperbaric bioreactor. *Biotechnology Progress* 19: 665–671.

Belo, I., Pinheiro, R. and Mota, M. 2005. Morphological and physiological changes in Saccharomyces cerevisiae by oxidative stress from hyperbaric air. *Journal of Biotechnology* 115: 397–404.

Bradford, M.M. 1976. A rapid and sensitive method for the quantitation of microgram quantities of protein utilizing the principle of protein-dye binding. *Analytical Biochemistry* 72: 248–254.

Díaz, A.B., Alvarado, O., Ory, I., Caro, I. & Blandino, A. 2013. Valorization of grape pomace and orange peels: Improved production of hydrolytic enzymes for the clarification of orange juice. *Food and Bioproducts Processing* 91(4): 580–586.

Lopes, M., Gomes N., Gonçalves, C., Coelho, M.A.Z., Mota M. & Belo, I. 2008. *Yarrowia lipolytica* lipase production enhanced by increased air pressure. *Letters in Applied Microbiology* 46(2): 255–260.

Lopes, M., Gomes, N., Mota, M. & Belo, I. 2009. *Yarrowia lipolytica* growth under increased air pressure: influence on enzyme production. *Applied Biochemistry and Biotechnology* 159(1): 46–53.

Mitchell, D.A., Krieger, N., Stuart, D.M., & Pandey, A. 2000. New developments in solid-state fermentation II. *Rational approaches to the design, operation and scale-up of bioreactors* 35: 1211–1225.

Oliveira, F., Moreira, C., Salgado, J.M., Abrunhosa, L., Venâncio, A., & Belo, I. 2013. Valorisation of olive pomace by solid-state fermentation with Aspergillus species for lipase production. In Proceedings of *2nd International Conference Wastes: Solutions, Treatment and Opportunities*, 283–288, 11–13 September 2013, Braga.

Pandey, A. 2003. Solid-state fermentation. *Biochemical Engineering Journal* 13: 81–84.

Pérez-Rodríguez, N., Oliveira, F., Pérez-Bibbins, B., Belo, I., Torrado Agrasar, A. & Domínguez, J.M. 2014. Optimization of xylanase production by filamentous fungi in solid-state fermentation and scale-up to horizontal tube bioreactor. *Applied Biochemistry and Biotechnology* 173(3): 803–825.

Serra, R., Cabañes, F.J., Perrone, G., Castellá, G., Venâncio, A., Mulè, G. & Kozakiewicz, Z. 2006. *Aspergillus ibericus*: a new species of section Nigri isolated from grapes. *Mycologia* 98(2): 295–306.

Singhania, R.R., Patel, A.K., Soccol, C.R. & Pandey, A. 2009. Recent advances in solid-state fermentation. *Biochemical Engineering Journal* 44(1): 13–18.

Vaseghi, Z., Najafpour, G.D., Mohseni, S. & Mahjoub, S. 2013. Production of active lipase by *Rhizopus oryzae* from sugarcane bagasse: solid state fermentation in a tray bioreactor. *International Journal of Food Science & Technology* 48(2): 283–289.

Wastes: Solutions, Treatments and Opportunities – Vilarinho, Castro & Russo (eds)
© *2015 Taylor & Francis Group, London, ISBN 978-1-138-02882-1*

New integrated polyphenols recovery and anaerobic digestion of Alpeorujo

M. Orive, B. Iñarra, M. Cebrián & J. Zufía
AZTI, Food Research Division, Derio, Spain

ABSTRACT: ABSTRACT: Solid-liquid extraction was used in order to recover polyphenols compounds from *alpeorujo*, a mixed olive mill liquid-solid waste with high pollution potential. The high extraction efficiency (70%) decreased the toxicity and enhanced the performance of anaerobic digestion when operating at OLRs ranging from 2.75 up to7.30 kg COD· m^{-3}·day^{-1}. The optimum biogas (4.26±0.59 m^3·m^{-3}·day^{-1}) and methane productions (2.68±0.91 m^3·m^{-3}· day^{-1}) were achieved at OLR of 7.30 kg COD·m^{-3}·day^{-1} and 24 days HRT with an average COD removal rate of 80.59±3.17%. The present paper also reports on the economic feasibility of these integrated processes in an industrial-scale bio-refinery plant which will treat 80,000 t·y^{-1}of *alpeorujo* in Andalusia region (Spain). Payback times range from 4.7 up to 8.7 years for Internal Results Rates among 8% and 16%.

1 INTRODUCTION

Olive oil manufacturing sector plays a relevant role in the economic development of the Mediterranean countries. Olive Mill Wastes (OMWs) produced in this area are estimated to be about 30 million m^3/year with considerable well–known harmful effects over the environment. As an example of the scale of the environmental impact of OMWs, 10 million m^3 year^{-1} of liquid effluent from three-phase systems corresponds to an equivalent load of wastewater generated from around 20 million people. The most extended technology for olive oil production in Spain is the two-phase process. This technology is environmentally friendly comparing to the traditional three-phase processes. However, problems related with the disposal of mixed solid-liquid waste from these systems (known as *alpeorujo* in Spanish) have not been completely solved due to the seasonality of the sector (primarily from early November to late February) and the geographical dispersion olive oil manufacturing industry.

The development of the bio-refinery concept as a sustainable approach for bio-based materials, biomolecules and biofuels production can be applied to *alpeorujo*, allowing a recovery of natural antioxidants such as polyphenols and other natural compounds and production of biofuels and biopolymers. The high economic value of these compounds makes their extraction one of the most promising alternatives to increase the benefits of waste management companies as can be exploited as ingredients for cosmetic and food industry. Several researchers have studied the techno-economic processes for recovering polyphenols from *alpeorujo* (Beccari et al. 1996; Azbar et al. 2008). Principal systems proposed for polyphenols recovery are extraction with solvents (solid-liquid, liquid-liquid and supercritical extractions), selective concentration by ultrafiltration and reverse osmosis. In previous studies, among all procedures employed for polyphenols recovery, solid-liquid solvent extraction represents a simple and convenient alternative, so it has been widely used in pilot-scale production and in commercial recovery (Bonazzi 1996). All these techniques resulted in a polyphenol rich extract and in a solid dephenolized fraction which is a complex mixture of water, sugars, lipids, nitrogenous substances, organic acids, residual polyphenols and solvents, tannins and inorganic compounds. Physicochemical or biological treatment of this exhausted fraction is necessary before discharging into the environment.

Many pollution disposal methods, such as concentration, evaporation, incineration, ultrafiltration/reverse osmosis, aerobic treatment or lagooning have already been tested on alpeorujo, but none of them led to industrial applications. However, anaerobic digestion seems to have some clear advantages that would make it the process of choice (Rincón et al., 2008). Indeed, this treatment process produces energy and a digested effluent with a significant reduction of the organic load. However, many problems concerning the high toxicity and inhibition of biodegradation of these effluents were encountered during anaerobic treatments, because some bacteria, such as methanogens, are particularly sensitive to high concentrations of polyphenols. Thus, the previous extraction of polyphenols from *alpeorujo* provides a double opportunity to obtain high-added value biomolecules and to reduce the phytotoxicity of *alpeorujo* over the fermentation processes. The additional advantages of anaerobic digestion processes comparing to other treatment methods are lower sludge production and less energy requirements for operation and easy restart of the process after several months of shutdown before seasonal waste production campaigns.

The aim of the present work was to develop an integrated chemical-biological process for polyphenols recovery from *alpeorujo* by liquid-solid extraction and the subsequent continuous anaerobic fermentation of the dephenolized fraction. Different Hydraulic Retention Times (HRTs) and Organic Loading Rates (OLRs) were tested to study the effects of feeding the diluted dephenolized *alpeorujo* over the anaerobic digestion. Finally, a technical-economic feasibility study for treating 80,000 $t \cdot y^{-1}$ of *alpeorujo* in an integrated bio-refinery process was performed for Jaén province in the Spanish Andalusia region.

2 MATERIALS & METHODS

2.1 *Preparation of samples*

The *alpeorujo* tested in the study resulted from the olive campaign of 2013. It was collected from *Gestor de Alpeorujos de Mágina, S.L.* located in Bedmar (Jaén, Spain). The average *alpeorujo* processed in the last four campaigns was around 80,000 $t \cdot y^{-1}$. All samples were frozen at $-5.0°C$ upon arrival and stored maximum for 3 months. The main physicochemical characteristics are illustrated in table 1.

2.2 *Batch polyphenols extraction from alpeorujo*

Batch extractions were performed at ambient temperature in 1 L stirred glass reactor at ambient temperature. The *alpeorujo* was mixed with ethanol and water solution (50/50 w/w) in 1:5 solid-liquid ratio and left under agitation at 200 r.p.m for 24 h. The composition of extracting solution and the solid-liquid ratios were determine and optimize in previous experiments (data not shown). Most of the stones were separated in this stage by decantation. Finally, the free-stone dephenolized *alpeorujo* was separated from the polyphenols extract by centrifugation at 2890 g for 10 min. The average batch extraction rates ranged between 80–85%.

2.3 *Biogas production*

In order to study the influence of extracting polyphenols before the anaerobic digestion, the reactor was initially fed with *non-dephenolized alpeorujo* until process became unstable.After reactor recovery,it was fed with *dephenolized alpeorujo* as unique substrate. Different OLRs and HRTs were tested (fig. 1). In order to reach the optimal moisture content for the digestion process, tap water was added to fix the volatile solids content around 9%.

The experiments were carried out in a 10 L semi-continuous stirred tank reactor (CSTR) with 9.5 L of working volume at mesophilic temperature ($37 \pm 1°C$). The digester was manually fed and purged with same volume daily to guarantee the same digestion volume. The operational temperature was maintained by circulating water from a heated water bath through a jacket system.

Table 1. Physicochemical parameters of *alpeorujo*, polyphenols extract and diluted dephenolized *alpeorujo* (± standard deviation).

Parameters	*Alpeorujo*	Polyphenols extract	Diluted dephenolized *alpeorujo*
pH	$5.55_{(0.35)}$		$5.00_{(0.03)}$
TS (%)	$37.61_{(5.81)}$	$2.51_{(0.73)}$	$9.43_{(2.26)}$
VS (%)	$36.45_{(4.52)}$	$2.11_{(0.64)}$	$9.07_{(2.28)}$
TKN (%)	$0.46_{(0.07)}$	$0.03_{(0.01)}$	$0.03_{(0.01)}$
TAN (mg$\cdot$100g^{-1})	$23.84_{(0.99)}$	$10.74_{(0.92)}$	$8.84_{(1.51)}$
COD (g $O_2\cdot L^{-1}$)	n.d	n.d.	$194.04_{(33.13)}$
TPh (g GAE$\cdot$ kg^{-1})	$5.01_{(0.74)}$	$3.64_{(0.28)}$	$0.91_{(0.25)}$
VFA (g$\cdot$kg^{-1} FM)	$4.90_{(0.12)}$	n.d.	$0.99_{(0.11)}$
Lipids (%)	$4.39_{(0.19)}$	$0.34_{(0.04)}$	$1.63_{(0.24)}$
Protein (%)	$2.89_{(0.42)}$	$0.18_{(0.05)}$	$1.32_{(0.17)}$
C/N ratio	$46.07_{(1.39)}$	n.d.	$20.50_{(3.42)}$

The digester was started with sludge from a Wastewater Treatment Plant (WWTP) located in San Sebastián (Gipuzkoa, Spain).

Biogas production and composition were measured daily. A sampling protocol was applied to digester to record data concerning to the following inhibitory parameters: free ammonium, polyphenols and volatile fatty acids. The flow of substrate to be treated was increased to determine the maximum OLRs and minimum HRTs at which inhibition by excess organic load occurred. The reactor was operated for 100 days.

2.4 *Analytical methods*

Total Solids (TS), Volatile Solids (VS), Total Chemical Oxygen Demand (COD), Total Kjeldahl Nitrogen (TKN), total ammonia nitrogen (TAN), free ammonia (NH_3) and lipids were determined by Standard Methods (APHA 2005). Alkalinity was evaluated as partial alkalinity (PA) and total alkalinity (TA) by titration to pH 5.75 and 4.3 with normalized 0.4 N H_2SO_4 respectively. Total phenolic compounds (TPh) into the extract and into the remaining dephenolized fraction were determined as gallic acid equivalents by modified Folin-Ciocalteau method.

All these variables were measured once a week. The biogas production and composition were determined daily by Milligas Counters provided by Ritter®(Dortmund, Germany) and by infrared equipment provided by Geothermical Instruments (Stuttgart, Germany). Volatile Fatty Acids (VFA) (acetic, propanoic, butyric, iso-butyric, valeric and iso-valeric) were determined by GC-FID fitted with INNOWAX 1909 1N-113 column (30 m × 0.32 mm × 0.25 μm film thickness). The operation temperatures for the injection port and flame injector detector were 260°C and 280°C respectively. Lipids content was determined by Blyer-Dyer method. The VS concentration in the feeding was maintained around 9% for the whole experimental period. The main physical-chemical parameters are summarized in table 1.

3 RESULTS & DISCUSSION

3.1 *Anaerobic digestion of non-dephenolized alpeorujo*

10 L semi-continuous reactor was initially fed with diluted raw alpeorujo (20%) at a first loading rate of 4.0 kg COD$\cdot$m$^{-3}\cdot$day^{-1} at HRT 50 days followed by higher loading rates and lower HRT. This toxicity was accompanied by a pH decrease and an accumulation of VFAs (data not shown). Several authors reported that inhibition of the methanisation of non-dephenolized *alpeorujo* occurred at mean OLRs of 1.5 kg COD$\cdot$m$^{-3}\cdot$day^{-1}. Previous reports concluded many problems such as high

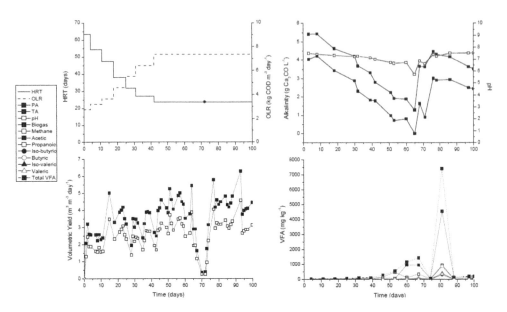

Figure 1. HRT, biogas and methane volumetric production yields, pH, alkalinity and VFAs.

toxicity, low biodegradability and the acidifications of reactors. Therefore, a co-digestion or a pre-treatment step was deemed necessary for decreasing the organic load and the phenolic compounds. This test will serve as a control for comparing the efficiency of the solvent extraction pre-treatment in the detoxification of *alpeorujo*.

3.2 *Anaerobic digestion of dephenolized alpeorujo*

Anaerobic digestion of diluted dephenolized *alpeorujo* fraction was found to be stable during the most the experimental period without toxicity phenomenon as shown in figure 1. Only when lower polyphenols extraction rates occurred (around 60%), some inhibitory phenomena were observed (days 65–70) due to polyphenols accumulation into the digester. This increasing of polyphenols concentrations correlated with high VFAs concentrations ($7.4 \text{ g} \cdot \text{L}^{-1}$) and the subsequent decreased of pH below 6.0.

Methane concentrations (62–74%) were preserved in a narrow range of the anaerobic process usual values. Concerning to total COD, the more concentrated influent corresponded to the highest removal percentages (80–83%).

3.3 *Industrial-scale process*

Before the polyphenols extraction, it is necessary to remove the maximum percentage of stones. The main advantages of removing stones are higher process performance and higher biomass availability as stones can be used for electric and/or for heat production. For the extraction process, *alpeorujo* is mixed with water and ethanol in steel continuous stirred tank reactors at ambient temperature for 24 h. In the extraction process, the limiting step is the consumption of the extracting solvent. Thus, a significant cost reduction is achieved as solvent is recycled after low temperature vacuum distillation which allows increasing the solids content in the extract from 2.5 to 5% TS. After the extraction process, the high-rich polyphenols extract is separated from the solid exhausted fraction by centrifuging. While the dephenolized *alpeorujo* is diluted with water to fix the volatile content of the anaerobic digesters fed, the high-rich polyphenols extract is pumped to drying process line.

Due to temperature limitations and high water initial content in the extract, the drying process is performed in three different stages. The first step allows increasing the extract from 5 to 30% DM

Table 2. Mass and energy balances for industrial-scale process sizing.

Mass balance				Energy balance		
INPUTS		OUTPUTS		INPUTS		OUTPUTS
Alpeorujo $(t \cdot y^{-1})$	80,000	Biogas $(Nm^3 \cdot y^{-1})$	11,977,000	Electric power (kWh)	4,439,976	–
Water $(m^3\ y^{-1})$	35,500	Compost $(t \cdot y^{-1})$	17,750	Thermal power (MW_t)	–	80,750
Ethanol $(t \cdot y^{-1})$	5,680 t	*Alpeorujo* stones $(t \cdot y^{-1})$	8,000	Natural Gas (MW_t)	15,000	–
		Polyphenols extract $(t \cdot y^{-1})$	3,360 (10% TS)			

through a triple-effect evaporator. As the viscosity increment affects the performance of evaporators, in the second-stage, a vacuum thin-layer evaporator is implemented to reach a final concentration of 50% TS. Finally, in order to commercialize the extract as powder, the final water content is reduced up to 10% through spray-drying technology. Table 2 shows the mass and energy balances used for industrial-scale process sizing.

3.4 *Economic feasibility*

The economic feasibility is based upon a number of the following assumptions:

1. Water is added with ethanol (50%/50%) for the polyphenols extraction in the maceration process. The expected water consumption considering solid-liquid ratio and the composition of extracting solution is 700 $L \cdot t^{-1}$ *alpeorujo*.
2. The annual ethanol consumption was calculated considering that 90% of the ethanol is recovered in the distillation process. Ethanol (96% v/v) cost was assumed to be 541 $€ \cdot t^{-1}$ (Tomsa Destil, S.L., personal communication).
3. The distillation, concentration and drying process of the polyphenols extract requires 80,750 $MWt \cdot y^{-1}$ (Falero & Laín personal communication). The annual biogas production covers the 85% process thermal energy needs. Therefore, additional natural gas consumption is required to cover the 100% thermal energy demand. It means an estimated natural gas consumption around 1,500,000 Nm^3 (15,000 MWt) per year.
4. Costs derived from operation and maintenance were assumed to be 138.450 €. Labour was assumed as 532,500 $€ \cdot y^{-1}$. The annual payment mortgage to pay the installation costs was calculated considering 5% interest during 15 years. Finally, transport costs of *alpeorujo* were estimated as follows: 110 $€ \cdot truck^{-1}$ (25 t capacity) which means a total cost of 352,000 $€ \cdot y^{-1}$ for 80,000 $t \cdot y^{-1}$.
5. Constructions cost are divided into civil works (1,810,000 €), the equipment purchasing (13.155.000 €) and taxes, building permits, etc. (789,300 €) (Falero & Laín, personal communication). Complementary costs are considered regarding civil engineering project (30,000 €) and other administrative and authorization requirements (95,000 €) (Falero & Laín, personal communication).
6. Due to the high variability in polyphenols extracts prices ranging from 7 up to 725 $€ \cdot kg^{-1}$, the economic feasibility is performed by calculating first the minimum sale price for polyphenols extract that gives zero profit for the total *alpeorujo* processed per year (80,000 $t \cdot y^{-1}$). Afterwards, different scenarios are calculated for different Internal Return Rates (IRRs).
7. The expected annual benefits are estimated based on polyphenols extract sale prices, for selling the compost as fertilizer (6 $€ \cdot t^{-1}$) and stones as biomass source for energetic valorization (60 $€ \cdot t^{-1}$) (Falero & Laín, personal communication).

Table 3. Economic feasibility for the industrial-scale plant.

Targeted % IRR	8	12	16	Minimum price for 0 benefit
Installation costs (€)	15,754,300	15,754,300	15,754,300	15,754,300
Annual costs (€·y^{-1})	5,343,450	5,343,450	5,343,450	5,343,450
Annual benefits (€·y^{-1})	6,321,913	6,472,812	9,055,403	0
Polyphenols extract sale price (€·kg^{-1})	2.14	2.32	2.52	1.77
Critical *alpeorujo* mass (t·y^{-1})	70,343	66,330	62,503	80,000
NPV (€)	3,279,501	8,144,231	13,364,801	−6,427,096
PBT (y)	8.7	6.8	4.7	>15
Shareholder's return (%)	16.7	25.1	34.1	−

Table 3 shows the sale price of the polyphenols extract, the minimum critical *alpeorujo* mass that gives zero profit, Net Present Value (NPV), the payback time (PBT) and the first year return on the shareholders' capital for the different IRRs.

4 CONCLUSIONS

The main advantage of polyphenols extraction prior to anaerobic digestion of diluted dephenolized fraction is the profitable use of *alpeorujo*'s polyphenols as a source of natural antioxidants for cosmetic and food industry. Additionally, the reduction of the toxicity improves around three-fold the biogas production comparing to non-depehenolized *alpeorujo* and enhances the process stability. After analyzing the techno-economic feasibility, it can be concluded that as the total processed *alpeorujo* (80,000 t ·y^{-1}) is higher than the critical mass calculated for the different targeted IRRs. In addition to the economic benefits for polyphenols extract and energy sales, the environmental benefits are the neutralization pollutant potential of the dephenolized *alpeorujo* and the transformation of digested sludge into compost.

ACKNOWLEDGMENTS

The present work was partially supported by SUDOE Interreg IV B research program (Provalue project SOE4/P1/E811).

REFERENCES

APHA, AWWA and WEF 2005. *Standard Methods for the Examination of Water and Wastewater*. 21st ed. Washington D. C.: American Public Health Association.
Azbar N., Keskin T., Yuruyen, A. 2008. Enhancement of biogas production from olive mill effluent (OME) by co-digestion. *Biomass and Bioenergy* 32(12): 1195–1201.
Beccari M., Bonemazzi F., Majone M., Riccardi C. 1996. Interaction between acidogenesis and methanogenesis in the anaerobic treatment of olive oil mill effluents. *Water Research* 30(1): 183–189.
Bonazzi, M. 1996. *Euro-Mediterranean policies and olive oil: Competition or job sharing. Executive Summary. EUR 17270 EN. 1996*.
Rincón B., Sánchez E., Raposo F., Borja R., Travieso L., Martín M.A., Martín A. 2008. Effect of the organic loading rate on the performance of anaerobic acidogenic fermentation of two-phase olive mill solid residue. *Waste Management* 28(5): 870–877.

Wastes: Solutions, Treatments and Opportunities – Vilarinho, Castro & Russo (eds)
© 2015 Taylor & Francis Group, London, ISBN 978-1-138-02882-1

Characterization of Construction and Demolition Wastes (C&DW)/geogrid interfaces

P.M. Pereira, C.S. Vieira & M.L. Lopes
Civil Engineering Department, Faculty of Engineering, University of Porto, Porto, Portugal

ABSTRACT: Over the last years the environmental sustainability has been demanding a progressive increase in the waste valorisation. The valorisation of Construction and Demolition Wastes (C&DW), reduces the use of natural resources and avoids congesting landfills with these inert materials. Although some studies have been carried out on the application of recycled C&DW, the valorisation of C&DW in geosynthetic reinforced embankments is almost an unexplored field. A research project on the assessment of the use of recycled C&DW as backfill material in geosynthetic reinforced structures is being developed at University of Porto. This work presents and discusses results of direct shear and pullout tests carried out to characterise the behaviour of C&DW/geogrid interfaces. The results have evidenced that C&DW, properly compacted, could exhibit shear strength similar to the natural soils commonly used in the construction of geosynthetic reinforced embankments. High values of the interaction coefficient for the geogrid/C&DW interfaces were achieved.

1 INTRODUCTION

Construction and Demolition Wastes (C&DW) are wastes derived from construction, reconstruction, cleaning of the work site and earthworks, demolition and collapse of buildings, maintenance and rehabilitation of existing constructions. Recycled C&DW have been considered as alternative materials for use as aggregates in concrete and pavement layers of transport infrastructures. As regards the application of C&DW in geotechnical works, it has been verified that the valorisation is performed mostly in road construction, particularly in base and sub-base layers of the infrastructures (Arulrajah et al., 2013a, Neves et al., 2013). Outside the scope of road infrastructures there are not many references to C&DW application in embankments. Apart from some recent studies (Santos et al., 2013, Arulrajah et al., 2013b, Vieira et al., 2014), the valorisation of recycled C&DW in geosynthetic reinforced structures is almost an unexplored field. Based on this evidence and bearing in mind the need of pursuing new ways to avoid landfilling of inert waste materials and to preserve the natural resources, a research project aiming to study the sustainable valorisation of recycled C&DW in geosynthetic reinforced structures has been under development.

In the design of geosynthetic reinforced structures, the interaction mechanisms between the reinforcement and the fill material have an utmost importance. These mechanisms depend on the fill properties, on the reinforcement characteristics and on the interaction between the two elements (fill and reinforcement).

Direct shear tests were carried out to characterize the shear strength of the C&DW and geogrid/C&DW interface shear strength. The interface geogrid/C&DW was also characterized through pullout tests. In the following sections the materials are described and test results are presented and discussed.

2 MATERIALS

Samples of recycled C&DW were collected from a Portuguese recycling plant, coming mainly from the demolition of single-family houses and cleaning of lands with illegal deposition of C&DW. The particle size distribution of the material is illustrated in Figure 1.

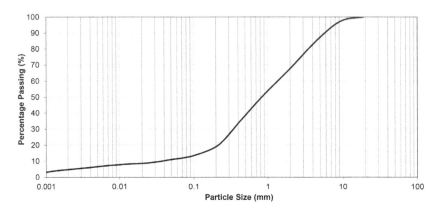

Figure 1. Particle size distribution of the recycled C&DW.

Table 1. Classification of recycled C&DW constituents.

Constituents	C&DW
Concrete, concrete products, mortar, concrete masonry units, R_c (%)	40.0
Unbound aggregate, natural stone, hydraulically bound aggregate, R_u (%)	36.5
Clay masonry units, calcium silicate masonry units, aerated non-floating concrete, R_b (%)	10.8
Bituminous materials, R_a (%)	0.5
Glass, R_g (%)	1.2
Soils, R_s (%)	10.8
Other materials, X (%)	0.1
Floating particles, FL (cm³/kg)	10.0

The constituents of the recycled C&DW, determined in accordance with the European Standard EN 933-11 (2009), are listed in Table 1. This C&DW consists mainly of concrete, unbounded aggregates, masonry and soil.

In order to evaluate the short term release of contaminants of the recycled C&DW, leaching tests were carried out. Table 2 presents the leaching test results, as well as, the acceptance criteria for leached maximum concentration for inert landfill define by the European Council Decision 2003/33/EC.

From the analysis of the results presented in Table 2, it can be concluded that only the value of sulphates exceeds the maximum values established by the European and Portuguese legislation. However the Directive 2003/33/EC states that "*if the waste does not meet these values for sulphate, it may still be considered as complying with the acceptance criteria if the leaching does not exceed 6000 mg/kg at L/S = 10 l/kg, determined either by a batch leaching test or by a percolation test under conditions approaching local equilibrium.*"

The geosynthetic used in this study is an extruded uniaxial polyester (PET) geogrid with aperture dimensions of 30 mm × 73 mm and nominal tensile strength of 80 kN/m, commonly used as reinforcement (Fig. 2).

3 DIRECT SHEAR TESTS

3.1 *Test procedures*

The direct shear tests were performed on a large scale direct shear device. The shear box comprises an upper box, fixed in the horizontal directions, with dimensions of 300 mm × 600 mm in plant and 150 mm in height. More details about this prototype can be found in Vieira et al. (2013).

Table 2. Leaching test results.

Parameter	Value (mg/kg)	Acceptance criteria for leached concentrations – Inert landfill
Arsenic, As	0.021	0.5
Lead, Pb	<0.01	0.5
Cadmium, Cd	<0.003	0.04
Chromium, Cr	0.012	0.5
Copper, Cu	0.10	2
Nickel, Ni	0.011	0.4
Mercury, Hg	<0.002	0.01
Zinc, Zn	<0.1	4
Barium, Ba	0.11	20
Molybdenum, Mo	0.018	0.5
Antimony, Sb	<0.01	0.06
Selenium, Se	<0.02	0.1
Chloride, Cl	300	800
Fluoride, F	6.1	10
Sulphate, SO_4	3200	1000
Phenol index	<0.05	1
Dissolved Organic Carbon, DOC	220	500
pH	8.2	–

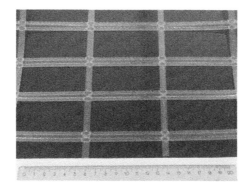

Figure 2. Visual aspect of the geogrid (ruler in centimetres).

The C&DW material was compacted inside each shear box in four layers (25 mm thick) at 90% of maximum modified Proctor dry density and at the optimum water content (12.5%). To characterize the interface geogrid/C&DW, the geogrid specimens were gripped with screws at the front edge of the lower box, outside the shear area. The direct shear tests were conducted with a constant displacement rate of 1 mm/min at normal stresses of 25, 50, 100 and 150 kPa.

3.2 *Direct shear strength of recycled C&DW*

The evolution of the shear stresses as a function of the imposed shear displacement, for the four values of the confining pressure is illustrated Figure 3(a). The shear stress–shear displacement curves have not revealed any peak of strength. Following the initial increase with the shear displacement, the shear stress has remained almost constant. The slight increase of shear stress for large displacements, more significant for the highest confining pressures, comes from the reduction of contact area during shear.

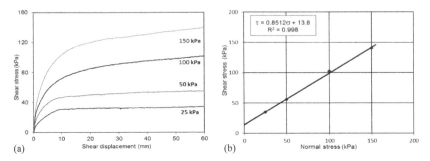

Figure 3. Results of direct shear tests for C&DW characterization: (a) shear stress-shear displacement curves; (b) failure envelope and direct shear strength properties.

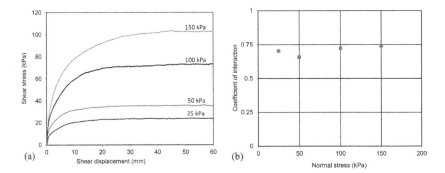

Figure 4. Results of direct shear tests carried out on geogrid/C&DW interfaces: (a) shear stress-shear displacement curves; (b) coefficients of interaction against normal stress.

Figure 3(b) presents the maximum shear stress for each value of the confining pressure (25, 50, 100 and 150 kPa), as well as, the corresponding linear best fit. Following the Coulomb failure criterion, the recycled C&DW revealed a friction angle of 40.4° and cohesion of 13.8 kPa. These values are frankly encouraging since they are similar (even higher) to the values for natural soils under similar conditions.

3.3 Characterization of geogrid/C&DW interface

Figure 4(a) illustrates the evolution of the shear stresses as function of the shear displacement for direct shear tests carried out on geogrid/C&DW interface. Comparing the direct shear test results of this interface with those of C&DW characterization (Fig. 3), it can be observed that the shear strength of the interface is lower than that of the C&DW material.

Following the Coulomb failure criterion, this interface revealed a friction angle of 33.1° and an adhesion of 6.6 kPa.

The shear strength of soil/geosynthetic interfaces is typically defined by the coefficient of interaction or friction ratio. The coefficient of interaction, f_g, is defined as the ratio of the maximum shear stress in a geosynthetic/C&DW direct shear test to the maximum shear stress in a direct shear test on C&DW under the same normal stress.

Figure 4(b) presents the coefficients of interaction for the interface under analysis. The coefficients of interaction are in the range 0.66–0.74. It should be highlighted that the coefficients of interaction achieved for these interfaces are higher than the value usually assumed (in the absence of test results) in the design of geosynthetic reinforced structures. On the other hand, they are similar to those presented by other researchers (Abu-Farsakh et al., 2007) for soil/geogrid interfaces.

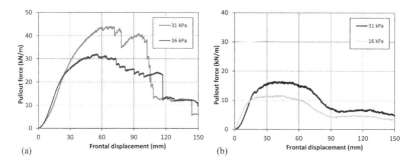

Figure 5.   Results of pullout test carried out for different confining stresses on: (a) geogrid as provided by manufacturer; (b) geogrid without transversal bars.

4   PULLOUT TESTS

4.1   *Test procedures*

In the upper part of a reinforced steep slope, the reinforcement can be pulled out from the soil, so the soil-geogrid interaction is better characterised through pullout tests. Laboratory pullout tests were also carried out to characterise the geogrid/C&DW interface. The test apparatus is composed by a pullout box (1000 mm × 1530 mm × 800 mm), a vertical load application system, a horizontal force actuator device and all the required instrumentation. More details about the device and test procedures can be found in Lopes & Ladeira (1996).

The C&DW was compacted inside the pullout device in four layers (150 mm thick) and, similarly to the direct shear tests, at 90% of maximum m odified Proctor dry density and at the optimum water content.

The geogrid was laid on the surface of the compacted C&DW and fixed to the clamps outside the box, when the fill level reached the pullout slot at the front of the box (300 mm). Two additional layers were placed, levelled, and compacted, resulting in a total C&DW thickness of 600 mm. The geogrid specimen inside the pullout box had 750 mm long and 250 mm wide.

The pullout tests were carried out with a constant displacement rate of 2 mm/min and normal stress at the interface level of approximately 16 kPa and 31 kPa.

4.2   *Pullout test results*

The effect of the confining pressure on the evolution of the pullout force per unit width with the front displacement is illustrated in Figure 5. Based on the evidence that, at the end of pullout tests, the junctions between the longitudinal and transversal bars were broken and, simultaneously, to assess the contribution of the transversal bars (less wide elements in Fig. 2) on the pullout resistance, pullout tests with geogrids without the bearing members were also carried out. Figure 5(a) presents the results of pullout tests carried out on intact geogrids (as provided by the manufacturer) and Figure 5(b) shows the results achieved with geogrids without the transversal bars.

Regardless of test conditions, the samples failed due to a deficient adherence with the surrounding material, i.e pullout failure occurred (failure due to lack of tensile strength can also occurs). As expected, the pullout resistance increases with the normal stress. For intact geogrids (Fig. 5a), the maximum pullout force for the normal stress of 16 kPa was 32.2 kN/m, mobilised at a frontal displacement of 55.1 mm. For the normal stress of 31 kPa the pullout resistance was 44.1 kN/m, mobilised at a frontal displacement of 67.5 mm.

Comparing the curves plotted in Figure 5(a) with those presented in Figure 5(b), it can be concluded that the maximum pullout force reduces approximately 64% and 62% when the geogrid transversal bars were removed, for confining pressures of 16 kPa and 31 kPa, respectively. This behaviour emphasizes the importance of the passive resistance mobilised on the geogrid bearing members (transversal bars).

The pullout interface coefficient, f_b, is defined as the ratio between the maximum shear stress mobilised at geogrid/C&DW interface during a pullout test and the direct shear strength of the C&DW under the same value of confining stress. For the interface under analysis, the pullout interface coefficient was 0.84 and 0.80 for confining pressures of 16 kPa and 31 kPa, respectively. These values are in the usual range of interface coefficients for soil/geogrid interfaces under similar conditions.

Comparing the pullout interface coefficients with the coefficients of interaction presented in Figure 4(b) for direct shear tests, it can be concluded that the former are slightly higher. Commonly this conclusion is also reached for interfaces with natural soils when the reinforcement elements are geogrids.

5 CONCLUSIONS

The results presented in this work show that the C&DW, properly selected and compacted, could exhibit shear strength similar to the backfill materials commonly used in the construction of geosynthetic reinforced structures.

The geogrid/C&DW interfaces have shown suitable shear strength and pullout resistance, with coefficients of interaction in the range of usual values for soil/geogrid interfaces.

This paper presents preliminary results of a broader research project under development. Even so, it is possible to state that the use of C&DW as backfill material in the construction of geosynthetic reinforced embankments seems to be an auspicious solution, able to balance the environmental and economic demands of current societies.

ACKNOWLEDGEMENTS

The authors would like to thank the financial support of Portuguese Science and Technology Foundation (FCT) and FEDER, through the Research Project: FCOMP-01-0124-FEDER-028842, RCD-VALOR – (PTDC/ECM-GEO/0622/2012).

REFERENCES

Abu-Farsakh, M., Coronel, J. & Tao, M. 2007. Effect of Soil Moisture Content and Dry Density on Cohesive Soil–Geosynthetic Interactions Using Large Direct Shear Tests. *Journal of Materials in Civil Engineering* 19: 540–549.

Arulrajah, A., Piratheepan, J., Disfani, M. & Bo, M. 2013a. Geotechnical and Geoenvironmental Properties of Recycled Construction and Demolition Materials in Pavement Subbase Applications. *Journal of Materials in Civil Engineering* 25: 1077–1088.

Arulrajah, A., Rahman, M.A., Piratheepan, J., Bo, M.W. & Imteaz, M.A. 2013b. Interface shear strength testing of geogrid-reinforced construction and demolition materials. *Advances in Civil Engineering Materials* 2: 189–200.

EN 933-11 2009. *Tests for geometrical properties of aggregates – Part 11: Classification test for the constituents of coarse recycled aggregate.* CEN.

Lopes, M.L. & Ladeira, M. 1996. Influence of the confinement, soil density and displacement rate on soil-geogrid interaction. *Geotextiles and Geomembranes* 14: 543–554.

Neves, J., Freire, A.C., Roque, A.J., Martins, I., Antunes, M.L. & Faria, G. 2013. Utilization of recycled materials in unbound granular layers validated by experimental test sections. *9th International Conference on the Bearing Capacity of Roads, Railways and Airfields, 2013 Trondheim, Norway.*

Santos, E.C.G., Palmeira, E.M. & Bathurst, R.J. 2013. Behaviour of a geogrid reinforced wall built with recycled construction and demolition waste backfill on a collapsible foundation. *Geotextiles and Geomembranes* 39: 9–19.

Vieira, C.S., Lopes, M.L. & Caldeira, L.M. 2013. Sand–geotextile interface characterisation through monotonic and cyclic direct shear tests. *Geosynthetics International* 20: 26–38.

Vieira, C.S., Pereira, P.P. & Lopes, M.L. 2014. Behaviour of geogrid-recycled Construction and Demolition Waste interfaces in direct shear mode. *Proceedings of the 10th International Conference on Geosynthetics, ICG 2014; ESTREL Convention Center Berlin; Germany; 21–25 September, 2014.*

Wastes: Solutions, Treatments and Opportunities – Vilarinho, Castro & Russo (eds)
© 2015 Taylor & Francis Group, London, ISBN 978-1-138-02882-1

Strategies for the treatment of metallurgical recycling slags

C. Pichler & J. Antrekowitsch

Christian Doppler Laboratory for Optimization and Biomass Utilization in Heavy Metal Recycling, Nonferrous Metallurgy, Montanuniversitaet Leoben, Leoben, Austria

ABSTRACT: The saving of the natural resources and the environment by different approaches is a global trend, which leads to so called "Zero Waste" concepts. To achieve this in different metallurgical processes, all generated by-products have to be reused instead of landfilling. Possible applications for residues could be the cement industry, the road construction or the fertilizer industry. However, they require different characteristics, like low amounts of leachable compounds or volume stability. To fulfil these demands, an additional treatment of such materials is needed.

A big variety of innovative processes is in development at the moment. Investigations were done in how recycling processes could be optimized to fulfil the mentioned requirements. Studies and designs of thermodynamic models, a necessary verification of the gained data in lab-scale trials as well as economic considerations for some special residues were done and partly presented in this paper.

1   GENERAL INSTRUCTIONS

Metallurgy often gets connected with environmental pollution. Such reproach may have been accepted in former times, but current produced residues from metallurgical processes are often used as secondary material for different applications, to avoid the landfilling option. This means in general the recovery of metals from slags, sludge or dust to produce an inert by-product without heavy metals and leachable compounds. For the flue dust recycling different processes are already developed and in use, e.g. the waelz kiln or the plasma technology. These facilities treat filter dust from the iron and steel industry to recover the zinc as saleable oxide. However, they produce partly huge amounts of slags, so called "recycling slags", which could be used as secondary resource for the cement production, the road construction or as railroad ballast. For such usability different requirements, like a defined amount of leachable compounds or a certain basicity are mandatory. To fulfil the requirements an additional treatment of these recycling slags or an optimization of the consisting recycling processes may be necessary. The target of the present research is to develop strategies for a removal of unwanted elements and compounds from recycling slags with a pyrometallurgical process and a utilization of the produced slag as secondary material for further use.

Because of the wide range of occurring residues a process which is suitable for all materials is not possible. The described developments in this paper are focused on zinc and lead containing residues. A characterization for the investigated slags was done by different methods to obtain useful information about the residue, which is important for a successful process development.

2   SLAGS FROM METALLURGICAL PROCESSES AND THEIR CHARACTERIZATION

Depending on the considered process, different slags occur in the metallurgy. If they do not fulfil the requirements regarding to elute behavior or mechanical properties for an usage as secondary

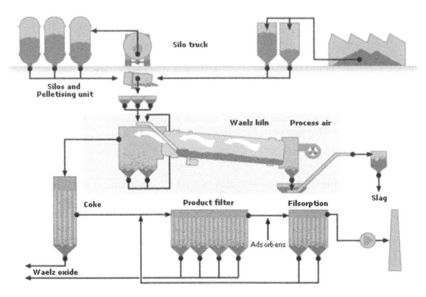

Figure 1. Waelz process of BEFESA Steel (BEFESA Steel, 2014).

material in another application, landfilling is the only possibility. To avoid this, some additional treatments are necessary.

2.1 *The waelz slag as example for a dumped slag in the metallurgy*

In Europe about 80% of the high zinc containing electric arc furnace dust is recycled by the waelz process. The waelz process with the SDHL-Technology in Figure 1 has been acknowledged as the best available technique for the recycling of steel mill dust. Beside electric arc furnace dust also other zinc bearing residues like acidic foundry dust and zinc bearing sludge are treated by this process. The whole waelz process can be divided into three process steps (Ruh and Krause, 2011):

- preparation of the raw material
- waelz line with slag granulation
- waelz oxide separation in the off gas cleaning step

In the first step, the raw materials are dosed, mixed and granulated to micro pellets, which are stored under roof for curing before they are transported to the kiln feeding pipe. In combination with defined quantities of reductants and fluxes the produced micro pellets are called self reducing pellets (SRP).

The following waelz line consists mainly of a rotary kiln with a typical length in the range of 40 to 65 meters and a diameter between 3 to 4.5 meters. During the SRP are continuously fed, the inclined kiln moves the micro pellets with a rotation speed of about 1.2 rpm to the discharge end. The retention time in the furnace is between 4 to 6 hours. After feeding to the kiln, the mixture is preheated by the countercurrent gas flow, produced by the air addition on the discharge end. In the following reduction zone temperatures between 1100 and 1200°C are reached. Zinc, lead and iron oxides are reduced in this reduction zone by carbon monoxide, which is built by the reaction of carbon with oxygen by the Boudouard-equation. While the iron is remaining in the kiln charge, zinc, lead, halogens and alkalis are mostly vaporized and afterwards reoxidized in the atmosphere of the rotary kiln and carried out at the feeding end by the off-gas stream. At the discharge end of the kiln, the air addition oxidizes the reduced iron to iron oxide and the produced heat is used for the chemical process in the middle of the kiln. For avoiding sticking it is essential, that there is no melting of the slag in course of the process.

Table 1. Chemical composition of crude and washed waelz oxide in weight-% (Rütten, 2008).

Weight-%	Crude waelz oxide
Zn	55.0–58.0
Pb	7.0–10.0
Cd	0.3–0.5
FeO	2.0–5.0
CaO	0.3–1.0
SiO$_2$	0.5–1.5
Cl	4.0–8.0
F	0.4–0.7
K$_2$O	1.5–2.0
Na$_2$O	2.0–2.5
C	0.5–1.5

The off-gas from the kiln with the valuable metal oxides reaches at first the dust settling chamber (DSC), where heavy particles – carry over – are separated. The off-gas is cooled by spraying water and/or addition of cooling air. The flue dust is collected by filter houses and impurities (such as Hg, As and Cd) and cyclic organic compounds are absorbed by activated carbon, which is added and caught by a second filtration step. After this procedure, the "TA-Luft" guidelines can be observed. The chemical analysis of crude waelz oxide can be seen in Table 1.

The advantages of this process are the comparative low investment costs, the low energy consumption and the simple process technology. The largest disadvantage is the high amount of remaining waelz slag after the process, which is about 700 kg per ton feeding material. The whole amount of this material goes to landfill (Bartusch et al., 2013). For the protection of the environment some additional treatment should be done to observe an inert material, which fulfils the requirements for the usage as secondary material in different applications.

For a successful process development, the properties of the considered slags must be known, which can be reached through a detailed characterization.

2.2 Characterization of metallurgical residues

To get detailed information from materials, an extensive characterization is necessary. Therefore different possibilities are available, like hot stage microscope, chemical analysis, scanning electron microscope, electron probe micro analysis or differential scanning calorimetry. One example of the huge variants, which are available for the characterization of different metallurgical residues is described in this paper. Figure 2 shows the element distribution within a waelz slag particle, done with a scanning electron microscope. In dependence on the formation of the residue, different zinc and lead containing phases can exist. They typically have influence on the parameters for a recycling concept. Such analysis can be informative in this case.

Figure 2 shows one area of the investigated sample with concentration values of different elements. As can be seen, zinc can be detected together with oxygen most of the time, sometimes it is combined with iron. Lead, which is present in low amounts, correlates also with oxygen. The slag components Ca and Si build together with oxygen their typical slag phases.

3 POSSIBLE TREATMENTS FOR METALLURGICAL SLAGS

Slags generated during the treatment of residues still contain heavy metals, which do not allow their utilization in the described fields. To reach a Zero-Waste concept an additional treatment is necessary. Therefore reaction models are available, depending on the physical conditions during the

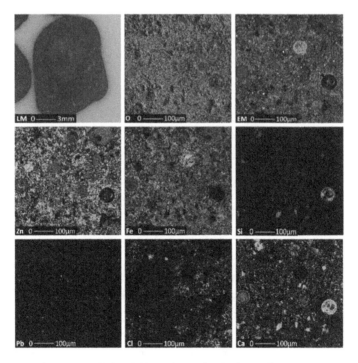

Figure 2.   Electron probe micro analysis (Pichler and Antrekowitsch, 2013b).

Figure 3.   Possible process models and utilizations (Pichler and Antrekowitsch, 2013a).

process. To gain an overview of ongoing investigations in metallurgical by-product treatment at the Christian Doppler Laboratory for Optimization and Biomass Utilization in Heavy Metal Recycling (CDL)/Dept. of Nonferrous Metallurgy, University of Leoben, Figure 3 gives a summary. Also some exemplary utilization possibilities are shown.

Table 2. Chemical composition of the used waelz slag.

Weight-%	ZnO	PbO	Fe_2O_3	FeO	$Fe_{met.}$	SiO_2	CaO	MgO
recycling-slag	9.9	1.1	5.7	24.4	11.0	7.2	22.8	3.4

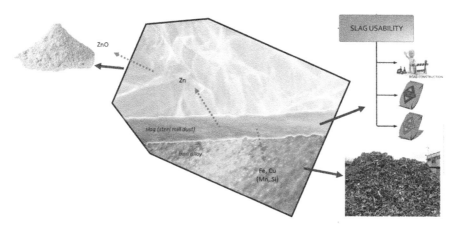

Figure 4. Principle of the metal bath process.

The term "recycling slag" mentioned in Figure 3 refers to the by-product which occurs during the recycling of zinc and lead containing residues. The valuable fraction is zinc- and leadoxide, recovered from the off-gas stream. Some of these processes are not able to remove the heavy metals to low contents, because of their defined parameters. For a possible utilization a low content of these components is mandatory.

In the following two reaction models are explained. One of these is already tested with practical investigations and for the other one a process simulation and design with FactSage™ 6.3 was done. Practical trials in this case are current under investigation and planned for the future.

3.1 Gas-Solid Reaction (carbon and slag(s))

First practical tests for the removal of zinc and lead out of recycling slags were done in a TBRC (Top Blown Rotary Converter) at the Chair of Nonferrous Metallurgy, University of Leoben. The input material for these trials was a waelz slag, like it is shown in Table 2.

Different preparation procedures and additives, like $CaCl_2$ were tested. The basic conditions were the same for all the different input materials, which were charged in the TBRC (1050°C process temperature and 180 min. residence time). The TBRC vessel is heated by a natural gas/oxygen burner and has a maximal rotating speed of 3 rpm. Pet coke and CO/H_2 (from a sub stoichiometric combustion) was used as reducing agent. During the whole treatment slag and carbon are solid, carbon reacts to CO and the reduction takes place. Because of the Boudouard-equation the resulting CO_2 reacts again to CO. Due to this reason the model can be described as gas-solid reaction, although solid carbon is used.

The results showed, that the reactions and formations, which take place during the treatment in the waelz kiln, have a significant influence on the possibilities to remove zinc and lead. This reason leads to investigations on a process in liquid state, where a total removing of heavy metals should be possible. This also includes a recycling of the containing iron.

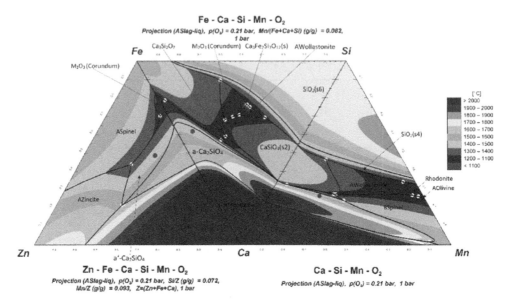

Figure 5. Iron bath process: thermodynamic calculation (Pichler and Antrekowitsch, 2013b).

3.2 *Liquid-Solid (Liquid-Liquid) Reaction (metal(l) and slag(s))*

Another option for the recovery of heavy metals and the production of an inert slag is the reduction of solid (liquid) slag on a liquid metal bath with solid and/or dissolved carbon. Such a process is known from the recycling of filter flue dust. The advantage is a simultaneous metal recovery. Some metals and oxides are vaporized at process temperature and collected via the off gas in the filter house. The non-volatile metals in the slag get collected in the liquid metal bath below the slag. The principle of this process is shown in Figure 4.

One important thing in this case is the knowledge about the behavior of the slag via the process. For the treatment of waelz slag, a process simulation was done with FactSage™ 6.3. The obtained result is shown in Figure 5.

Because of the iron oxide removing the melting point could increase which leads to problems during the treatment. With thermodynamic programs like FactSage™ it is possible to predict such effects and find solutions to do such a process in liquid state.

Figure 5 shows the procedure of the reduction in different phases. It is constructed out of three ternary systems. During the first step (Zn-Ca-Fe) the whole zinc oxide is reduced and by this, the process enters the next step (Fe-Ca-Si). Under the present condition the iron oxides also get reduced. After the reduction processes a final slag with a high melting point results. To lower it, different additives like blast furnace slag or foundry sand can be added.

First practical trials with this reaction model are done in a TBRC and showed promising results. Further investigations with different residues with this concept are planned for the future.

4 CONCLUSION

Different technologies allow the recycling of residues from the metallurgical industry. Examples are the recycling of dust from the steel mill, the secondary copper production and secondary lead industry. These residues contain certain amounts of zinc, lead, copper, tin and iron (Piret, 1995; Stubbe et al., 2008). Especially the treatment of flue dust from the iron and steel industry to recover the containing zinc is common but still shows an optimization potential. Various processes like the waelz kiln, the Rotary Hearth Furnace, the PRIMUS technology, the Plasma Arc Fuming and the

PIZO process are developed and partly in industrial use. However, the most important and dominant one is the waelz process, which produces again high amounts of slag (Rütten, 2009). This slag is called recycling slag, which means a residue resulting from a recycling process. Most of the time landfilling is the only option for such materials. To satisfy the Zero-Waste idea, an additional treatment must be performed, to generate a product for the application in other areas, like the cement industry or the road construction. Within this research some possible reaction models were evaluated within practical and theoretical investigations. Promising results were received with a liquid metal bath process, which is known from the recycling of different dust from the copper and iron- and steel industry.

REFERENCES

BEFESA Steel 2014. Available at: www.befesa-steel.com, retrieved 2014.

Bartusch, H., Fernández Alcalde, Ana María and Fröhling, M. 2013. Erhöhung der Energie- und Ressourceneffizienz und Reduzierung der Treibhausgasemissionen in der Eisen-, Stahl- und Zinkindustrie (ERESTRE). *Erhöhung der Energie- und Ressourceneffizienz und Reduzierung der Treibhausgasemissionen in der Eisen-, Stahl- und Zinkindustrie (ERESTRE)* 2.

Pichler, C. and Antrekowitsch, J. 2013a. Innovative Recyclingslag-Treatment. In *WASTES: solution treatment opportunities, Braga*. Guimaraes: CVR. 329–335.

Pichler, C. and Antrekowitsch, J. 2013b. *Slags from metal recycling processes as innovative secondary resource*. Dubai.

Piret, N. 1995. *Criteria for Optimization of Recycling Processes of Primary and Secondary Copper*.

Ruh, A. and Krause, T. 2011. The Waelz Process in Europe. In *3. Seminar Networking between Zinc and Steel, Austria, Leoben*. GDMB, Clausthal-Zellerfeld.

Rütten, J. 2008. *Application of Waelz and Other Processes on Treatment of Steel Mill (EAF) Dust*.

Rütten, J. 2009. The Waelz Kiln. *ValoRes GmbH* 10.

Stubbe, G., Harp, G., Hillmann, C. and Scholl, W. 2008. Schließung von Stoffkreisläufen beim Einsatz von verzinktem Schrott im Oxygenstahlwerk. *Stahl und Eisen* 128: 55–60.

Wastes: Solutions, Treatments and Opportunities – Vilarinho, Castro & Russo (eds)
© 2015 Taylor & Francis Group, London, ISBN 978-1-138-02882-1

Alkaline hydrolysis applied to animal tissues treatment

S.C. Pinho, M.F. Almeida & O.P. Nunes
LEPABE, Faculty of Engineering, University of Porto, Porto, Portugal

ABSTRACT: In this study, the minimum temperature and NaOH concentration for the total destruction of animal tissues were determined using either pork or beef as surrogates. The alkaline hydrolysis trials were performed with 1 M and 2 M NaOH solutions at different temperatures and different hydrolysis times. The pork and beef showed a similar behavior when subjected to alkaline hydrolysis. The destruction of meat was faster than the bone. The meat was totally hydrolyzed above 90°C in less than 60 minutes. The effluents obtained after alkaline treatment are hazardous due to their very high pH. Although with very high organic load, the effluents produced in alkaline hydrolysis were biodegradable after neutralization. Therefore, it may be acceptable to discharge the neutralized effluents in a domestic wastewater treatment plant.

1 INTRODUCTION

One of the fundamental policies for waste management is based on waste quantities reduction. For this purpose the Best Environmental Practices (BEP) include source reduction, segregation, resource recovery and recycling. Likewise, the selection of the treatment system takes into account the waste characteristics, the volume and mass reduction obtained in the treatment, among many others (Chartier et al., 2014). Animal by-products are a type of waste that has increased over the last years due mainly to the growing consumption of meat. Animal by-products are animal carcasses, parts of animals or other materials which come from animals but are not meant for humans to eat. Of the 47 million tons of animals slaughtered for meat production in Europe every year, 17 million tons of by-products, such as minus hides, skins and bones for gelatin production are handled by the animal by-products industry. However, considerable numbers of carcasses are also left to rot or are illegally dumped (BREF, 2005). This illegal disposal poses a potential risk to the public health and to the environment.

In 2002, with the appearance of BSE emerged restrictions that have led to an increased proportion of solid material being disposed of to landfill and by incineration. The limits placed on the traditional uses for animal by-products have led to further alternative uses and to new methods for disposal. Indeed, animal proteins have a very high biological value which opens wide possibilities for their use for generating of energy. Disposal by incineration and co-incineration is an advantageous energetic valorization process however, due to its high organic matter content, there is a great potential for anaerobic digestion. Nevertheless, this technology requires long process time and large facilities due to slow anaerobic processes involved (Ro et al., 2007).

Therefore, the alkaline hydrolysis emerges as a waste treatment option, mainly because it is able to significantly reduce the volume of animal wastes and to produce sterile by-products (Tracker, 2004) which can be used for soil fertilization (Kalambura et al., 2011). The efficacy of alkaline hydrolysis on the destruction of tissue wastes, including anatomical parts, organs, placenta, blood, body fluids, specimens, human cadavers and animal carcasses has been proven in some studies (Tracker, 2004). However, conditions for destruction are reported to be 150°C with a time of contact between three and eight hours. This work intended to study the minimal conditions for destruction of animal tissues. The behavior of animal tissues, such as pork and beef, when treated by alkaline hydrolysis was studied. The efficiency of the treatment was assessed by determination of weight losses on the materials and characterization of total organic carbon (TOC), chemical oxygen demand (COD) and biochemical oxygen demand after five days (BOD$_5$) in the effluents resulting from the alkaline hydrolysis treatment.

2 MATERIALS AND METHODS

The treatments were carried out in a Parr batch reactor with a titanium vessel of 450 mL capacity under temperature control and with a pressure gauge, with a heating rate of 10°C/min. The samples were heated at 80°C, 90°C, 95°C, 100°C and 110°C, during 30 to 240 minutes. A liquid/solid ratio of 5:1 (w/w) was used in all the tests. The tests were made using samples of 20 to 30 g of pork meat including bones. The alkaline solutions used were 1 M and 2 M NaOH, the selection of these concentrations was based on studies (Taguchi et al., 1991; Taylor et al., 1997). To compare the behavior of pork and beef samples, both containing bone, tests were performed at 90°C using 1 M or 2M NaOH solutions over 150 minutes. To compare the hydrolysis efficiency of meat versus bone, pork meat or bone, samples were hydrolyzed at 90°C, with 1 M or 2 M NaOH, at different times. After cool down to ambient temperature, the resulting solid product was filtered, washed with distilled water in order to remove all of the sodium hydroxide, after which it was dried at room temperature, further held for 48 hours in a desiccator and finally weighted. The pH, TOC, COD and BOD$_5$ were determined in the solutions resultant from the treatments. TOC in solutions were determined with a Shimadzu TC analyzer model TOC-VCSH, according to EN 1484 (1997). Measurements of pH were made with a pH-meter model 632 of Metrohm. COD was determined following 5220 D: Closed reflux – colorimetric method; and BOD$_5$ following 5210 B: 5-Day BOD method as described by the Standard Methods for Examination of Water and Wastewater (APHA, 1998). Also, the main chemical compounds, in hydrolyzate at 110°C and 1 M of NaOH solution, were analyzed by gas chromatography with mass detector (GCMS) using an Agilent HP 6890/MSD 5793N from HP, 30 m × 0.25 mm I.D., 0.5 μm P/N 19091S-133 column; and using as carrier gas He at constant flow of 1.2 mL/min. Tests were carried out in the following conditions: split-splitless injector at 280°C; oven 1 minute at 50°C, followed by heating at 10°C/min till 300°C; transference line at 290°C; and MSD scan mode. The separated compounds were identified using NIST 1998 library match. Additionally, thermogravimetric (TG) and differential scanning calorimetric (DSC) analyses were performed on the samples before and after alkaline hydrolysis with 1 M NaOH solution. The method used in TG and DSC analyses were performed using two equipments (Setaram, model 92-16.18 and model Labsys, respectively). Briefly, samples of 40 mg were heated at a rate of 10°C/min up to 135°C, held 3600 s at this temperature, after which they were heated up to 200°C at the same rate.

3 RESULTS AND DISCUSSION

For the temperature and NaOH concentrations used in this work, it was verified that when temperature was raised, the time required to hydrolyze the pork and beef samples decreased. At 100°C and using 1 M NaOH, 45 min were sufficient to destroy all the samples. Accordingly, the percentage of solid residue decreased with the increase of the temperature (Table 1). The increase of NaOH concentration, under specific conditions, namely at 90°C and 95°C for 1 M and 2 M NaOH, did not influence the hydrolysis time. Indeed, small differences were found in all the analyzed parameters for tests carried out at 1 M and 2 M NaOH. Nevertheless, differences in TOC, COD and BOD$_5$ were observed, which may be due to variation on the proportion of meat and bone in each tested sample. In the most aggressive condition of temperature and NaOH concentration, the proteins were hydrolyzed and esterified as showed in the chromatographic analysis of the hydrolyzate of 110°C and 1 M NaOH solution. The molecular ions more frequently detected and the ions with higher relative abundance had molecular masses of 28, 44, 72, 86 and 117. The main chemical structures identified include alkyl group and chains, amides and esters. NIST 1998 data base proposes the

Table 1. Alkaline hydrolysis conditions of samples composed by pork meat including bones versus hydrolysis time, residue amount, TOC, COD and BOD$_5$ of the effluent produced.

Experimental conditions	Time min	Residue w/w %	Effluent		
			TOC (g/kg)	COD (g/kg)	BOD$_5$ (g/kg)
80°C, 1 M	240	25.0	97.5	n.d.	n.d.
90°C, 1 M	90	12.3	101.0	345.8	316.6
90°C, 2 M	90	17.5	100.0	307.0	220.8
95°C, 1 M	50	15.0	122.5	360.2	319.2
95°C, 2 M	50	4.9	101.6	312.0	271.2
100°C, 1 M	45	3.2	110.6	351.5	302.2
100°C, 2 M	40	3.6	83.0	287.4	204.7
110°C, 1 M	35	3.8	115.7	356.9	260.9
110°C, 2 M	35	3.1	116.1	345.6	230.3

n.d. not determined

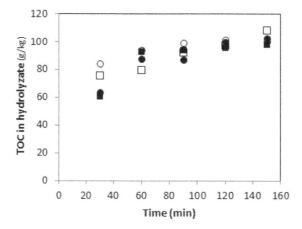

Figure 1.   Alkaline hydrolysis of pork ((□) 1 M NaOH; (■) 2 M NaOH) and bovine ((○) 1 M (●); 2 M NaOH) meat at 90°C over time.

presence of propanamide, 4-methylphenol, 4-methylpentanamide, ethyl ester L-isoleucine, indole, and triethyl phosphate.

When bovine and pork samples, both containing meat and bone, were compared small differences were observed only for short times of contact (up to 60 min). For higher times of hydrolysis, the TOC values obtained for both type of samples were similar, independently of the NaOH concentration (1 M or 2 M) as shown in Figure 1.

When meat and bone were hydrolyzed separately at 90°C, it was observed that meat requires less time of contact than bone, as expected due to their composition. The meat is composed mainly by water followed by proteins and lipids, which are easily hydrolyzed. The bone is composed predominantly by mineral components (carbonated hydroxyapatite), organic components (mainly type I collagen, 22% by weight and 36% by volume) and water (Collins et al., 2002; Figueiredo et al., 2012). For the same temperature and NaOH concentration, higher TOC values and lower solid residue values were obtained for increasing hydrolysis time, i.e., the treatment efficiency increased for both type of samples (Table 2). As expected, for the same time of contact and temperature, treatment with 2 M NaOH showed a better efficiency, for both type of samples (Table 2). Even

Figure 2. Hydrolyzed bone and hydrolysate.

Table 2. Alkaline hydrolysis conditions versus hydrolysis time, residue amount of the meat and bone and TOC in hydrolyzate.

	Experimental conditions	Time Min	Residue %	TOC (in hydrolyzate) g/kg
Meat	90°C, 1 M	30	6.7	90.2
		60	0.0	94.0
	90°C, 2 M	30	1.0	89.1
		45	0.0	92.0
Bone	90°C, 1 M	30	66.5	77.9
		45	38.2	85.6
		60	18.2	104.4
	90°C, 2 M	30	52.7	90.0
		45	23.4	92.0

under the harsher conditions tested (90°C, 2 M NaOH and 60 minutes), it was not possible to completely destroy the bone samples, since solid residues constituted 20% of the initial weight. This resultant solid residue could be crushed with reduced pressure to powder size fragments, probably because hydrolysis promotes the digestion of the structural collagen, which is needed to strengthen the bone structure (Collins et al., 2002) but not the mineral fraction, mainly constituted by calcium phosphate (Kaye et al., 1998) (Fig. 2). In contrast, 90°C, 1 M NaOH and 60 minutes were sufficient to achieve complete meat hydrolysis as shown in Table 2.

The analysis of the TG profiles of bone showed that the highest weight loss occurred during the first heating step, up to 135°C (Figure 3). Untreated bone showed a weight loss of 5% due, essentially, to water loss. In opposition, TG profiles of hydrolyzed bone at 80°C, 90°C and 100°C with 1 M NaOH showed weight losses of 10%. Most probably, these weight losses were due to the increased amount of water absorbed by the matrix during the treatment. The fact that hydrolyzed bone showed similar TG profiles independently of the temperature, suggest that this parameter does not seem to influence the hydrolysis of the bone inorganic matrix. Similar conclusions were obtained through the analysis of the DSC profiles. These profiles showed an endothermic peak corresponding to the evaporation of water for both untreated and hydrolyzed bone, which occurred at a lower temperature for the hydrolyzed ones. The TG analysis of untreated meat showed a high weight loss of 74% essentially due to water evaporation, the main component of meat (data not shown). As expected, it was not possible to obtain TG profiles of treated meat, since no residues were obtained.

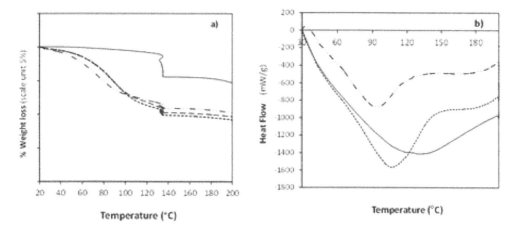

Figure 3. TG (a) and DSC (b) profiles of bone; (—) untreated; (– · · –) alkaline treated at 100°C (- - -) alkaline treated at 90°C; (· · · ·) alkaline treated at 80°C.

The effluents obtained from alkaline hydrolysis of animal tissues showed a brownish color and were gelatinous, suggesting that the collagen was solubilized (Collins et al., 2002). The effluents showed also a higher value of TOC, COD and BOD5 with average values of about 20 000 mg/L, 70 000 mg/L and 50 000 mg/L, respectively. Although with very high pH, approximately 13, and organic load, the effluents produced in alkaline hydrolysis are, presumably, biodegradable, since a BOD_5/COD ratio above 0.70 was obtained after neutralization. Therefore, these effluents might be discharged and treated in a wastewater treatment plant.

4 CONCLUSIONS

Higher NaOH concentrations and longer times of contact favored the hydrolysis of animal tissues. When tested separately, the meat was hydrolyzed faster than the bone when subjected at same treatment conditions. Meat hydrolysis was complete while bone generated a solid residue, which was easily crushed into a powder. Temperature at which hydrolysis was carried out influenced considerably the time required to destroy animal tissues as well as the solid residue obtained. Effluents resultant from alkaline hydrolysis of animal tissues, although with very high pH and organic load, were, presumably, biodegradable after neutralization.

REFERENCES

APHA – American Public Health Association. 1998. Standard Methods for the Examination of Water and Wastewater. 20th edition, Public office: American Public Health Association Washington DC.
BREF – European Commission 2005. Integrated Pollution Prevention and Control Reference Document on Best Available Techniques in the Slaughterhouses and Animal By-products Industries. Joint Research Center, Institute for Prospective Technological Studies (IPTS). Available at: http://eippcb.jrc.ec.europa.eu/reference/BREF/sa_bref_0505.pdf
Chartier, Y., Emmanuel, J., Pieper, U., Pruss, A., Rushbrook, P., Stringer, R., Townend, W., Wilbum, S., Zghondi R. 2014. Safe Management of Wastes from Health Activities. 2nd edition, World Health Organization, Geneva.
Collins, M.J., Nielsen-Marsh, C.M., Hiller, J., Smith, C.I., Roberts, J.P. 2002. The survival of organic matter in bone: A review. *Archaemetry* 44 (3): 383–394.

Figueiredo, M.M., Gamelas, J.A.F., Martins, A.G. 2012. Characterization of Bone and Bone-Based Graft Materials Using FTIR Spectroscopy, Infrared Spectroscopy – Life and Biomedical Sciences, Prof. Theophanides Theophile (Ed.).

Kalambura, S., Voća, N., Krička, T., Sindrak, Z., Spehar, A., Kalambura, S. 2011. High-risk biodegradability waste processing by alkaline hydrolysis. *Archives of Industrial Hygiene and Toxicology* 62, 249–253.

Kaye, G., Weber, P., Evans, A., & Venezia, R. 1998. Efficacy of alkaline hydrolysis as an alternative method for treatment and disposal of infectious animal waste. *Contemporary topics in laboratory animal science* 37 (3): 43–46.

Ro, K.S., Cantrell, K., Elliott, D., Hunt, P.G. 2007. Catalytic Wet Gasification of Municipal and Animal Wastes. *Ind. Eng. Chem. Res.* 46 (26): 8839–8845.

Taguchi F., Tamai Y., Uchida K., Kitajima R., Kojima H., Kawaguchi T., Ohtani Y., Miura S. (1991) Proposal for a procedure for complete inactivation of the Creutzfelt-Jakob disease agent. *Archives of Virology* 119, 297–301.

Taylor D.M., Fernie K., McConnell I. (1997). Inactivation of the 22A strain of scrapie agent by autoclaving in sodium hydroxide. *Veterinary Microbiology* 58, 87–91.

Thacker, L.H. 2004. Carcass Disposal: A Comprehensive Review. National Agricultural Biosecurity Center, Kansas State University.

Wastes: Solutions, Treatments and Opportunities – Vilarinho, Castro & Russo (eds)
© 2015 Taylor & Francis Group, London, ISBN 978-1-138-02882-1

Separation of plastics: Jigging versus froth flotation

F.A. Pita & L.A. Menezes
Geosciences Centre, Department of Earth Sciences, Faculty of Sciences and Technology, University of Coimbra, Coimbra, Portugal

A.M. Castilho & N. Rodrigues
CEMUC – Centre for Mechanical Engineering, Department of Earth Sciences, Faculty of Sciences and Technology, University of Coimbra, Coimbra, Portugal

J.A. Pratas
MARE – Marine and Environmental Sciences, Department of Earth Sciences, Faculty of Sciences and Technology, University of Coimbra, Coimbra, Portugal

ABSTRACT: This paper compares the separation of three bi-component plastics mixtures (PS + PMMA; PS + PET; PS + PVC) by jigging and froth flotation. The effect of density of plastics, size and shape of the particles was analysed. Separation of PS + PMMA and PS+PVC mixtures by jigging and froth flotation improved with the increase of the particles size. Jigging provided better results for the fine fractions. The PS+PET mixtures showed the worst results and froth flotation worsened with the increase of the particles size. Froth flotation provided better results for fine fractions. Jigging separation of a bi-component mixture presents better results if the light plastic, that overflows, has lamellar shape and the heavy plastic, that sinks, has regular shape. The efficiency of separation of two plastics by froth flotation improved when they have different hydrophobicity and when the least hydrophobic plastic has a higher density and a rounder shape.

1 INTRODUCTION

Plastics are used in more and more applications and they have become essential to our modern economy. This wide-ranging application is consequence of its versatility and excellent properties. The increase in plastics production and consumption has been largely responsible for the increased in municipal solid wastes (MSW) production. Plastics waste has become one of the larger categories in municipal solid waste. In Portugal plastics account for approximately 11% of MSW by weight, however, for more than 30% of MSW by volume (because of the low density of plastic waste). Over time plastic has acquired a negative reputation, due to its difficult decomposition. When properly collected and processed, plastics can be made useful through recycling and/or energy recovery. However, in order to recycle this material it is necessary to separate the several different types of plastics.

Several separation methods used in mining industry are employed to separate a mixed plastic based on plastic type. These included the application of gravity methods (Pascoe, 1999; Ferrara et al., 2000; Hori et al., 2009; Ito et al., 2010; Carvalho et al., 2009; Gent et al., 2011; Richard et al., 2011), froth flotation (Shibata et al., 1996; Shen et al., 2002; Agante et al., 2004; 2004; Burat et al., 2009; Kangal, 2010), and electrostatic separation (Dodbiba et al. 2005, Bedeković et al., 2011).

Jigging is a gravity separation process that occurs in a pulsating bed where the separation is based upon density difference of materials. A mixture of solid-water is placed in a perforated vessel at the bottom (jig), through which are forced vertical currents of water, ascending and descending. The action of those currents either expands (ascending currents) or compacts (descending currents) the pulsating bed. The particles stratification is based on density differences between the constituent particles of the mixture. The densest particles are in the base of that stratum, being kept inside the jigging cell while the particles of lower density are in the superficial layers, being overflowed.

Froth flotation is the most important separation method used in mining industry. It is a physical-chemical process based on selective adhesion of specific particles to the air (hydrophobic) and the others to the water (hydrophilic) in a solid/water pulp. The separation takes place in a container (flotation cell), where the water and the particles are put together, and then air is continuously injected giving rise to the formation of air bubbles. This pulp is previously conditioned with small quantities of specific chemicals reagents to promote the selective formation of aggregates between solid particles of a given composition and air bubbles. Hydrophobic particles adhere, after collision, to the air bubbles, which move upwards to the top of flotation cell where they are recovered as the floated product. Hydrophilic particles settle in the pulp, become the non-floated product or sunken.

Plastics are naturally hydrophobic, so selective wetting of one or more components is necessary for the plastic mixture separation by froth flotation.

This study aims to compare the separation of three different plastics mixtures through the use of jigging and froth flotation, and to analyse the influence of size and shape of the particles.

2 EXPERIMENTAL MATERIALS AND METHODS

2.1 Materials

Four different types of post-consumer plastic: PS (Polystyrene), PMMA (Polymethyl methacrylate), two PET (Polyethylene Terephthalate) and PVC (Polyvinyl Chloride) have been employed in this study. In order to analyse the influence of the particle size the floated and non-floated products of jigging and flotation were dried and sieved, being obtained the following size fractions: +1–1.4 mm, +1.4–2 mm, +2–2.8 mm, +2.8–4 mm and +4–5.6 mm.

The densities of these plastics, measured in the laboratory, are: PS – 1.047 g/cm^3; PMMA – 1.204 g/cm^3; PET – 1.364 g/cm^3 and PVC – 1.209 g/cm^3.

The shape of the particles was described by the shape factor $F = D_{min}/D_{max}$ (where D_{min} and D_{max} represent the minimum and the maximum size diameters). Plastic particles have different shapes according to the plastic type and with the particles size. PMMA particles have more regular shapes and PET presents more lamellar shape, with a medium thickness of 0.3 mm. Particles of PMMA, PS and PVC have an F-shape that increases with increasing particle size and PET has an F-shape that decreases with particle size.

Separation by jigging and froth flotation was carried out on bi-component mixtures, with each type of plastic representing 50% of the weight. The mixtures were the following: PS+PMMA, PS+PET and PS+PVC.

2.2 Flotation experiments

Flotation tests were carried out in a Denver cell with a capacity of 3 dm^3 at a rotational speed of 600 rpm. Each test used 80 g of material that was conditioned with tannic acid (wetting agent) and with methyl isobutyl carbinol (MIBC) as frothing agent at a constant concentration of 30×10^{-3} g/L in all flotation experiments. The suspension of plastics was conditioned with the wetting agent for about 5 minutes and later with the frother for about 2 minutes before flotation experiments. The floated product was collected during 6 minutes, dried, screened and weighed. The non-floated (sink) product remaining in the flotation cell after the experiment was also dried, screened and weighed.

The ideal tannic acid concentration that led to the most efficient separation of each type of bi-component mixture was selected from previous tests.

2.3 Jigging experiments

The jigging tests were carried out in a Denver jig with a rectangular section with 10×15 cm. In each test it was used 0.5 kg of material, i.e. 0.25 kg of each type of plastic. Jigging time for separation was about 6 minutes. Since PS plastic has a lower density than the other three plastics, PS is overflowed and the others are kept inside the jigging cell. After the experiments, the sink was removed from the jig. The floated and sink products were dried, screened and weighed. For writing easiness the product that overflows in the jigging operation will be designated by floated; the one that remains inside will be designated by sink.

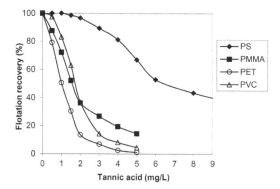

Figure 1. Floatability of four plastics versus tannic acid concentration (MIBC: 30×10^{-3} g/L).

The control of flotation and jigging separation was carried out from recovery and grade of each type of plastic in the floated and in the sink. These parameters were obtained after manual sorting and weighing of the two types of plastics in the floated and in the sink products. This was possible because plastics have different colours and different shapes. The recovery of each plastic in the floated and in the sink was calculated dividing the mass recovered in the floated and in the sink by the initial mass.

3   RESULTS AND DISCUSSION

Plastics are naturally hydrophobic, requiring a wetting agent that promotes their wettability, in order to prevent the attachment to air bubbles and their subsequent flotation. Figure 1 shows the recovery of four plastics in the floated versus tannic acid concentration.

The flotation recovery decreased with the increase of the tannic acid concentration. The PS was the plastic with higher floatability, since it was only depressed with high tannic acid concentrations. Concentration of tannic acid below 1.5 mg/L had no effect on PS floatability, while the other three plastics were depressed at very low tannic acid concentrations. These plastics showed similar variation of the floatability with tannic acid concentration. The floated recovery decreased sharply with the addition of the tannic acid. When increasing the concentration of tannic acid up to 5 mg/L, with the exception of PS, the recovery of plastics decreased to almost zero. The PET was the plastic that presented lower floatability.

It is possible to separate the four plastics into two groups. The first group constituted only by the PS, which has a high floatability, and the second group that includes the other three plastics, with similar floatability.

3.1   *PS+PMMA Mixture*

In order to determine the tannic acid concentration that led to the best separation of the PS+PMMA mixture, flotation tests were developed using four tannic acid concentrations: 1.5, 2, 3 and 4 mg/L. It was verified that the concentration of 3 mg/L led to the most efficient separation for the finer fraction, while the concentration of 2 mg/L led to the most efficient separation for the coarser fraction. In this study are only presented the results corresponding to the tannic acid concentration of 2 mg/L, because for this concentration the PS recovery in floated is similar to the PS recovery obtained by jigging.

In jigging tests, the recovery of PS in the floated of all fractions is close to 100% (Table 1). However, in the froth flotation, PS recovery in the floated decreased slightly with the particles size. In both separation methods, PMMA recovery in the sink increased with the particle size, with a marked increase in froth flotation.

Jigging presented better separation than froth flotation, in particular for the three finer fractions. Froth flotation showed poor results for the two finest fractions. The two separations improved with the increase of the particles size but this improvement is more pronounced in froth flotation. The separation

Table 1. Results of separation by jigging and flotation of the PS+PMMA mixture, for five size fractions.

| | Size fraction (mm) | FLOATED | | | | SINK (%) | | | | Separation Efficiency (%) |
| | | Recovery (%) | | Grade (%) | | Recovery (%) | | Grade (%) | | |
		PS	PMMA	PS	PMMA	PS	PMMA	PS	PMMA	
Jigging	1–1.4	99.6	32.6	75.4	24.6	0.4	67.4	0.6	99.4	67.0
	1.4–2	99.6	27.6	78.3	21.7	0.4	72.4	0.5	99.5	72.0
	2–2.8	99.6	11.8	89.4	10.6	0.4	88.2	0.4	99.6	87.8
	2.8–4	99.7	7.3	93.1	6.9	0.3	92.7	0.3	99.7	92.4
	4–5.6	99.7	5.4	94.9	5.1	0.3	94.6	0.3	99.7	94.3
Flotation	1–1.4	98.3	82.9	54.2	45.8	1.7	17.1	8.9	91.1	15.4
	1.4–2	97.9	62.1	61.2	38.8	2.2	37.9	5.4	94.6	35.8
	2–2.8	97.2	28.6	77.3	22.7	2.9	71.4	3.8	96.2	68.6
	2.8–4	94.8	5.0	95.0	5.0	5.2	95.0	5.2	94.8	89.8
	4–5.6	93.4	0.00	100.0	0.00	6.6	100.00	6.2	93.8	93.4

efficiency of PS+PMMA mixture improved with particle size: from 67.0% for the finer fraction, to 94.2% for the coarser in the jigging and from 15.4% for finer fraction, to 93.4% for coarser fraction in the froth flotation.

The fine particles of PMMA led to worst results in jigging since they are easily dragged to the overflow. Over the five fractions, the jigging sink material was almost pure PMMA. It should be noted that this also occurred with the other two plastics mixtures later analysed. This was the result of extended jigging tests that aimed to overflow almost all the PS.

In froth flotation the decrease of PMMA recovery on the floated associated with the increase particle size is due to the greater weight of the coarse particles, because they require the attachment of many air bubbles to form particles-bubbles aggregates with a density lower than one.

Thus, the particle size control is important for separation. For particles with size coarser than 2.8 mm, good separations by jigging and froth flotation were obtained.

3.2 PS+PET Mixture

The effect of particle size on the separation efficiency for the PS+PET mixture using jigging and froth flotation is shown in Table 2. In the case of froth flotation, the results correspond to tests with 2 mg/L of tannic acid concentration. For the three finer fractions froth flotation produced better quality separation than jigging, with an efficiency rate of about 83%. For the coarser fraction, jigging showed lightly greater separation efficiency than froth flotation.

In both separation processes, the PET recovered in the floated increased with particle size, leading to a decrease in the PS grade of the floated material. This is most evident in the froth flotation separation. Therefore, separation efficiency, in particular froth flotation, worsens with the increase of the particles size.

Jigging showed a greater recovery of PS in the floated than froth flotation, but also greater recoveries of PET. For the five fractions, froth flotation produced a richer floated in PS than the one produced by jigging.

In jigging, a practically pure PET product was obtained with a grade of approximately 99% for the five size fractions.

Unlike to the PMMA, no significant effect was observed due to the PET particle size on jigging separation. The separation efficiency was slightly worst for the coarser fractions. So, despite greater weight of coarser particles, the lamellar shape of the particles was responsible for greater difficulty in penetrating and remaining in the jigging bed. Also Pascoe (1999) and Ferrara et al. (2000) verified that, in the separation of plastics with lamellar shape by media separation cyclones, the underflow recovery of the heavy plastic decreased with the increasing of the particle size.

Table 2. Results of separation by jigging and flotation of the PS+PET mixture, for five size fractions.

	Size fraction (mm)	FLOATED				SINK (%)				Separation Efficiency (%)
		Recovery (%)		Grade (%)		Recovery (%)		Grade (%)		
		PS	PET	PS	PET	PS	PET	PS	PET	
Jigging	1–1.4	99.5	26.3	79.1	20.9	0.5	73.7	0.7	99.3	73.2
	1.4–2	99.5	26.9	78.7	21.3	0.5	73.1	0.7	99.3	72.6
	2–2.8	99.6	27.8	78.2	21.8	0.4	72.2	0.6	99.5	71.8
	2.8–4	99.5	29.9	76.9	23.1	0.5	70.1	0.7	99.4	69.6
	4–5.6	99.6	31.5	76.0	24.0	0.4	68.5	0.6	99.4	68.1
Flotation	1–1.4	97.4	14.1	87.4	12.6	2.6	85.9	2.9	97.1	83.3
	1.4–2	96.5	12.7	88.4	11.6	3.5	87.3	3.8	96.2	83.8
	2–2.8	94.3	12.5	88.3	11.7	5.8	87.5	6.2	93.8	81.8
	2.8–4	92.7	18.1	83.6	16.4	7.3	81.9	8.2	91.8	74.6
	4–5.6	90.6	24.0	79.0	21.0	9.5	76.0	11.1	88.9	66.6

Table 3. Results of separation by jigging and flotation of the PS+PVC mixture, for five size fractions.

	Size fraction (mm)	FLOATED				SINK (%)				Separation Efficiency (%)
		Recovery (%)		Grade (%)		Recovery (%)		Grade (%)		
		PS	PVC	PS	PVC	PS	PVC	PS	PVC	
Jigging	1–1.4	99.6	35.9	73.5	26.5	0.4	64.1	0.6	99.4	63.7
	1.4–2	99.5	30.5	76.5	23.5	0.5	69.5	0.7	99.3	69.0
	2–2.8	99.5	22.2	81.8	18.2	0.5	77.8	0.7	99.3	77.3
	2.8–4	99.5	12.0	89.2	10.8	0.5	88.0	0.6	99.4	87.5
	4–5.6	99.5	5.9	94.4	5.6	0.5	94.1	0.6	99.4	93.6
Flotation	1–1.4	98.2	47.2	67.6	32.5	1.8	52.8	3.3	96.7	51.0
	1.4–2	97.5	40.3	70.8	29.3	2.5	59.7	4.0	96.0	57.2
	2–2.8	96.8	18.5	83.9	16.0	3.2	81.5	3.8	96.2	78.3
	2.8–4	96.0	8.9	91.6	8.4	4.0	91.2	4.2	95.8	87.1
	4–5.6	94.1	3.4	96.5	3.5	5.9	96.6	5.7	94.3	90.7

3.3 PS+PVC Mixture

Table 3 shows the recovery and grade of the floated and sink materials using jigging and froth flotation, for both types of plastics over five size fractions. In the case of flotation, the results correspond to tests using 2mg/L of tannic acid concentration.

In both separation processes, the separation efficiency of PS+PVC mixture improved clearly with the increase of the particles size (Table 3). The floated recovery of the two plastics decreased with the particle size, with a more evident decrease of the PVC recovery: from 35.9% for the finer fraction, to 5.9% for the coarser fraction in jigging and from 47.2% for the finer fraction, to zero for the coarser fraction in froth flotation. Thus, for two separation processes the PS grade in the floated increased with the particle size.

Jigging produced better separation than flotation for the two finer fractions. It led to larger recoveries of PS and smaller recoveries of PVC in the floated.

For the coarse fraction (4–5.6 mm), both processes led to the similar separation efficiency. In jigging was obtained a floated with a PS grade of 94.4% and sink material with a PVC grade of 99.4%; in flotation was obtained a floated with a PS grade of 96.5% and sink material with a PVC grade of 94.3%.

239

Whilst in jigging was obtained a sink material almost pure PVC, in froth flotation was also obtained a floated almost pure PS.

Thus, to achieve a better separation by jigging and froth flotation, should be used a mixture with particles with greater size than 2.8 mm.

4 CONCLUSIONS

The quality of the separation, in jigging and in froth flotation, changes according to the plastic mixture, varying with both size and shape of the particles.

In the PS+PMMA mixture, jigging produced better results with the finer size fractions. In the two coarser fractions, both processes led to similar separations. In this mixture, the separation quality of the two processes, and mainly in froth flotation, clearly improved with an increase in the particles size.

The PS+PET mixture produced the worst results and neither of the separation methods could be seen to predominate. In this mixture froth flotation worsened as the particle size increased. As jigging is a gravity process, based on density differences, it was expected that the best separations would be obtained in the PS+PET mixture. However, PS+PET mixtures produced the worst jigging results for particle sizes greater than 2 mm. This was a consequence of the lamellar shape of PET particles, which makes them more difficult to penetrate the jigging bed.

In the PS+PVC mixture, both processes led to similar quality separations. In both cases, separation improved as the particles size increased.

Better jigging and froth flotation separations were obtained for mixtures where the denser plastic have regular shapes and for particles greater than 2.8 mm.

REFERENCES

Agante, E., Carvalho, M.T. Durão, Bártolo, T. 2004. Separation of PET, PVC and PS using gravity concentration and froth flotation. *Proceedings of the Global Symposium on Recycling, Waste Treatment and Clean Technology (REWAS'04)*. Madrid, Spain: 1691–1700.

Bedeković, G., Salopek, B., Sobota, I. 2011. Electrostatic separation of PET/PVC mixture. *Technical Gazette* 18(2): 261–266.

Burat, F., Güney, A., Kangal, M.O. 2009. Selective Separation of Virgin and Post-Consumer Polymers (PET and PVC) by Flotation Method. *Waste Management* 29: 1807–1813.

Carvalho, M.T., Ferreira, C., Portela, A., Santos, J.T. 2009. Application of fluidization to separate packaging waste plastics. *Waste Management* 29: 1138–1143.

Dodbiba, G., Sadaki, J., Okaya, K., Shibayama, A., Fujita, T. 2005. The use of air tabling and triboelectric separation for separating a mixture of three plastics. *Minerals Engineering* 18: 1350–1360.

Ferrara, G., Bevilacqua, P., Lorenzi, L., Zanin, M. 2000. The influence of particle shape on the dynamic dense medium separation of plastics. *International Journal Mineral of Processing* 59: 225–235.

Gent, M.R., Mario, M., Javier, T., Susana, T. 2011 – Optimization of the recovery of plastics for recycling by density media separation cyclones. *Resources, Conservation and Recycling* 55: 472–482.

Hori, K., Tsunekawa, M., Ueda, M., Hiroyoshi, N., Ito, M., Okada, H. 2009. Development of a New Gravity Separator for Plastics a Hybrid-Jig. *Materials Transactions* 50(12): 2844–2847.

Ito, M., Tsunekawa, M., Ishida, E., Kawai, K., Takahashi, T., Abe, N., Hiroyoshi, N. 2010. Reverse jig separation of shredded floating plastics – separation of polypropylene and high density polyethylene. *International Journal Mineral of Processing* 97: 96–99.

Kangal, M.O. 2010. Selective flotation technique for separation of PET and HDPE used in drinking water bottles. *Mineral Processing & Extractive Metall Rev.* 31: 214–223.

Pascoe, R.D., Hou, Y.Y. 1999. Investigation of the importance of particle shape and surface wettability on the separation of plastics in a LARCOMEDS separator. *Minerals Engineering* 12: 423–431.

Richard, G.M., Mario, M., Javier, T., Susana, T. 2011. Optimization of the recovery of plastics for recycling by density media separation cyclones. *Resources, Conservation and Recycling* 55: 472–482.

Shen, H., Forssberg, E. and Pugh, R.J. 2002. Selective Flotation Separation of Plastics by Chemical Conditioning With Methyl Cellulose. *Resources, Conservation and Recycling* 35: 229–241.

Shibata, J., Matsumoto, S., Yamamoto, H. and Lusaka, E. 1996. Flotation Separation of Plastics Using Selective Depressants. *International Journal Mineral of Processing* 48: 127–134.

Full-scale vermicomposting of sludge from milk plants with pulp and paper mill sludge

M. Quintern & M. Morley
Quintern Innovation Ltd., Noke Ltd., Tauranga, New Zealand/Germany

B. Seaton
Carter Holt Harvey Pulp & Paper, Tokoroa, New Zealand

ABSTRACT: Noke Limited is currently vermicomposting 21,000 tonnes per year of dewatered Waste Activated Sludge (WAS) and Dissolved Air Flotation (DAF) sludge from milk processing plants in New Zealand. The sludge is blended with 30,000 tonnes of several pulp and paper mill sludge from Carter Holt Harvey Pulp & Paper Mills to adjust pH, C/N ratio and other nutrient requirements for successful vermicomposting. Approximately 13,000 tonnes of vermicast is produced and is land applied to pasture and maize to improve soil functions. Productivity, soil water efficiency, nutrient uptake increased. Root depth will reduce the risk for nutrient leaching of nitrate. The new vermicomposting operation will be integrated into the overall farm management system of the 1,600 ha farm. The Noke process is a holistic approach where organic resources from primary sector industries, are converted to high quality soil conditioner, while producing positive environmental and economical outcomes for industries and agriculture.

1 ORGANIC WASTE RESOURCES

Within the central North Island region of New Zealand, the primary economic drivers are based on dairying and the timber industries. The resultant organic 'waste' resources originating from milk processing and pulp and paper plants have traditionally, mostly been disposed of in landfills until recently when these valuable organic resources have been treated through proprietory vermicomposting technologies and are presented and discussed in this paper.

1.1 *Milk Plants*

Several milk plants in the central North Island region of New Zealand are producing some 30,000 tonnes of various organic wastes, which are only partially land applied. The majority of the organic waste sludge has been disposed at landfills. Waste Activated Sludge (WAS) is thickened by decanters at two sites to approximately 15% solids. Typical analysis of the WAS from the two sites are given in Table 1. Wastewater is introduced into continuous flow pools where larger particles including fat and proteins are extracted through a dissolved air flotation (DAF) process. The particles attach to the air bubbles and float to the surface as a DAF sludge and is mechanically skimmed off. The sludge is decanted prior to transportation to the vermicomposting operation. Nutrient contents (Table 1) are lower than those of the WAS.

1.2 *Pulp & Paper Plants*

Carter Holt Harvey operates three Pulp & Paper mills located in New Zealand's North Island. All three produce organic wastes generally described as pulpmill solids or sludge's. The fibrous pulpmill solids will be used as a sources of blending with the milk plant sludge. The characteristics of pulpmill solids can be highly variable depending upon the particular pulp mill wastewater treatment and sludge separation technology.

Table 1. Characteristics of Waste Activated Sludge (WAS) and Dissolved Air Flotation (DAF) sludge from various milk plants in Central North Island – New Zealand.

| Parameter | Unit | Waste Activated Sludge | | Dissolved Air Flotation Sludge | Limits organic certification |
		Waitoa	Te Rapa	Tirau	
Dry Matter	(%)	15	16	13	
Carbon	(%)	41	43	5.2	
Nitrogen	(%)	6.9	6.6	1.0	
C/N ratio	–	5.9	6.5	5.2	
pH	–	6.8	5.8	3.9	
Phosphorus	(ppm)	15,300	18,700	2,100	
Potassium	(ppm)	7,800	9,800	1,100	
Calcium	(ppm)	15,600	13,500	1,900	
Magnesium	(ppm)	3,200	2,700	510	
Sulphur	(ppm)	8,500	9,000	1,100	
Sodium	(ppm)	4,200	4,900	930	
Boron	(ppm)	128	54		
Arsenic	(ppm)	4	<5		20
Cadmium	(ppm)	<0.10	<0.3		1
Chromium	(ppm)	10	7		15
Copper	(ppm)	16	7		60
Lead	(ppm)	5.4	0.8		250
Mercury	(ppm)	<0.10	<0.3		1
Nickel	(ppm)	9	<5		60
Zinc	(ppm)	100	67		300

Table 2. Characteristics of pulp and paper mill sludge and recycled paper solids from various Carter Holt Harvey Pulp & Paper Mills in Central North Island – New Zealand.

| Parameter | Unit | Tasman | | Kinleith | | Limits |
		Primary Solids	Secondary Solids	Secondary Solids	Recycled paper Solids	organic certification
Dry Matter	(%)	58.9	17.8	18.0	26.9	
Carbon	(%)	34.2	37.6	17.0	37	
Nitrogen	(%)	0.43	0.5	0.53	0.17	
C/N ratio	–	80	75	32	217	
pH	–	7.4	7.4	8.5	6.9	
Phosphorus	(ppm)	717	509	1,203	143	
Potassium	(ppm)	2,060	1,060	1,203	240	
Calcium	(ppm)	78,300	24,200	42,000	58,900	
Magnesium	(ppm)	1,260	2,240	863	1,900	
Sulphur	(ppm)	1,292	3,200	5,000	<5,000	
Sodium	(ppm)	2,030	1,130	1,582	820	
Boron	(ppm)	9	0.28	<20	<6	
Arsenic	(ppm)	2.1	1.0	20	<5	20
Cadmium	(ppm)	0.16	0.1	1.0	0.2	1
Chromium	(ppm)	34	2.9	100	7	15
Copper	(ppm)	16	9	80	46	60
Lead	(ppm)	4.9	1.7	33	15.2	250
Mercury	(ppm)	0.04	0.04	–	<0.3	1
Nickel	(ppm)	5.5	1.3	11	5	60
Zinc	(ppm)	43	43	320	77	300

Losses to the Pulp & Paper mill's sewer systems from routine cleaning of the pulp & paper machine's, average around 15 t/d (dry basis) of rejected fibre material. This is referred to as 'primary solids' and is removed in the primary clarifier. The pulp & paper mill's wastewater treatment system produces approximately 12 t/d (dry basis) of secondary biological solids from the oxidation ponds which are combined with the primary solids prior to being pumped into the two gravity-dewatering ponds (Quintern et al. 2013). By blending the different solids the limits for organic certification are met and an organic certified product is achieved.

The recycled paper solids show similar carbon and nitrogen concentrations to the primary solids. The slightly higher metal concentrations of the recycled paper solids (Table 2) compared to the primary pulp and paper mill sludge are still well below the limits for safe land application (NZWWA 2003). At Tasman Mill, the primary pulp and paper solids are separated in the wastewater treatment plant clarifier by adding a polymer to coagulate the solids which are then de-watered mechanically resulting in a much dryer product than the sludge from the sedimentation ponds at Kinleith Mill (Table 2). The pulp and paper solids are generated in a continuous flow process with consistent supply to the vermicomposting operation.

2 VERMICOMPOSTING

From the 1990th scientist started using pulp and paper mill sludge also referred as pulpmill solids for vermicomposting as bulking agent (Butt 1993; Elvira et al. 1996; Elvira et al. 1997; Elvira et al. 1998; Lazcano et al. 2008). Vermicomposting of purely pulpmill solids has not been conducted at any significant scale because the wide C/N ratio of pulpmill solids is generally not suitable for vermicomposting. Laboratory trials have shown that best reproduction of earthworms is achieved when the C/N ratio is adjusted to 25 with nitrogen richer waste streams (Aira et al. 2006). Paper wastes with a C/N ratio of up to 200 and higher were used as a carbon rich blending agent for nutrient rich wastes such as biosolids, food wastes (Arancon et al. 2005; Edwards 1988), manure (Arancon et al. 2005), and other industrial wastes. In recent years Quintern et al. has demonstrated that a sub-optimal C/N ratio in the earthworm feedstock can be successful in commercial vermicomposting operations and nitrogen sources captured from the wastewater treatment plant of a pulpmill can be used as nitrogen source for blending with primary pulpmill solids (Quintern et al. 2009; Quintern 2011). Vermicomposting of DAF sludge from milk processing industries in combination with pulpmill solids has first been studied and published by Quintern (2009).

Industrial scale vermicomposting of sludge from Carter Holt Harvey Pulp & Paper mills and other industrial and municipal organic waste streams commenced in 2007 and currently totals 150,000 t/a (Quintern et al. 2009; Quintern 2011; Quintern et al. 2013; Quintern 2014a; Quintern 2014b). In 2010 Noke Ltd., undertook feasibility studies on commercial vermicomposting of the organic waste stream from pulp and paper industries in combination with dewatered WAS and DAF sludge. The aim was to produce valued added products, vermicast, for higher beneficial land utilisation. In 2014 the project established the efficacy of operating as a full scale vermicomposting operation and a new site is being developed to receive only these products and will be operational in 2015.

2.1 *Vermicomposting process*

Vermicomposting is a mesophilic process, which is using *epigeic* compost earthworms to ferment and decompose organic matter producing a specific earthworm casting called vermicast or vermicompost (Edwards & Neuhauser 1988). Compost worms can 'operate' in moist conditions between 60 to 90% depending on the structure of the feedstock, so no pre drying is required for vermicomposting. Vermicomposting requires no mechanical turning of the feedstock as the feedstock waste is applied in smaller quantities as earthworms are mixed with the feedstock through their activity. This action is described as 'bio-turbation' in soil science or could be described as 'vermi-turbation' in the process of vermicomposting. During the digestion of the organic waste a volume reduction of up to 80% can be achieved, which will reduce transportation costs of the vermicast to its end users.

Table 3. Characteristics of the Quintern Windrow Vermicomposting Systems in comparison to conventional Windrow Vermicomposting Systems described by Edwards (2011).

	Windrow Vermicomposting Systems	
	Quintern	Conventional
Benefits	• Low CAPEX* • Easily managed • Labour extensive (mechanised) • Reduced footprint • Retaining of nutrient through wider C/N ratio • Harvesting vermicompost without earthworms	• Low CAPEX* • Easily managed
Drawbacks	• 6 to 12 months processing time	• Labour-intensive • Large footprint • 6 to 16 months processing time • Loss of nutrient through leaching and volatilisation • Impossible to harvest vermicompost without earthworms

*Capital expenditure.

2.2 Vermicomposting technology

The vermicomposting technology developed by Quintern Innovation Ltd. is based on the Windrow Vermicomposting System described as low-technology vermicomposting systems (Edwards 2011). Windrow vermicomposting technology in general is characterised as low capital expenditure (CAPEX) and easy to manage. The drawbacks of conventional Windrow Vermicomposting System are a large footprint, labour-intensive, slow processing time, considerable nutrient losses, and need for separation of earthworms from vermicompost. Most of these drawbacks have been resolved by Quintern Innovation Ltd. (Table 3). The key improvements have been achieved by avoiding continuous feeding practices in order to reduce labour costs and processing time. The footprint/net production ratio for vermicomposting has been significantly reduced by minimising the non-productive areas of the vermicomposting site. The Quintern Windrow Vermicomposting Technology has been developed and proven over the last decade and is successfully operating at an economic industrial scale.

To manage the nutrient losses, such as leaching and gaseous losses from vermicomposting sites, the C/N ratio of the feedstock is carefully adjusted, which will reduce the potential of enriching the underlying topsoil with nutrients and humic acids. Operating at the same location over multiple years could create a hot spot of nutrient intake in the soil ecosystem. To address this potential risk the author adopted the strategy of integrating vermicomposting into farm and forest management practice. Similar to integrating outdoor pig ranging, with extreme high nutrient accumulation in hot spots, the vermicomposting sites will rotate either on the farm (Quintern 2005; Quintern & Sundrum. 2006) or within the forest. After a certain time of vermicomposting on one specific site the vermicomposting windrows will be placed on a new paddock or forest block. On the former vermicomposting site the vermicast will be harvested and a so-called 'catch crop' with high nutrient demand, such as maize, will be planted. In forest management systems the new plantation benefits from the vermicast residues as the demand for nutrient of trees is highest in the first years of plant growth.

2.3 Vermicast

Vermicast and vermicompost are used in the literature often synonymously. In New Zealand vermicast is defined as mature pure earthworm casting (NZS 2005). There are hundreds of references characterising vermicompost in relation to the degree of earthworm activity, vermicomposting technology used, and the parent wastes or 'feedstock' used. Arancon & Edwards (2011) provides a review of the recent findings of beneficial use of vermicast.

In comparison to commonly advised vermicomposting technology, the vermicast produced in this commercial operation has received a much larger portion of fibrous and carbon rich feedstock source as a blending agent. As a result the vermicast has a more advanced peaty structure with a lower nitrogen content and therefore a slightly wider C/N ratio. This provides the MyNOKE vermicast with various advantages when applied in nutrient sensitive ecosystems, which are lacking of soil humus for maintaining a high nutrient and water holding capacity.

3 BENEFICIAL LAND UTILISATION OF VERMICAST

Vermicast is applied in bulk to various crops in rates from 2.5 t/ha to pasture, 10 t/ha to kiwi-fruit orchards, and 20 t/ha to maize and other crops according to the specific soil quality and nutrient demand of the crop. Screened vermicast can be applied with standard compost or manure spreaders.

Vermicast produced by compost earthworms (Eisenia foetida) contains humic substances described as humic acids, humates, gibberelins, auxins, 3-indole acetic acid, and various other substances. These humic substances promote plant growth in multiple ways starting with faster germination, increased root development of more and longer lateral and vertical roots, increased area of root hairs and even higher root activity (Zandonadi et al. 2010; Roy et al. 2010; Lazcano et al. 2011; Canellas et al. 2013). Increased root area, depth, and activity leads to increased nutrient uptake and access to more available soil water during dryer seasons. Positive effects were measured in higher numbers of blossoms, flowers, and fruits, increased photosynthesis which overall leads to higher yields (Roy et al. 2010), and fruit harvests. Recent studies are showing that vermicast has the potential to suppress plant diseases and control pests such as insects and nematodes.

Little is known on continuous application of vermicast in intensive dairying, especially on pasture and intensive monoculture maize cropping systems. Intensive dairy farming is currently criticised for an increase in nitrate leaching and phosphate runoff. Improved topsoil functions in regards to water and nutrient retention would most likely mitigate nutrient losses, such as nitrate in the groundwater. Higher humus content and more intense root systems would improve these soil functions, as well as increase nutrient uptake including nitrogen.

4 OUTLOOK

Vermicast application to soil has the potential to mitigate nitrate losses, increase soil organic matter directly as a carbon source and indirectly by increasing root production.

Vermicomposting of combined industrial and municipal organic wastes offers multiple benefits on the carbon footprint for industry, the community and for the Agribusiness sectors. Of significance is the reduction in greenhouse gases emissions originating from land filling of organic wastes and the avoidance of the high economic penalties associated with the design, construction and ongoing monitoring of landfills. For the farming and horticulture sectors soil carbon would be increased and higher carbon sequestration by increasing root mass and crop yields. Potential fertiliser reduction would reduce the carbon footprint and a better soil structure could reduce laughing gas emissions from pastoral soils.

ACKNOWLEDGEMENTS

The authors would like to thank Jonathan Stevenson and his team for their endless support in establishing the commercial vermicomposting. Ed Mercer and Craig Andrews for supporting all trials allowing this project to come so far. Barry Campbell from Waikato Regional Council and his team for the fruitful discussion around implementing industrial scale vermicomposting into safe environmental practice.

REFERENCES

Aira, M., Monroy, F. & Domínguez, J. 2006. C to N ratio strongly affects population structure of Eisenia fetida in vermicomposting systems. *European Journal of Soil Biology* 42: 127–131.

Arancon, N.Q. & Edwards, C.A. 2011. The use of vermicomposts as soil amendments for production of field crops. In C.A. Edwards N.Q. Arancon & R. Sherman (eds), *Vermiculture Technology: Earthworms, Organic Wastes, and Environmental Management:* 129–151. Boca Raton, London, New York: CRC Press, Taylor & Francis Group.

Arancon, N.Q., Edwards, C.A., Lee, S.S. & Yardim, E. 2002. Management of plant parasitic nematode populations by use of vermicomposts. *Brighton crop protection conference pests and diseases.* 705–710.

Arancon, N.Q., Galvis, P.A. & Edwards, C.A. 2005. Suppression of insect pest populations and damage to plants by vermicomposts. *Bioresource Technology* 96(10): 1137–1142.

Butt, K.R., 1993. Utilisation of solid paper-mill sludge and spent brewery yeast as a feed for soil-dwelling earthworms. *Bioresource Technology* 44(2): 105–107.

Canellas, L.P., Balmori, D.M., Médici, L.O., Aguiar, N.O., Campostrini, E., Rosa, R.C.C., Façanha, A.R. & Olivares, F.L. 2013. A combination of humic substances and Herbaspirillum seropedicae inoculation enhances the growth of maize (Zea mays L.). *Plant and Soil* 366(1–2): 1–14.

Canellas, L.P., Teixeira Junior, L.R.L., Dobbss, L.B., Silva, C.A., Medici, L.O., Zandonadi, D.B. & Façanha, A.R. 2008. Humic acids crossinteractions with root and organic acids. *Annals of Applied Biology* 153(2): 157–166.

Edwards, C.A. 1988. Breakdown of animal, vegetable and industrial organic wastes by earthworms. In C.A. Edwards & E.F. Neuhauser (eds), *Earthworms in Waste and Environmental Management:* 21–31. The Hague, The Netherlands: SPB Academic Publishing BV.

Edwards, C.A. 2011. Low-technology vermicomposting systems. In *Vermiculture Technology: Earthworms, Organic Wastes, and Environmental Management:* 79–90. CRC Press.

Edwards, C.A., Arancon, N.Q., Emerson, E. & Pulliam, R. 2007. Suppressing plant parasitic nematodes and arthropod pests with vermicompost teas. *BioCycle* 48(12): 38–39.

Edwards, C.A., Arancon, N.Q., Vasko-Bennett, M., Askar, A., Keeney, G. & Little, B. 2010. Suppression of green peach aphid (Myzus persicae)(Sulz.), citrus mealybug (Planococcus citri)(Risso), and two spotted spider mite (Tetranychus urticae)(Koch.) attacks on tomatoes and cucumbers by aqueous extracts from vermicomposts. *Crop Protection* 29(1): 80–93.

Elvira, C., Goicoechea, M., Sampedro, L., Mato, S. & Nogales, R. 1996. Bioconversion of solid paper-pulp mill sludge by earthworms. *Bioresource Technology* 57(2): 173–177.

Elvira, C., Sampedro, L., Benitez, E. & Nogales, R. 1998. Vermicomposting of sludges from paper mill and dairy industries with Eisenia andrei: a pilot-scale study. *Bioresource Technology* 63(3): 205–211.

Elvira, C., Sampedro, L., Dominguez, J. & Mato, S. 1997. Vermicomposting of wastewater sludge from paper-pulp industry with nitrogen rich materials. *Soil Biology and Biochemistry* 29(3/4): 759–762.

Lazcano, C., Revilla, P., Malvar, R.A. & Domínguez, J. 2011. Yield and fruit quality of four sweet corn hybrids (Zea mays) under conventional and integrated fertilization with vermicompost. *Journal of the Science of Food and Agriculture* 91(7): 1244–1253.

Lazcano, C., Sampedro, L., Zas, R. & Domínguez, J. 2008. Enhancement of pine (Pinus pinaster) seed germination by vermicompost and the role of plant genotype. *Compost and digestate: sustainability, benefits, impacts for the environment and for plant production*: 253–254.

NZS, 2005. *Composts, Soil Conditioners and Mulches.* New Zealand Standard, NZS 4454:2005.

NZWWA, 2003. *Guidelines for the safe application of biosolids to land in New Zealand.* Wellington, New Zealand: New Zealand Water & Wastes Association.

Quintern, M., 2005. Integration of organic pig production within crop rotation: Implications on nutrient losses. *Landbauforschung Völkenrode.*(281): 31–40.

Quintern, M. 2009. *Vermicomposting of lake weeds and pulp & paper solids for carbon resource recovery for primary sectors.* Wellington, New Zealand: Ministry of Agriculture and Forestry.

Quintern, M., Wang H., Magesan G. & Slade A. 2009. *Vermicomposting primary and secondary solids from the pulp and paper industry.* Report Scion, Rotorua, New Zealand: Scion.

Quintern, M., 2011. Organic waste free pulpmill through vermicomposting – The Kinleith way. *New Zealand Land Treatment Collective: Proceedings for the 2011 Annual Conference.* Rotorua: New Zealand Land Treatment Collective. 84–88.

Quintern, M., 2014a. Full scale vermicomposting and land utilisation of pulpmill solids in combination with municipal biosolids (sewage sludge). *7th International Conference on Waste Management and the Environment.* Ancona: Wessex Institute. 65–76.

Quintern, M., 2014b. Industrial scale vermicomposting of municipal biosolids by blending with fibrous industrial wastes. *Eurasia 2014 Waste Management Symposium.* Istanbul, Turkey: 1–9.

Quintern, M. & Sundrum, A. 2006. Ecological risks of outdoor pig fattening in organic farming and strategies for their reduction – Results of a field experiment in the centre of Germany. *Agriculture, ecosystems & environment* 117(4): 238–250.

Quintern, M., Seaton, B., Mercer, E. & Millichamp, P. 2013. Industrial scale vermicomposting of pulp and paper mill solids with municipal biosolids and DAF sludge from dairy industries. *Appita* 66(4): 290–295.

Zandonadi, D.B., Santos, M.P., Dobbss, L.B., Olivares, F.L., Canellas, L.P., Binzel, M.L., Okorokova-Façanha, A.L. & Façanha, A.R. 2010. Nitric oxide mediates humic acids-induced root development and plasma membrane H^+-ATPase activation. *Planta* 231(5): 1025–1036.

Wastes: Solutions, Treatments and Opportunities – Vilarinho, Castro & Russo (eds)
© 2015 Taylor & Francis Group, London, ISBN 978-1-138-02882-1

Risk management in landfills. A public health perspective

B. Rani-Borges & J.M.P. Vieira
University of Minho, Braga, Portugal

ABSTRACT: Currently, sanitary landfill is an appropriate option for disposal of waste, mainly because it is a technology with little aggressive environmental impact and associated to low costs and simple operation. However, the deficit in best practices in solid waste management is recognized in scientific literature as the cause of adverse effects on the environment and public health. Since landfill has a relevant social, economic and environmental impact as a sanitary urban infrastructure, new control approach based on public health and environmental integrity should be developed in order to prevent diseases propagation and negative environmental impacts. This research work presents a novel concept of landfill safety plan, using similar approach as the well established water safety plan methodology, structured on risk assessment and risk management throughout the waste collection and disposal system.

1 INTRODUCTION

The continuous increase on urbanization and people standards of life has a high influence on waste collection and disposal policies. The accumulation of waste has become a real urban problem, causing serious impacts on public health and environmental integrity. Currently, landfill as an appropriate technological option for municipal solid waste (MSW) disposal, is used especially because it has little environmental impact, with small costs and simple operation (Cossu, 2013). However, the deficit in best practices of solid waste management systems, especially the absence of a public health protection policy and environmental pollution control are recognized in the scientific literature as the cause of adverse effects associated to MSW management (Giusti, 2009).

Quality assurance of the MSW management system is an essential element of public health policies since human health can be affected by solid waste in different ways (Karak et al., 2012). The negative aspects presented by sanitary landfills are caused by inadequate waste disposal and unstructured management systems which can cause contamination of water, air and soil. Moreover bad practices contribute to the appearance of vectors and epidemiological agents which involve various impacts on the environment and human health (Brown et al., 2011). Thus, grows the need to seek improvements to the technology for MSW collection and disposal (Pereira & Maia, 2012).

A special issue to be considered in landfills operation is the emission of pollutants (liquid and gaseous: leachate and biogas) originated from degradation of the organic matter fraction of MSW. The environmental impacts of these emissions essentially depends on the characteristics of the residues brought to the landfill, the geological features of the site, the applied technology and other conditions (Gomes, 2009).

Today, more and more people are concerned about the risks to health and the environment associated to landfills and are increasingly demanding definitive solutions not just only for the present but also for the future generations. Also according to the World Health Organization (WHO) it is necessary to identify and to assess the health risks to prevent illness and injury. Hence, it is recommended to implement an integrated risk management strategy covering all the steps off MSW management system with special focus on landfills.

2 BACKGROUND

2.1 *Public health risks related to landfills*

The disposal of MSW in landfills is an adequate technologic solution when best practices and operational standards are followed. However this solution can cause environmental concerns (mainly related to waste biodegradation) and public health. The pollution can reach different levels of the environment and, consequently, affect the public health in different ways (Butt et al., 2008).

Health impacts can be associated to all the steps of MSW management system by direct or indirect contact (Forastiere et al., 2011). Human exposure to toxic chemicals that exist in landfills can occur through inhalation of pollutants in the air, by contact with contaminated soil and water, leachate, or micro and macro-vectors like birds, insects, rodents.

Several authors (EA, 2011, Elliott et al., 2001, Giusti, 2009, Jarup et al., 2002, Palmiotto et al., 2014, Perez et al., 2006) have investigated the negative impacts on human health associated with landfills by describing the potential causes of disease according to exposure routes and substances formed during and after the process of MSW biodegradation as shown in Figure 1.

During the process of treatment and disposal of MSW in landfills, the environment can be found in many situations of potential risks, including impacts on the quality of air, soil and groundwater. Although there are no ways to completely avoid the negative environmental impacts, risk assessment

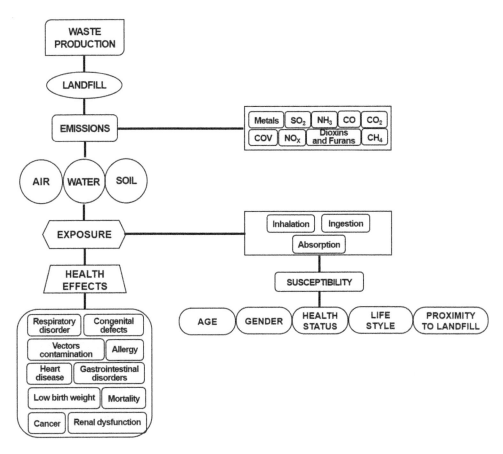

Figure 1. Diagramatic scheme of solid waste emissions and exposure routes.

and risk management can significantly reduce them so as to allow the maintenance of environmental health (Butt et al., 2014) and consequently protecting the public health quality.

Most of the environmental problems related to landfills include, among other problems: contribution to the greenhouse effect, resulting from emissions of CH_4 and CO_2; generation of several toxic substances and volatile organic compounds; odor; noise; depletion of the ozone layer; damage to vegetation; contamination of air, water and soil with leachate and heavy metals; affect the animals health and contributes to the proliferation of insects and rodents (Cossu, 2013, Forastiere et al., 2011, Giusti, 2009).

With the development of science and technology it is possible to control or even eliminate certain hazards and undesirable hazardous events, such as those triggered due to the operation of the landfill, by applying risk management approach (Proag & Proag, 2014).

2.2 *Risk management*

Risk management is a systematic process consisting of a set of analytical steps and activities required for control of hazard and hazardous events for a specific system or project (Aven, 2009, Irimia-Dieguez et al., 2014). Risks and hazardous events evaluated by this tool can be positive or negative, since the risk management study any situation that influences the project in question, regardless of the nature and may also be identified inside or outside of organization's environment (PWC, 2008).

Although it is a tool that needs improvements (Serpella et al., 2014), is widely used in different areas of science and industry because it is essential for good project management since it is applied to identify, analyze and define strategies to deal with the risks (Dey, 2012). Besides it providing a good foundation for decision-making that help to reduce the expected impacts. The risk management process is indispensably multidisciplinary, bringing together knowledge and data as necessary for a good risk assessment, including socio-economic and environmental data (Travis & Bates, 2014).

This process consists of several steps, which are outlined in Figure 2. Enforcement of sequence of steps that make up the risk management allows decision making according to the risks and likely

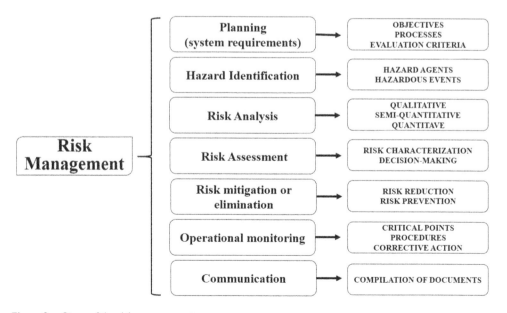

Figure 2. Steps of the risk management process.

hazardous events, in order to guarantee only the occurrence of acceptable events (Tohidi, 2011). What is possible because of the prioritization and measures to eliminate or mitigate risks that reduce the magnitude of hazardous events (Struik et al., 2015).

2.3 *Risk management in landfills*

The adoption of methodologies for risk assessment and risk management must comply with the environmental and economic requirements and consequently reflect on safety and ensuring the quality of management and treatment of solid waste service. The result of analyzes shall indicate the performance of risk management and may thus be a measure of the evolution of the damage suffered by the public health and the environment (Butt et al., 2014).

Risk analysis for landfill is based on characterization of the sources of danger, and the landfill risks must be established from emissions for several toxic and carcinogenic substances and their amplitude varies according to the exposure levels. That is, the risks are increased according to inhalation of gases or particles, water intake and food contaminated with leachate, direct contact with contaminated soil and other situations (Davoli et al., 2010). The hazard identification process depends on understanding all the steps and system operation of MSW management, which makes it possible to analyze all potential biological, physical, chemical and radiological hazards that may be associated (Forastiere et al., 2011).

Besides pollution, the risk analysis must include less impacting issues such as odor, risks of failures in engineering processes and characterization of chemical and biological components in the system (Pollard et al., 2006).

The impacts on human health and environment integrity by the components of landfill could not be validated because the results of the studies are considered insufficient due to the large variables that result in the absence of concrete evidence (Giusti, 2009, Palmiotto et al., 2014, Porta et al., 2009). Therefore it is essential to carry out future research work that include a risk management for the disposal process of MWS and that may elucidate the risks present on this system and any eventual consequences as well as notify the population affected by this technological solution.

3 LANDFILL SAFETY PLAN

The World Health Organization Guidelines for Drinking-water Quality (WHO, 2004, WHO, 2011), introduced a new concept of risk assessment and risk management in drinking-water systems through water safety plans (WSP). This integrated and preventive methodology starts from the establishment of health-based targets and provides a systematic and comprehensive risk assessment and risk management approach for consistently ensuring the safety of a drinking-water supply system from the catchment to the consumer. Generous bibliography has been published on this subject and this is a well-established methodology around the world (Vieira & Morais, 2005; Vieira, 2011).

A keynote lecture at the 2nd International Africa Sustainable Waste Management Conference (Vieira, 2014) recommended the development of a concept of landfill safety plan for a similar purpose. This novel concept should comprise three essential steps: (i) *system and exposure assessment*, referring to mapping the system, assessing the risks along the waste collection and disposal system, and the exposure level of different vulnerable groups; (ii) *operational monitoring* establishing control measures for previously identified and prioritized hazards and exposures at control points; (iii) *management plans* describing actions and control measures to be taken during normal operation or extreme and incident conditions. This concept of landfill safety plan must be built on the structure of a water safety plan, but also with significant differences due to the specificity nature of a waste collection and disposal system.

4 CONCLUSION

Human activities generate big quantities of solid waste being its management in urban scale difficult due to the implications in public health and environment. Science and technology developments permit the implementation of different methods for collection and disposal of MSW. The use of landfilling technology for waste disposal presents some risks to public health and environmental integrity although it presents a broad variety of benefits at the social, economic and environmental dimensions.

Hazards and hazardous events associated to landfills can cause severe damages in human health and in environment integrity, through the contamination of air, water and soil and contribute to the proliferation of pathogenic agents and consequently trigger serious consequences for exposed populations. These aspects show the necessity to develop a new approach to MSW management systems based on risk assessment and risk management methodologies in order to identifying hazards and hazardous events, assessing and prioritizing risks and applying control measures to eliminate or mitigate those risks and promote the values of public health and environmental protection.

Based on a successful approach of risk management for drinking-water supply systems applying the water safety plan methodology it is recommended to develop a novel concept of landfill safety plan for a similar purpose. This new approach must consider the specific nature of a waste collection and disposal system.

REFERENCES

Aven, T. 2009. Risk Analysis and Management. Basic Concepts and Principles. *Reliability & Risk Analysis: Theory & Applications* 57–73.

Brown, C., Milke, M., & Seville, E. 2011. Disaster waste management: A review article. *Waste Management* 1085–1098.

Butt, T.E., Gouda, H.M., Balochd, M.I., Paul, P., Javadi, A.A., & Alam, A. 2014. Literature review of baseline study for risk analysis — The landfill. *Environment International* 149–162.

Butt, T.E., Lockley, E., & Oduyemi, K.O. 2008. Risk Assessment Of Landfill Disposal Sites – State Of The Art. *Waste Management* 952–964.

Cossu, R. 2013. Groundwater contamination from landfill leachate: when appearances are deceiving! *Waste Management* 1793–1794.

Davoli, E., Fattore, E., Paiano, V., Colombo, A., Palmiotto, M., Rossi, A.N., & Fanelli, R. 2010. Waste Management Health Risk Assessment: A case Study of a Solid. *Waste Management* 1608–1613.

Dey, P.K. 2012. Project risk management using multiple criteria decision-making technique and decision tree analysis: a case study of Indian oil refinery. *Production Planning & Control: The Management of Operations* 903–921.

EA. 2011. *Environment Agency. Additional guidance for H4 odour management.* Bristol.

Elliott, P., Briggs, D., Morris, S., Hoogh, C., Hurt, C., & Jensen, T.K. 2001. Literature review of baseline study for risk analysis — The landfill leachate case. *BMJ* 363–368.

Forastiere, F., Badaloni, C., Hoogh, K., Kraus, M.K., Martuzzi, M., Mitis, F., & Briggs, D. 2011. Health Impact Assessment of Waste Management Facilities in Three European Countries. *Environmental Health.*

Giusti, L. 2009. A review of waste management practices and their impact on human health. *Waste Management* 2227–2239.

Gomes, L.P. 2009. *Estudos de caracterização e tratabilidade de lixiviados de aterros sanitários para as condições brasileiras.* Rio de Janeiro: ABES/PROSAB.

Irimia-Diéguez, A.I., Sanchez-Cazorla, A., & Alfalla-Luque, R. 2014. Risk Management in Megaprojects. *Procedia – Social and Behavioral Sciences* 407–416.

Jarup, L., Briggs, D., Hoogh, C., Morris, S., Hurt, C., Lewin, A., & Elliott, P. 2002. Cancer risks in populations living near landfill sites in Great Britain. *British Journal of Cancer* 1732–1736.

Karak, T., Bhagat, R.M., & Bhattacharyya, P. 2012. Municipal Solid Waste Generation, Composition, and Management: The World Scenario. *Critical Reviews in Environmental Science and Technology* 1509–1630.

Palmiotto, M., Fatore, E., Paiano, V., Celeste, G., Colombo, A., & Davoli, E. 2014. Influence of a municipal solid waste landfill in the surrounding environment: Toxicological risk and odor nuisance effects. *Environment International* 16–24.

Pereira, A.L., & Maia, K.M. 2012. Contribuição da gestão de resíduos sólidos e educação ambiental na durabilidade de aterros sanitários. *Sinapse Múltipla* 68–80.

Perez, H.R., Frank, A.L., & Zimmerman, N.J. 2006. Health Effects Associated With Organic Dust Exposure During the Handling of Municipal Solid Waste. *Indoor and Built Environment* 207–212.

Pollard, S.J., Smith, R., Longhurst, P.J., Eduljee, G.H., & Hall, D. 2006. Recent developments in the application of risk analysis to waste technologies. *Environment International* 1010–1020.

Porta, D., Milani, S., Lazzarino, A.I., Perucci, C.A., & Forastiere, F. 2009. Systematic review of epidemiological studies on health effects associated with management of solid waste. *Environmental Health*.

Proag, S.L., & Proag, V. 2014. A framework for risk assessment. *Procedia Economics and Finance* 206–213.

PWC 2008. *A Pratical Guide for Risk Assessment – How principles-based risk assessment enables organizations to take the right risks*. Price WaterhouseCoopers.

Serpella, A.F., Ferrada, X., Howarda, R., & Rubio, L. 2014. Risk management in construction projects: a knowledge-based approach. *Procedia – Social and Behavioral Sciences* 653–662.

Struik, L.C., Jong, S.V., Shoubridge, J., Pearce, L.D., & Dercole, F. 2015. *Risk-based Land-use Guide*. Geological Survey of Canada.

Tohidi, H. 2011. The Role of Risk Management in IT systems of organizations. *Procedia Computer Science* 881–887.

Travis, W.R., & Bates, B. 2014. What is climate risk management? – Editorial. *Climate Risk Management* 1–4.

Vieira, J.M.P., & Morais, C. M. 2005. *Planos de Segurança da Água para Consumo Humano (Drinking Water Safety Plans)*. Lisbon, Portugal: IRAR Edition. ISBN: 972-99354-5-9. P. 173.

Vieira, J.M.P. 2011. A Strategic Approach for Water Safety Plans Implementation in Portugal. *Journal of Water and Health*, IWA Publishing, 09.1: 107–116.

Vieira, J.M.P. 2014. Landfill Safety Plan. A Methodology to Guarantee Safe Waste Management. *2nd International Africa Sustainable Waste Management Conference*. Luanda, Angola.

WHO 2004. *Guidelines for Drinking-water, third edition*. Geneva: World Health Organization.

WHO 2011. *Guidelines for Drinking-water, fourth edition*. Geneva: World Health Organization.

Wastes: Solutions, Treatments and Opportunities – Vilarinho, Castro & Russo (eds)
© 2015 Taylor & Francis Group, London, ISBN 978-1-138-02882-1

Lead (II) adsorption by modified eggshell waste

A. Ribeiro & J. Carvalho
CVR – Centre for Waste Valorisation, Guimarães, Portugal

C. Vilarinho
Mechanical Engineering and Resources Sustainability Center, Mechanical Engineering Department,
UMinho, Guimarães, Portugal

ABSTRACT: The current global trend towards more stringent environmental standards, technical applicability and cost-effectiveness, became key factors in the selection of adsorbents. After demonstrating the performance of eggshell-derived adsorbent under a range of lab operating conditions, this work focused the adsorption efficiency of thermic activated EGGIF (Eggshell Inorganic Fraction) in the adsorption of lead (II) ions. The effects of pH, contact time, adsorbent concentration and initial concentration were studied in batch experiments. Maximum uptake of lead (II) ion $(18.35\,\mathrm{mg\,g^{-1}})$ was obtained at pH 6.0 and initial lead concentration of $100\,\mathrm{mg\,L^{-1}}$. It was also obtained removal rates higher than 90% for all experimental conditions tested. These suggest that EGGIF is a promising adsorbent for lead (II) ions present in wastewater effluents.

1 INTRODUCTION

Water is an essential natural resource for animals, plants and mankind. Therefore the contamination of water by toxic contaminants through the discharge of industrial wastewater is a significant environmental problem (Gupta et al., 2009). The rapid industrialization, the increase of world population and the modern methods of agricultural and domestic activities led to an increasing demand for water, resulting in the generation of large volumes of wastewater with a number of pollutants that are harmful to both human and animal life (Wang et al., 2003).

The most common chemicals found in wastewater in a dissolved state and considered as potential pollutants are heavy metals, dyes, phenols, detergents, pesticides, polychlorinated biphenyls (PCBs), and others organic substances, such as organic matter (Gupta et al., 2009). Unlike organic contaminants, heavy metals are not biodegradable and tend to accumulate in living organisms and many heavy metal ions are known to be toxic or carcinogenic. Toxic heavy metals of particular concern in treatment of industrial wastewaters include zinc, copper, nickel, mercury, cadmium, lead and chromium (Ahmaruzzaman, 2011).

Lead (Pb) is a rare metal on the earth's crust. It is not known any benefit or essential character of Pb to any living organism. Its accumulation by humans can occur through inhalation, ingestion, dermal absorption and placental transfer, resulting then in the poisoning of metabolic pathways and consequent affection of several systems, like hematopoietic, nervous or even reproductive (Eisler, 2000). In contraposition to its harmful effects, Pb is widely used by humans, bringing them great comfort through the production of several useful products. The deposition of old paints, batteries, pesticides, solder, glass, brass, bronze, pigments and ammunition, but also activities like industrial smelting, coal burning, mining and traffic exhaustion are examples of anthropogenic sources of Pb, greatly responsible for its excessive accumulation (Eisler, 2000; Cameron, 2009).

Various methods are available for the removal of heavy metals from water and wastewater including: filtration; ultrafiltration; reverse osmosis; ion exchange; electrodialysis; electrolysis and adsorption (Ahmaruzzaman, 2011; Wang et al., 2003). Many of the enumerated methods are only employed for the control of specific pollutants. By contrast, adsorption is by far the most versatile and widely used method for the removal of a large spectrum of pollutants due to its high removal capacity and ease of operation at large scale (Meski et al., 2010). Adsorption is a physic-chemical technique which involves mass transfer between liquid and solid phase to remove or at least reduce chemical residues from wastewaters (Ahmaruzzaman, 2011). Activated carbon has been the most and widely used adsorbent in wastewater treatment technologies. Despite of its efficiency in adsorption process, this type of adsorbent has a high cost which makes it no longer attractive to be used in small-scale industries. Therefore, many researchers have applied regenerated natural wastes to treat aqueous solutions (Park et al., 2007). In this particular, eggshell from egg-breaking operations constitutes significant waste disposal problems for the food industry, so the development of value-added by products from this waste is to be welcomed (Carvalho et al., 2011). The porous nature of eggshell makes it an attractive material to be employed as an adsorbent (Tsai et al., 2006; Carvalho et al., 2011). Eggshell has been studied and applied successfully in many researches as low cost adsorbent in wastewaters treatment. These studies demonstrated the effectiveness of this adsorbent in the removal of heavy metals (Otun et al., 2006; Tsai et al., 2006; Park et al., 2007; Ahmaruzzaman, 2011), dyes (Kourmanova et al., 2002) and pesticides (Elwakeel et al., 2010). Nevertheless, in order to increase the adsorption capacity of this material, it is necessary to proceed to their thermal activation.

Considering previous studies, the aim of the present work was to evaluate the adsorption mechanism, adsorption capacity and removal rate of lead (II) by modified eggshell waste through the analyses of different operational conditions: influence pH, initial lead concentration, adsorbent concentration and operation time.

2 MATERIALS AND METHODS

2.1 EGGIF (Eggshell Inorganic Fraction) adsorbent preparation and characterisation

The adsorbent used in the present paper was calcined eggshell inorganic fraction powder and was collected from a local hatchery waste. Shells were washed successively with distilled water and eggshell membrane (Organic Fraction) was separated manually. After complete removal of the organic fraction, shells (Inorganic Fraction) were washed again. After that, material was dried at 105°C for 24 h, milled and calcined at 1000°C for 2 h. Chemical-physical characterization of EGGIF comprised the analysis of the surface area and particle porosity, through mercury porosimetry analysis. Chemical composition of EGGIF was obtained by X-Ray Fluorescence (XRF) and Scanning Electron Microscopy (SEM) with X-Ray Microanalysis.

2.2 Adsorption experiments

To determine the optimised operating conditions for lead uptake, batch studies were carried out with EGGIF adsorbent in a 1000 mL beaker of contaminated solution with constant agitation and pH control. Synthetic solutions were all prepared by diluting Pb (II) standard stock solutions (concentration 1000 mg L^{-1}) obtained by dissolving Pb(NO$_3$)$_2$ in distillate water. Real lead concentrations in the beginning of each assay were determined by atomic absorption spectrometry (Clescer et al., 1998). All batch adsorption studies were conducted at room temperature and the effect of contact time, initial Pb concentration, optimal pH and adsorbent dosage on adsorption capacity and removal rate, were monitored.

Effect of pH on lead adsorption capacity was investigated at pH values of 2.0 or 6.0 by using Pb solutions with 50 mg L^{-1} and an EGGIF concentration of 5.0 g L^{-1}. Desired pH of used solutions – 4 or 6 – was maintained by adding HCl or NaOH at the beginning and during the experiments. The effect of contact time was studied in the time range of 2–120 min. Effect of EGGIF concentration on lead adsorption capacity and removal rate was studied using different concentrations of EGGIF (1, 5 and 10 g L^{-1}) under the determined optimum pH value and lead (II) concentration of 50 mg g^{-1}. Effect of contact time was studied in the time range of 2–120 min. During the contact time of each assay, pH was maintained at value of 6 by the continuous addition of HCl 0.1 N. Effect of initial Pb (II) ion concentrations on adsorption process by EGGIF was undertaken using lead (II) solutions with the concentration range of 10–100 mg L^{-1} and an adsorbent dosage of 5 g L^{-1}. Effect of contact time was studied in the time range of 2–120 min. During the contact time of each assay, pH was again maintained at value of 6 by the continuous addition of HCl 0.1 N.

At the end of the adsorption process in each assay, EGGIF was separated from the solution through centrifugation and supernatant was analyzed for residual Pb (II) concentration by atomic absorption spectrometry (Clescer et al., 1998). Adsorption capacity was calculated by determining the final concentration at equilibrium, according to Equation 1.

$$Adsorption\ capacity\ (qt) = \frac{(C_0 - C_e) \times V}{m} \tag{1}$$

where qt (mg g^{-1}) is the solute adsorption capacity per gram of eggshell, C_0 and C_e (mg L^{-1}) are the initial an equilibrium concentration, respectively V (L) is the volume and m (g) is the mass of EGGIF used.

The percentage of pollutant removed (R/%) from the solutions were calculated using Equation 2.

$$Removal\ rate\ (R) = \frac{[(C_0 - C_e) \times 100]}{C_0} \tag{2}$$

where R (%) is the removal rate, C_0 and C_e (mg/L) are the initial an equilibrium concentration, respectively.

3 RESULTS

3.1 EGGIF adsorbent preparation and characterisation

Data in Table 1 indicates the BET surface area, pore volume and chemical composition of calcined eggshell particles. It is possible to verify that EGGIF have a surface area of 3.469 m^2 g^{-1} and a pore volume of 9.364 m^3 g^{-1}. According to the classification of Brunauer, Deming, Deming and Teller (BDDT), this type of material belong to a typical type II, which is frequently encountered when adsorption occurs on materials with macropores or open voids. Table 1 also shows the chemical composition of EGGIF.

This determination showed that EGGIF is mostly constituted by calcium oxide (CaO) and others elements in residuals traces. According to Tsai et al. (2006), this occurs because in the process of calcination, calcium carbonate (CaCO$_3$) is converted in calcium oxide releasing carbon dioxide (CO$_2$). These results were confirmed through scanning electron microscopy determination as shown in Figure 1(b). The textural structure examination of calcined eggshell particles can be observed from the SEM photographs in Figure 1. It is possible to observe a well-defined pore structure and a crystalline structure showing an angular pattern of fracture.

3.2 Effect of pH and contact time in adsorption capacity and removal rate of lead (II)

The acidity of solutions (pH) is one of the most important parameters controlling uptake of heavy metals from aqueous solutions. In this evaluation, the concentration of Pb (II) (50 mg L^{-1}), the amount of EGGIF adsorbent (5.0 g L^{-1}), the contact time (120 min) and the desirable pH were kept

Table 1. Physical-chemical composition of EGGIF adsorbent.

Sample	Surface area (m^2 g^{-1})	Pore volume (m^3 g^{-1})	Chemical composition (%)				
			CaO	MgCO$_3$	CaCO$_3$	P$_2$O$_5$	SO$_3$
EGGIF after thermic activation	3.469	9.3639	95.42	1.13	2.75	0.58	0.21

255

Figure 1. (a) SEM images for calcined eggshell powder; (b) diagram of the composition of calcined eggshell obtained from scanning electron microscopy.

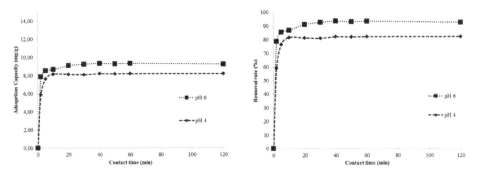

Figure 2. Effect of pH and contact time in adsorption capacity and removal rate of lead (II) on EGGIF adsorbent (Contact time: 120 minutes; Initial [Pb]: $50 \, mg \, L^{-1}$; [EGGIF]: $5 \, g \, L^{-1}$; Volume: 1 L).

constant. Figure 2 illustrate the effect of pH and contact time in adsorption capacity and removal rate of lead (II) on EGGIF adsorbent.

Results demonstrate that adsorption of lead (II) is more effectiveness at pH 6. At this value it was achieved a maximum adsorption capacity of $9.36 \, mg \, g^{-1}$ and a removal rate of 94%. The effect of pH on adsorption can vary substantially due to the different functional groups on the adsorbent surfaces, and the sorption capacity may therefore increase or decrease as a function of pH (Liao et al., 2010). At low pH levels, EGGIF adsorbent is protonated, which results in the poor adsorption of metal cations. An optimum pH leaves the binding sites un-protonated and the adsorption capacity is consequently maximized. Above this optimum range, most metals tend to precipitate out of the solution in hydroxide form (Liao et al., 2010, Riao et al., 2010). According to Pourbaix diagram of lead in aqueous matrix, this metal at this concentration and pH value (6.0) is in soluble state (Pb^{2+}). According to the same diagram, above pH 7, lead assumes the state of stable metal hydroxide ($Pb(OH)_2$). For that reason, the contact of these metals with EGGIF even in low concentration, leads to their adsorption by chemisorption processes.

3.3 Effect of EGGIF concentration and contact time in adsorption capacity and removal rate of lead (II)

Experiments were carried out using different concentrations of EGGIF (1, 5 and $10 \, g \, L^{-1}$) under the determined optimum pH value and contact time of 120 minutes. During the contact time of each assay, pH was maintained at value of 6 by the continuous addition of HCl 0.1 N. Figure 3 illustrate the effect of EGGIF concentration and contact time in lead (II) uptake by EGGIF.

Results showed that uptake of lead (II) enhanced with increasing the concentration of EGGIF adsorbent. At the concentration of $1 \, g \, L^{-1}$ of EGGIF it was achieved a maximum removal rate of 87% of lead (II). In contrast, at the concentration of $10 \, g \, L^{-1}$ of EGGIF it was achieved a maximum removal rate of 95%. Nevertheless it's also verified that the difference in lead (II) removal rate with $5 \, g \, L^{-1}$ to $10 \, g \, L^{-1}$ of EGGIF is insignificant, since there was obtained a removal

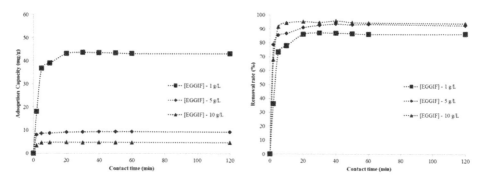

Figure 3. Effect of EGGIF concentration and contact time in adsorption capacity and removal rate of lead (II) on EGGIF adsorbent (contact time: 120 minutes; initial [Pb] 50 mg L^{-1}; pH: 6.0; Volume: 1 L).

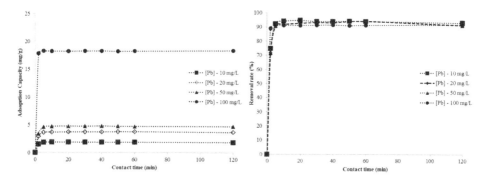

Figure 4. Effect of initial lead (II) concentration and contact time in adsorption capacity and removal rate of lead (II) on EGGIF adsorbent (Contact time: 120 minutes; initial [EGGIF] 5 mg L^{-1}; pH: 6.0; Volume: 1 L).

rate of 94%. Therefore, optimal conditions for lead (II) uptake were achieved using 5 g L^{-1} of EGGIF adsorbent. Rao et al. (2010) performed a study in lead adsorption on eggshells. In this study the authors report a maximum removal of lead of approximately 96% at a concentration of 20 mg metal L^{-1} solution. At a concentration of 50 mg L^{-1} lead in solution, these authors report a removal rate of approximately 89%.

3.4 Effect of initial lead (II) concentration and contact time in adsorption capacity and removal rate of lead (II)

Experiments were carried out using various initial concentrations of Pb (II) solution under the determined optimum pH value and contact time of 120 minutes. The effect of initial lead ion concentration was studied in the range of 10–100 mg L^{-1}. Figure 4 shows the effect of initial lead (II) concentration and contact time in adsorption capacity and removal rate of lead (II) on EGGIF adsorbent.

According to the results expressed in figure 4, it is possible to verify that increasing metal concentration slightly reduce the removal rate of lead (II) by EGGIF adsorbent (95 to 91%). At lower concentrations, lead ions could interact with all binding sites of the adsorbent, which increase the effectiveness of adsorption. In opposite, at higher concentration the binding sites of the adsorbent are more saturated due to decrease of available surface area which decreases the adsorption yield. Nevertheless, in all tested lead concentrations were achieved adsorptions yields higher than 90%. These confirm the high affinity of EGGIF adsorbent with lead (II) ion. Naturally, adsorption capacity (qt) obtained highest values on the high concentrated solution (18.35 mg L^{-1}) due to saturation of the binding sites.

4 CONCLUSIONS

Results confirmed that thermal activation of inorganic fraction of eggshell waste enhances its capability on uptake heavy metals from wastewaters. This process it's very promising, since it will allow the separation and valorisation of both fractions of eggshell wastes (organic and inorganic). Results showed that EGGIF (Eggshell Inorganic Fraction) may be a promising adsorbent for lead (II) ions present in wastewater effluents. Regarding to technical aspects of lead (II) ions uptake by EGGIF it is possible to conclude that:

– Lead uptake by EGGIF adsorbent is strongly dependent by pH. Results demonstrate that optimal pH value is 6.
– Adsorption of lead (II) ions enhanced with increasing the concentration of EGGIF adsorbent. Optimal conditions for lead (II) uptake were achieved using $5\,g\,L^{-1}$ of EGGIF adsorbent been obtained a removal rate of 94% and a adsorption capacity of $9.27\,mg\,g^{-1}$.
– Regarding to initial lead concentration, results proved that the increase of metal concentration slightly reduce the uptake of lead (II) by EGGIF adsorbent. Nevertheless, in all tested lead concentrations were achieved adsorption yields higher than 90%. Highest adsorption capacity value ($18.35\,mg\,L^{-1}$) was achieved testing a high contaminated lead solution ($100\,mg\,L^{-1}$).

REFERENCES

Ahmaruzzaman, M. 2011. Industrial wastes as low-cost potential adsorbents for the treatment of wastewater laden with heavy metals. *Advances in Colloid and Interface Science*, 166: 36–59.

Cameron, R.E. 2009. *Guide to Site and Soil Description for Hazardous Waste Site Characterization, EPA/600/4-91/029*. Environmental Protection Agency, United States of America.

Carvalho, J., Araújo, J. & Castro, F.P. 2011. Alternative low-cost adsorbent for water and wastewater decontamination derived from eggshell waste – an overview. *W. Bio*. Vol. 2: 157–163.

Clescer, L., Greenberg, A. & Eaton, A. 1998. *Standard methods for the examination of water and wastewater"*. 20th ed, American Public Health Association, Washington.

Eisler, R. 2000. *Handbook of chemical risk assessment: health hazards to humans, plants, and animals*. CRC Press LLC, Boca Raton.

Elwakeel, K.Z. & Yousif, A.M. 2010. Adsorption of malathion on thermally treated egg shell material. *Inter. Water Tech. Conf* (2010).

Gupta V.K., Carrott, P.J.M., Ribeiro, Carrott, M.M.L. & Suhas. 2009. Low-Cost Adsorbents: Growing Approach to Wastewater Treatment—a Review. *Critical Reviews in Environmental Science and Technology*, 39: 783–842.

Koumanova, B., Peeva, P., Allen, S.J., Gallagher, K.A. & Healy, M.G. 2002. Biosorption from aqueous solutions by eggshell membranes and Rhizopus oryzae: equilibrium and kinetic studies. *J. Chem. Tech. Biotech*. 77: 539–545.

Liao, D., Zheng, W., Li, X. & Yan, Q. 2010. Removal of lead(II) from aqueous solutions using carbonate hydroxyapatite extracted from eggshell waste. *Journal of Hazardous Materials* 177: 126–130.

Meski, S., Ziani, S. & Khireddine, H. 2010. Removal of lead Ions by hydroxyapatite prepared from the egg shell. *J Chem. Eng*. 55: 3923–3928.

Otun, J., Oke, I., Olarinoye, N., Adie, D. & Okuofu, C. 2006. Adsorption isotherms of Pb (II), Ni (II) and Cd (II) ions onto PES. *Journal of Applied Sciences* 6(11): 2368–2376.

Park, H.J., Jeong, S.W., Yang, J.K., Kim, B.G. & Lee, S.M. 2007. Removal of heavy metals using waste eggshell. *J. Envir. Sci*. 19: 1436–1441.

Rao, J., Kalyani, G., Rao, V. & Anu, T. 2010. Kinetic studies on biosorption of lead from aqueous solutions using egg shell powder. *International Journal of Biotechnology and Biochemistry*, 957–968.

Tsai, W.T., Yang, J.M., Lai, C.W., Cheng, Y.H., Lin, C.C. & Yeh, C.W. 2006. Characterization and adsorption properties of eggshells and eggshell membrane. *Bioresh. Tech*. 97: 488–493.

Wang, Y., Lin, S. & Juang, R. 2003. Removal of heavy metal ions from aqueous solutions using various low-cost adsorbents. *Journal of Hazardous Materials*. B102: 291–302.

Wastes: Solutions, Treatments and Opportunities – Vilarinho, Castro & Russo (eds)
© 2015 Taylor & Francis Group, London, ISBN 978-1-138-02882-1

Irregular disposal of unserviceable tires in Brazil and management proposals

M.C. Rizk & A.C.C. Gomes
Departamento de Planejamento, Faculdade de Ciência e Tecnologia, Universidade Estadual Paulista "Júlio de Mesquita Filho", São Paulo, Brasil

ABSTRACT: The unserviceable tires do not have more usefulness after some time of use. If these wastes are abandoned in rivers, lakes and land, it can cause the proliferation of vectors and rodents, posing risks to public health. Therefore, it is necessary to identify the irregular disposal areas of unserviceable tires to can be traced the necessary and reasonable measures to reduce the impacts. Thus, this paper aims to identify the irregular disposal of unserviceable tires that occurs in the city of Presidente Prudente – São Paulo – Brazil. The study allowed presenting some proposals to improve the current management of unserviceable tires in order to minimize the risks and composing the guidelines of unserviceable tires management in the municipality.

1 INTRODUCTION

Tires are principally composed of vulcanized rubber, rubberized fabric containing reinforcing textile cords, steel or fabric belts, and steel wire-reinforced rubber beads. To strengthen the rubber and to increase abrasion resistance is used the carbon black (Hazarika & Yasuhara, 2008).

Since the tires are manufactured to withstand severe impacts, its structure needs to be strong and therefore the eliminating of waste tire becomes difficult.

In the USA, each year, are generated over 5 billion tons of non-hazardous solid waste materials. Of these, more than 270 million scraptires (approximately 3.6 million tons) are generated each year (Siddique & Naik, 2004).

Lagarinhos & Tenório (2013) related that in Brazil 67.3 million of tires were produced in 2010; 18.1 million of tires were imported and 23.9 million tires were exported. The tires output is given to the replacement and sale market (45%); automobile industry and manufacturers (30%) and exportation (25%). Between the years 2001 and 2012 it was produced 538.7 million tires in Brazil.

Because of the huge generation of tires and the several environment impacts that are caused, the tire recycling has become a necessity. There is about one worn tire produced per year and per person in the developed countries (Roy et al. 1990).

Sasikumar et al. (2010) related that, approximately, 800 million unserviceable tires are disposed around the world every year and this amount is expected to increase 2% each year.

The irregular discard of unserviceable tires in inadequate locals can cause harm to public health and the environment (Souza & D'Agosto, 2013).

The unserviceable tires can provide breeding sites for mosquitoes, which can spread diseases and often constitute fire hazards. Utilization of discarded tires minimizes environmental impact and maximizes conservation of natural resources (Jang et al., 1998).

Sasikumar et al. (2010) related that due to increasing environmental impacts, government regulations, social responsibilities, resource reduction and economic factors, many companies are now engaged in the product recovery business.

Logistics reverse associates activities with the handling and management of tires. It is accompanied by a series of steps as collection, transport and final destination: remanufacturing or recycling.

The logistics reverse chain can be affected by collection step. How much more locations use, the collection is more difficult. This occurs because the tires are not discarded in the same places of use or purchase, for example (Brito & Dekker, 2002).

Lagarinhos & Tenório (2008) related that in Brazil, about 4.000 tire's sellers make part of the unserviceable tires collection process. Besides, there are 270 ecopoints, which are sites provided voluntarily by municipalities through cooperation agreements with tires manufacturers, where the unserviceable tires collected by public service or discarded voluntarily by the population are taken and temporarily stored.

In Brazil, it is required by CONAMA Resolution 416/2009, that the manufacturers and distributors collect its tires, but there is still no payment associated with the return of the unserviceable tires as occurs in the deposit-return systems.

The environmental policies developed in Brazil are relatively new and still need to be structured aiming to minimizing impacts to the environmental and public health caused by the improper disposal of unserviceable tires. It is necessary to develop a management plan for unserviceable tires together with studies that diagnose the main obstacles in the unserviceable tires management.

In this context, the aim of this paper is to study the irregular areas of unserviceable tires disposal in the city of Presidente Prudente, São Paulo, Brazil and present proposals to reducing these areas and the impacts generated from the irregular destination of tires.

2 MATERIALS AND METHODS

The irregular areas of tires disposal were recognized by collecting data from the Municipal Department of the Environment of Presidente Prudente – SP, along with field works. The points of irregular areas were demarcated using GPS.

Fieldworks were realized to know each area and taking photos.

During the fieldworks, it was observed as the management of unserviceable tires influences the irregular discard.

Thus, through these data, it was traced the chain up until the tire destination step after one unserviceable tire be generated.

The analysis of the irregular disposal sites allowed to check the generation of environmental impacts caused by the unserviceable tires and the risk that its offer to security of heath population.

3 RESULTS AND DISCUSSION

According to Censo 2010 (IBGE, 2010), the projected population of Presidente Prudente in 2013 was 218,960 inhabitants, with a population density of 368.89 inhabitants/km^2. The population increased 15.74% between the years 2000 and 2013.

Due to the growth of the city and elsewhere, the vehicle's fleet consequently also increased with a total in 2013 of 143,910 vehicles, including cars, motorcycles, bus, trucks and tractors (IBGE, 2010).

With the data of the Municipal Department of the Environment of Presidente Prudente – SP and field surveys about the irregular areas of unserviceable tires disposal, it was identified 12 critical areas of irregular disposal unserviceable tires in this city.

Of the 12 areas, seven of them are located along one of the major city avenues by focusing in eastern urban area of Presidente Prudente. While other irregular disposal points are more spaced with two points far away from urban area, nearly to a rural area. All points are in peripheral areas of the urban perimeter.

Of the total, 11 discard areas of tires are close to rivers, which can harm the drainage system and can cause visual and sanitary pollution.

Figure 1.   Tires disposed in lateral area of urban road system.

The irregular disposal of unserviceable tires is found along the urban road system (Figure 1), in vacant lots and even in places which there are information prohibiting the disposal of unserviceable tires.

Solid waste that require different forms of treatment and disposal are disposed in open at the same place, difficulty the municipal management process of wastes and rendering more expensive the public sanitation service. In the areas, also it was founded old furnishings, clothes, pieces of wood, building waste, plastics, etc. as seen by Figure 2.

The amount of waste found in irregular areas are significant and require, monthly, to the public sanitation sector the collection of solids wastes.

In order to reduce the tires amount in discard areas, it is necessary that the focus be on the generation step of unserviceable tires. For this, the local government, including the dissemination of environmental educational programs, could deploy support programs about tires management to tire's repairmen.

The waste tires, which are in the open, constitute potential breeding grounds for mosquitoes, particularly *Aedes aegypti*, the transmitter of diseases such as dengue and yellow fever.

Dengue, in particular, is among the most reemerging diseases in the world and is one of the diseases transmitted by arthropods considered more important. In Brazil, dengue has a seasonal pattern, with the highest incidence of cases in the first five months of the year, warmer and wetter period, typical of tropical climates (Braga & Valle, 2007).

The overcrowded cities and the consequent poor sanitation conditions favor the proliferation of the mosquito *Aedes aegypti*. Moreover, the practice of improper disposal of unserviceable tires aggravates this situation, resulting in situations that can pose risks to the health and welfare of the population.

In the case of an epidemic, the irregular areas of unserviceable tires disposal presented deserve to be constantly monitored by local sanitation agents that should together with the population require proper management of this type of waste, considering the danger that it may entail for aggravation of scenario the dengue epidemic in Brazil.

Figure 2. Disposal of several waste solid in valley bottom.

The necessary measures are public campaigns with sanitary focus to draw the attention of consumers, tire's repairmen and the population for the losses that may be caused by irregular disposal unserviceable tires.

The consumers can also be influenced by campaigns that show the importance of keeping the tires always calibrated and balanced beyond keeping the car lined up, increasing the life of the tire. This practice is extremely important in environment awareness divulgations, which provides for the reduction of the generation of tire waste, the first step in the hierarchy political of the 3Rs (reduce, reuse and recycle).

In order to reduce the irregular disposal in inappropriate areas of the city, cleaning task forces can be realized. The municipal government can develop projects to collect tires and spoils in the own homes of the city. The collect can be organized in accordance with existing neighborhoods in the city. Each region of the city can be attended during one week with hours of pre-released collection to residents, who need only leave their tires and spoils on the sidewalk, so that the truck collects the disposed materials.

Partnerships can be made so that more and more tires be collected and properly forwarded to shredded, avoiding the irregular disposal.

It is clear that a routine inspection of the areas is extremely important. Cleaning should be maintained periodically in order to prevent further proliferation of rodents and vectors of diseases, with special attention to the areas of valley bottom.

Based on the results obtained the application of better management of waste tires by municipal government in Presidente Prudente is necessary to ensure the reduction of irregular disposal of tires.

Also, if the tires could be treated at the city of Presidente Prudente, the quantity of tires that is stored will decrease, reducing the number of tires that can be disposed irregularly.

4 CONCLUSIONS

Due of concern about public health and environment, the identification of irregular discard of unserviceable tires is of huge importance for the management of pneumatic waste. From this identification, it becomes possible to draw the necessary measures to reduce the impacts of irregular disposal of unserviceable tires, such as environmental education campaigns to consumers and generators of tires and legal measures to management of unserviceable tires by local government.

Legal measures should be taken by the government also in relation to the risk of spread of dengue and yellow fever due to poor disposal of waste tires. Public campaigns can be made to detect dengue cases and to eliminate outbreaks of *Aedes aegypti*, ensuring the health of the population and minimizing disposal of unserviceable tires in the open sky.

Thus, the results obtained can contribute to the establishment of guidelines that will assist in the development of an integrated management of unserviceable tires.

ACKNOWLEDGEMENTS

The authors would like to thank FAPESP (Fundação de Amparo à Pesquisa do Estado de São Paulo) for financial support.

REFERENCES

Braga, I. A. & Valle, D. 2007. Aedes aegypti: history of control in Brazil. *Epidemiology and Health Services* 16: 113–118.

Brasil. Conselho Nacional do Meio Ambiente (CONAMA). Resolution n. 416/2009. Provides for the prevention of environmental degradation caused by waste tires and environmentally sound disposal.

Brito, M. & Dekker, R. 2002. Reverse Logistics – a framework. *Econometric Institute Report*. hdl.handle.net/1765/543. Online publication date: 10-Oct-2002.

IBGE – Brazilian Institute of Geography and Statistics. Censo 2010. Available in: www.ibge.gov.br. Acess in 10s-ago-2014.

Jang, J.-W., Yoo, T.-S., Oh, J-H., & Iwasaki, I. 1998. Discarded tire recycling practices in the United States, Japan and Korea. *Resources, Conservation and Recycling* 22 (1–2): 1–14.

Hazarika, H. & Yasuhara, K. 2008. *Scrap tire derived geomaterials – Opportunities and Challenges*. London: Taylor & Francis Group.

Lagarinhos, C.A.F. & Tenório, J.A.S. 2008. Technologies used in the reuse, recycling and energy recovery of tires in Brazil. *Polymers 8*: 106–118.

Lagarinhos, C.A.F. & Tenório, J.A.S. 2013. Reverse logistics of unserviceable tires in Brazil. *Polymers* 23: 49.

Roy, C., Labrecque, B. & Caumia, B. 1990. Recycling of scrap tires to oil and carbon black by vacuum pyrolysis. *Resources, Conservation and Recycling 4*: 203–213.

Sasikumar, P., Kannan G. & Haq, A.N. 2010. A multi-echelon reverse logistics network design for product recovery – a case of truck tire remanufacturing. *International Journal of Advanced Manufacturing Technology* 49: 1223–1234.

Siddique, R. & Naik, T.R. 2004. Properties of concrete containing scrap-tire rubber – an overview. *Waste Management* 24 (6): 563–569.

Souza, C.D.R. & D'Agosto, M.A. 2013. Value chain analysis applied to the scrap tire re-verse logistics chain: An applied study of co-processing in the cement industry. *Resources, Conservation and Recycling* 78: 15–25.

Wastes: Solutions, Treatments and Opportunities – Vilarinho, Castro & Russo (eds)
© 2015 Taylor & Francis Group, London, ISBN 978-1-138-02882-1

Diagnosis of food waste generation in a university restaurant

M.C. Rizk & B.A. Perão

Departamento de Planejamento, Faculdade de Ciência e Tecnologia, Universidade Estadual Paulista
"Júlio de Mesquita Filho", São Paulo, Brasil

ABSTRACT: The restaurants produce meals in large scale, so the management of the solid waste is an environmental requirement and a hygienic-sanitary issue in this kind of establishment. Considering the need to reduce the generation of wastes, it was done an evaluation in a university restaurant, which were quantified the solid wastes generated at the meal production. During the 14 days of analysis, it was produced the amount of 3410.10 kg of food and it was generated 605.98 kg of solid wastes, as food remains. It was observed failures in the process of cutting and/or removal of excessive peel during the handling of food. The average of leftovers was 5.04% (prepared food that is not served on plates) and the serving losses average was 18.52%. Thus, to minimize the waste generation is necessary an awareness campaign focusing on reducing food waste involving employees and users of the university restaurant.

1 INTRODUCTION

Katajajuuri et al. (2014) related that during recent years, there has been increasing international interest in the amount of food the world wastes. It is argued that globally roughly one-third of food produced is lost or wasted, which amounts to about 1.3 billion tonnes per year. Therefore, politicians are interested in food waste and are seeking ways to reduce it. In most studies, food waste has been explored by conducting waste compositional analyses, measuring food waste through the analysis of waste streams.

According to Engström & Carlsson-Kanyama (2005), food produced but not effectively used in human consumption is a mismanagement of natural resources. Food losses occur throughout the food system that starts with farm production and ends when the food is consumed. Lowering food losses is one of the potential measures for overcoming hunger. Increasing the efficiency with which food is handled will also reduce the ecological side effects from increasingly intensive agriculture and will help to reduce the demand for land.

Jeong et al. (2014) presented that as customers become more environmentally conscious and their demands for eco-friendly products and services increase, many industry professionals have engaged in developing and promoting ecologically sustainable products and services while striving to strengthen their business commitment to sustainability. Green practices enable companies to save on long-term operational costs while gaining a competitive advantage by developing or enhancing a positive image and reputation.

Wang et al. (2013) summarized the fundamental green restaurant concepts in sustainability (eco-friendly production and service, using organic food and environmentally friendly products, green building), low-carbon (energy conservation and carbon reduction, energy efficiency and water-saving, using local food and resources), environmental conversation (waste reduction, low-pollution, and saving resources through concepts such as recycle, reduce and reuse), environmental management policies, green education for employees and consumers, health, sanitation and safety, and social responsibility. Implementing the green restaurant philosophy thus involves incorporating environmental conversation into each management process, and increasing the emphasis on green food, green production and service to provide customers with healthy, safe and eco-friendly meals.

In this context, the aim of this paper was to determine the quantity and types of food wasted generated in a university restaurant and analyzing the reasons for food waste, aiming to improve the waste management.

2 MATERIALS AND METHODS

The study covered a university restaurant located at FCT/UNESP campus in Brazil. The restaurant served around 300 meals per day. The study period was in January of 2015, during 14 days.

The routine activities of the restaurant were followed to identify and characterize the stages of meal's production and to identify the areas where the solid waste is generated, as well as the forms of segregation and disposal of the solid waste generated.

The losses were divided into different types: preparation losses, which are mostly seeds, peel, etc. from fruits and vegetables; leftovers, which are prepared food never served; and serving losses, which are what are left on serving dishes.

The restaurant personnel weighed the sorted waste. In addition, the personnel completed forms with the daily amounts of food prepared, and the amounts of food waste from cooking and leftovers.

With the data, it was determined three loss's indicators in the restaurant; the correction factor related to the preparation losses and the index of serving losses and leftovers.

According to Abreu et al. (2009), the correction factor was obtained by Equation 1:

$$\text{Correction factor} = \left(\frac{\text{Gross weight}}{\text{Net weight}} \right) \qquad (1)$$

To determine the percentage of leftovers, it was used the Equation 2 (Abreu et al. 2009):

$$\text{Leftovers (\%)} = \left(\frac{\text{Total produced} - \text{Total distributed}}{\text{Total produced}} \right) \times 100 \qquad (2)$$

Prepared foods in the distribution counter, but that were not distributed also were considered serving losses. To determine the serving losses it was used the Equation 3 (Abreu et al. 2009):

$$\text{Serving losses(\%)} = \left(\frac{\text{Weight of rejected meal}}{\text{Weight of distributed meal}} \right) \times 100 \qquad (3)$$

3 RESULTS AND DISCUSSION

The studied university restaurant provides meals during the week (Monday to Friday), at lunchtime.

During the period of this study, it was served an average of 287 meals. The menu composition is basic, consisting of a base plate (rice/beans), main course, trim, salad, dessert and juice. The provision of meals is by the centralized mode, in which the restaurant personnel serve the trim and the main course and the base plate (rice/beans) and the salad are self-serviced.

The solid waste generated is separated into organic and dry. The location of temporary waste disposal is appropriate, organic are stored in a cold room, as dry are stored after the service on the outside of the restaurant. The disposal of the organic material is for feeding pigs and other animals and third parties collect it daily. The dry material is disposed in the municipal landfill.

During the research, it was served 3994 meals totaling 3410.10 kg of food produced, of which 3242.88 kg were distributed and 605.98 kg of food were discarded as waste.

To assess food waste in the pre-preparation of vegetables step it was used the correction factor. Table 1 shows the results of the correction factor found in this study and the values showed by Nunes (2010).

Table 1. Correction factor of greenery in the pre-preparation.

Greenery	Samples	Average gross weight (kg)	Average net weight (kg)	Correction factor	
				Results	Nunes (2010)
Chard	4	18.5	16.95	1.09	1.54–1.66
Carrot	3	10.3	8.7	1.18	1.17
Lettuce	4	19.5	16.04	1.22	1.09–1.33
Tomato	1	20	15.3	1.31	1.25
Cabbage	5	11.4	10.36	1,10	1.72
Collard green	3	13	9.53	1.36	1.06–2.22
Broccoli	1	36	30	1.20	2.12
Zucchini	1	34	30	1.13	1.33–1.38
Beet	1	24	21	1.14	1.61–1.88

Table 2. Determination of the leftovers.

Analyzed days	Weight of produced meals (kg)	Weight of distributed meals (kg)	Leftovers (%)
1st	246.10	246.10	0
2nd	250.78	246.08	1.87
3rd	316.50	312.30	1.33
4th	242.90	242.90	0
5th	250.60	222.20	11.33
6th	263.60	240.10	8.92
7th	248.70	236.70	4.83
8th	245.30	234.90	4.24
9th	240.50	240.50	0
10th	225.50	199.10	11.71
11th	211.00	195.10	7.54
12th	223.80	204.30	8.71
13th	216.90	204.60	5.67
14th	228.00	218.00	4.39
Total	3410.18	3242.88	–
Average	243.58	231.63	5.04
Standard deviation	26.36	29.59	4.12

These results show that the losses of non-edible parts of these foods are as expected. Only the carrot and tomato showed higher values than those in literature, although close to those found in the literature. Such values could be even smaller if there was the full use of food.

Table 2 shows the results obtained to the leftovers determination.

It was observed that the average percentage of leftovers was 5.04%. In some days, there were no leftovers, however, in some days the leftover's level was above the average. According to Vaz (2006), the acceptable values are at most 3%, so the value found in this study is higher than the acceptable.

During the study developed by Katajajuuri et al. (2014) the authors found that the restaurants evaluated discarded 19% of all food produced and served. Of that, 6% was kitchen waste, 5% service waste and 7% leftovers.

Table 3 shows the results obtained to the serving losses determination.

With regard to the serving losses index, the value obtained was superior to the literature. Teixeira et al. (2006) indicate an acceptable value below 10% and the value obtained was 18.52%. Several factors may influence the food waste, such as food preferences of each user, training of personal

Table 3. Determination of the serving losses.

Analyzed days	Weight of distributed meals (kg)	Weight of rejected meals (kg)	Number of meals served	Serving losses (%)
1st	246.10	44.06	287	17.90
2nd	246.08	44.95	292	18.27
3rd	312.30	66.30	294	21.23
4th	242.90	46.20	287	19.02
5th	222.20	34.30	273	15.44
6th	240.10	52.80	270	21.99
7th	236.70	49.40	287	20.87
8th	234.90	36.77	290	15.65
9th	240.50	61.10	290	25.41
10th	199.10	43.20	284	21.70
11th	195.10	35.70	280	18.30
12th	204.30	31.50	294	15.42
13th	204.60	23.80	286	11.63
14th	218.00	35.90	280	16.47
Total	3242.88	605.98	3994	–
Average	231.63	43.28	285.29	18.52
Standard deviation	29.59	11.57	7.30	3.53

responsible for distribution, users who use larger amounts than they can consume in the self-service counter, and the lack of user awareness about food waste and the impacts that these wastes are disposed of in the environment.

It is important to note that the generation of food waste, in addition to harming the environment with its final disposal, also causes impacts of agriculture, and economic and social impacts.

According to Kantor et al. (1997) while food recovery and recycling technologies may help to utilize food that would otherwise be discarded, programs designed to prevent food loss in the first place may be particularly useful in reducing consumer and foodservice food losses. Education programs that help consumers change their food discard behavior may also be effective in preventing food loss. For instance, educational programs that help meal planners determine appropriate portion sizes and distinguish between spoiled and safe food can help consumers reduce plate waste and better utilize leftovers. Educational programs that increase the awareness of food loss by manufacturers, retailers, and consumers may reduce the amount of food loss and in turn the environmental and economic costs of waste disposal.

4 CONCLUSIONS

It was observed that the university restaurant studied demonstrates a concern for environmental issues, already adopting some practices that aim to reduce the negative impact of their activities on the environment. However, some measures should be incorporated and other must be appropriate, as concern about the disposal of recyclable.

The percentage of leftovers can be because the quantity of food is produced based on the number of meals sold, this control is performed daily and the number of tickets sold is always 300, but the daily frequency is never the same. Another factor that may have influenced is the fact that the research was conducted in summer, with high temperatures, where it was observed that the diners preferred the cold dishes (salads) rather than hot dishes, which can cause large amounts of rice leftovers.

Regarding the serving losses, the nutritionist responsible assesses what preparations are better accepted by users, this measure prevents preparations were not very well accepted are repeatable, contributing to reduction of waste.

With the results obtained it is concluded that awareness is required of employees and diners to avoid the food waste, by conducting environmental education campaigns, so there is reduction of waste.

ACKNOWLEDGEMENTS

The authors would like to thank PIBIC/UNESP (Reitoria) for financial support.

REFERENCES

Abreu, E.S., Spinelli, M.G.N. & Pinto, A.M.S. 2009. *Gestão de unidades de alimentação e nutrição: um modo de fazer*. São Paulo: Metha.
Engström, R. & Carlsson-Kanyama, A. 2004. Food losses in food service institutions: Examples from Sweden. *Food Policy* 29: 203–213.
Jeong, E., Jang, S-C.(S)., Day, J. & Ha, S. 2014. The impact of eco-friendly practices on green image and customer attitudes: An investigation in a café setting. *International Journal of Hospitality Management* 41: 10–20.
Kantor, L., Lipton, K., Manchester, A. & Oliveira, V. 1997. Estimating and addressing America's food losses. *Food Review* 20: 2–12.
Katajajuuri, J-M., Silvennoinen, K., Hartikainen, H., Heikkilä, L. & Reinikainen, A. 2014. Food waste in the Finnish food chain. *Journal of Cleaner Production* 73: 322–329.
Nunes, R.M. 2010. Table of per capita, correction factor and performance. Available at: http://www.ufjf.br/renato_nunes/files/2010/08/apostila-de-fator-de-corre%C3%A7%C3%A3o-dos-alimentos.pdf. Accessed on: 16 Feb. 2015.
Teixeira, S. et al. 2006. *Administração aplicada às unidades de alimentação e nutrição*. São Paulo: Atheneu.
Vaz, C.S. 2006. *Restaurantes – controlando custos e aumentando lucros*. Brasília: Editora Metha.
Wang, Y-F., Chen, S-P., Lee, Y-C. & Tsai, C-T.(S). 2013. Developing green management standards for restaurants: An application of green supply chain management. *International Journal of Hospitality Management* 34: 263–273.

Wastes: Solutions, Treatments and Opportunities – Vilarinho, Castro & Russo (eds)
© 2015 Taylor & Francis Group, London, ISBN 978-1-138-02882-1

Anaerobic co-digestion of fruit and vegetable wastes with different substrates

M.C. Rizk
Universidade Estadual Paulista "Júlio de Mesquita Filho" – Faculdade de Ciências e Tecnologia,
Departamento de Planejamento, Urbanismo e Ambiente, Presidente Prudente, São Paulo, Brazil

R. Bergamasco & C.R.G. Tavares
Universidade Estadual de Maringá – Departamento de Engenharia Química, Maringá, Paraná, Brazil

ABSTRACT: Anaerobic biological treatment of organic solid wastes can be an acceptable alternative to current disposal strategies. Therefore, the aim of the present study was to evaluate the anaerobic biodegradability potential of fruit and vegetable wastes (FVW) using different co-substrates. The FVW were shredded, blended and diluted with water, domestic sewage, sewage sludge and swine waste in concentrations that varied from zero to 40%. The experiments were conducted in glasses bottles of 100 ml, 25°C, in a shaker at 150 rpm during 60 days. After analyzing the monitored parameters, it can be said that the best results had been obtained when the FVW had been diluted with high proportions of domestic sewage and swine waste. In these conditions, the chemical oxygen demand removal was superior to 60%, the final C/N ratio had significant reduction and the biogas production was close to $0.40 \, m^3 \, kg^{-1}$ of total volatile solids.

1 INTRODUCTION

Problems of solid waste management are prevalent worldwide, but they are conspicuously more serious in developing than developed countries. This situation can be explained by resource and technology constraints, which plague most developing countries, and poor governance. In addition, most developing countries started tackling solid waste management problems seriously relatively recently (Mbuligwe & Kassenga, 2004).

Sustainable waste treatment concepts that favor waste recycling and the recirculation of nutrients back to the soil represent benefit for the environment. Anaerobic digestion of organic solid waste offers the advantage of both a net energy gain, by producing methane, as well as the production of a fertilizer from the residuals (Hartmann & Ahring, 2005).

Anaerobic digestion involves the degradation and stabilization of organic materials under anaerobic conditions by microbial organisms. As one of the most efficient waste and wastewater treatment technologies, anaerobic digestion has been widely used for the treatment of municipal sludge and limited application in the treatment of organic industrial wastes including fruit and vegetable processing wastes, packinghouse wastes, and agricultural wastes (Chen et al., 2007).

Anaerobic degradation of carbohydrates, protein and fat from biowastes proceeds via hydrolysis, acidogenesis, acetogenesis and methanogenesis in wet or dry biowaste fermentation systems. For wet fermentation, the dry matter content is adjusted to 8–16% by addition of process water, whereas for dry-fermentation no or only little process water is added to the moist material and the dry matter content should be higher than 30% (Gallert et al., 2003).

An interesting option for improving yields of anaerobic digestion of solid wastes is co-digestion. The benefits of co-digestion include dilution of potential toxic compounds, improved balance of nutrients, synergistic effect of microorganisms, increased load of biodegradable organic matter and better biogas yield. In anaerobic digestion, co-digestion is the term used to describe the combined treatment of several wastes with complementary characteristics, being one of the main advantages of anaerobic technology (Ağdağ & Sponza, 2007).

Therefore, the aim of the present study was to evaluate the anaerobic biodegradability potential of fruit and vegetable wastes (FVW) using different solids concentrations, which had been obtained by the FVW dilution with different co-substrates.

2 MATERIALS AND METHODS

2.1 Substrate characteristics

The fruit and vegetable wastes were collected in a central distribution market for food. The collected waste was shredded, blended and diluted with water, domestic sewage, sewage sludge and swine waste. The pure FVW had been evaluated and it had been studied the FVW diluted with 10, 20, 30 and 40% (mass:mass) of each co-substrate.

2.2 Experimental design

The experiments were conducted in glasses bottles of 100 ml filled with 90% of its volume. It had been studied the mixing condition, so the tests had been done at 25°C in a shaker at 150 rpm. The bottles were sealed with butyl rubber stopper. A syringe was connected to each bottle for collecting and measuring the biogas. The experiments run for 60 days.

2.3 Analytical procedure

The waste was characterized and evaluated in terms of pH (digital pH meter Digimed – DM2); total alkalinity (TA), volatile fatty acids (VFA) (Silva, 1977); organic carbon (C) (Kiehl, 1985); Kjeldahl nitrogen (N) (IAL, 1985); chemical oxygen demand (COD), total volatile solids (TVS) (APHA, 1998) and biogas production.

3 RESULTS AND DISCUSSION

The characterization of the FVW diluted in different concentrations of co-substrates is shown in the Table 1.

It can be observed that the pH has not showed significant variation in the different dilutions. The values of pH with the addition of water, domestic sewage, sewage sludge and swine waste were, respectively, around 3.5, 3.8, 4.0 and 4.4. Although it had been observed an increase in the pH values in some samples, these values were still low and this fact could cause prejudices because the optimum value of this parameter for the methanogenic bacteria is between 6.5 and 7.2 (Appels et al., 2008).

The VFA/TA ratio of the pure FVW and the VFA/TA ratio of the samples diluted with 10, 20, 30 and 40% of water and domestic sewage and diluted with 10 and 20% of sewage sludge couldn't be determined, because of the middle pH was very low and this fact hasn't allowed the determination of the total alkalinity. The samples with 30 and 40% of sewage sludge and with the swine waste (all percentages) had shown very high VFA/TA ratio, since the VFA concentrations were so high.

The increase of the FVW dilution with water promoted a consecutive increase in the total volatile solids concentration. The increase of the FVW dilution with domestic sewage, sewage sludge and swine waste had reduced the TVS concentration. This fact could be explained because these co-substrates are also constituted of inorganic compounds.

The initial COD concentration had been around 40,000 mg/L in almost all studied conditions. The increase of the FVW dilution with water and swine waste had promoted a reduction in the COD concentration. The COD of the swine waste was inferior to the FVW and the dilution had contributed

Table 1. Fruit and vegetable wastes characteristics.

Essays	pH	AV/AL	STV (%)	DQO (mg L^{-1})	C/N
100% FVW	3.5	– [*]	92.0	42,031	58.5/1
10% water + 90% FVW	3.6	– [*]	96.9	41,467	57.6/1
20% water + 80% FVW	3.6	– [*]	97.3	36,639	53.5/1
30% water + 70% FVW	3.6	– [*]	97.6	33,028	53.4/1
40% water + 60% FVW	3.6	– [*]	97.7	32,260	53.4/1
10% domestic sewage + 90% FVW	3.9	– [*]	89.6	61,286	51.2/1
20% domestic sewage + 80% FVW	3.8	– [*]	89.4	51,362	44.4/1
30% domestic sewage + 70% FVW	3.8	– [*]	85.3	45,641	45.9/1
40% domestic sewage + 60% FVW	3.8	– [*]	83.5	40,037	39.6/1
10% sewage sludge + 90% FVW	4.0	– [*]	85.9	46,320	44.5/1
20% sewage sludge + 80% FVW	4.0	– [*]	86.3	45,675	44.8/1
30% sewage sludge + 70% FVW	4.1	27	85.8	42,867	38.4/1
40% sewage sludge + 60% FVW	4.2	30	82.6	35,734	30.1/1
10% swine waste + 90% FVW	4.4	4.8	85.4	36,799	53.5/1
20% swine waste + 80% FVW	4.4	5.0	83.2	33,242	40.7/1
30% swine waste + 70% FVW	4.4	4.8	82.2	28,233	41.0/1
40% swine waste + 60% FVW	4.6	3	81.0	28,157	34.6/1

[*]Values not determined because of the low pH.

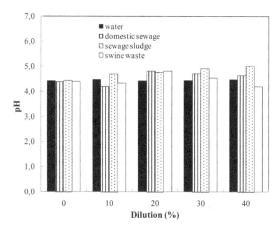

Figure 1. Final pH in the tests varying the dilution and the co-substrates.

to the COD reduction. The addition of domestic sewage and sewage sludge had increased the COD concentration.

The C/N of the wastes had been very high, 40-60/1, mainly because the studied wastes were highly organic.

The Figure 1 shows the final pH in the tests varying the dilution and the co-substrates. The pH values indicated that the VFA, produced during the organic compounds decomposition, had been little consumed, because the final values for almost all tested conditions had been around 4.5. After the experiments, the pH wastes still had been low for the anaerobic digestion.

The final VFA/TA ratio in the tests varying the dilution and the co-substrates is shown in the Figure 2 and it can be said that after the consumption of the VFA it could be possible to determine the VFA/TA for all the samples. The VFA for the pure FVW was around to 5.5, indicating a low reduction of the VFA/TA. The VFA/TA of the FVW diluted in 20, 30 and 40% with water, domestic

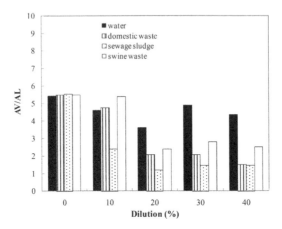

Figure 2.　Final VFA/TA ratio in the tests varying the dilution and the co-substrates.

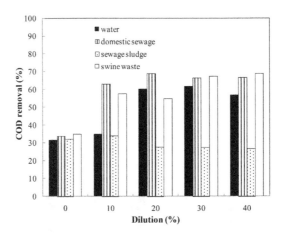

Figure 3.　COD removal in the tests varying the dilution and the co-substrates.

sewage, sewage sludge and swine waste were close to 3.5, 2.0, 1.5 and 1.4, respectively. Therefore, it can be said that there were a significant reduction in the VFA/TA ratio and that the best values had been obtained in dilution of the FVW superior to 20%, mainly for the co-substrates domestic sewage, sewage sludge and swine waste.

Since it had been occurred reductions in the VFA/TA, these values were still distant to ideal conditions that need to be inferior to 0.4 (Callaghan et al., 2002).

The COD efficiency removal is presented in the Figure 3. It can be said that COD removal of the pure FVW was around 30%. In the 20, 30 and 40% of the FVW dilution with water, the COD removal was very close and around 60%. In all dilutions with domestic sewage and in the 30 and 40% dilutions with swine waste the COD removal were close to 70%. In all dilutions with sewage sludge the COD removal was around 30%.

Therefore, it can be concluded that the FVW diluted in concentrations superior to 20% can facilitate its degradation and the most appropriated to the degradation increase is the use of water, domestic sewage and swine waste.

The final C/N ratio is showed in the Figure 4. It can be said that the C/N had been considerably reduced during the experiment, because the initial values were around 40-60/1 and the final values were inferior.

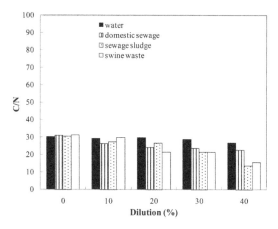

Figure 4.   Final C/N ratio in the tests varying the dilution and the co-substrates.

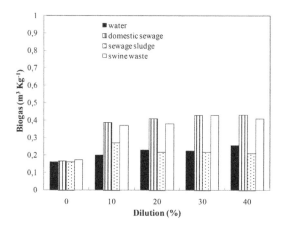

Figure 5.   Biogas production in the tests varying the dilution and the co-substrates.

It also can be said that the dilution of the FVW hadn't exerted influence in the C/N final ratio of the pure FVW and the FVW diluted with water, which were around 30/1 and 28/1, respectively.

The increase of the FVW dilution with domestic sewage, sewage sludge and swine waste had promoted a low reduction in the C/N ratio. In the dilution of 40% the C/N were 23, 14 and 16/1 for the domestic sewage, sewage sludge and swine waste, respectively.

C/N ratio is most often used to indicate both the stability of organic matter and the quality of the digested substrate for its further use. C/N ratio of digested substrate in the range 15–17/1 is considered stable and high quality compost (Rao & Singh, 2004). In this study, only the 40% dilution of the FVW with sewage sludge and swine had promoted this condition. In a general way, the efficiency of C/N reduction was superior to 50%.

The biogas production obtained in the in the tests varying the dilution and the co-substrates is shown in the Figure 5.

It can be observed that the higher productions of biogas had been obtained when it was added domestic sewage and swine waste to the FVW. The dilution almost had not exerted influence in the biogas production that was around 0.40 m^3 kg^{-1} of total volatile solids, close to the values obtained by Rao et al. (2000) that were between 0.429 and 0.568 m^3 kg^{-1} of total volatile solids. In the other tested conditions, the biogas production was close to 0.20 m^3 kg^{-1} of total volatile solids.

It also can be said that the higher biogas production, as the high COD removal and the low final C/N ratio, had been obtained when the FVW had been diluted with domestic sewage and swine waste. This fact can be explained by the high amount of microorganisms present in these co-substrates, which had accelerated the degradation.

4 CONCLUSIONS

The results obtained in this experiment permit to conclude that the best results had been obtained when the FVW had been diluted with high proportions of domestic sewage and swine waste. Therefore, in these conditions, the pH and VFA/TA had run to stabilization, but still there were distant to the optimum conditions. The COD removal was superior to 60%, the final C/N ratio had significant reduction and the biogas production was close to $0.40 \, m^3 \, kg^{-1}$ of total volatile solids.

ACKNOWLEDGEMENTS

The authors would like to CAPES (Coordenação de Aperfeiçoamento de Pessoal de Nível Superior) for financial support.

REFERENCES

Ağdağ, O.N. & Sponza, D.T. 2007. Co-digestion of mixed industrial sludge with municipal solid wastes in anaerobic simulated landfilling bioreactors. *Journal of Hazardous Materials* 140: 75–85.

APHA – American Public Health Association 1998. *Standard methods for the examination of water and wastewater*. Washington D.C., 20th ed.

Appels, L., Baeyens, J., Degrève, J. & Dewil, R. 2008. Principles and potential of the anaerobic digestion of waste-activated sludge. *Progress in Energy and Combustion Science* 34: 755–781.

Bouallagui, H., Touhami, Y., Cheikh, R.B. & Hamdi, M. 2005. Bioreactor performance in anaerobic digestion of fruit and vegetable wastes. *Process Biochemistry* 40: 989–995.

Callaghan, F.J., Wasea, D.A.J., Thayanithya, K. & Forster, C.F. 2002. Continuous co-digestion of cattle slurry with fruit and vegetable wastes and chicken manure. *Biomass and Bioenergy* 27: 71–77.

Chen, Y., Cheng, J.J. & Creamer, K.S. 2008. Inhibition of anaerobic digestion process: A review. *Bioresource Technology* 99: 4044–4064.

Gallert, C., Henning, A. & Winter, J. 2003. Scale-up of anaerobic digestion of the biowaste fraction from domestic wastes. *Water Research* 37: 1433–1441.

Hartman, H. & Ahring, B.K. 2005. Anaerobic digestion of the organic fraction of municipal solid waste: Influence of co-digestion with manure. *Water Research* 39: 1543–1552.

IAL – Instituto Adolfo Lutz 1985. *Métodos Químicos e Físicos para Análises de Alimentos*. São Paulo: Editoração Débora D. Estrella Rebocho.

Kiehl, E. J. 1985. *Fertilizantes Orgânicos*. São Paulo: Editora Agronômica – CERES.

Lastella, G., Testa, C., Cornacchia, G., Notornicola, M., Voltasio, F. & Sharma, V.K. 2002. Anaerobic digestion of semi-solid organic waste: biogas production and its purification. *Energy Conversion and Management* 43: 63–75.

Mbuligwe, S. & Kassenga, G.R. 2004. Feasibility and strategies for anaerobic digestion of solid waste for energy production in Dar es Salaam city, Tanzania. *Resources, Conservation and Recycling* 42: 183–203.

Rao, M.S. & Singh, S.P. 2004. Bioenergy conversion studies of organic fraction of MSW: kinetic studies and gas yield–organic loading relationships for process optimization. *Bioresource Technology* 95: 173–185.

Silva, M.O.S.A. 1977. *Análises Físico-Químicas para Controle das Estações de Tratamento de Esgotos*. São Paulo: CETESB.

Vavilin, V.A., Lokshina, L.Y., Jokela, J.P.Y. & Rintala, J.A. 2004. Modeling solid waste decomposition. *Bioresource Technology* 94: 69–81.

Wastes: Solutions, Treatments and Opportunities – Vilarinho, Castro & Russo (eds)
© *2015 Taylor & Francis Group, London, ISBN 978-1-138-02882-1*

Waste management state plans – The tip of the iceberg on waste management solutions

C. Santiago, E. Pugliesi, A. Santi & C. Milano
Federal University of São Carlos, São Paulo, Brazil

ABSTRACT: Waste Policies in Brazil are recent and struggle to develop fast to adequate waste management situation. In this sense, the main tool for waste management presented on Brazilian policies is the Waste Plan. These plans analyze and guide development of waste management. São Paulo state, the biggest economy in Brazil, approved its Plan in 2014, with the intention of following policies and guiding the cities on how to develop waste management in an adequate and optimized way. Therefore, this paper aims at analyzing São Paulo's waste plan, evoking its importance as a tool for waste management, the analysis of the state's panorama and of the direction of the plans resolutions. The situation is critical, and the plan serves as a first shot to implement the delineated measures, but it is necessary to keep in mind that actions must be developed in an integrate way, involving society and monitoring the process.

Waste generation is presented as a major issue for our society nowadays due to the increase of the urban population all over the globe, along with the increase of consumerism and obsolescence. This situation is especially delicate in developing countries, which are still struggling to adequate their national patterns, approve laws and coordinate integrated management in many fields, such as waste management. Brazil, for instance, produced an amount of waste 6.8% higher in 2010 than in 2009. In addition, this amount is six times superior to the population growth for the same period, according to ISWA's (International Solid Waste Association) representative in Brazil, Abrelpe (Abrelpe, 2011).

In order to adequate waste management in national level, Brazilian politicians approved the National Waste Policy in 2010, after 21 years of debate. Strongly based on the German waste policies, Brazil's National Waste Policy focus on optimized locations for waste management. Besides, this policy incorporates principles of the environmental law such as precaution, prevention and polluter pays principle. The main objectives of Brazilians' National Waste Policy are to protect both the public and environmental health; adequate disposal of rejects; encouragement of the recycling industry; integrated waste management and cooperation among the government, industry and society (Brazil, 2010).

Brazilian National Waste Policy approval comes with delay comparing to the international scenario. The European Union, for instance, deals with this subject since the 90s and Germany particularly, which was the main base for the Brazilian Policy, mentioned 'waste' in law for the first time in 1957, approving the *Waste Law* later in 1986, focusing the precaution principle (European Comission, 2010; Juras, 2012; Schmidt, 2005).

The Policy proposes nineteen tools to assist waste management. In this paper, we analyze the first of them, *waste management plans*. These plans consist essentially of a reference document to guide waste management in a territorial context – national, state, municipal and regional. This tool is crucial to organize the integrated management and its logistics, as well as develop strategies according to the diagnosis presented, making the management suitable for each reality, which is well outlined in these documents.

The state of São Paulo, known for its representative economy in national and international horizons, approved its own Waste State Policy in 2006, with the same goal of the National Policy, and similar principles, focusing on the cooperation and systemic vision to apply integrated management. The objectives of the State Policy are especially orientated to the universe of recycling, the social issues involved in it and the need to encourage the development and appreciation of this chain. This Policy also presents "waste management plans" as an essential tool for successful management (São Paulo, 2006; Góes & Silva, 2012).

As a result of National and State laws, the State of São Paulo presented its State Waste Management Plan in October 2014. Despite a certain delay on the presentation of this document, observing that the laws were approved in 2006 (state) and 2010 (national), the Plan comes to initiate a major process of change in São Paulo's society, regarding its attitudes, and consumption habits, according to São Paulo's environmental agency (São Paulo, 2014).

A team from the major environmental agencies in São Paulo (CETESB and SMA) was responsible for the development of the State Plan and the environmental planning coordinating body led the project.

The result is a four-piece-document: the first piece presents an overall on the current situation of waste management in São Paulo, exploring the processes of generation, collection, sorting, treatment, disposal and management. The main goal of this first piece is to present a tangible diagnosis, enabling a deep analysis of this outlook and proposition of viable actions and achievements on the way to a better waste management.

This panorama brings information of the most diverse branches, from related laws and patterns to a panorama of waste pickers and waste picker's associations in the state. Therefore, this piece has become an essential reference for cities and regions of the state intending to produce a more specific plan. In addition, since this document has open access to the population, it contributes for environmental education and shaping of a joint conscience concerning waste (São Paulo, 2014).

The next pieces are based on the previous diagnosis. The second piece is a regionalization study and proposal of intermunicipal arrangements, the third is the scenario proposal and the fourth proposes guidelines, goals and actions. The second piece, concerning regionalization, is the main reference of the German waste policy absorbed in our National Waste Policy. This kind of study is crucial especially when it comes to a continental country, such as Brazil, and São Paulo, a country-sized-state, which contains a large percentage of Brazil's population. Besides, in Brazil and in São Paulo the major amount of cities is of small cities, with less than 20 thousand inhabitants. These cities often cannot afford to build an adequate landfill. Under such circumstances, the best option concerning waste management is to gather other cities and join resources to build adequate landfills and manage their waste together. The National Waste Policy encourages this type of arrangement, prioritizing financial resources to cities consortia instead of individual cities by themselves.

In the state of São Paulo, there is another territorial arrangement to feature – the metropolitan regions. The most famous is the state's capital metropolitan region, which covers São Paulo and several conurbated cities, but the state has four other metropolitan regions. These conglomerated centers often have major waste management issues due to lack of integrated management and cooperation among cities.

In the third piece, four scenarios are proposed based on projections concerning waste generation, population growth and GDP. The first scenario represents an optimistic situation regarding the economy; the second scenario represents a less advantageous situation regarding the economy; the third scenario is pessimistic for the economy; and the forth or reference scenario is the most likely, adopting a yearlong average GDP growth rate of 3.5% for the period between 2005 and 2035. These scenarios can also be used by cities and other regional agreements to analyze and compare their situation with the state expectations (São Paulo, 2014).

It is important to declare that the plan projected waste generation growth only for municipal waste (household waste and commercial waste) and demolition waste, due to the lack of data and methodology on waste generation. Estimations were made for the other types of waste.

Considering population growth expectations, São Paulo would increase waste generation in 13.6% by 2030. However, when the plan considered population growth rates and an average GDP

Table 1. Urban waste generation variation for four scenarios (São Paulo, 2014).

Administrative region	Waste generation variation 2012–2030 (%)			
	Reference Scenario	Scenario 1	Scenario 2	Scenario 3
Araçatuba	100.2	142.1	80.3	56.7
Barretos	93.9	134.5	74.6	51.7
Bauru	102.9	145.4	82.7	58.8
Campinas	114	158.9	92.8	67.5
Central	104.3	147.1	84	59.9
Franca	104.3	147.1	84	59.9
Marília	97.2	138.6	77.6	54.4
Presidente Prudente	94.8	135.7	75.5	52.5
Registro	95.5	136.4	76	53
Ribeirão Preto	115	160.1	93.7	68.3
Santos	114.2	159.1	92.9	67.6
São José do Rio Preto	101.4	143.6	81.4	57.6
São José dos Campos	112.6	157.1	91.4	66.4
Sorocaba	112.3	156.8	91.2	66.2
RMSP (São Paulo Metropolitan Region)	105.8	149	85.4	61.1
São Paulo State	107.6	151.1	86.9	62.4

Table 2. Goals established for Guideline 1 (Data from São Paulo, 2014).

Goal number	Guideline 1 (G1)	Number of related actions
1	Institute yearly waste declaration system	5
2	Implement State waste inventory	5
3	Monitor waste management quality indicators	5

increase rate of 3.5%, municipal waste generation would double in the state between 2012 and 2030, going from 36.4 thousand ton/day to 75.4 thousand ton/day (São Paulo, 2014). The following table, adapted from the State Waste Plan, presents the generation growth situation for all the administrative regions in São Paulo's state in the four scenarios.

Finally, the fourth piece of the State Plan is the conclusion of the previous work, proposing guidelines for waste management in a state level, goals for the improvement of waste management in the state and actions, which are the means to put the plan into practice. The development of these guidelines, goals and actions occurred based on the ones proposed by Brazil's National Waste Plan, according to São Paulo's specific situation (São Paulo, 2014).

This piece proposed four guidelines: improvement of waste management planning (G1); promotion of sustainable solutions for waste management (G2); improvement of waste management in the State of São Paulo (G3); encouragement of the increase of efficiency on natural resources use (G4) and promotion of awareness, communication, and environmental education for waste management (G5). The following tables present goals stablished for each of the guidelines.

From the guidelines and goals, it is clear that São Paulo, just like Brazil, lacks a solid base to build an efficient system for waste management, so its primary needs are basic: coordinating systems to collect data; eliminating inappropriate and illegal sites of disposal; encouraging solutions and assisting cities to organize and coordinate their waste management systems.

However, considering the current situation and development of waste management around the world, it is imperative to develop not only basic needs on the field, but also mature and contemporary measures and tools. These mature measures are also present on the goals and guidelines already

Table 3.　Goals established for Guideline 2 (Data from São Paulo, 2014).

Goal number	Guideline 2 (G2)	Number of related actions
1	Encourage input of financial and other resources for implementing the State Waste Policy	3
2	Improve economic tools for implementing the State Waste Policy	4

Table 4.　Goals established for Guideline 3 (Data from São Paulo, 2014).

Goal number	Guideline 3 (G3)	Number of related actions
1	Municipal and Intermunicipal Plans (%)	3
2	Promotion of cities association in regional arrangements for optimizing waste management	4
3	Elimination of all illegal dumping	2
4	Improvement of existing landfills	1
5	Promote sustainable solutions for waste management	3
6	Recovery of degraded areas (outdated inappropriate landfills) (%)	4
7	Reduction of recycling waste disposed on landfills (%)	7
8	Reduction of humid waste disposed on landfills (5)	5
9	Promote landfill gas utilization for energy generation (MW)	3
10	Social inclusion and empowerment of waste-pickers and organizations (%)	5
11	Recycling of sludge following principles of cleaner production	2
12	Rural sanitation	1
13	Adequate treatment for biomedical waste	5
14	Adequate disposal of biomedical waste	1
15	Implement systems for treating ports, airports, customs, bus stations and train stations (%)	1
16	Adequate disposal of industrial rejects	2
17	Implementing Policy for generation-reduction of industry rejects	6
18	Coordinate development of agricultural waste inventory	1
19	Implementing voluntary deliver stations (PEV) in rural areas	1
20	Implementing Mining Waste Management Plans (%)	1
21	Elimination of all illegal dumping of demolition waste by 2019	3
22	Implementing voluntary deliver stations (PEV), transfer and screening areas, landfill (if necessary) for demolition waste in every city	4
23	Reuse and recycling of demolition waste (% of cities)	5
24	Development of demolition waste plans by large generators	1
25	Encourage measures for demolition waste generation reduction and demolition rejects	4

Table 5.　Goals established for Guideline 4 (Data from São Paulo, 2014).

Goal number	Guideline 4 (G4)	Number of related actions
1	Implement Reverse Logistics in São Paulo	5
2	Encourage initiatives of good practices for waste generation reduction and incentive to use of recycled materials	2
3	Use of São Paulo's government purchasing power to stimulate technological innovation and development of a less-waste-generator market	1

Table 6. Goals established for Guideline 5 (Data from São Paulo, 2014).

Goal number	Guideline 5 (G5)	Number of related actions
1	Implementing an environmental education program focused on waste management	6

exposed. Among them, we can highlight guidelines 4 and 5, since G4 focus on preventing waste generation, trying to solve the source of the problem; and G5 promotes educational measures, commonly forgotten or not set as priorities in Public Policies.

Along with other environmental policies, waste-related policies and plans focus on the participating planning which means including the population in the process of developing plans and even the policies themselves. In this sense, São Paulo's population (NGOs, civil society, government as a whole and industries) participated for validation of the State Waste Plan in public audiences and consultations. Furthermore, the diagnosis (first piece of the Plan) became available online through a four-month period and the preliminary version of the document was available for one month for public consultation. In the same period, five public audiences happened in five different regions of the state (São Paulo, 2014a).

These steps of public participation are crucial to ease implementation of the plan as well as making society feel like they are a part of public policies. It is clear that this kind of policy development and monitoring, including society as a major part with the power to make the plan work efficiently, is incipient in Brazil and other developing countries. We have much to learn but the initiatives are happening little by little, successfully.

Therefore, it is crucial for São Paulo to coordinate the development of basic needs in waste management along with mature measures on the field, once there is no time to waste, the environment is already responding to several impacts in several ways. This particular state of Brazil struggles with its large population and occupation of the land, causing problems like the lack of areas for landfill building and misbalanced waste generation on the state area, which demand innovating solutions (São Paulo, 2014).

Waste Plans, like the one analyzed in this paper, are essential to guide and define where waste management needs to go, as well as the fragile and potential aspects. However, the definitions of the plan have to be implemented for an efficient waste management, because once the plan is done all the work should begin. Along with that, it is imperative to invest in the source of the problem, and this applies for the whole world – consumption patterns must be revised, and the population and society must stop acting as if the resources were infinite.

REFERENCES

Associação Brasileira de Empresas de Limpeza Pública e Resíduos Especiais (ABRELPE) 2011. *Panorama dos resíduos sólidos no Brasil 2010*. São Paulo: Abrelpe. Disponível em: <http://www.wtert.com.br/home2010/arquivo/noticias_eventos/Panorama2010.pdf>. Acesso em 10 nov 2013.

Brasil 2010. *Lei nº 12.305 de 02 de agosto de 2010*. Política Nacional de Resíduos Sólidos. Brasília.

European Commission 2010. *Being wise with waste: the EU's approach to waste management*. Luxembourg: Publications Office of the European Union.

Góes, L. & Silva, R. C. 2012. A Experiência da Política de Resíduos Sólidos do Estado de São Paulo. In: Jardim, A.; Yoshida, C.; Machado Filho, J.V. (org.) *Política Nacional, gestão e gerenciamento de resíduos sólidos*. Barueri: Manole. 589-598.

Juras, I. A. G. M. 2012. *Legislação sobre Resíduos Sólidos: Comparação da Lei 12.305/2010 com a Legislação de Países Desenvolvidos*. Consultoria Legislativa da Câmara dos Deputados. Estudo. Abr 2012.

São Paulo 2006. *Lei Estadual Nº 12.300, de 16 de Março de 2006*. Institui a Política Estadual de Resíduos Sólidos e define princípios e diretrizes. São Paulo.

São Paulo 2014. *Plano Estadual de Resíduos Sólidos do Estado de São Paulo*. Governo do Estado de São Paulo, Secretaria do Meio Ambiente, Cetesb – Companhia Ambiental Do Estado De São Paulo. São Paulo.

São Pauloa 2014. *Plano de Resíduos Sólidos do Estado de São Paulo é lançado*. Disponível em: <http://www.ambiente.sp.gov.br/blog/2014/10/29/plano-de-residuos-solidos-do-estado-de-sao-paulo-e-lancado/>. Acesso em fev 2014.

Schmidt, T. 2005. *Planos de gestão integrada de resíduos sólidos urbanos: avaliação da arte no Brasil, comparação com a situação na Alemanha e proposições para uma metodologia apropriada*. Recife.

Wastes: Solutions, Treatments and Opportunities – Vilarinho, Castro & Russo (eds)
© 2015 Taylor & Francis Group, London, ISBN 978-1-138-02882-1

A waste rock and bioash mixture as a road stabilization product

M. Sarkkinen, T. Luukkonen & K. Kemppainen
KAMK University of Applied Sciences, Kajaani, Finland

ABSTRACT: The economic and ecological utilization of biomass based fly ashes (bioashes) generated in energy production is a growing problem. In addition, waste rock forms the highest amount of unused waste material. On the other hand, the structural load bearing capacities of roads increasingly require repair and strengthening partly due to the negative effects of climate change. This study investigates the use of a bioash and waste rock mixture as a stabilization material in road construction. The experimental phase compares the effects of different additives with bioash. The characteristics of the materials and mixtures are studied using X-ray fluorescence (XRF), X-ray diffraction (XRD) and thermogravimetric analysis (TGA). In addition, unconfined compressive strength, optimum moisture content and freeze-thaw resistance are studied. The results show that bioash and dolomite based waste rock react together, forming a strong and durable material suitable, for example, for base course stabilization.

1 INTRODUCTION

Biofuels are increasingly replacing coal in energy production. This has stimulated a need for more active research into the utilization of biofuel based ashes (bioashes). Moreover, the economic and ecological utilization of bioashes is a growing problem due to increasing legal demands as well as higher waste handling costs. For example, the combustion of biofuels (wood and peat) generates 500 000 tpy ashes in Finland (Korpijärvi et al. 2009). In addition, waste rock is roughly 50% of all excavated rock material, being approx. 34 Mty in Finland (Statistics Finland 2012).

This study concerns the use of a bioash and waste rock mixture as the stabilization material in road construction. The insufficient load bearing capacity of gravel roads is a typical problem in geographical areas prone to continuous freezing and thawing periods and such problems are expected to increase due to the negative effects of climate change, such as heavy rain and flooding. The aim of the study was to develop an economically and ecologically viable stabilization product for the strengthening of private roads prone to bearing heavy vehicle traffic.

To evaluate the optimal mix design for the road upgrade, various additives with bioash were tested to optimize the mix proportions. XRF, XRD and TGA were used to analyze the material's characteristics. In addition, unconfined compression, proctor density and freeze-thaw durability were tested to study the performance of the materials.

1.1 *Base course stabilization*

The purpose of stabilization is generally to improve load bearing capacity, repair damage or reduce sensitivity to frost. Base course stabilization is suitable for the construction of new roads and the repair of old roads. In base course stabilization, a binder material is mixed with new aggregate or old road surface structure by using a stabilization cutter (Ramboll 2012). In principle, stabilization can be realized mechanically or chemically either on or outside the site. Chemical stabilization means the addition of binders to soil i.e. the chemical reaction between the stabilizing binder and soil minerals (Makusa 2012).

1.2 Bioashes as stabilization materials

There are extensive research results on the use of coal fly ash as a stabilization material for soils and roads. However, there are still only a few studies on the use of bioash as a stabilization material. The use of bioash in stabilization is more economical and ecological than using cement or lime. Bioashes have a typically high CaO-content, which makes it possible to use them as a replacement for burnt lime for the stabilization of silt and clay soils. The addition of CaO into the soil reduces humidity which improves solidification and carbonization which in turn increases strength. In addition to CaO, the fine particle size and composition of bioash also improves its binding properties i.e. fly ash is a better binding material than bottom ash. However, the CaO-content of bioash is typically lower than in burnt lime resulting in a higher relative amount of bioash required for binding (Suspancic & Obernberger 2012). Active CaO also improves the strength of bioash which can further be improved by the addition of cement (Ramboll 2012).

Bioashes are challenging due to the high variety in chemical composition of fly ashes, which depend on the fuels used and combustion processes of energy plants. Bioash composition also varies depending on the sampling point in the process. The utilization of bioash is limited due to heavy metal content, which can be controlled for example by using fractioning, sampling from different process phases and using additives. It has been stated that all heavy metals restricting earth construction (e.g. As, Ba, Cd, Cr, Cu, Pb, Mo, Zn and V) tend to concentrate in smaller particles (Raiko et al. 1995). According to some calculations it is more economical to classify bioash by particle fractioning in order to improve its quality compared to disposal costs (Korpijärvi et al. 2009). Based on some environmental assessments, the risks from using bioash in earth construction are small when harmful substance content is low (Suspancic & Obernberger 2012). According to some studies, leaching from a 40 cm thick bioash based gravel road (gravel 10 cm) is low and reduces over time before reaching ground water level (Thurdin et al. 2006). However, the use of ash when it exceeds permitted heavy metal content is not recommended (Hottenroth et al. 2003).

There are positive examples of using bioash as an alternative for burnt lime for example in the stabilization of silt and clay soils in Austria (Suspanic & Obernberger, 2012). Another good example is the strengthening of gravel roads in Sweden (Vestin et al. 2012; Macsik 2012; 2006). Trial areas containing 5% cement and 5% bioash mixed with 90% of gravel provided the best strength values (6.5 MPa) and 20–30% bioash was sufficient to gain acceptable strength for light load bearing roads (>2 MPa) (Macsik 2012). The quality of the road and its substructure (gravel), the water content of gravel and binder, compaction depth and degree, in addition to binder content and quality control were identified as critical success factors in the test. In Finland, Huttunen & Kujala (2001) have reported good results on the use of peat ash in road structures where it has also prevented frost damage.

Particle size distribution, optimal water content, proctor density, shear strength and the unconfined compression test provide information on the usability of bioash in different applications. The high shear strength of bioash (>8 MPa) indicates potential usability without additional cement. In addition, the storage of bioash in dry conditions and its use when fresh reduce the need for additives and improve performance (Macsik ym. 2012.). It has been stated that fly ashes already lose their binding capacity after storage in damp conditions within two weeks (Lagerlund & Jansing 2012).

2 MATERIALS AND METHODS

2.1 Bioash

The bioash used in the study is a by-product of an energy plant burning wood, peat and paper mill waste as fuel. Figures 1 and 2 and Tables 1 and 2 describe XRD, TGA, XRF and particle size distribution of the bioash.

The XRD, XRF and TGA analyses indicate, for example, the presence of lime, silicon oxide, anhydrite and iron oxide in the bioash. As the values show, the CaO content is relatively high

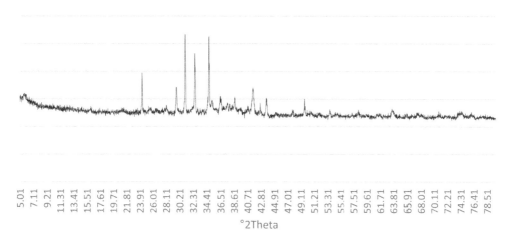

Figure 1.   XRD-analysis of the bioash.

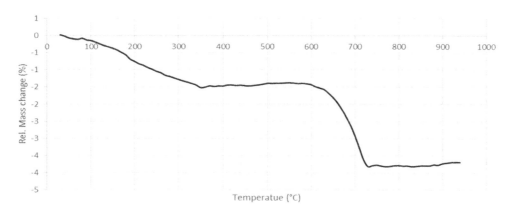

Figure 2.   TGA-analysis of the bioash. Mass loss (%) in relation to temperature (°C).

Table 1.   XRF analysis of the bioash.

CaO	SiO$_2$	Fe$_2$O$_3$	Al$_2$O$_3$	SO$_3$	MgO	P$_2$O$_5$	K$_2$O	Na$_2$O	Mn	Cl
28.528	21.946	17.731	6.365	5.227	3.241	3.063	2.46	0.63	0.476	0.198
In	Ti	Eu	Zn	Ag	Ba	Sr	Tot.			
0.163	0.162	0.123	0.116	0.107	0.102	0.076	90.817			

Table 2.   Gradation of the bioash measured with Alpine wind sieve.

Sieve (μm)	20	25	32	45	63	75	90	100	150
Passing (%)	43.81	53.93	63.45	72.77	80.56	84.02	87.95	89.48	94.87

Table 3. Main components of the waste rock.

CaO	SiO$_2$	MgO	Fe$_2$O$_3$	Al$_2$O$_3$	K$_2$O	Ti	In	Ba
36.611	19.744	11.701	8.103	6.115	1.343	0.377	0.266	0.175
Mn	SO$_3$	Ag	Eu	V	Sr	Cr	Zn	Rb
0.151	0.105	0.102	0.042	0.015	0.014	0.012	0.008	0.006

(28.528%). In addition, leaching of heavy metals in the bioash were analyzed according to the SFS EN 12457-3. The identified total quantities were below the recommended limit values defined for the paved soil constructions in Finland.

2.2 Waste rock material

The aggregate used was a byproduct from a quarry close to the road to be stabilized. It was dolomite based crushed rock generated during dolomite production. The maximum grain size of the aggregate was 16 mm. The main components of the aggregate according to XRF-analysis are depicted in Table 3.

2.3 Methodology

The experimental program was carried out in two phases. The aim of the first test phase was to find out the optimal binder combination by comparing different additives with bioash/waste rock mixture. The used binder/aggregate mass ratio was 20:80. Various bioash/waste rock mixtures were tested by using the unconfined compressive test (UCT) and optimum moisture content (proctor density).

The used specimens were cylinders, 100 mm in diameter and 100 mm in height. The specimens were compacted with ICT (impact compactor tester) to a pressure of 6 bars and 100 rounds/specimen. Three specimens per one sample were prepared. The specimens were stored on a steel mesh above the bottom of the box, which was partly filled with water. In the second phase, the chosen option was tested with three different binder/waste rock ratios by using UCT and freeze-thaw testing.

3 RESULTS

3.1 Unconfined compression test

The results of the unconfined compression tests with different binder combinations are given in Table 4.

Dolomite filler, calcined dolomite filler with and without potassium monophosphate, and metakaolin were used as additives with bioash. According to the results, plain bioash when combined with the aggregate (test 2) provided a sufficiently good result compared to other results and it was decided that it would be used in further tests.

3.2 Proctor density

The aim of the Proctor test was to find out the optimal water content i.e. maximum density providing the best compaction. The higher value indicates higher material density. Proctor tests were executed by using the plain aggregate and a mix of aggregate:bioash 80:20 with three different water contents (Table 5).

The highest water content (8.15%) was too high leading to the leakage of water from the cylinder. Thus, the optimum water content according to the test was 7.58% corresponding to the proctor density of 21.37 kN/m^3.

Table 4. Unconfined compression test (UCT) and densities after 7d and 28d.

	Binder composition	component ratios (%)	binder: aggregate (%)	total water (%)	UCT 7d (MPa)	UCT 28d (MPa)	Density (g/l)
1	waste aggregate	–	0:100				
2	bioash	100	20:80	7.3	2.85	7.30	1819
3	bioash: calcined dolomite: potassium monophosphate	90:5:5	20:80	7.8	3.64	5.95	1796
4	bioash:dolomite	80:20	20:80	7.1	2.55	7.33	1836
5	bioash: metakaolin	90:10	20:80	7.2	4.37	5.53	1899
6	bioash:calcined dolomite	80:20	20:80	7.9	1.6	5.1	1885
7	bioash: metakaolin: calcined dolomite	90:5:5	20:80	7.4	3.7	6.37	1863

Table 5. Proctor density tests.

Mix	Ratio (m-%)	Total water (%)	Proctor density (kN/m^3)
Aggregate 0–16 mm	100	1.7	20.92
Aggregate:bioash	80:20	4.62	21.11
		7.58	21.37
		8.15	22.82

Table 6. UCT values of specimens before and after freeze-thaw test.

Binder: aggregate (%)	UCT 28d, R_B (MPa)	Density 28d (g/l)	UCT after freeze-thaw test, test, R_A (MPa)	Density after freeze-thaw test (g/l)	R_A/R_B
10:90	5.0	1882.5	3.6	1872.3	0.71
20:80	9.1	1874.3	9.5	1865.3	1.04
15:85	10.0	1909.8	7.9	1900.0	0.79

3.3 *Freeze-thaw resistance*

In the second phase, the chosen binder (bioash 20%) and optimized water content were used as a basis for optimizing the aggregate:binder ratio with UCT after 28d and consequent freeze-thaw testing with three different aggregate:binder ratios. The specimens were prepared with the ICT machine as in the first test phase. The water content (m-%) was the same in all three tests. Six specimens were prepared with each combination. The specimens were prepared and stored as in the first test phase.

At the age of 28d, 3×3 specimens were moved to freeze-thaw testing into a weather simulator test machine. A freeze-thaw test was performed according to FprCEN/TS 13286-54 (Unbound and hydraulically bound mixtures- Part 54: Test method for the determination of frost susceptibility-Resistance to freezing and thawing of hydraulically bound mixtures). Table 6 depicts the UCT values after 28d (R_B) and after 10 freeze-thaw cycles (R_A). Value R_A/R_B indicates freeze-thaw resistance. Figure 3 describes the temperature change during one freeze-thaw cycle.

The bioash/waste rock mixture consisting of bioash 20% with total water content of 7.3% provided the highest freeze-thaw resistance value (R_A/R_B 1.04), and it was decided that this combination would be used in the pilot test during summer 2015. The increase in strength may be possible due to higher relative humidity (RH up to 100%) in the test cabin during the thawing period compared to the RH during the 28d curing period before the freeze-thaw test (<100%).

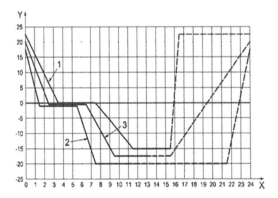

Figure 3. Freeze-thaw test according to CEN/TS 13286-54:2014. The figure presents one test cycle. Y = temperature (°C), X = time (h), 1 = upper limit value, 2 = lower limit value, 3 = control.

4 CONCLUSIONS

In the study, the characteristics of the stabilization mixture consisting of bioash with various additives and waste rock were investigated. The optimum bioash content was found to be 20% considering unconfined compressive strength and freeze-thaw resistance. The strength values were adequate for the bioash/ waste rock mixture without any additional additives indicating a positive reaction between the bioash and dolomite based waste rock. The results show the mixture is well suited for use as a material for base course stabilization.

REFERENCES

Huttunen, E. & Kujala, K. 2001. *The use of peat ash in earth construction*, Geological survey of Finland, Special paper 32, Espoo.

Korpijärvi, K., Mroueh, U-M., Merta, E., Laine-Ylijoki, J., Kivikoski, H., Järvelä, E., Wahlström, M. & Mäkelä, E. 2009. *Energiatuotannon tuhkien jalostaminen maarakennuskäyttöön*. VTT tiedotteita 2499 (in Finnish).

Lagerlund, J. & Jansing, C. 2012. *Long term effects on wet stored calcium rich fly ash with bearing on ground improvement work*. Värmefors report 1226.

Macsik, J. 2006. *Flygaska som förstärkninslager i grusväg. Miljöriktig användning av askor*. Värmeforsk rapport 949.

Macsik, J., Edeskär, T., Rogbeck, Y. & Ribbing, C. 2012. *Stabilization of road structures with fly ash as binder component – through demo projects to full scale use*. ASH 2012. Stockholm, Sweden.

Makusa, G.P. 2012. *Soil stabilization methods and materials in engineering practice, State-of- the-art review.* Department of civil, environmental and natural resources engineering, Division of mining and geotechnical engineering. Luleå University of Technology. Luleå, Sweden, pp. 35.

Raiko, R., Saastamoinen, J., Hupa, M., Kurki-Suonio, I. 1995. *Poltto ja palaminen*. International Flame Research Foundation. Gummerus Kirjapaino Oy. Jyväskylä (in Finnish).

Ramboll 2008. *UUMA-materiaalien ja –rakenteiden inventaari (raportti)*, pp. 113 (in Finnish).

Ramboll 2012. *Tuhkarakentamisen käsikirja, verkkojulkaisu*. 13.1.2012, pp. 65 (in Finnish).

Ministry of employment and the economy. 2012. http://www.tem.fi, read 27.2.2015.

Suspancic, K. & Obernberger, I. 2012. *Wood ash utilization as a binder in soil stabilization for road construction – first results of large-scale tests*. ASH 2012. Stockholm, Sweden.

Thurdin, R.T., van Hees, P.A.W., Bylund, D. & Lundström, U. 2006. *Bio fuel ash in a road construction: Impact on soil solution chemistry*. In Waste Management, vol. 26 no. 6, pp. 599–613.

Vestin, J., Arm, M., Nordmark, D., Lagerkvist, A., Hallgren, P. & Lind, B. 2012. *Fly ash as a road construction material*. WASCON 2012 Conference proceedings.

Wastes: Solutions, Treatments and Opportunities – Vilarinho, Castro & Russo (eds)
© *2015 Taylor & Francis Group, London, ISBN 978-1-138-02882-1*

'*Ecofreguesias*' experience: Some contributions for local sustainable waste management

L. Schmidt & S. Valente
Instituto de Ciências Sociais da Universidade de Lisboa, Lisboa, Portugal

ABSTRACT: The aim of this paper is to present the results of a research-action project – SEPARA®(Awareness and Behavioural Change Relating to Selective Collection of Urban Solid Waste) – which tested a community-based research model. Tratolixo.IMC (Inter Municipal Company) for Urban Solid Waste (USW), had developed a Strategic Plan based on investments in infrastructure and technological solutions with the aim of maximising recycling in its Lisbon area of operation. The 'ecoFreguesias' ('ecoParishes') initiative was created to promote direct cooperation between the USW management body, the Parish Councils and a group of local organizations, adopting a bottom-up and top-down communication model. Through the implementation of the initiative, based on a double component of training and social participation, recognition was given to the efforts of the Parish Councils, local organizations and the general public itself in meeting the criteria set out in a special regulation designed to be adaptable to different types of Parish Councils.

1 INTRODUCTION

Social participation is a crucial factor for the effective success of public environmental policies mostly in specific questions such as the Urban Solid Waste (USW) sorting. Good technologies and sophisticated management systems can be jeopardised if they are not received by people who are able to learn and use them. Action has to be taken on the processes of raising awareness and social behaviour change. This involves a high level of complexity and has been the subject of very little study in Portugal.

The fact that Portugal arrived late to a market and consumer economy, having maintained closed protectionism practically to the end of the 70s, explains the late emergence of both the environmental USW problem and the scarcity of sociological studies on this topic (Schmidt, Truninger & Valente 2004; Valente, Schmidt & Truninger 2012; Valente 2013). There are pilot schemes and awareness campaigns to encourage recycling, but in general there has been no serious and ongoing assessment of the results of the actions carried out with the public (Valente 2001, 2013).

In other Western European countries, where the whole problem of an excess of USW and the respective solutions of management and handling arose earlier, studies analysed the participation of the general public in recycling programmes. These studies identified some of the most relevant factors in joining waste sorting practices: factors such as the use of penalties or incentives; the involvement of local leaders; the preparation and distribution of information; the impact of feedback on the results given to participants; or the type of infrastructures available (Porter, Leeming & Dwyer 1995; Scott 1999; Vicente & Reis 2008).

Such studies, however, focus their analysis on an operative vision, rarely making reference to the various factors that influence the willingness to sort waste. These factors are, amongst others, the social contexts and local spaces, routines, a sense of local identity, the relationship between the public and official organizations, the implementation of policies for public spaces (Derksen & Gartell 1993; Chappels & Shove 1999; Barr, Ford & Gilg 2003).

This paper aims at presenting the findings of a research-action project – SEPARA®, Sensibiliza-ção e Mudança Comportamental Relativamente à Recolha Selectiva de Resíduos Sólidos Urbanos RSU (Awareness and Behavioural Change Relating to Selective Collection of Urban Solid Waste USW) –, undertaked over a period of 3 years which tested a community-based research model (Pereira, Vaz & Togetti 2006). Prior to the end of this research project, a key-achievement was the development of the ecoFreguesias Initiative, in itself a reference case-study and a model of intervention with a strong potential for implementation elsewhere.

A Strategic Waste Plan developed by Tratolixo.IMC (Inter Municipal Company) encompassed a set of investments in infrastructure and technological solutions aimed at optimising recycling in its catching area – Cascais, Oeiras, Mafra and Sintra (four boroughs in the Lisbon Metropolitan Area). Yet, it turned out that the attempt, which was restricted to the areas of technology and infrastructure, was not enough to realize the goal that had been set and decisive to this was the general public's take up of the practice of sorting of USW.

2 METHODOLOGY AND FIELDWORK TOOLS

The objectives of the project were set out on 3 levels (1) to increase the quality and quantity of selective sorting of USW to move towards meeting domestic and European sorting objec-tives; (2) to identify the social factors that influence take up of USW sorting; (3) to develop an action/intervention model together with the public, identifying ways of speeding up the processes for the general public's take up of USW sorting.

Given the lack of studies on the social reality in Portugal, one of the distinctive features of this project was to use a multitude of methodologies, both quantitative and qualitative, to create solid knowledge regarding the relationship between the general public and USW.

The start of the project involved choosing 6 pilot areas, selected in specific neighbourhoods with different socio-economic-urban types – Consolidated Urban; New (sub)urban; Qualified (sub)urban; Unqualified (sub)urban; Uncertain and Rural (INE, 2003). In these 6 areas, direct observation of the use of recycling containers and their surroundings took place over 7 days between 9:00 and 21:00. During this period a survey of the users of the recycling containers under observation was carried out. In-depth interviews with a panel representing the various social groups were also held. The aim of these interviews was to allow greater understanding of the relationship between the public and the USW theme. Afterwards, a door-to-door research-action questionnaire was undertaken to a representative sample of residents from each of the 6 areas. This was combined with the distribution of an information leaflet and the clarification of doubts about sorting. This research-action methodology allied a component of gathering, processing and analysis of information to a component of relay of information, training and triggering change of practice of the target public in the pilot areas.

After this exploratory phase focussing on the 6 areas, the project was extended to the whole of the company's geographical area of operation through the implementation of a model – the ecoFreguesias initiative – based on a double component of training and social action. Each of the 53 parishes of the four town councils – Cascais, Mafra, Oeiras and Sintra – made up the units of analysis and intervention. Direct observation took place in the parishes and in depth interviews were carried out with their leaders. Additionally, 2 surveys in the form of questionnaires were put to a representative sample from the 4 council areas – one before the implementation of the initiative and the other after, in order to monitor changes in attitude.

3 RESULTS

The intervention developed on the ground by the SEPARA® project, as well as the actions for communication and strengthening of the network of recycling points carried out by Tratolixo.IMC

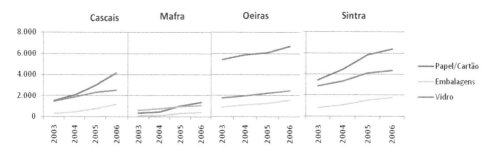

Source: Tratolixo IMC
[Right side of chart: **Paper/Cardboard** – Packaging – **Glass**]

Figure 1.   Evolution of selective collection 2003–2006 (ton.).

contributed to the increase in sorting levels. Furthermore, the sociological analysis on the ground provided clues and precise indications to the preferred location of these infrastructures from the user's point of view. If a positive evolution in selective collection was noted between 2004 and 2006, in 2007, the total values exceeded 2006, most of all in the case of packaging. One of the results of the monitoring survey carried out after the initiative revealed that a percentage of those surveyed who had not previously sorted waste started to sort in that year (Figure 1).

With regard to social factors, a huge disparity between the low level of awareness of the existence of waste and the enormous quantity of waste produced daily was noted immediately. This led to problems in understanding the dimension of the problem and the urgency of taking action. In fact, not only did people in general not think about the problem of waste systemically, but also the way in which they see waste – or put better, "they do not see" waste – is miscalculated. They ignore it, keeping their distance from the reality of "rubbish", which they simply want to get rid of. For example, there is no recognition of the economic dimension of waste as raw material and they are unaware of its final destination (Schmidt & Valente 2005; Schmidt & Martins (ed.) 2006).

Taking into account the level of disinformation, various factors crucial to stimulating a greater take up of USW sorting by the general public were identified.

The quality of the public space and the state of conservation and cleanliness of the recycling containers are crucial aspects which have an influence upon people's attitude to "waste" in general, and the practice of USW sorting in particular. The better the care of the urban context, the public space and the state of conservation and cleanliness of the recycling points are, the better the accomplishment in terms of depositing and sorting of USW is.

Also, the existence of urban facilities for local social conviviality linked to the identity of the neighbourhood has proven to be crucial in mobilizing the general public to sort USW.

Another pertinent question was the need for a coherent, regular and interactive communicational model. Direct communication (personal interaction), which implies the effective and personalised interpretation and relay of information, showed positive effects, most of all in the context of a great lack of information and, at times, contradictory information generated from mistakes and interference (Ferreira & Valente 2014).

As a result the need arose to turn to innovative methods of communication through direct technical support capable of bringing dynamism to the processes of acquiring know-how associated with change and activating the dynamics of participation.

Also, in terms of communication, what stood out was the importance of beginning processes to make connections at a local government level, bringing together the citizens' political leaders, creating confidence and making decisions easier to understand. In this context the Parish Councils

291

Low density population socio-urban context	**High density population socio-urban context**
(rural, undefined and consolidated urban)	(new suburban, qualified and unqualified)

Continuity between house-street Valuing of place	Little use of public space "social void"

Appropriation of the public space as their own	Lack of identification as public space "no man's land"

Involvement of the general public in maintenance of the public space	Lack of interest in maintenance of the public space – lack of sensitivity to degradation and loss of identity

Greater hygiene and cleanliness Greater sorting of USW "friendlier" process	Greater amount of dirt (dirt attracts more dirt) Less generalised sorting of USW Less "friendly" process

S. Pedro de Sintra Parish.	Rio de Mouro Parish.
Source: Separa®/ICS-UL	Source: Separa®/ICS-UL

proved to be the organizations, which in the opinion of the general public, can generate greater local involvement in the promotion of waste sorting (Schmidt, Seixas & Baixinho 2014).

4 THE "ECOFREGUESIAS" [ECOPARISHES] INITIATIVE AS A MODEL FOR ACTION

The knowledge and experience acquired from the results of the research pointed to the need to create a model for action based on direct methods of communication for the relay of information, as well as in the involvement and linking of the various social actors on a local level.

This became a reality in the "zero" edition of the ecoFreguesias Initiative, in which direct cooperation between the USW management company, the Parish Councils and also a diverse group of local organizations was promoted, adopting a bottom-up and top-down communication model with the benefit of external mediation by the research team.

In this pilot experiment recognition and appreciation are given to the efforts of the Parish Councils, local organizations (companies, charities, non-governmental organizations and schools) and of the general public itself in meeting the criteria set out in a special regulation designed to be adaptable to the different sizes of Parish Councils and the resources available to them. In-depth interviews were held with all the leaders of the Parish Councils so as to give them an opportunity to be heard and involved in an initiative which adopted them as central players.

Action on two fronts made the participation of the Parish Councils and local organizations a reality:

(1) the implementation of good sorting practices in their headquarters, allowing all the Parish Councils and the participating organizations the opportunity to equip their premises and so begin the process of good practices for USW sorting. In the same way, the possibility for Parish Councils to develop actions for awareness and requalification of public spaces with the direct support of Tratolixo.IMC was facilitated (e.g. cleaning woodland areas and riverbanks; rebuilding gardens; making use of abandoned spaces).

(2) the frequency of training sessions on USW for at least one member of the executive and one of staff in the case of the Parish Councils; and of one or more representatives from each of the other organizations involved. These training sessions operated as a venue for the provision of information, the raising of awareness and the promotion of behavioural change in relation to the USW sorting, its life cycle and its potential to provide raw materials or sources of energy. At this level, the visit to the Trajouce Ecoparque shows clearly what the training provided, allowing direct contact with the reality of the handling and destination of USW, and promoting understanding of the cycle of waste beyond simply depositing it in containers.

The use of videos about the problems of USW also stood out in bringing awareness to the trainees, not only through the use of shocking images, but also through the evocation of memories provoked by certain images. Besides this, the creation of the "sorting game" clarified doubts about the placement of the different USWs and the cycle of the materials in a fun way.

The whole emphasis of training was placed on the transition from a pattern of a "vicious cycle of waste" (useless, unthinkable and undesirable) to a "virtuous cycle of waste" which, when sorted, is cleaner, can take on new useful forms and become a new product – including the production of local energy.

These two fronts of action – implementation of good practices and training – were always interlinked by ongoing communication in the form of monitoring and mediation by the project team which, in addition to training, created an interpersonal dynamic between all those involved. This component of the project also contributed to promoting the link between the general public and the institutions, thereby short-circuiting the formality of institutional relationships and narrowing the distances between the government and the people.

The intention was to avoid lapses in communication, promote information on the decisions of Tratolixo.IMC and strengthen the relationship of trust and partnership between the company and local organizations. The presence of an external team working to reduce remoteness and lack of trust also helped to give credibility to the Initiative and drew attention to the need to use different communication strategies according to each area's different populations and realities.

In sum, 35 Parish Councils participated in the initiative, with 32 Parish Councils being candidates to the title ecoFreguesias 2007. The number of local organizations involved, between schools, civic associations, services and business, rose to 229. The initiative provided training directly to around 400 people and led to the distribution of around 4000 indoor recycling points. What also became clear was the potential for involvement this intervention model and its expansion. The initiative indirectly impacted on a wider public: as various studies proved, the participants extended their sphere of influence to their respective family and/or work contexts (Bartiaux & alli 2006; Garabuau-Moussaoiu & alli 2009; Schmidt, Horta, Correia & Fonseca 2014).

5   DISCUSSION AND CONCLUSION

Throughout this project, which culminated in the ecoFreguesias Initiative, in addition to the more practical aspects of raising awareness and starting sorting practices in the communities and of the dynamics of qualification of public spaces, it was possible to account for a set of factors which acted interdependently in the processes of improvement of sorting (Ferreira & Valente 2014). The strong points of the model applied were:

1. Identifying the Parish Councils as the key actors in the processes of USW sorting through the capacity they showed in mobilizing the local organizations and the public. The Parish Councils are a close level of government with an enormous potential to stimulate change on various levels. Specifically, the leaders of the Parish Councils played an active role, attending the training sessions along with their staff, making proposals for public and private organizations to get involved in the whole initiative, and taking on the implementation of sorting in their own Parish Councils thereby giving an "example" to the general public.
2. Evidencing the need to anchor USW sorting practices to other actions for the qualification of local public space. It is this integrated approach that leads people to adopt sorting, not merely as an isolated act, but as a basic issue in the improvement of the environment and quality of local life. Public participation in the initiative was not merely instrumental but committed, and this was due, to a great extent, to the simultaneous actions in improving the recycling points and in the qualification of the local public space.
3. Relaying the knowledge and the information with experience and in a clear form are central aspects in the process of behavioural change. Specifically, the direct experience stood out – seeing, smelling and witnessing the waste cycle –, as well as having the benefit of effective communication tools, such as videos and games and the interaction between people throughout the process.
4. Activating the relationship between the various players involved – Parish Councils, companies, local organizations, the general public – through the mediation of an external team capable of closing the gap eliminating the lack of communication. A restructuring of the company began during the project with the aim of creating positions for skilled technicians in the area of communication to give continuity to the mediation processes.
5. Debating the environmental problems of local USW in a participative and constructive way, bringing up ideas to resolve the situations that affect the quality of local life. In this sense, the residents, instead of being part of the problem, became part of the solution, getting involved in concrete actions and generating dynamics of collective civic responsibility.

In sum, a new culture of waste involves a closer relationship between government and the public on a local scale. This in turn implies greater visibility, presence and level of action from those responsible in local government – more particularly the Parish Councils – in the direct management of public spaces, taking on the responsibility they have over them in an effective way and putting an end to a double anonymity: the lack of identity of places and the invisibility of those responsible for public administration.

It also implies the re-appropriation of public space as "common property" by communities, involving them and creating a feeling of belonging and identity in relation to the space or neighbourhood where they live.

It implies ultimately policies of continuity – the absence of strategic memory of public policies on the environment in general is, furthermore, a classic factor in the lack of success in its implementation.

Communication, trust, common responsibility and cooperation evidenced in this study are key-components in a more creative model for sustainable waste management.

ACKNOWLEDGEMENTS

The authors would like to thank Tratolixo.IMC for the opportunity and support to carry out this research. The authors are also grateful to the Parish Councils of the municipalities of Cascais, Mafra, Oeiras and Sintra and all the local organizations that took part in the field work of Separar, especially in the ecoFreguesias Initiative. Last but not least, the authors wish to thank to the team that contributed to the data collection and analysis, consisting of a group of researchers from Instituto de Ciências Sociais da Universidade de Lisboa (ICS-ULisboa) and Centro de Estudos em Economia da Energia, dos Transporte e do Ambiente (CEEETA). Namely: Álvaro Martins (co-coordinator), Carlos Laia, Francisco Lima, Vanessa Pereira, Inês Carneiro, Alexandra Figueiredo, Ana Vicente and Alexandra Pereira.

REFERENCES

Barr, S., Ford, N.J. & Golg, A. 2003. Attitudes towards recycling household waste in Exeter Devon: quantitative and qualitative approaches. *Local Environment* 8(4): 407–421.

Bartiaux F., Vekemans G., Gram-Hanssen K., Maes D., Cantaert M., Spies B. & Desmedt J. 2006. *Socio-technical factors influencing residential energy consumption*, SEREC, Final Report. Belgian Science Policy Office, Brussels. (www.belspo.be/belspo/home/publ/pub_ostc/CPen/rappCP52_en.pdf)

Chappels, H. & Shove, E. 1999. The dustbin: a study of domestic waste, household practices and utility services. *International Planning Studies* 4(2).

Derksen, L. & Gartell, J. 1993. The social context of recycling. *American Sociological Review* 58(3): 434–442.

Ferreira, J.G & Valente, S. 2014. *A Água e os resíduos: duas questões-chave*. In Schmidt, L. & Delicado, A. (eds.). Ambiente, alterações climáticas, alimentação e energia: Portugal no contexto europeu. Imprensa de Ciências Sociais, Lisboa.

Garabuau-Moussaoui I., Bartiaux F. & Filliastre M. 2009. *Entre école, famille et médias, les enfants sont-ils des acteurs de transmission d'une attention environnementale et énergétique ? Une enquête en France et en Belgique*. In N. Burnay et A. Klein (eds) Figures Contemporaines de la Transmission. Presses Universitaires de Namur, Collection Transhumances IX, pp. 105–120.

INE 2003. *Tipologia Sócio-económica da Área Metropolitana de Lisboa*, INE, Lisboa.

Pereira, A., Vaz, S.G & Togetti, S. (ed.) 2006. *Interfaces Between Science and Society*. Sheffield: Greenleaf Publishing.

Porter, B.E., Leeming, F.C. & Dwyer, W.O. 1995. Solid waste recovery. A review of behavioral programs to increase recycling. *Environment and Behavior* 27(2):122–152.

Schmidt, L., Trüninger, M. & Valente, S. 2004. *Problemas ambientais, prioridades e quadro de vida*, pp. 65–168. In: Almeida, J. F. (ed.). Os Portugueses e o Ambiente: II Inquérito Nacional às Representações e Práticas dos Portugueses sobre o Ambiente. Oeiras: Celta Editora.

Schmidt, L. & Valente, L. 2005. *O lixo: uma história residual. O contributo das ciências sociais, Relatório Final*. Lisboa: FCT-UNL.

Schmidt, L. & Martins, A. (ed.) 2006. *Separa®1 – Sensibilização e mudança comportamental relativamente à recolha selectiva de resíduos sólidos urbanos, Relatório Final*. Lisboa: ICS-UL e CEEETA.

Schmidt, L., Horta, A., Correia, A. & Fonseca, S. 2014. *Generational Gaps and Paradoxes Regarding Electricity Consumption and Saving*. Nature and Culture 9(2), Summer 2014: 183–203 © Berghahn Journals, doi:10.3167/nc.2014.090205.

Schmidt, L., Seixas, J. & Baixinho, A. 2014. *Governação de Proximidade – As Juntas de Freguesia de Lisboa*. Lisboa: INCM.

Scott, D. 1999. Equal opportunity, unequal results: determinants of household recycling intensity. *Environment and Behavior* 31(2): 267–290.

Valente, S. 2001. *Campanhas pelo Ambiente, Processos de (In)Comunicação. Tese de Mestrado*. Lisboa: ISCTE.

Valente, S. 2013. *Hábitos Privados, Práticas Públicas – o lixo no quotidiano. Tese de Doutoramento em Ciências Sociais*. Lisboa: ICS-Universidade de Lisboa.

Valente, S., Schmidt, L. & Trüninger, M. 2012. *Consumo e Lixo na sociedade portuguesa: uma evolução histórica-ambiental (1960-2010)*. In VII Congresso Português de Sociologia, APS, Porto.

Vicente, P. & Reis, E. 2008. Factors influencing households' participation in recycling. *Waste Management Research*. April 1, 26(2): 140–146.

Wastes: Solutions, Treatments and Opportunities – Vilarinho, Castro & Russo (eds)
© 2015 Taylor & Francis Group, London, ISBN 978-1-138-02882-1

Evaluation of cork as a natural sorbent for oil spill treatments

A. Sen & H. Pereira

Centro de Estudos Florestais, Instituto Superior de Agronomia, Universidade de Lisboa, Lisboa, Portugal

ABSTRACT: The oil sorption capacity of raw cork was examined and compared to that of heat-treated commercial cork sorbent (Corksorb), *Q. cerris* cork and wood. The investigated oils were gasoline, diesel, vegetable oil and motor oil. Oil retention, oil recovery and reusability of cork after sorption were evaluated. Oil sorption capacities (g/g) of cork were 5.9, 8.4, 9.3, and 9.5 for gasoline, diesel, vegetable oil and motor oil respectively. The oil sorption capacity of cork is in the lower range of synthetic sorbents but higher than that of the commercial Corksorb granulate, and of *Q. cerris* cork and wood. Oil sorption was significantly reduced in water-soaked cork. Particle size plays an important role on oil sorption. Oil can be recovered from cork by mechanical pressing up to 70% and cork previously used for oil adsorption can be reused several times whilst oil sorption yield decreases after each sorption cycle.

1 INTRODUCTION

Oil pollution of water is one of the principal threats to nature. Oil spills generate huge pollution in the open sea and shorelines: the well-known examples of Exxon Valdez (1989) and Deep Water Horizon (2010) spills released into the open sea approximately 40 and 800 million liters of crude petroleum respectively. A wide range of physical, chemical and biological methods are applied for oil removal and degradation, such as application of booms, skimmers, adsorbents, solidifiers, dispersants, in-situ burning and bioremediation. The most applied methods to remove spilled oil in the open sea are skimming with vessels and use of dispersants in larger areas (Ventikos et al. 2004). However the success is limited e.g. the estimated skimming and dispersant removal yields of the spilled oil from Deep Water Horizon Spill were 4% and 8%, respectively (Lubchenco et al. 2010). The accumulation of spilled oil in low-mobility areas such as beaches and wetlands poses an additional threat to the environment since the oil may remain for decades in these areas (Thibodeaux et al., 2011). In such cases adsorbent booms are generally applied to remove the oil because other methods are not appropriate to these relatively small areas (ITOPF, 2011). Adsorbents are also applied in the open sea to remove the remaining thin oil layer after skimming (Fingas, 2010). Synthetic polypropylene and polyurethane sorbents are the most commonly used commercial sorbents. These sorbents have oil uptake capacities ranging from 4.5 to 34 (wt:wt) (Adebajo et al., 2003). The usage of synthetic sorbents increased in the recent years and various sorbents have been tested including natural sorbents, organo-clays, expanded perlite, silica aerogels, zeolites, activated carbon, magnetic sorbents, modified polyurethane sponges and carbon nanotube sponges. Oil recovery and cost efficiency play an important role in sorbents applications. Synthetic polypropylene, polyurethane and inorganic sorbents are highly efficient in oil recovery but they are costly compared to natural sorbents. Natural sorbents, on the contrary, have relatively lower sorption capacity and higher water uptake values than synthetic and inorganic sorbents. They have also a sinking problem. Synthetic polypropylene or polyurethanes in the form of strips, foams or "pom poms" are therefore the most effective sorbents but their secondary contamination must also be considered. Since the oil containing sorbent is usually burned or buried, the synthetic plastic sorbents will release toxic gases by burning and they degrade slowly compared to natural sorbents. Among the natural sorbents, cork offers a promising opportunity: it is a renewable

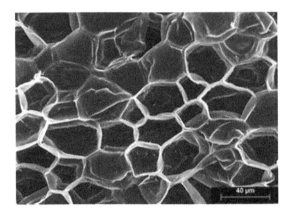

Figure 1.   SEM image of a cork sample showing the compact structure of small, thin walled and closed cells.

resource of biological origin (it is part of the outer bark of a tree, the cork oak, *Quercus suber*) and a rather inexpensive product, with hydrophobic character and good buoyancy properties because of its closed and air-filled cells (Pereira 2007) (Fig 1). In addition, the surface tension of cork ($32 \pm 3 \, \mathrm{mNm^{-1}}$) is in the range of that of polypropylene (Gomes et al., 1993). The surface tensions of commercial oils vary between 20 to $34 \, \mathrm{mNm^{-1}}$ which imply that cork can be used to adsorb a wide range of oils. Yet another advantage of cork is that it is possible to modify cork performance towards different oils by changing cork particle size e.g. smaller particles for light oils and coarser particles for heavy oils. The residual cork materials from the well-established cork industrial chain are the obvious source for cork-based sorbents: discarded refuse cork, low quality cork particles, expanded cork granulates and recycled cork materials. A commercial cork sorbent has successfully been used recently in oil recovery (Corticeira Amorim, 2009; Pintor et al., 2012). The aim of the present study is to study the oil sorption of granulated cork of different types with various oils with differing properties e.g. density and viscosity.

2   MATERIALS AND METHODS

2.1   *Materials*

The following spherical granular materials were used as sorbents in the sorption assays: Raw cork (from *Quercus suber*), commercial cork sorbent granulate (Corksorb) obtained from Corticeira Amorim, *Q. cerris* cork obtained from Turkey, pine (*Pinus pinaster*) and eucalypt (*Eucalyptus globulus*) woods from Portugal. Four common oils were selected as sorbates: gasoline, diesel oil, engine lubricating oil (termed as motor oil) and vegetable sunflower oil (termed as oil). Diesel oil (Oil e+) and gasoline (Efitec 95) were purchased at a Repsol Service Station, motor oil (BP Visco 2000 A3/B3 15W-40) was provided from a BP Service Station and vegetable sunflower oil (Fula) was obtained from a local market in Lisbon (Table 1). Sea water was collected from the Atlantic coast at Costa de Caparica beach, Portugal. Ultra-pure Milli Q water was provided from our laboratory.

2.2   *Methods*

Oil sorption capacity was determined following the ASTM 726-12 standard method. Approximately 1 g of sorbent was weighted into a stainless-steel spoon-shaped tea infuser and submerged into the oil (100 ml) at 20°C during 15 minutes, in triplicate experiments. After removal from the oil and 1 minute of draining time, the oil sorption capacity was determined as follows:

Table 1. Important properties of oils used in sorption tests.

	Gasoline	Diesel	Oil	Motor oil
Specific gravity at 15.6°C	0.70	0.80	0.90	0.85
Kinematic viscosity (mm^2s^{-1}) at 40°C	1.0	3.0	30	165

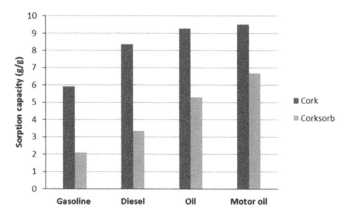

Figure 2. Oil sorption by cork and Corksorb particles (0.42–2.00 mm).

$A = (P2 - P1) \div P1$ where A is the sorbent oil sorption capacity (g/g), P1 is the sorbent oven-dried weight and P2 the sorbent weight after oil adsorption.

The oil retention test was carried out using the same method by successive weighing along 12 time intervals (from 0.25 to 1440 min) at 20°C. For the determination of oil sorption of water-soaked cork, samples were submerged into ultra-pure water and sea water during 18h. After water treatment, cork granules were drained for 30 minutes and submerged into 100 ml of the oil contained in a beaker. After 15 minutes sorption and 1 minute draining time the sorption capacity was calculated. The possibility of oil recovery after sorption was also determined. The oil-soaked cork granules were either mechanically pressed between metal wires using a laboratory hand press or a laboratory centrifugation method was applied (3000 rpm and 15 min centrifugation time) (Lim and Huang, 2007). The oil recovery was calculated as follows: $B = (P5 \div (P4 - P3)) \times 100$ where B is the oil recovery (%), P3 is the weight of moist cork, P4 is the weight of oil-sorbed moist cork and P5 is the weight of oil (g) obtained after mechanical press or centrifugation. The pressed or centrifuged cork was later tested for sorption capacity as indicated above to determine recycling possibilities of cork.

3 RESULTS AND DISCUSSION

Cork sorbed oil 6 to 10 times its weight (Fig. 2). This value is in the range of polypropylene fibers and higher than most of the vegetable sorbents (Adebajo et al., 2003; Bayat et al., 2005; Payne et al., 2012). The highest and lowest sorption values were attained with motor oil and gasoline respectively (Fig 4), showing that sorption increased with increasing oil density and viscosity. Wei et al. (2003) also obtained higher oil sorption capacity in a polypropylene sorbent with increased oil viscosity. It seems that oil sorption onto cork occurs mainly by surface adsorption. The raw cork particles adsorbed higher amount of oil than the heat-treated cork particles of the commercial cork sorbent (Fig. 2). The heat treatment of cork used in the production process of Corksorb leads to the decomposition of extractives and hemicelluloses, modification of lignin structure at lower

299

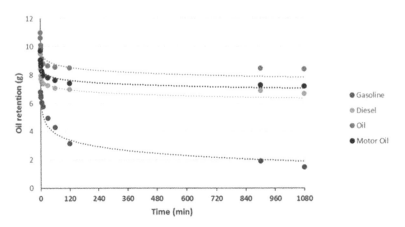

Figure 3. Oil retention patterns of gasoline, diesel oil, vegetable oil and motor oil in cork particles after adsorption.

temperatures and decomposition of lignin, cellulose and suberin at higher temperatures. The lower oil adsorption values obtained with the heat-treated cork should mainly be due to the chemical modification of the cork surface. Because the main cork component responsible for adsorption, i.e. lignin, is chemically modified and cork extractives are totally removed. The adsorption of heavy metals on cork also showed the similar trend (Sen et al., 2012). Following the oil sorption, and left to drain, the cork particles lost approximately 20% of the adsorbed oil in 2 hours (Fig. 3). Oil losses were not significant after this time. All oil types showed logarithmic oil retention losses over time. The gasoline retention pattern was different from others with a faster and greater loss because of its relatively higher volatile character. The oil retention after 1 day was around 75% in diesel, oil and motor oil and 30% in gasoline. These results imply that the oil-sorbed cork should be treated immediately after adsorption either by oil recovery methods or burning to prevent oil leaching to the environment.

Since the cork particle size distribution was somewhat different in cork and Corksorb particles (Fig. 2), with a higher proportion of larger particles in the Corksorb, differences in the oil sorption shown in Fig. 2 could be the result of the differences in particle size. The effect of sorbent on oil sorption was therefore studied on fractions with the same particle sized corks (Fig. 4). Comparing the oil adsorption of 1–2 mm particles of cork and Corksorb, it was found that cork adsorbed higher amount of oil than Corksorb although the difference was small (e.g. 8.6 and 8.1 g/g adsorption of vegetable oil for cork and Corksorb respectively). When comparing oil adsorption of two granulometric fractions of Corksorb (Fig. 4) it was found that increasing particle size resulted in lower oil uptake as expected e.g. 5.2 g/g vs. 8.1 g/g adsorption of vegetable oil onto 2–4 mm and 1–2 mm Corksorb particles. *Q. cerris* cork adsorbed less oil than heat treated cork of Corksorb with the same particle size (Fig. 4). The low oil sorption of *Q. cerris* cork is probably because of the high phloem (lignocelulosic non-cork component of bark) content of this bark fraction (Sen et al., 2011). In fact the wood samples showed similar and lower oil sorption values than cork. The macroscopic and microscopic differences between the hardwood (eucalypt) and softwood (pine) therefore do not seem to play an important role in oil sorption. The differences in their respective chemical composition should explain this adsorption difference: the wood cell wall is composed of 60–80% polysaccharides, 20–30% lignin and 1–10% extractives while cork is formed by 40–50% suberin, 20–25% lignin, 20% polysaccharides and 14–18% extractives (Pereira, 2007). Suberin should therefore play an important role in oil sorption, while the presence of woody, lignocellulosic components in the sorbent will lower its sorption capacity, as it was found for the *Q. cerris* cork fraction.

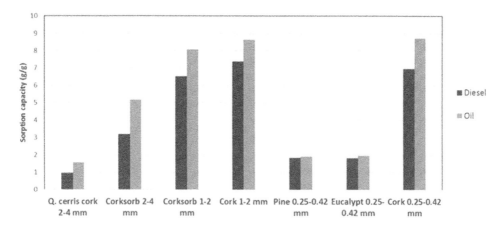

Figure 4.   Effect of sorbent types and granulometry on oil sorption.

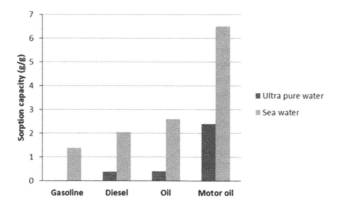

Figure 5.   Oil sorption of water-soaked cork after 18 hour.

Oil sorption was considerably reduced in water-soaked cork with reductions of 63%, 81%, 84% and 100% respectively for motor oil, diesel, vegetable oil and gasoline sorptions (Fig. 5). The surface groups of cork become saturated with water (water-soaked cork increased 4 to 7 times in weight, data not shown) and further oil adsorption was prevented. Interestingly, oil sorption was significantly higher in sea water-soaked cork than in ultra-pure water-soaked cork. The alkaline character and NaCl content of sea water must have increased oil sorption on cork. This result imply the importance of pH and presence of inorganic ions on oil sorption. A similar result was obtained by Chubar et al. (2004) who tested heavy metals sorption on cork after NaCl and NaOH treatments.

One of the important advantages of using cork as adsorbent of oil is its compression properties which might be utilized for oil recovery after the sorption. Cork recovers largely its dimensions even after substantial deformations and compression occurs without cell rupture (Pereira, 2007; Anjos et al. 2014). Diesel oil and vegetable oil were recuperated from oil-adsorbed cork samples using mechanical pressing and centrifugation methods (Lim and Huang, 2007). The oil recuperation rates were 70% for Diesel oil and 60% for vegetable oil using mechanical pressing. The centrifugation method did not yield a significant oil recovery. In a second cycle of oil adsorption using the recuperated cork after pressing, the oil sorption yields were reduced to approximately 50% of those obtained when the original cork was used for the sorptions of Diesel oil and vegetable oil, but the oil recuperation rates increased to 90% and 65% respectively. In a third cycle, the same trend was continued: the adsorption yields were reduced 60% relative to first cycle while oil recuperation rates were approximately 80%. These results imply that after each sorption cycle the cork surface is increasingly saturated with oil particles that may be difficult to remove by pressing, and further adsorption is blocked. Despite the decreasing oil sorption yields, cork can be reused up to three times after oil recuperation where the adsorbent capacity reaches approximately 2 g/g oil/dry cork after 3rd cycle. Since the tested oils are mixtures of different compounds, their adsorption on cork can be compared also by taking into consideration of their carbon numbers (gasoline: 4–12, diesel: 7–25, oil: 16–18, and motor oil: 12–50). It is clear from the results that higher sorption values obtained with higher carbon containing oils.

4 CONCLUSIONS

The oil sorption capacity of cork was studied for the first time and compared to that of commercial heat-treated cork (Corksorb), *Q. cerris* cork and wood. Cork has an oil sorption capacity in the lower range of the synthetic sorbents, a little above that of Corksorb, and considerably higher than that of *Q. cerris* cork and wood. Oil sorption was significantly reduced in water-soaked cork, although less in sea water-soaked cork than in ultrapure water-soaked cork. These results direct to some application recommendations e.g. cork as a sorbent should be applied dry and over the spillage. Cork chemical components and particle size play an important role on oil sorption. Oil saturated cork particles showed significant oil losses by draining within the first 2 hours and there-fore oil-adsorbed cork should be removed immediately after adsorption to avoid secondary contaminations. The adsorbed oil can be recovered from cork by mechanical pressing up to 70% and cork after oil adsorption can be reused several times.

REFERENCES

Adebajo, M. O., Frost, R. L., Kloprogge, J. T., Carmody, O., Kokot, S. 2003. Porous materials for oil spill cleanup: a review of synthesis and absorbing properties. *Journal of Porous Materials*, 10 (3), 159–170.

Anjos, O., Rodrigues, C., Morais, J., Pereira, H. 2014. Effect of density on the compression behavior of cork. *Materials and Design*, 53, 1089–1096.

ASTM 726-12. *Standard methods of testing sorbent performance of adsorbents*.

Bayat, A., Aghamiri, S. F., Moheb, A., Vakili-Nezhaad, G. R. 2005. Oil spill cleanup from sea water by sorbent materials. *Chemical engineering & technology*, 28 (12), 1525–1528.

Chubar, N., Carvalho, J. R., Correia, M. 2004. Heavy metals biosorption on cork biomass: effect of the pre-treatment. *Colloids and Surfaces A: Physicochemical and Engineering Aspects*, 238 (1), 51–58.

Corticeira Amorim, S.G.P.S. 2009. *CORKSORB – Sustainable Absorbents*.

Fingas, M. 2010. (Ed.). *Oil spill science and technology*. Gulf professional publishing.

Gomes, C. M. C. P. S., Fernandes, A. C., de Almeida, B. D. J. V. S. 1993. The surface tension of cork from contact angle measurements. *Journal of colloid and interface science*, 156 (1), 195–201.

ITOPF. 2011. The Use of Sorbent Materials in Oil Spill Response. *Technical Information Paper 8*. ITOPF-The International Tanker Owners Pollution Federation Limited. London. United Kingdom.

Lim, T. T., & Huang, X. 2007. Evaluation of kapok (Ceiba pentandra (L.) Gaertn.) as a natural hollow hydrophobic–oleophilic fibrous sorbent for oil spill cleanup. *Chemosphere*, 66 (5), 955–963.

Lubchenco, J., McNutt, M., Lehr, B., Sogge, M., Miller, M., Hammond, S., Conner, W. 2010. Deepwater Horizon/BP Oil Budget: what happened to the oil? *National Oceanic and Atmospheric Administration Report*. Silver Spring, MD. National Oceanic and Atmospheric Administration.

Payne, K. C., Jackson, C. D., Aizpurua, C. E., Rojas, O. J., Hubbe, M. A. 2012. Oil spills abatement: Factors affecting oil uptake by cellulosic fibers. *Environmental science & technology*, 46 (14), 7725–7730.

Pereira, H. 2007. *Cork: Biology, Production and Uses: Biology, Production and Uses*. Elsevier.

Pintor, A., Ferreira, C. I., Pereira, J. C., Correia, P., Silva, S. P., Vilar, V. J., Boaventura, R. A. 2012. Use of cork powder and granules for the adsorption of pollutants: A review. *Water research*, 46 (10), 3152–3166.

Sen, A., Quilhó, T., Pereira, H., 2011. Bark anatomy of *Quercus cerris* L. var. cerris from Turkey. *Turk J. Bot.* 35, 45–55.

Sen, A. U., Olivella, A., Fiol, N., Miranda, I., Villaescusa, I., Pereira, H. 2012. Removal of chromium (VI) in aqueous environments using cork and heat-treated cork samples from Quercus cerris and Quercus suber. *BioResources*, 7(4), 4843–4857.

Thibodeaux, L. J., Valsaraj, K. T., John, V. T., Papadopoulos, K. D., Pratt, L. R., Pesika, N. S. 2011. Marine oil fate: Knowledge gaps, basic research, and development needs; A perspective based on the Deep-water Horizon spill. *Environmental Engineering Science*, 28(2), 87–93.

Ventikos, N. P., Vergetis, E., Psaraftis, H. N., Triantafyllou, G. 2004. A high-level synthesis of oil spill response equipment and countermeasures. *Journal of hazardous materials*, 107 (1), 51–58.

Wei, Q. F., Mather, R. R., Fotheringham, A. F., Yang, R. D. 2003. Evaluation of nonwoven polypropylene oil sorbents in marine oil-spill recovery. *Marine Pollution Bulletin*, 46 (6), 780–783.

Wastes: Solutions, Treatments and Opportunities – Vilarinho, Castro & Russo (eds)
© 2015 Taylor & Francis Group, London, ISBN 978-1-138-02882-1

The mounds of Estremoz marble waste: Between refuse and reuse

C.F. Silva & L.D. Esteves
School of Architecture, University of Minho, Guimarães, Portugal

ABSTRACT: Waste is commonly, considered an effect separated from the economic processes of production, construction, and consumption. This is seen both in the politics of recycling and in the legal apparatus regarding the reintegration of industrial wastelands, in the aftermath cessation of the exploitation. Consequently, this management represents merely the 'after' attempts that try to minimize the effects of waste negatively affecting our lands and lives. In contrast, this paper addresses the interdependence between production and waste by making visible the 'mounds of marble waste' in Estremoz Anticline (Alentejo, Portugal). These mounds of marble waste allow us to demonstrate that the intervals of 'artificial time', as explained by Cedric Price (1996), that is, that use, reuse, mis-use, dis-use, and refuse, are not only successive but coexistent (Kümmel 1968).

1 INTRODUCTION

To reveal the interdependence between production and waste, this paper's argument is structured in three parts: (a) in 'marble waste, between refuse and use' we show the mis-use of this natural resource, as confirmed by a simple statistic. From all the matter extracted from Estremoz Anticline, only 10 to 20% is used in construction; the remainder (80 to 90%) (Ribeiro 2011) is refused and accumulated in mounds of marble waste close to the extraction quarries; (b) in 'marble waste, between dis-use and reuse' we highlight that the found matter has a strong economical and social potential through reintegration into the productive system, namely by creating innovative products for construction; and finally, (c) we argue that 'waste in transit' reclaims the urgency of considering a cyclical and coexistent strategic approach to waste, through the interconnection of the intervals of artificial time, as an open-ended process.

2 MARBLE WASTE: BETWEEN REFUSE AND USE

The Estremoz Anticline mining industry is predominately one of extraction and manufacture of marble. The stone exploitation in this region recalls ancient civilizations. In fact, the first known reference to the use of marble refers to a tombstone ordered in 370 BC by the Carthaginian Capitain Maarbal (Luz 2005). Nevertheless, it was mainly between the decades of 1930–90 that the level of production grew substantially to "concentrate more than 99% of the total active marble quarries in the country" (Luz 2005). Consequently, this industrialization process manifested a strong economic, social, and environmental impact on this territory, continuously reshaping its landscape.

The extraction areas are predominantly surrounded by removed stone that is considered worthless from a commercial point of view. These rock fragments that jointly represent 80–90% (Ribeiro 2011) of the extracted marble is usually set aside to contiguous areas and accumulated in piles, thus erecting artificial mounds. Furthermore, refused marble amounts to an expectant 50 million tons of still-waiting raw material (Ribeiro 2011). Considered together, this unused marble represents approximately 178 heaps in this region. As a result, the matter transferred from deep beneath the earth to the surface creates a "second nature" (Beigel & Christou 1996) landscape of holes and mountains. This landscape mutation is represented in three case-study territorial sections (Fig. 2)

Figure 1.   The marble waste mounds of Estremoz anticline.

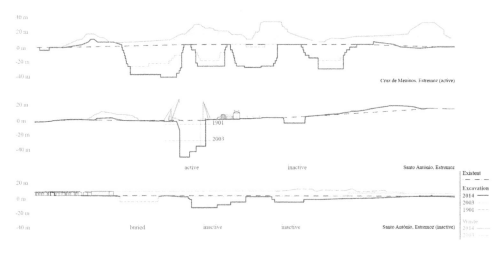

Figure 2.   Between refuse and use, landscape in mutation.

that make visible the noticeable topographic variation (registered between 2003 and 2014) (Esteves 2015), caused by the transition between the excavation and deposition processes and the soaring prominence of marble waste production. These mounds of marble waste are essentially elements characterized by a strong volumetric presence, reaching heights above dozen of meters, depending on the level of stone waste.

The considerable dross of marble matter reveals the questionable sustainability of this natural resource exploitation. For example, in 2001 the Anticline of Estremoz reported a rubble production of 1,485,100 m^3, a very significant value, taking into account the production of marble blocks (ready for trade) of that year (222,700 m^3). Perhaps if this earth-place had not been so generous, both in the quality of this geologic resource and in its dimensions — 40 km long per 7 km wide, men would have already thought about how to waste less matter, imagining alternatives to reintegrate it into the economic cycle of production. In that sense, in 21st century ecological concern, refused marble is not only an egregious waste of a natural matter; but simultaneously a waste of land and water, which leads to serious implications in agriculture production and obviously in all living

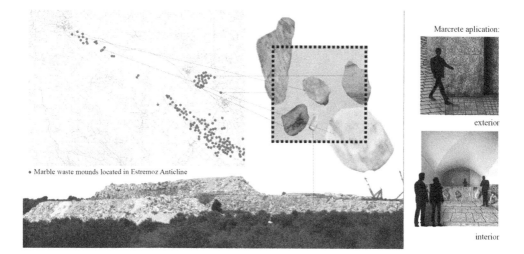

Figure 3. Marcrete.

system, including human cycles of life. As Kevin Lynch points out: "wasting is useful where it supports life and its development, and wasteful where wasting is blocked, accumulates in toxic form, or causes a loss of organic material" (Lynch 1990).

But what to do with this waste?

3 MARBLE WASTE: BETWEEN DIS-USE AND REUSE

The current economic context reflects the need to ensure and develop new approach strategies to reuse or use better existent resources. In the Estremoz Anticline region, the potential of waste as a construction material has been undervalued. Due to this fact and to the major scale of this dis-use we explored (Esteves 2015) in partnership with the Department of Civil Engineering of Minho University and CVR [Centro para a Valorização de Resíduos], the hypothesis to create a new construction material, *Marcrete*, mainly fabricated with marble waste and concrete. Moulded with existing raw materials, this newly created material reintroduces refused matter into the cycle of production.

Fundamentally, this solution aims to demonstrate the opportunity and priority of considering the huge waste produced by this mining industry as valid matter for the manufacture of new materials, with applications in the architectural context. Due to its environmental approach, *Marcrete* can also be integrated into the Environmental Landscape Recovery Plan [PARP – Plano Ambiental de Recuperação Paisagística]. Although the reintegration of industrial wastelands, after the cessation of the exploitation, is compulsory, examples of such practice are lacking in this territory. Thus, in the case of Estremoz, the demand for new industrial approaches based on the reuse of the waste produced by the marble industry and the reintegration of abandoned extractive areas is urgent. Moreover, *Marcrete* incites a strategy based on permanent acting-cycles, creating a cross-protection system that ensures greater productive incomes and less environmental dis-use.

Marcrete is one of several possibilities that need to be explored to resolve the waste issue. More than finding 'the' solution, the aim is to demonstrate the potential of reusing this existent dis-used matter, to act simultaneously as a stimulus for a reconfiguring of local companies. Strategically, we also point out the relevance of adopting synchronic strategies to reuse the refused matter, rather than the embedded one that considers waste in its linear succession: that is, reuse follows dis-use which follows refuse. This would not only prevent the escalation of the spoiled land and matter, but more importantly it would integrate the interdependence between production and waste.

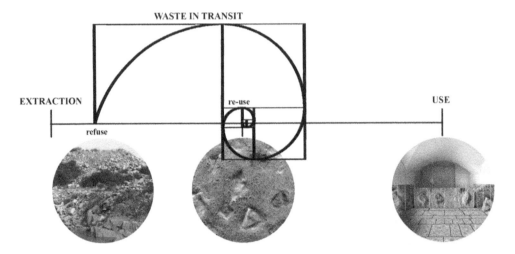

Figure 4.   Waste in transit.

4   WASTE IN TRANSIT

As long as we keep looking at the question of quarry wastelands as something that needs to be taken care of 'after' the end of the production cycle we are loosing sight of the big picture and only implementing superficial and aesthetical measures. To date, this remains the anachronous discourse and law apparatus regarding these areas of land. To hold onto an ideal image based on a naïve belief in the reversibility of long-time processes of exploitation of the earth by filling the hole of years of extraction, is no longer possible.

Jeremy Till (2009) states that "Architecture is no more than waste waiting to happen," pointing out that "demolition and construction, waste and order are kept apart through disciplinary policing of the boundary between the two." This implies that building and decay are not opposites but a transitory state 'in-between,' through which waste becomes. Applying Cedric Price (1996)'s intervals of 'artificial time' – use, reuse, mis-use, dis-use and refuse – to the marble exploitation in Estremoz Anticline reveal the paradoxical nature of time – not only successive, but coexistent (Kümmel 1968) – that has been generating this second nature landscape.

If this is the case, is it not time for protection and recovering landscapes policies to be more than an 'after action', one that re-acts to the inaction of a dis-used matter caused by the mis-use of this precious geologic resource, that represents 80 to 90% of extracted matter?

Production and waste are interdependent, and a strategic approach is also needed. As Sir Geoffrey Vickers (1968 cited by Price 1996) states, "There are many situations in which to be systematically late, is to be systematically wrong." This is the case of Estremoz Anticline wastelands: they have been waiting for a long time for something to be done to "Maintain reversibility in the short run and open-endedness in the long; don't put out more than the context can absorb. The greatest wastes occur when species and cultures are extinguished. Decline and death on the other hand, are normal and life-enhancing. (…) We should value a connected flow: of matter, of energy (Lynch 1990).

5   CONCLUSIONS

What to do with the mounds of waste in Estremoz Anticline? In this paper, we have brought visibility to the mounds of marble waste in Estremoz Anticline, generated by the continuous flow between extraction-deposition, between use and refuse, between refuse and mis-use, and between dis-use and reuse. Furthermore, by highlighting the principle of interdependence between production and waste we have reclaimed the urgency for cross-actions between the two, coexistent in time.

There are eight practical benefits of *Marcrete*, the proposed innovative material that could to solve waste in Estemoz Anticline: (a) the reintroduction of refused matter into the cycle of production; (b) the ability to be moulded with existing raw materials; (c) it is simple to produce; (d) the use of pre-existent industrial infrastructures and tools; (e) low-cost production; (f) the contribution to the increase in the productivity generated by the extracted mineral resources; (g) the addition of diversification to the business segments where this sector competes; and, (h) the reduction of the environmental impact of this dis-use waste.

Lastly, the *Marcrete* example has been introduced as a stimulus to trigger the active reuse of these huge piles of waste; but we argue that what is more important is to incite a more interconnected way of taking care of the earth, one that creates less wasteful waste.

REFERENCES

Beigel, F. & Christou, P. 1996. Brikettfabrik Witznitz: specific indeterminancy – designing for uncertainty. *arq: Architectural Research Quarterly* 2(2): 18–38.
Esteves, L. D. 2015. *O Marmorear de Estremoz: Paisagem em Movimento* [The Estremoz Marbleize: Landscape in Motion]. Master Thesis supervised by C. F. Silva. Guimarães: School of Architecture, University of Minho.
Kümmel, F. 1968. Time as Succession and the Problem of Duration. In J. T. Fraser (ed.), *The Voices of Time. A cooperative survey of mans views of time as understood and described by the sciences and by the humanities*: 31–55. London: Allen Lane The Penguin Press.
Luz, L. B. 2005. *Análise Crítica ao Modelo de Desenvolvimento do Sector das Pedras Naturais: O Caso dos Mármores no Triângulo de Estremoz-Borba-Vila Viçosa, 1890–2003*. Master Thesis supervised by H. C. P. Faustino. Lisboa: Universidade Técnica de Lisboa.
Lynch, K. 1990. *Wasting Away. An Exploration of Waste: What it is, How it Happens, Why we Fear it, How to do it Well*. M. Southworth (ed.). San Francisco: Sierra Club Books.
Price, C. 1996. Anticipating the Unexpected. *Architects journal* 204(9): 27–39.
Ribeiro, T. G. M. 2011. *Valorização de escombreiras da indústria extractiva de mármores no Alentejo*. Master Thesis supervised by P. Caetano. Lisboa: Faculdade de Ciências e Tecnologia, Universidade Nova.
Till, J. 2009. *Architecture Depends*. Cambridge: MIT Press.

ILLUSTRATION CREDITS

Esteves, L. D. 2015. *O Marmorear de Estremoz: Paisagem em Movimento* [The Estremoz Marbleize: Landscape in Motion]. Master Thesis supervised by C. F. Silva. Guimarães: School of Architecture, University of Minho.

Wastes: Solutions, Treatments and Opportunities – Vilarinho, Castro & Russo (eds)
© 2015 Taylor & Francis Group, London, ISBN 978-1-138-02882-1

New valorization strategies for Eucalyptus spp. bark extracts

F.S. Silva
Centro de Biotecnologia Agrícola e Agro-Alimentar do Alentejo (CEBAL)
Instituto Politécnico de Beja (IPBeja), Beja, Portugal

A.R. Guerra
Centro de Biotecnologia Agrícola e Agro-Alimentar do Alentejo (CEBAL)
Instituto Politécnico de Beja (IPBeja), Beja, Portugal
CICECO and Department of Chemistry, University of Aveiro, Aveiro, Portugal
University of Aveiro, Aveiro, Portugal

M.F. Duarte
Centro de Biotecnologia Agrícola e Agro-Alimentar do Alentejo (CEBAL)
Instituto Politécnico de Beja (IPBeja), Beja, Portugal

B. Soares, S.R. Freire & A.J.D. Silvestre
CICECO and Department of Chemistry, University of Aveiro, Aveiro, Portugal
University of Aveiro, Aveiro, Portugal

C. Calçada, C. Pereira-Wilson & C.F. Lima
Centre for the Research and Technology of Agro-Environment and Biological Sciences (CITAB),
Department of Biology, University of Minho, Campus de Gualtar, Braga, Portugal

ABSTRACT: The forest-based industry is important industrial sector in Europe, and particularly in Portugal. The increasingly higher demands for competitiveness, in forest-industry prompted the exploitation of different side-streams, such as bark residues. Within a biorefinery-based concept, these residues can be further refined to high value-added chemicals, with relevance for chemical and pharmaceutical industries. Results from our research team reveal that Eucalyptus globulus bark lipophilic extracts contain high quantities of triterpenoids (up to $12\,g\,kg^{-1}$ bark), predominantly triterpenic acids (TAs), namely betulinic, betulonic, ursolic and oleanolic acids. Similar composition has also been detected on Eucalyptus nitens outer bark. In the present work we report the chemical characterization of Eucalyptus spp. bark derived extracts, rich in TAs and the assessment of their antitumoral potential, using two distinct in vitro human cancer cell models. Our results represent an important step forward in terms of economic valorization of Eucalyptus spp bark residues derived from forest industry.

1 INTRODUCTION

Eucalyptus species are the most important fiber source for pulp and paper production in south-western Europe (Portugal and Spain) and south America (Brazil and Chile), where this sector has witnessed a fast growing during the last few years (Celpa, 2008). In the temperate and Mediterranean zones, E. globulus and E. nitens are the most common planted species (Pye-Smith & Cossalter, 2003). In the Portuguese context, *E. globulus* ranks third in terms of forest area (about 672.000 ha), representing nearly 31% of *E. globulus* world production (Potts et al. 2004) with 2 million tons/year of derived pulp, and the main raw material for pulp and paper production in Iberian Peninsula (Potts et al. 2004).

The agro-industrial exploitation of eucalyptus for pulp production generates large amounts of biomass residues, particularly bark, which, on average represents 11% of the stem dry weight, and that can be recovered in the pulp mill in the debarking line. Thus, a pulp mill with a production capacity of 0.5 Mtons/year of bleached kraft pulp can generate around 1 Mtons/year of bark. The exploitation of pulp industries by-products is one of the most promising examples of the implementation of biorefinery concept (Huang et al. 2008), which has been attracting increasing interest of agro-forest industries in recent years, from the perspective of promoting integrated exploitation of agro-forest biomass resources, taking the maximum value out of their crops, while searching for new alternatives to the non-renewable sources and, minimizing the waste streams without affecting the current most important outputs of the existing mills. In this perspective, the valorization of eucalyptus bark components, which nowadays is mainly burned for energy production, can represent an important contribution for the integrated valorization of this species.

Members of our research team, had demonstrated that *Eucalyptus spp.*, barks, and particularly their outer fractions (Domingues et al. 2010), are abundant sources (up to $\sim$120 g kg^{-1}) of several triterpenoids and particularly of triterpenic acids (TAs) namely, betulonic (BoA), betulinic (BA), 3-acetylbetulinic, ursolic (UA), 3-acetylursolic, oleanolic (OA) and 3-acetylbetulinic acids. The development of new applications for these compounds namely in the nutraceutical or in the pharmaceuticals domains will represent an important contribution for the valorization of this natural resource.

TAs have been associated to a broad spectrum of pharmacological activities coupled with a low toxicity profile, sparking renewed interest with regard to human health and disease, especially in what concerns their potential in cancer treatment [for review (Domingues et al. 2014)]. Based on our research team interest in valorization of eucalyptus barks as a natural source of TAs, as well as in the search for new modulators of cellular growth, the aim of the present study is to evaluate the activity of *Eucalyptus* bark derived extracts as modulators of cancer development, to determine the extracts effect on controlling cellular viability, cell cycle regulation and apoptosis using two distinct cellular models, and finally to relate this results with the specific composition of each extract, as well as with the individual pure TAs.

2 MATERIAL AND METHODS

2.1 *Chemicals and solvents*

Nonacosan-1-ol (98% purity) and β-sitosterol (99% purity) were purchased from Fluka Chemie (Madrid, Spain); UA (98% purity), BA (98% purity) and OA (98% purity) were purchased from Aktin Chemicals (Chengdu, China); BoA (95% purity) was purchased from CHEMOS GmbH (Regenstauf, Germany); palmitic acid (99% purity), dichloromethane (99% purity), pyridine (99% purity), bis(trimethylsilyl)trifluoroacetamide (99% purity), trimethylchlorosilane (99% purity), and tetracosane (99% purity) were supplied by Sigma Chemical Co (Madrid, Spain). 3-(4,5-dimethylthiazol-2-yl)-2,5-diphenyltetrazolium bromide (MTT) were purchased from Calbiochem (Merck, Millipore, Billerica, USA).

2.2 *Extraction*

E. globulus and *E. nitens* bark samples were taken from 12-year-old and 10-year-old trees, respectively, randomly sampled from clone plantations cultivated in Center region of Portugal.

Bark outer part was separated by hand as described elsewhere (Freire et al. 2002), air dried until a constant weight and ground to a granulometry lower than 2 mm prior to extraction. The outer bark samples of *E. globulus* and *E. nitens* (15 g each) were then Soxhlet extracted with dichloromethane for 7 h. The solvent was evaporated to dryness, the extracts were weighed and the results were expressed in percent of dry outer bark. Dichloromethane was chosen because it is a fairly specific solvent for lipophilic extractives.

2.3 *GC–MS analysis*

Before GC–MS analysis, nearly 20 mg of each dried sample were converted into trimethylsilyl (TMS) derivatives according to the literature (Ekman 1983, Freire et al. 2002). GC–MS analyses were performed using a Trace Gas Chromatograph 2000 Series equipped with a Thermo Scientific DSQ II mass spectrometer, using helium as carrier gas ($35\,cm\,s^{-1}$), equipped with a DB-1 J&W capillary column ($30\,m \times 0.32\,mm$ i.d., $0.25\,\mu m$ film thickness). The chromatographic conditions were as follows: initial temperature: $80°C$ for 5 min; temperature rate of $4°C\,min^{-1}$ up to $260°C$; $2°C\,min^{-1}$ till a final temperature of $285°C$; maintained for 10 min; injector temperature: $250°C$; transfer-line temperature: $290°C$; split ratio: 1:50. The MS was operated in the electron impact mode with electron impact energy of 70 eV and data collected at a rate of $1\,scan\,s^{-1}$ over a range of m/z 33–700. The ion source was maintained at $250°C$.

For quantitative analysis, GC–MS was calibrated with pure reference compounds, representative of the major lipophilic extractives components (namely, palmitic acid, nonacosan-1-ol, β-sitosterol, BA, UA and OA), relative to tetracosane, the internal standard used. The respective multiplication factors needed to obtain correct quantification of the peak areas were calculated as an average of six GC–MS runs. Two aliquots of each extract were analyzed. Each aliquot was injected in triplicate. The presented results are the average of the concordant values obtained for each part (less than 5% variation between injections of the same aliquot and between aliquots of the same sample).

2.4   Cell line and culture conditions

The human breast and colon cancer cell lines, respectively, MDA-MB-231, HCT116 and HCT15, were purchased from American Type Cell Culture (ATCC, Manassas, VA, USA). Cell lines were maintained at $37°C$ in a 5% CO_2 humidified atmosphere, and MDA-MB-231 cultured in DMEM supplemented with 10% heat-inactivated fetal bovine serum (iFBS) and 1% penicillin-streptomycin mixture, being the colon cells cultured in RPMI-1640 medium supplemented with 10 mM HEPES, 0.1 mM pyruvate, 1% antibiotic/antimycotic solution and 6% iFBS. All tested compounds were added to culture medium to the desired concentration ensuring that vehicle concentration (ethanol or DMSO) did not exceed 0.5% (v/v); controls received vehicle only.

2.5   Viability assay

The effects of *Eucalyptus spp.* bark derived extracts and pures TAs (UA, OA, BA and BoA) on cell viability were determined using 3-(4,5-dimethylthiazol-2-yl)-2,5-diphenyltetrazolium bromide (MTT) reduction assay. Cells were seeded, at a density of 2×10^5 cells/mL (for MDA-MB-231cells) and 1.6×10^5 cells/ml (for HCT116 and HCT15 cells), in 96-well plates, and incubated for 24 h at $37°C$. Then, the cells were incubated with different concentrations of Eucalyptus spp. bark derived extracts or with TAs individually during 24, 48 or 72 h. The number of viable cells in each well was estimated by the cell capacity to reduce MTT, determining the absorbance at 540 nm. The results were expressed as percentage of cell viability relative to the control (cells received vehicle only), and the concentration providing 50% cell growth inhibition (IC_{50}) was calculated by plotting the percentage of cell viability against the different compound concentrations logarithm.

2.6   Apoptosis assessment by the nuclear condensation assay

After treating cells for 48 h with different concentrations of pure TAs or Eucalyptus spp.extracts, cells were collected (both floating and attached cells) and fixed with 4% of paraformaldehyde for 20 min at room temperature. Cells were then attached into polylysine-treated slide using a Shadon Cytospin for estimation of cell death by apoptosis by the nuclear condensation assay as previously described (Lima et al. 2015). After staining cells with the nuclear Hoechst dye for 10 min, cells were observed under a fluorescent microscope and the percentage of apoptotic cells was calculated from the ratio between apoptotic (cells presenting condensed and/or fragmented DNA) and the total number of cells. For each condition, more than 500 cells were counted.

2.7   Cell cycle analysis

The analysis of cell cycle was done by flow cytometry as previously described (Lima et al., 2015). In brief, after treatment of cells for 24 h with different concentrations of extracts/pure compounds, permeabilized cells were stained with propidium iodide and treated with RNAse A for 15 min at 37°C. Analysis of cell cycle progression was done using a Coulter Epics XL Flow Cytometer) counting, at least, 40,000 single cells per sample. Data was obtained using the FlowJo Analysis Software, through the mathematical Watson Pragmatic model.

3 RESULTS AND DISCUSSION

The yields of the dichloromethane extractives from the *E. globulus* and *E. nitens* outer bark samples investigated Table 1 are in good agreement with previous reported data (Domingues et al. 2010, Domingues et al. 2011, Freire et al. 2002).

The chemical compositions of the lipophilic extract of *E. globulus* and *E. nitens* varied between the two Eucalyptus species Figure 1, as previously reported (Domingues et al. 2010, Domingues et al. 2011, Freire et al. 2002), however in both extracts the main components are TAs (about 4.7 and 19.9 g kg^{-1} of outer bark of *E. globulus* and *E. nitens*, respectively) with lupane, oleanane and ursane structures, namely BoA, BA, 3-O-acetylbetulinic, OA, 3-O-acetyloleanolic, UA and 3-O-acetylursolic acids, also in good agreement with literature (Freire et al. 2002). Specifically, BA, BoA and OA acids are the most abundant compounds in the outer bark extract of *E. nitens* Table 2, while 3-O-acetylursolic, UA and 3-O-acetyloleanolic are the main components of *E. globulus* lipophilic extract Table 2.

Regarding the effect of *Eucalyptus spp* outer bark lipophilic extracts on the modulating tumor cell growth, using a breast and colon cancer cells models, respectively, MDA-MB-231, HCT116 and HCT15, it was observed that both extracts were capable of reducing cellular viability Table 3. *E. nitens* extract had effect upon MDA-MB-231 cell line only after 72 h of incubation, instead both colon cancer cells presented themselves more sensible to the same extract. On the other hand, and according to the results *E. globulus* extract just had effect upon cellular viability after 72 h of incubation, at concentrations higher than what observed for *E. nitens* extracts Table 3.

Corroborating the MTT results, HCT116 cells were more susceptible to apoptosis induced by *E. nitens* extracts as compared to MDA-MB-231 cells. In addition, as shown by the nuclear

Table 1. Extraction yields (% w/w) of the eucalyptus outer bark fractions characterized.

Sample	Extraction yield (% w/w)
E. nitens	3.24 ± 0.76
E. globulus	2.17 ± 0.19

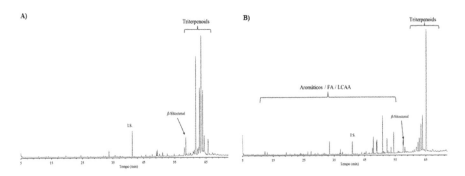

Figure 1. GC–MS chromatogram of the dichloromethane extracts of *E. nitens* (A) *and E. globulus* (B) outer bark. IS, internal standard (tetracosane), fatty acids; LCAA, long chain aliphatic alcohols; and IS, internal standard (tetracosane).

condensation assay and by the fraction of cells at the sub-G1 cell cycle phase, *E. nitens* extract was more active than the *E. globulus* counterpart. Induction of cell cycle arrest by *Eucalyptus spp.* extracts after 24 h of incubation was also more pronounced in HCT116 cells as compared with the breast cancer cells. As well, *E. nitens* extract was more effective showing significant cell cycle arrest at G1 phase in HCT116 cells.

Regarding the major triterpenic acids, present in the extracts, cell cycle arrest was only observed in HCT116 cells after 24 h of incubation, and more significantly for BoA. On the other hand, *E. nitens* bark lipophilic extracts (25 µg/mL) seem to be induce, in HCT 116 cancer cells, more apoptosis than *E. globulus* bark extracts Figure 2. BA, the major TA presented in *E. nitens* bark extracts, was the most potent inducing apoptosis showing about 5 fold induction at 2.5 µM. Interestingly, UA was the most potent inducing apoptosis in MDA-MB-231 cells, but significantly only after 72 h of incubation with about a 15 fold induction.

In conclusion, we observed that HCT116 cells were more susceptible to *Eucalyptus* extracts enriched in triterpenic acids than MDA-MB-231 cells. Whereas BoA was the most active compound for HCT116 cells, UA was for the MDA-MB-231 cells. These results underlie the biological potential of TA rich extracts, demonstrating that isoprenoids seem to carry some selectivity towards induction of cell death in cancer cells from different origins that may be further exploited for personalized target therapies.

Table 2. Compounds identified in the dichloromethane extracts of the outer bark of *E. nitens* and *E. globulus*.

Retention time (min)	Compounds	*E. nitens* (mg/mg$_{total\ extract}$)	*E. globulus* (mg/mg$_{total\ extract}$)
	Sterols		
58.09	β-Sitosterol	1.01 ± 0.12	1.54 ± 0.16
	Triterpenoids		
58.03	β-Amirin	0.32 ± 0.06	0.85±0.04
61.75	Betulonic acid	14.72 ± 0.69	1.01±0.12
62.94	Oleanolic acid	10.71 ± 0.56	0.78±0.08
63.47	Betulinic acid	20.02 ± 0.90	1.48±0.16
63.97	Ursolic acids	9.90 ± 0.63	2.80±0.33
64.42	3-O-Acetyloleanolic acid	3.16 ± 0.36	2.72 ± 0.38
65.64	3-O-Acetylursolic acid	2.54 ± 0.28	11.88 ± 1.47
	Total quantified compounds	86.0	32.5
	Total triterpenoids	61.0	22.4

Table 3. Cell viability at 24, 48 and 72 h of human breast cancer (MDA-MB-231) and human colon cancer cells (HCT115 and HCT16) incubated with *E. nitens* and *E. globulus* outer bark lipophilic extracts, as well as pure TAs (ursolic, betulinic and betulonic acids).

IC$_{50}$Values

Cell line	Time (h)	*E. Nitens* (µg/mL)	*E. globulus* (µg/mL)	Ursolic Acid (µM)	Betulinic Acid (µM)	Betulonic Acid (µM)
MDA-MB-231	24	–	–	41.35 ± 2.48	–	–
	48	–	–	Nd*	10.22 ± 0.45	–
	72	1.23 ± 0.41	16.45 ± 1.63	Nd*	28.01 ± 2.68	–
HCT116	24	4.66 ± 0.47	–	–	5.43 ± 0.25	43.27 ± 4.06
	48	3.49 ± 0.46	–	24.22 ± 0.49	4.00 ± 1.27	1.65 ± 0.47
	72	1.67 ± 0.13	6.29 ± 0.80	3.26 ± 0.55	1.33 ± 0.31	0.79 ± 0.28
HCT15	24	13.14 ± 3.45	–	21.32 ± 4.63	–	39.96 ± 6.28
	48	6.56 ± 2.51	–	17.26 ± 1.83	5.27 ± 1.49	14.45 ± 6.06
	72	3.67 ± 0.79	7.59 ± 1.65	12.17 ± 2.51	4.86 ± 2.09	1.59 ± 0.59

Values are expressed as mean ± SEM (standard error of mean). *Nd – not determined.

Control   E. Nitens 25µg/mL

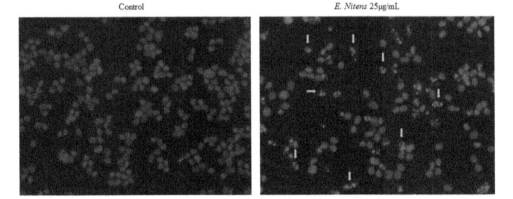

Figure 2.   Effect of *E. nitens* bark extracts (25 µg/ml) on cellular death by apoptosis. Representative images of the induction of apoptosis in HCT116 cells after 48 h incubation, arrows indicate apoptotic cells with condensed and/or fragmented DNA. Photos were taken with objective 20X.

ACKNOWLEDGMENTS

The authors acknowledge Fundação para a Ciência e a Tecnologia for the financial support through the project NEucBark– New valorization strategies for Eucalyptus spp. Bark Extracts (PTDC/AGR-FOR/3187/2012), award of PhD grant to A. R. Guerra (SFRH/BD/98635/2013), CICECO (UID/CTM/50011/2013), and CITAB (UID/AGR/04033/2013). Carmen S.R. Freire acknowledges FCT/MCTES (Portugal) for a contract under *Investigador FCT 2012*.

REFERENCES

Celpa. 2008. Boletim estatístico. (http://www.celpa.pt/Default.aspx?PageId=207&ContentId=28&ChannelId =112).

Domingues, R. M. A., Guerra, A. R., Duarte, M., Freire, C. S. R., Neto, C. P., Silva, C. M. S. & Silvestre, A. J. D. 2014. Bioactive triterpenic acids: from agroforestry biomass residues, to promising therapeutic tools. *Mini-Reviews in Organic Chemistry* 11:382–399.

Domingues, R. M. A., Patinha, D. J. S., Sousa, G. D. A., Villaverde, J. J., Silva, C. M., Freire, C. S. R., Silvestre, A. J. D., & Neto, C. P. 2011. Eucalyptus biomass residues from agro-forest and pulping industries as sources of high-value triterpenic compounds. *Cellul. Chem. Technol.,* 45:475–481.

Domingues, R. M. A., Sousa, G. D. A., Freire, C. S. R., Silvestre, A. J. D., & Neto, C. P. 2010. Eucalyptus globulus biomass residues from pulping industry as a source of high value triterpenic compounds. *Industrial Crops and Products,* 31:65–70.

Ekman, R. 1983. The Suberin Monomers and Triterpenoids from the Outer Bark of Betula verrucosa Ehrh. *Holzforschung,* 37:205–211.

Freire, C. S. R., Silvestre, A. J. D., Neto, C. P. & Cavaleiro, J. A. S. 2002. Lipophilic Extractives of the Inner and Outer Barks of Eucalyptus globulus. *Holzforschung,* 56:372–379.

Huang, H. J., Ramaswamy, S., Tschirner, U. W. & Ramarao, B. V. 2008. A review of separation technologies in current and future biorefineries. *Separation and Purification Technology,* 62:1–21.

Lima, C. F., Costa, M., Proenca, M. F. & Pereira-Wilson, C. 2015. Novel structurally similar chromene derivatives with opposing effects on p53 and apoptosis mechanisms in colorectal HCT116 cancer cells. *Eur J Pharm Sci,* 72:34–45.

Potts, B. M., Vaillancourt, R. E., Jordan, G., Dutkowski, G. W., Mckinnon, G., Steane, D., Volker, P., Lopez, G. A., Apiolaza, L. A., Li, Y., Marques, C., Borralho, N. M. G. & Costa E. Silva, J. 2004. Exploration of the Eucalyptus globules gene pool, in Eucalyptus in a Changing World. *Proc. of IUFRO Conf. 2004.* Aveiro.

Pye-Smith, C. & Cossalter, C. 2003. Fast-wood Forestry: Myths and Realities. *Centre for International Forestry Research.*

Wastes: Solutions, Treatments and Opportunities – Vilarinho, Castro & Russo (eds)
© *2015 Taylor & Francis Group, London, ISBN 978-1-138-02882-1*

Fixed bed adsorption dynamics of CO_2/CH_4 mixtures in zeolite 13X for biogas upgrading

J.A.C. Silva
Escola Superior de Tecnologia e Gestão, Instituto Politécnico de Bragança, Bragança, Portugal

A.E. Rodrigues
Laboratory of Separation and Reaction Engineering, Departamento de Engenharia Química,
Faculdade de Engenharia, Universidade do Porto, Portugal

ABSTRACT: The separation performance of CO_2/CH_4 mixtures in binderless beads of zeolite 13X has been studied in a fixed bed adsorption system for biogas upgrading strategies to biomethane. Zeolite 13X proved to be very efficient for the separation leading to breakthrough curves that show a plateau of pure CH_4 of approximately 4 minutes depending of the operating conditions chosen. The separation performance for CO_2/CH_4 measured by the selectivities range from 37 at a low pressure of 0.1 MPa to approximately 5.0 at the high temperature of 423 K. A mathematical model for fixed bed adsorption studies was used and validated through the experimental data predicting with good accuracy the behavior of the transient breakthrough experiments being a valuable tool for the analysis and design of cyclic adsorption processes (PSA) for biogas upgrading strategies to biomethane and CO_2 capture using 13X zeolite.

1 INTRODUCTION

Biogas is a raw gaseous stream produced by the anaerobic decomposition of organic matter in units called digesters. It is mainly composed by CH_4 (50 to 70%) and CO_2 (30 to 40%) and to obtain a high energy content CO_2 needs to be separated from CH_4. Major sources of biogas production are landfills, waste-water treatment plants, manure fermentation, fermentation of energy crops and coal-bed methane. To increase the heating value of the biogas it's necessary to remove CO_2 and the upgraded biogas can be used as a vehicle fuel in cars and buses or injected in the natural gas grid of a city or power plant.

Two recent reviews discuss this matter with great detail concerning the use of adsorbents (porous solids) based technologies to handle CO_2 capture and CO_2/CH_4 separations (Férey et al., 2011; D'Alessandro et al., 2010). For this purpose many solid physical adsorbents have been considered including molecular sieve zeolites and a new class of adsorbents named Metal-Organic Frameworks (MOFs). Zeolite 13X is still one promising adsorbent to capture CO_2 because of the strong adsorbate-adsorbent interactions (D'Alessandro et al. 2010). One of the technologies for biogas upgrading to biomethane using adsorbents is called Pressure Swing Adsorption (PSA). With this technique, carbon dioxide is separated from the biogas by adsorption in a proper adsorbent under elevated pressure. The adsorbent material, is regenerated by a sequential decrease in pressure before the column is reloaded again, hence the name of the technique. A review about the use of PSA technology for biogas upgrading is described in detail by Grande, 2011.

The modelling of fixed bed adsorption dynamics is of fundamental importance for the design of industrial adsorbers due to the complexity of these systems, that involve several mechanisms for mass and heat transfer coupled with thermodynamic models that describe the equilibrium adsorption between gas and solid phases.

In this work, we present fixed bed adsorption transient experimental data with single and multicomponent mixtures of CO_2 and CH_4 on binderless beads of zeolite 13X at 313, 373 and 423 K and pressures ranging from 0.1 to 0.5 MPa. A transient mathematical fixed bed adsorption model taking into account mass and energy conservation laws is used to capture the fixed bed adsorption experimental data, which could be used in the implementation (simulation) of cyclic adsorption processes (PSA, TSA) for the purification of biogas or CO_2 sequestration.

2 THEORETICAL

2.1 *Mathematical model to study fixed bed adsorption dynamics*

The modelling of fixed bed adsorption dynamics is very important to the design of cyclic adsorption processes to operate at industrial scale. The model should take into account the conservation laws of mass and energy in the fixed bed coupled with the thermodynamic (adsorption equilibrium isotherms) and kinetics (mass transfer rate of adsorbable species to solid adsorbent) of sorption characteristic of a certain system.

Let us consider that at time zero a mixture of known composition is introduced at the inlet of a fixed bed containing a suitable adsorbent for the separation of the mixture. If the following assumptions are made:

1. Ideal gas;
2. There is no pressure drop in the column;
3. The flow pattern is described by the axial dispersed plug flow model;
4. The mass transfer between bulk gas phase and adsorbent particle is accounted by a Linear-Driving-Force model (LDF);
5. The system is non-isothermal and non-adiabatic;
6. A resistance to heat transfer between solid adsorbent and bulk gas phase could exist in the external fluid film around the solid;
7. There is no temperature gradients inside the porous adsorbents (the temperature is homogeneous in the solid).

Silva (1998) according to these assumptions developed a mathematical model to simulate the fixed bed adsorption of gaseous mixtures and PSA cycles in adsorbent materials.

2.2 *Thermodynamic adsorption model*

Coupled with the mathematical fixed bed adsorption model we must have a suitable thermodynamic adsorption equilibrium description for the system since the transient response of the bed at the outlet will depend dramatically of the efficacy of such model. There are in literature several idealized adsorption isotherm models to describe the equilibrium distribution of guest molecules and host porous adsorbents. In a previous work it has been shown that an appropriate model to describe the binary sorption behavior of CO_2 and CH_4 on binderless zeolite 13X is the Fowler model (Silva et al., 2012) which is described by the following equations,

$$\frac{1}{p_1}\frac{\theta_1}{1-(\theta_1+\theta_2)} = b_1\exp\left(-\frac{w_{11}\theta_1}{RT} - \frac{w_{12}\theta_2}{RT}\right) \tag{1}$$

$$\frac{1}{p_2}\frac{\theta_2}{1-(\theta_1+\theta_2)} = b_2\exp\left(-\frac{w_{22}\theta_2}{RT} - \frac{w_{12}\theta_1}{RT}\right) \tag{2}$$

where $\theta = q/q_m$ is the degree of filling of sites, b is an equilibrium constant, p the pressure, q the amount adsorbed and q_m is the amount adsorbed at the saturation of the adsorbent, w is the extra energy when sorbate molecules occupy adjacent sites, R the ideal gas constant and T the temperature. The subscripts 1 and 2 refer to the adsorbable species (1) (CO_2) and 2(CH_4) along the manuscript. Table 1 show the adsorption equilibrium model parameters.

Table 1. Isotherm model parameters for single and binary sorption of CO_2 and CH_4 in binderless beads of 13X zeolite (data from Silva et al. 2012).

		$CO_2(1)$		$CH_4(2)$
q_m	(mol/kg$_{ads}$)	7.4		7.4
ΔH	(kJ/mol)	-43.1		-8.9
w_{11}	(kJ/mol)	12.3		$-$
w_{22}		$-$		$-$
		313 K		
b	(atm^{-1})	21.3		0.0643
$-w_{12}/RT$	$(-)$		1.39	
		373 K		
b	(atm^{-1})	1.49		0.0374
$-w_{12}/RT$	$(-)$		1.25	
		423 K		
b	(atm^{-1})	0.286		0.0256
$-w_{12}/RT$	$(-)$		1.10	

3 EXPERIMENTAL

3.1 *Adsorbent and adsorbates*

The powder of 13X from which the binderless beads were formed is from Chemiewerk Bas Kostritz GmbH (Germany) with a Si/Al ratio of 1.18. Metakaolin is used to manufacture the beads. The synthesis and characterization procedure is described in detail elsewhere (Schumann et al., 2012). Briefly, the beads formed consist in spherical particles with a diameter ranging from 1.2 to 2.0 mm. The size of the zeolite crystals are around 2 μm.

The sorbate and inert gases were furnished by Air Liquid with the following purities: methane N35 (99.95%), carbon dioxide N48 (99.998%), and helium ALPHAGAZ 2 (99.9998%).

3.2 *Single and multicomponent fixed bed experiments*

To study of the adsorption of CO_2 and CH_4 in the fixed bed was performed in a stainless steel column of 4.6 mm i.d. with 80 mm length containing the zeolite 13X beads which were placed inside a chromatographic oven with automatic temperature control. A typical experiment consists in measuring continuously the transient concentration histories at the outlet of the column using a thermal conductivity detector (TCD) and a mass spectrometer (MS) after feeding the column with a single component or mixtures of CO_2 and CH_4 of known composition. When the saturation is reached, the column is regenerated being prepared for another run. Details of the apparatus and experimental procedure can be found in detail elsewhere (Bastin et al., 2008).

4 RESULTS AND DISCUSSION

Breakthrough curves or the transient response at the outlet of the bed to an input of a single component or mixture feed at inlet is the more realistic way to evaluate the performance of an adsorbent for a specific separation or encapsulation of compounds. In the present study, several breakthrough curves were performed by feeding a fixed bed column containing beads of zeolite 13X with CO_2 or CH_4 or a mixture of both to evaluate the separation performance of the fixed bed adsorption system.

4.1 *Single component fixed bed adsorption experiments of CO_2 and CH_4*

Figure 1 shows two typical experimental breakthrough experiments obtained for the sorption of CO_2 (Figure 1a) and CH_4 (Figure 1b) at 313 K and total pressure in the column of 0.1 MPa. The

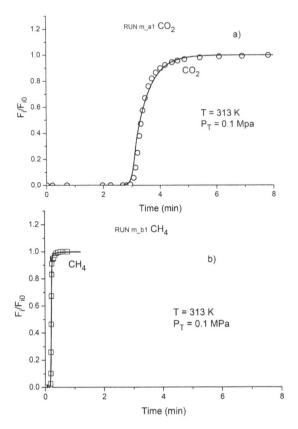

Figure 1. Single component experimental breakthrough curves (points) on binderless beads of 13X zeolite at 313 K and total pressure 0.1 MPa and comparison with the mathematical model (lines). (a) CO_2 feed at 21.7 Nml/min diluted in helium at 21.8 Nml/min. (b) CH_4 feed at 21.8 Nml/min diluted helium at 19.3 Nml/min. The simulation of the experimental runs with the mathematical model and respective parameters and correlations were adapted from the work of Silva (1998).

breakthrough time of CO_2 starts at 3 min spreading to the final input concentration at near 6 min. On the contrary, CH_4 (Figure 1b) breaks the column in much less than 1 min (practically with no adsorption) with a sharpening front. These results show that zeolite 13X has a strong affinity for CO_2 at 313 K being the sorption of CH_4 practically negligible. This means that we can predict that in mixture sorption zeolite 13X is able to separate with great efficiency CO_2 from binary mixtures of CO_2/CH_4 with a very high selectivity. At the same time, we can also conclude that the capacity of zeolite 13X to sequestrate CO_2 is very high being a good alternative to be used in systems where it is necessary to capture it. For the experiment reported in Figure 1a the amount adsorbed of CO_2 is around 3.8 mol/kg which can be considered a very high value at a total pressure in the column around 0.1 MPa. The lines in the figures represent model predictions and we can also conclude that the mathematical model capture with good accuracy the concentration profile of the transient breakthrough curves.

4.2 Binary CO_2/CH_4 fixed bed adsorption experiments

In strategies for biogas upgrading we wish to separate CO_2 from CH_4 by feeding the fixed bed column with mixtures of known composition and separate CH_4 from CO_2. When in contact with the adsorbent, the mixture is selectively adsorbed in the fixed bed due to differences in the adsorption strengths between the compounds, giving rise to the formation of different mass travelling waves

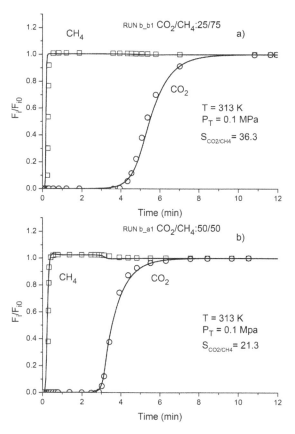

Figure 2. Binary CO_2/CH_4 experimental breakthrough curves (points) on binderless beads of 13X zeolite at 313 K and 0.1 MPa and comparison with the mathematical model (lines). (a) Binary mixture with 25(CO_2)/75(CH_4) feed at 9.2 Nml/min CO_2, 26.9 Nml/min CH_4 diluted in helium at 19.2 Nml/min. (b) Binary mixture with 50(CO_2)/50(CH_4) feed at 16.5 Nml/min CO_2, 16.5 Nml/min CH_4 diluted in helium at 16.5 Nml/min. The simulation of the experimental runs with the mathematical model and respective parameters and correlations were adapted from the work of Silva (1998).

along the bed, resulting in a transient breakthrough curve at the outlet with a different composition of the one at bed inlet until it is completely saturated. To demonstrate this, we present two typical binary breakthrough curves obtained after feeding the column with mixtures of CO_2/CH_4. Figure 2a shows an experimental breakthrough curve after feeding the column with a 50(CO_2)/50(CH_4) and Figure 2b with a 25(CO_2)/75(CH_4) mixture diluted with helium (inert) at the temperature of 313 K and total pressure in the column of 0.1 MPa. Figure 2a clearly shows a complete separation between CO_2 and CH_4 with a long plateau of pure CH_4 of almost 4 min for the 25/75 mixture ratio. The selectivity measured by the ratio of the amounts adsorbed of the two components is $S_{CO2/CH4} = 36.3$. Also interesting is the fact that Figure 2a,b shows that CH_4 breaks the column practically at the beginning of the experiment. Remarkable also is that the mass transfer zone for CH_4 is very sharp which very interesting for separation purposes. In Figure 2b for the 50/50 mixture ratio the selectivity decrease ($S_{CO2/CH4} = 21.3$) but it can be considered also very high. At the same time, the lines in the figures show that the mathematical model developed is also capable to capture with good accuracy the concentration profiles of both compounds as well as the plateaus of pure CH_4 observed experimentally in mixture sorption, being a valuable tool for the design of cyclic industrial processes (PSA).

Table 2. Selectivities as a function of temperature and total mixture pressure for several binary 50/50 and 25/75 binary, obtained from several breakthrough experiments.

Total Pressure (MPa)	50(CO_2)/50(CH_4)			25(CO_2)/75(CH_4)		
	313 K	373 K	423 K	313 K	373 K	423 K
0.5	10.4	6.4	5.4	14.4	8.7	6.8
0.3	11.3	7.9	5.5	19.7	10.8	7.1
0.1	21.3	16.6	7.4	36.3	19.8	10.3

4.3 Selectivities for several CO_2/CH_4 fixed bed experiments

Table 2 resumes the selectivities obtained for several 50/50 and 25/75 CO_2/CH_4 mixture breakthrough experiments as a function of total pressure. Table 2 shows that the selectivities are high at the low temperature of 313 K and total pressure 0.1 MPa being 36.3 (Experiment – Figure 2a) and 21.3 (Experiment – Figure 2b) for the 25/75 and 50/50 mixtures, respectively. As the pressure increases the selectivities decrease but the values are still considerable high at 313 K and total pressure of 0.5 MPa ranging from 14.4 to 10.4 for the 25/75 and 50/50 mixtures ratios, respectively. These results show that binderless beads of 13X zeolite can be considered an excellent separator of mixtures CO_2/CH_4 when appropriate operating conditions are chosen.

5 CONCLUSIONS

Through this work we show an experimental study of the sorption of single and binary mixtures of CO_2 and CH_4 in binderless beads of 13 X zeolite. The single component transient fixed bed adsorption experiments performed with CO_2 show that zeolite 13X in very efficient to encapsulate it. However, with CH_4 we see a complete different picture where the compound practically breaks the columns at beginning of the experiment practically with no adsorption. The efficiency of the separation of mixtures CO_2/CH_4 in the binderless beads of 13X zeolite starting from a fresh column is illustrated through two binary breakthrough curves where it is clearly seen the ability of zeolite 13X to separate CO_2 from CH_4 . These results show that zeolite 13X is a good choice for strategies of upgrading biogas to biomethane. For other conditions, Table 2 resumes the selectivities obtained from several mixture breakthrough experiments with values ranging from 37 at the low pressure of 0.1 MPa and temperature of 313 K to approximately 5 at the high temperature of 423 K.

A mathematical fixed bed adsorption dynamic model coupled to the thermodynamic model of adsorption of Fowler was used and validated through the experimental data proving to be a valuable tool for the design of cyclic adsorption processes for biogas upgrading to biomethane (by PSA) and CO_2 capture.

REFERENCES

D'Alessandro, D. M., Smit, B., Long, J. R., 2010. Carbon Dioxide Capture: Prospects for New Materials. *Angew. Chem. Int. Ed.* 49, 6058–6082.
Bastin, L., Bárcia, P. S., Hurtado, Silva, J.A.C., Rodrigues, A. E., Chen, B., 2008, A Microporous Metal-Organic Framework for Separation of CO2/N2 and CO2/CH4 by Fixed-Bed Adsorption, *J. Phys. Chem. C.* 112, 1575–1581.
Férey, G., Serre, C., Devic, T., Maurin, G., Jobic, H., Llewellin, P.L., Weireld, G., Vimont, A., Daturi, M., Chang, J. S., 2011. Why hybrid solids capture greenhouse gases?. *Chem. Soc Rev.* 40, 550–562.
Grande, C. A., 2011. *Biogas Upgrading by Pressure Swing Adsorption, Biofuel's Engineering Process Technology*, M. A. S Bernardes (Ed.), ISBN: 978-953-307-480-1.
Silva, J.A.C., Schumann, K., Rodrigues, A. E., 2012. Sorption and kinetics of CO2 and CH4 in binderless beads of 13X zeolite. *Microporous and Mesoporous Materials* 158, 219–228.
Silva, J.A.C. 1998. *Separation of n/iso -paraffins by adsorption processes*, Ph. D. Thesis (pages 105-117), University of Porto (http://repositorio-aberto.up.pt/handle/10216/11499).

Wastes: Solutions, Treatments and Opportunities – Vilarinho, Castro & Russo (eds)
© 2015 Taylor & Francis Group, London, ISBN 978-1-138-02882-1

Sewage Sludge Ash (SSA) as a phosphate fertilizer in the aspect of legal regulations

M. Smol & A. Henclik
The Mineral and Energy Economy Research Institute of the Polish Academy of Sciences, Kraków, Poland

J. Kulczycka
AGH University of Science and Technology, Kraków, Poland

B. Tarko, K. Gorazda & Z. Wzorek
Cracow University of Technology, Institute of Chemistry and Inorganic Technology, Kraków, Poland

ABSTRACT: Incinerated Sewage Sludge Ash (SSA) is an ideal base for inorganic fertilizer production due to its relatively high phosphorus (P)-content. The aim of paper was to determine the method of SSA utilization in the agricultural purposes. In this study, samples of biological swage sludge were obtained from three currently operational minucipal wastewater treatment plants in Poland: Częstochowa, Rzeszów and Tarnów. The sludge samples were incinerated in electrically heated laboratory chamber kiln in 950°C. The product of sludge combustion was fly ash with a good fertilizer properties, especially in terms of the content of phosphorus and calcium. However, in accordance to the Polish legislation, direct utilization of ash as a phosphate fertilizer in agriculture is not possible. Therefore there is necessary to look for other ways of recovery of this waste, eg. by develop processing system appropriate for the recovery of valuable phosphate from SSA and making a recycled fertilizer from it.

1 INTRODUCTION

Nowadays, the handling of waste is one of the most significant challenges in waste management over the world. The European Union (EU) emphasizes an issue of the more efficient use of waste under "A zero waste programme for Europe" [COM 398, 2014]. This visionary concepts for solving waste problems, suggests that moving towards a more circular economy (CE) is an essential way to deliver the resource efficiency agenda established under the Europe 2020 Strategy for smart, sustainable and inclusive growth. Higher and sustained improvements of resource efficiency performance can bring large economic benefits. Due to this facts, researchers have attempted reuse and recycle waste to enhance a sustainable environment [Kowalski & Makara 2010].

In the last years, an increase in volume of municipal wastewater caused by acceleration of population growth and urbanization process was observed [Lin & Ma, 2012]. Due to large quantity of generated municipal sewage sludge [Xu et al. 2012], one of the most important target of the EU is to achieve sustainable development in sewage sludge (SS) management. In accordance to the Council Directive 98/15/EEC amending Directive 91/271/EEC concerning the urban wastewater treatment 'sludge arising from wastewater treatment shall be re-used whenever appropriate'. Its aim is to promote pro-ecological management of sewage sludge. Therefore, it is expected that by developing and adopting appropriate treatment methods, a significant amount of materials and energy will be recovered from sludge [Kliopova & Makarskienė, 2015]. In the last years, the European Project SUSAN (Sustainable and Safe Re-use of Municipal Sewage Sludge for Nutrient Recovery) was aimed to develop a sustainable and safe strategy for nutrient recovery from sewage sludge using thermal treatment [Adam et. al., 2009b]. The large benefit of sewage sludge incineration is 80–90% decline of stored sludge amount and this solution should be more frequently applied in modernization and investment projects [Chen et al., 2013]. Product of sewage sludge combustion is sewage sludge ash (SSA), which could be used as a raw material in engineering applications, eg. as fertilizers. The aim of this paper is an overview of the possible application of SSA in the agricultural purposes.

2 WASTEWATER SEWAGE SLUDGE AS FERTILIZERS

Wastewater sludge contains significant amounts of nitrogen (N), phosphorus (P), organic matter (OM) and other trace elements, which are a good soil conditioner to improve physical and chemical properties of soil and a good source of nutrients for plant growth. Moreover, due to the high organic nitrogen content, sewage sludge may be evaluated as a slow release fertilizer. Therefore, sewage sludge is recycled in many countries for agricultural purposes [Topaç et al., 2008]. Reuse of sludge must be performed under conditions limiting the risks liable to the pathogenic microorganisms, which are presented in sludge. Among the several methods, such as biological digestion, composting and lime stabilisation, thermal treatment has been used to destroy pathogens from sludge. The incineration residues are ashes (bed ash and fly ash, depending on the location at which the ash is collected) with high phosphorus content (15–25% P_2O_5), which is an essential element for all organisms and it is fundamental in fertilizers used for food production [Adam et al., 2009a]. The combustion technology, variations in fuel composition and the conditions in the particle separation equipment affect the composition and properties of the various ash flows leaving the combustion unit. In general, the phosphorus is enriched in the fly ash together with trace elements. It need to be also mentioned that an important issue in the development of sewage sludge ash (SSA) utilization strategies is whether the concentration of trace elements [Pettersson et al., 2009]. The problem is toxic heavy metals such as cadmium (Cd), chromium (Cr), copper (Cu), lead (Pb), manganese (Mn), zinc (Zn) and nickel (Ni), as well as high soluble salt [Xu et al. 2012], present in municipal sewage sludge. This contaminants should be also effectively removed from sewage ash, below the limits for agricultural use according to EU and national fertilizer ordinances.

3 PHOSPHORUS RECOVERY

Phosphorus is an essential plant nutrient that cannot be substituted. Currently, some 80% of the world's phosphate production is used for fertilizer production. A continuous input of P-fertilizers to soils is required to sustain plant production for the growing world population. Most P-fertilizers are gained from sedimentary phosphate rock deposits that are non-renewable resource. If the presently recoverable reserves will be used in future at today's rates, they could be finite in roughly one century [Mattenberger et al. 2008]. Due to this, major producer countries are attempting to secure domestic raw materials and have designated phosphorous as a strategic resource [Sano et al. 2012]. Sewage sludge and incinerated sewage sludge ashes are significant sinks for P and the recovery of P from these waste streams, and its recycling could contribute to decelerating global P-flows [Petzet et al. 2012]. Therefore, the development of technologies aimed to phosphorus recovery from different waste streams is observed. Moreover, among the methods for P recovery, incineration has some advantages, such as reduction, stabilization, and harmless treatment [Li et al., 2014], which makes it be the most attractive method for thermal processing of sludge and should be more frequently applied in investment projects [Adam et al. 2009b]. Treating the SSA through a thermal process (850–1100°C) with a chlorine addition ($MgCl_2$, $CaCl_2$, KCl) in order to remove heavy metals and transform phosphorus into an available mineral form was studied [Donatello et al. 2010, Franz 2008]. In the previous study, SSA was leached with sulfuric [Franz 2008], nitric [Wzorek et al. 2006] or phosphoric acid [Gorazda et al. 2012]. However, the leaching processes of SSA with mineral acids are firmly developed and high content of phosphorus can be extracted to make an adequate phosphate fertilizer.

4 MATERIALS AND METHODS

Samples of biological swage sludge were obtained from each of the three currently operational municipal wastewater treatment plants in Poland: Częstochowa, Rzeszów and Tarnów.

The conditions of thermal treatment process of sewage sludge were described previously by the authors of this publication and colleagues [Wzorek et al. 2006]. In current study, the sludge was incinerated in optimal temperature determined empirically ($950°C$) in electrically heated laboratory chamber kiln. The product of the sewage sludge combustion was sewage sludge ash with code 19 01 14 (fly ash).

In accordance with generally accepted methodology, for the determination of quantitative and qualitative concentrations of selected components and heavy metals present in the SSA, X-ray Fuorescence Spectrometry (XRF) and Inductively Coupled Plasma Spectroscopy (ICP) were used.

5 RESULTS AND DISCUSSION

In accordance to the EU legislation, finding the most effective waste management method is necessary. In Poland, Act on Waste [Jurnal of law 2013, item. 21] indicates that generated waste should be subjected to the possible greatest extent recovery processes, and in a situation where it can not be subject to recovery method meeting environmental protection requirements, waste should be disposed of. Analysis of possible ways of waste management should be begun by reviewing the authorized recovery methods provided for this type of waste. Annex 5 to the Act on Waste includes the lists of the most likely recovery processes [Białowiec et al.2009]. These are:

R10 – Distribution on the earth surface in order to fertilizer or soil improvers.

R14 – Other activities involving the use of waste in whole or in part,

R15 – Waste processing, in order to prepare it for recovery, including recycling.

In the R10 process, the use of waste substances (in accordance with the Act on Waste) directly as a fertilizer or soil conditioners substances is provided. The possibility of waste using as fertilizer or improve soil properties formulation is determined by Regulation the Minister on R10 recovery process [Jurnal of law 2015, item. 132]. The fly ash studied in this paper has a good fertilizer properties, especially in terms of the content of phosphorus and calcium (Table 1). The PO_4^{3-} content of these ashes was equal to 16.45% in Tarnów, 22.9% in Rzeszów, and the highest contennt of PO_4^{3-} was obtained in ash from Częstochowa – 28.32%. The content of iron ranged from 3.31% in ash from Rzeszów to 16.16% in ash from Częstochowa. The content of calcium was equal to 6.07% in ash from Rzeszów, 11.1% in ash from Częstochowa and 16.72% in ash from Tarnów. It contains also magnesium, which is a valuable plant nutrient as well. Due to the high phosphorus content, this ash could be applied to acidic soils, poor in phosphorus. Unfortunately, although ash contains amounts of phosphorus comparable to commercial superphosphate, its direct utilization as a phosphate fertilizer in agriculture is not possible because the regulation on the R10 recovery process [Jurnal of law 2015, item. 132] does not allow for the recovery of fly ash with code 19 01 14 by the distribution on the earth surface.

Sewage sludge ashes from three Polish wastewater plants are not only carriers of valuable nutrients like phosphorus, calcium, magnesium and iron, but also contain considerable amounts of heavy metals. The content of heavy metals in the tested ash and standardized values is presented in Table 2. The content of heavy metals in SSA that apply to fertilizers is standarized for cadmium (50 mg/kg DW), lead (140 mg/kg DW), mercury (2 mg/kg DW) and arsenic (50 mg/kg DW). The values obtained in this study indicate that the ash from the three wastewater meets the Polish standards in order to cadmium and lead. The highest content of cadmium was observed in ash from Częstochowa – 29 mg/kg DW. The content of cadmium in ash from Rzeszów was equal to 17 mg/kg DW, and from Tarnów – 11 mg/kg DW. The highest content of lead was obtained in fly ash collected from Tarnów – 51 mg/kg DW. The content of lead in ash from Częstochowa was equal to 44 mg/kg DW. The lowest value was observed in fly ash from Rzeszów – 30 mg/kg DW.

Table 1. Chemical composition of average SSA samples after combustion of sewage sludge from three wastewater treatment plants in Poland.

	% dry weight of sample Częstochowa	Rzeszów	Tarnów
PO_4^{3-}	28.32	22.90	16.45
Fe	16.16	3.31	6.82
Ca	11.10	6.07	16.72
Mg	1.52	2.26	1.62

Table 2. Content of heavy metals in SSA from three wastewater treatment plants in Poland compared with Polish heavy metals limiting values that apply to fertilizers.

	mg/kg dry weight of sample Częstochowa	Rzeszów	Tarnów	Maximum levels according to Polish fertilizer regulations*
Cd	29	17	11	50
Cu	810	460	414	ns.
Ni	833	148	216	ns.
Zn	3775	1479	993	ns.
Cr	1293	198	518	ns.
Pb	44	30	51	140

ns. – not standarized
* Jurnal of law 2008, no. 119, item. 765

As it was mentioned, Polish regulation on the R10 recovery process [Jurnal of law 2015, item. 132] does not allow for the recovery of fly ash with code 19 01 14 by the distribution on the earth surface. However, it is possible to preparation of mixtures of various components, including waste, in order to obtain a final mixture meeting the specified requirements. For example, it is possible to prepare fertilizer granules having commercial value [Mattenberger et al. 2010]. The norm in this regard is the Regulation the Minister of Agriculture and Rural Development on implementation of certain provisions of the Act on fertilizers and fertilization [Jurnal of law 2008, no. 119, item. 765].

Therefore, preparation of fertilizer mixtures and substances improving soil properties is permitted process of recovery, defined by Act on Waste as R15 - waste processing in order to prepare it for recovery, including recycling. However, on the market, there is not always appropriate demand for this kind of products, hence other ways of waste management need to be seek. A further possibility of this waste recovery must be analyzed – the process of R14 – other activities involving the use of waste in whole or in part. This includes actions specified in the Regulation of the Minister of the Environment for the recovery or disposal of waste outside of installations and equipment [Jurnal of law 2006, no. 49, item. 356]. This Regulation provides the following ways for using of waste (outside installations where there is potential possibility for use of the ash): fill areas adversely transformed (landslides, non-operated pit excavation); hardening of the surface area; use in underground mining techniques; use for protection against water and wind erosion surface of the closed landfill; biological reclamation of closed landfills; construction of embankments, road and railway embankments, substructures of roads and highways. Unfortunately, in each of these cases, the Regulation does not allow the use of fly ash waste with code 19 01 14. It is therefore necessary to look for other ways of recovery of this waste. The research works in that range are implemented under the project PBS1/A1/3/2012 'Environmentally-friendly technology for sewage sludge ash utilization as a source of fertilizers and construction materials' ('ECOPHOS'). Due to the fact that fly ash is a valuable phosphate resource, in this project, investigations to develop a processing system appropriate for the recovery of valuable phosphate from ISSA and making a recycled fertilizer from it, are conducted.

6 CONCLUSSION

Sewage sludge ashes taken from three wastewater plants in Poland (Częstochowa, Rzeszów and Tarnów) exhibit high phosphorus contents of approximately 17-28% PO_4^{3-}. However, due to the Polish principles of environmental protection and waste management, the use od ash direct in agriculture is not possible because the regulation on the R10 recovery process does not allow for the recovery of fly ash with code 19 01 14 by the distribution on the earth surface. Therefore, the preferred method for utilization discussed fly ashes should be their use for the production of fertilizer granules.

ACKNOWLEDGEMENTS

This work was funded by the The National Centre for Research and Development in Poland under project PBS1/A1/3/2012 "Environmentally-friendly technology for sewage sludge ash utilization as a source of fertilizers and construction materials" ("ECOPHOS").

REFERENCES

Act of Waste of 14 December 2012 (Jurnal of law 2013, item. 21) (in Polish).
Adam, C., Peplinski, B., Michaelis, M., Kley, G., & Simon, F. G. 2009. Thermochemical treatment of sewage sludge ashes for phosphorus recovery. *Waste Management*, 29(3), 1122–1128.
Adam, C., Vogel, C., Wellendorf, S., Schick, J., Kratz, S., & Schnug, E. 2009. Phosphorus recovery by thermochemical treatment of sewage sludge ash–Results of the European FP6-project SUSAN. In International conference on nutrient recovery from wastewater streams. IWA Publishing, London, 417–430. b
Atienza–Martínez, M., Gea, G., Arauzo, J., Kersten, S. R., & Kootstra, A. M. J. 2014. Phosphorus recovery from sewage sludge char ash. *Biomass and Bioenergy*, 65, 42–50.
Białowiec, A., Janczukowicz, W., & Krzemieniewski, M. 2009. Możliwości zagospodarowania popiołów po termicznym unieszkodliwianiu osadów ściekowych w aspekcie regulacji prawnych. *Środkowo-Pomorskie Towarzystwo Naukowe Ochrony środowiska*, 11, 959–971. (in Polish).
Chen, M., Blanc, D., Gautier, M., Mehu, J., Gourdon, R., 2013. Environmental and technical assessments of the potential utilization of sewage sludge ashes (SSAs) as secondary raw materials in construction. *Waste Management* 33(5), 1268–1275.
Commission of European Communities. Communication No. 398, 2014. Towards a circular economy: A zero waste programme for Europe.
Donatello, S., Tong, D. & Cheeseman, C.R. (2010). Production of technical grade phosphoric acid from incinerator sewage sludge ash (ISSA). *Waste Management*, 30 (8–9), 1634–1642.
Franz, M. (2008). Phosphate fertilizer from sewage sludge ash (SSA). *Waste Management*, 28(10), 1809–1818.
Gorazda, K., Kowalski, Z., & Wzorek, Z. 2012. From sewage sludge ash to calcium phosphate fertilizers. *Polish Journal of Chemical Technology*, 14(3), 54–58.
Li, S., Li, Y., Lu, Q., Zhu, J., Yao, Y., & Bao, S. 2014. Integrated drying and incineration of wet sewage sludge in combined bubbling and circulating fluidized bed units. *Waste Management*, 34(12), 2561–2566.
Lin, H., Ma, X., 2012. Simulation of co-incineration of sewage sludge with municipal solid waste in a grate furnace incinerator. *Waste Management*, 32(3), 561–567.
Kliopova, I., & Makarskienë, K. 2015. Improving material and energy recovery from the sewage sludge and biomass residues. *Waste Management* 36, 269–276.
Kowalski Z., Makara A. 2010. Methods of ecological and ecomonic evaluation of technology, *Chemik*, 64(3), 158–167.
Mattenberger, H., Fraissler, G., Brunner, T., Herk, P., Hermann, L., & Obernberger, I. 2008. Sewage sludge ash to phosphorus fertiliser: Variables influencing heavy metal removal during thermochemical treatment. *Waste Management*, 28(12), 2709–2722.

Mattenberger, H., Fraissler, G., Jöller, M., Brunner, T., Obernberger, I., Herk, P., & Hermann, L. 2010. Sewage sludge ash to phosphorus fertiliser (II): influences of ash and granulate type on heavy metal removal. *Waste Management*, 30(8), 1622–1633.

Pettersson, A., Åmand, L. E., & Steenari, B. M. 2008. Leaching of ashes from co-combustion of sewage sludge and wood—Part I: Recovery of phosphorus. *Biomass and Bioenergy*, 32(3), 224–235.

Petzet, S., Peplinski, B., & Cornel, P. 2012. On wet chemical phosphorus recovery from sewage sludge ash by acidic or alkaline leaching and an optimized combination of both. Water Research, 46(12), 3769–3780.

Regulation of the Minister of Agriculture and Rural Development of 18 June 2008 on implementation of certain provisions of the Act on fertilizers and fertilization (Jurnal of law 2008, no. 119, item. 765) (in Polish).

Regulation of the Minister of the Environment of 20 January 2015 on R10 recovery process (Jurnal of law 2015, item. 132) (in Polish).

Regulation of the Minister of the Environment of 21 March 2006. for the recovery or disposal of waste outside of installations and equipment (Jurnal of law 2006, no. 49, item. 356) (in Polish).

Sano, A., Kanomata, M., Inoue, H., Sugiura, N., Xu, K. Q., & Inamori, Y. 2012. Extraction of raw sewage sludge containing iron phosphate for phosphorus recovery. *Chemosphere*, 89(10), 1243–1247.

Topaç, F. O., Başkaya, H. S., & Alkan, U. 2008. The effects of fly ash incorporation on some available nutrient contents of wastewater sludges. *Bioresource Technology*, 99(5), 1057–1065.

Wzorek, Z., Jodko, M., Gorazda, K., & Rzepecki, T. 2006. Extraction of phosphorus compounds from ashes from thermal processing of sewage sludge. *Journal of Loss Prevention in the Process Industries*, 19(1), 39–50.

Xu, J. Q., Yu, R. L., Dong, X. Y., Hu, G. R., Shang, X. S., Wang, Q., & Li, H. W. 2012. Effects of municipal sewage sludge stabilized by fly ash on the growth of Manilagrass and transfer of heavy metals. *Journal of Hazardous Materials*, 217, 58–66.

Wastes: Solutions, Treatments and Opportunities – Vilarinho, Castro & Russo (eds)
© 2015 Taylor & Francis Group, London, ISBN 978-1-138-02882-1

Effect of amendment of urine on clayey sandy soil salinity

M. Sou/Dakouré, F. Kagabika, D. Sangaré, B. Sawadago & A.H. Maïga
International Institute for Water and Environmental Engineering (2iE), Burkina Faso

R. Lahmar
Centre de Coopération Internationale en Recherche Agronomique pour le Développement (CIRAD), France

N. Hijikata
Department of Environmental Engineering, University of Hokkaido, Japan

ABSTRACT: By-products coming from resource recovery sanitation (i.e. urine, compost or treated greywater) are commonly used in agriculture as alternative source fertilizers. The present study aims to (1) assess global salinity progress in soil amended with urine; (2) evaluate a mixed urine/compost amendment effect on soil salinity and (3) identify the main salt accumulated after experiment. Three treatments were tested in pot experiment: (i) control (dam water); (ii) 100 % urine and (iii) 50% urine + 50% compost. The pots were filled with clayey sandy soil applying the fertilizers at three times. Experiment lasted 2 months from the end of the dry season to the beginning of rainy season. Salinity in pots amended with urine increased after the first rain event meaning that important salt dissolution occurred. Salinity is mainly due to high sodium accumulation coming from urine. Indeed, salinity value in pots treated only with urine was $273 \pm 20\,\mu s\,cm^{-1}$ higher than control pots salinity ($173 \pm 5\,\mu s\,cm^{-1}$). Mixing urine and compost has a slight mitigation effect on salinity augmentation with a final salinity of $222 \pm 6\,\mu s\,cm^{-1}$.

1 INTRODUCTION

In rural areas of Sahel regions, it is vital to provide water and fertilizers as essential inputs to sustain a small farming activity, which in turn is vital as subsistence agriculture. However, conventional source of water as well as chemical fertilizers are respectively rare and too expensive for these poor farmers, so alternative sources of water and fertilizers must be found. Concerning alternative water, greywater have a promising potential. Coming mainly from showers, dishwashers, and hand washbasins, it is estimated at between 70 to 85% of total wastewater generated, which represents an important source of water for small farming (Abu Ghunmi et al., 2010). Concerning alternative source of fertilizers, an average person excretes annually 2.8 kg of nitrogen (N), 0.45 kg of phosphorous (P) and approximately 1.3 kg of potassium (K) (Dagerskog and Bonzi, 2010). Nitrogen mainly comes from urine. However, urine is also very rich in sodium useless for plant so that it could accumulate in soil, leading to damaging effect on soil quality (Sou/Dakouré et al., 2013) The present study aims to (1) assess global salinity progress in soil amended with urine; (2) evaluate a mixed urine/compost amendment effect on soil salinity and (3) identify the main salt accumulated after experiment.

2 MATERIEL

The study was carried out in a typical sub-Saharan city located in the savannah region characterized by a hot and dry climate. Average temperature is 28°C and annual average rainfall is 773 mm, as observed over a 97-year period (Ouédraogo et al., 2007), whereas reference evapotranspiration is $1900\,mm \cdot year^{-1}$ and real evapotranspiration about 785 mm·year-1 (Ahmed and Al-Hajri, 2009).

Table 1. Chemical values of initial soil and inputs.

	Soil BP	Dam Water	Urine	Compost
pH (1:2.5)	6.80	7,54	8.10	
EC (1:5) (mS/cm)	0.081	0.189	21	8.35
SAR	0.21	0.3	64.2	10.9
	mg/g	mg/l	g/l	mg/kg
Total Nitrogen (N)	0	0,84	2.7	54.7
Total Phosphorus (P)	48.78	2,8	0.425	194
Potassium (K)	0.57	11.5	3.2	1848
Sodium (Na)	2.56	6.4	2.8	565
Calcium (Ca)	5.6	20.4	0.06	160
Magnesium (Mg)	2.9	4	0.05	96
Chloride (Cl)	4	3.15	2.6	497
LAS	ND	ND	ND	ND

BP: Before planting; ND: Not Detected; SAR: Sodium Adsorption Ratio; EC: Electrical Conductivity; LAS: Linear alkylbenzene sulphonate (as surfactant)

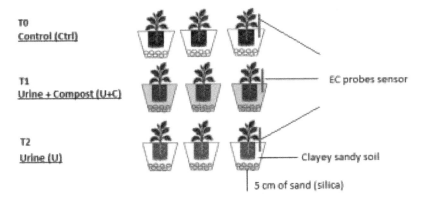

Figure 1. Experimental design.

The rainy season lasts from May to October; the study was performed from the end of dry season to the beginning of rainy season (april to may).

The table 1 presents the chemical values of soil (before planting), irrigation water (coming from a dam) and the bio fertilizers (as urine and compost) Values of the total forms of N, P, K (as the macronutrients necessary for plant growth) indicate that the soil and the irrigation water are relatively poor compared to urine (which provides the most important source of nitrogen) and compost, particularly rich in potassium. Concerning chemical likely to induce toxicity in soil and/vegetables, urine contains 5 times higher sodium than compost and then, represent a risk (as mentioned by the SAR value) of sodium accumulation on amended soil.

2.1 Experimental design

Experiment was performed in pots of 50 L of capacity each. They were filled with clayey sandy soil preceded by a layer of 5 cm of sand (figure 1).

Three treatments were tested in three repetitions per treatment, leading to the nine pots: treatment [T0] control (no urine, no compost); treatment [T1] compost and urine (50% N from urine and 50% N from compost) and treatment [T2], urine (100% N from urine)

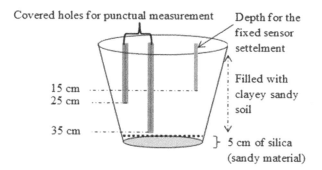

Figure 2. EC probes for salinity assessment.

Tomato was the experimental crop. Its needs in nitrogen, phosphorus and potassium are 200–150 kg N/ha, 60–80 kg P/ha and 190–200 kg K/ha respectively. Provision of urine and compost in treatment [T1] and [T2] were done based on the nitrogen (N) needs of tomatoes. Nitrogen mean concentration in urine and compost are presented in table 1.

2.2 *Data collection*

Soil salinity was evaluated proceeding by measurements of Electrical Conductivity (EC) of the soil using permanent sensors fixed in one pot of each treatment. Sensors were set at the depth of 15 cm (considered as root zone) and were set to record data each half hour from one day before the first application of urine to the end of the trial. An EC probe was also used to take punctual measurements of electrical conductivity at 25 and 35 cm depth (figure 2).

After harvest, soil samples were collected from the three different depths of measures by treatment and were taken to laboratory for chemical analysis. At laboratory level, electrical conductivity of the soil was measured using electrodes connected to a WTW 350i multi-parameters device after extraction 1:5 soil-deionized water. Determination of cations was done after extraction 1:10 ratio. Na and K were determined by flame emission spectrophotometer; Ca and Mg were determined by titration with EDTA and chlorides were determined by titration with $AgNO_3$. The SAR value of each sample was calculated using formula 1. Same analyses of soil were also done before the trial (table 1).

$$SAR = \frac{[Na]}{\sqrt{\frac{[Ca]+[Mg]}{2}}} \qquad (1)$$

Where concentrations are in milliequivalent per liter.

Urine was applied through 3 phases indicated on figures 4 and 5 by vertical arrows. The first application was 3 weeks after transplanting, the second was 2 weeks after the first application and last one, 2 weeks after the second application. Compost was applied the same day that urine was applied for the first time. Harvest happened almost 2 weeks after the last urine application.

3 RESULTS & DISCUSSION

3.1 *Electrical conductivity at soil surface (0-5 cm) during dry and rainy seasons*

The electrical conductivity monitoring results are presented on figures 3 to 5. Observation of the three curves trends indicates two main parts: the first part starts from the beginning of the experiment (10/04/2014) until the date of 10/05/2014, which corresponds to the first rain event. This first part (indicated by dotted lines on the curves) represents the dry period of the experiment, in contrast to the rainy period (from 10/05/2014 until the end of the experiment). This second part

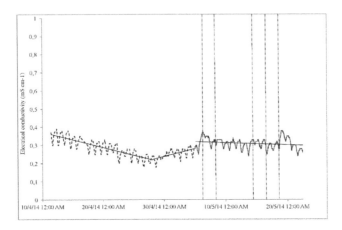

Figure 3. Salt monitoring in control pot (T0).

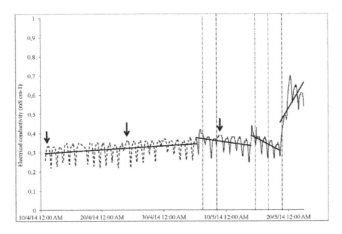

Figure 4. Salt monitoring in urine + compost amended pot (T1).

(indicated by full lines on the curves) is characterized by four rain events (indicated by vertical dotted lines).

During the dry season period of the experiment, EC exhibits a slight increase (from 0.3 to 0.4 mS/cm) at T1 and T2 despite of the two applications of urine already done in these pots. In contrast, during the rainy season period of the experiment, important increases of EC were observed mainly after the first rain event (for T2) and after le last rain event (both for T1 and T2). Such increases were not observed in the control treatment (T0) where EC trend to keep a mean value comparable to the initial EC. EC augmentation on pots amended with urine and compost (T1) or only urine (T2) means that salinity induced by urine applications produces a very sensitive reaction when in contact with rainwater. Salinity augmentation is delayed in urine+compost treatment probably because this treatment receives only 50% of the total volume of urine applied to T2. So the degree of salinity augmentation depends on the previous amount of urine amended in the pot.

EC augmentation in urine and urine/compost treatments after rain events, results from salts dissolution by rainwater. These salts have been precipitated during dry season period of experiment. Due to a relative small amount of irrigation water provided during dry period and the rate of evapotranspiration, a major part of the salts kept the solid form until the rainy period where higher amount of water allowed more dissolution of precipitated salts.

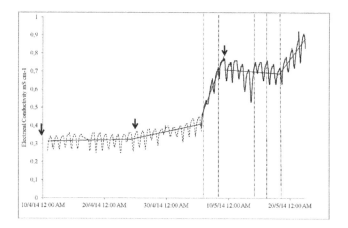

Figure 5. Salt monitoring in urine amended pot (T2).

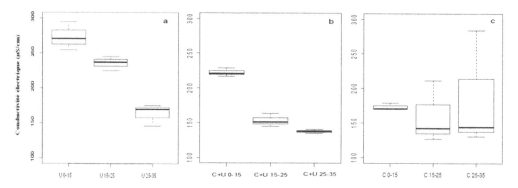

Figure 6. Soil electrical conductivity at 0–15, 15–25 and 25–35 cm depth.

3.2 *Electrical conductivity at 15, 25 and 35 cm depths*

Figure 6 presents final salinity at the three different depths of each treatment measured at the end of experiment. Electrical conductivity in control pot did not reach $200 \, \mu S \, cm^{-1}$ at 15 cm depth and kept its value below $150 \, \mu S \, cm^{-1}$ at 25 and 35 cm depths. Knowing that EC mean value of the soil before experiment was $81 \, \mu S \, cm^{-1}$, we can consider that dam water did not induce high salinity after experiment and the slight increase of EC is limited at the first 15 cm depth. Salinity induced by Urine/compost amendment is between 200 and $250 \, \mu S \, cm^{-1}$ at 15 cm depth and comparable to those of control treatment at deeper layers (25 and 35 cm). Urine/compost amendment limits salinity augmentation at 15 cm depth whereas only urine amendment induces salinity augmentation until 25 cm depth.

Box is 25th percentile (lower quartile); 75th percentile (upper quartile) and the median (50th percentile) in bold line. Bars are min and max values.

Figure 7 presents SAR values for the three treatments at three different depths. These data highlight a significant accumulation of sodium in urine and urine/compost treatments, at 0–15 cm depth. Such accumulation, in accordance with the highest values of salinity found 15 and 25 points measurement (figure 6) indicates that salinity is mainly caused by sodium accumulation. That also traduces a poor drainage in the pots.

Inside each treatment, values with different letters are significantly different (Tukey's test $p < 0.05$)

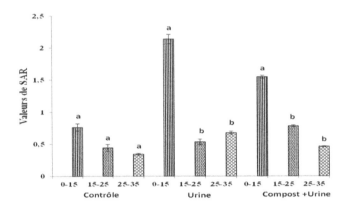

Figure 7. Sodium Absorption Ratio after experiment.

3.3 *Biomass production*

Tomato was used as experimental crop because it is very sensitive to salinity (Brady et Weil, 1999). At the end of experiment, the plants did not produce fruits in any treatment. At control treatment, this is due to a lack of nutrients (not add to better assess residual salinity after irrigation). For urine and urine/compost treatment, the plants growth was delayed and some leafs even burnt in urine treatment. This symptoms indicate that both urine and urine/compost treatment induced too high salinity to allow a suitable growth of tomato.

4 CONCLUSION

The present study mainly focused on salinity assessment when urine and urine/compost are applied in clayed sandy soil as bio fertilizer. The amount of urine applied represents the equivalent needs in nitrogen for tomato. But such amendment induces salt dissolution (mainly sodium) during rainy season, leading to significantly increase of global salinity until 25 cm depth. However, when urine total amount is halved and completed with compost, salinity augmentation tends to be delayed and globally decreased to 20% compared to urine treatment. In addition, salinity progression is limited at the 15 cm depth.

These first results need to be completed with further investigations focusing on the best ratio of urine/compost suitable to get a good productivity of tomato, but also a deeper investigation of pots drainage system in order to have a global assessment of salts balance sheet.

REFERENCES

Abu Ghunmi, L., Zeeman, G., Fayyad, M. & van Lier, J.B. 2010 Grey water treatment in a series anaerobic–aerobic system for irrigation. *Bioresour Technol* 101(9), 41–50

Ahmed, T. A. & Al-Hajri, H. H. 2009. Effects of Treated Municipal Wastewater and Sea Water Irrigation on Soil and Plant Characteristics. *Int. J. Environ. Res.*, 3 (4), 503–510

Brady, N., Weil, C. & Ray, R. 1999. The nature and properties of soils. Twelfth edition, Upper Saddle River. *NJ: Prentice Hall. Technology & Engineering*, 881p.

Dagerskog, L. & Bonzi, M. 2010. Opening Minds and Closing Loops–Productive Sanitation Initiatives in Burkina Faso and Niger. *Sustainable Sanitation Practice*. Issue 3.

Ouédraogo, E., Mando, A., Brussaard, L. & Stroosnijder, L. 2007. Tillage and fertility management effects on soil organic matter and sorghum yield in semi-arid West Africa. *Soil and Tillage Research*, 94 (1), 64–74.

Sou/Dakouré, M., Mermoud, A., Yacouba, H. & Boivin P. 2013. Impacts of irrigation with industrial treated wastewater on soil properties. *Geoderma*, 200-201, 31–39.

Wastes: Solutions, Treatments and Opportunities – Vilarinho, Castro & Russo (eds)
© *2015 Taylor & Francis Group, London, ISBN 978-1-138-02882-1*

The co-processing operation in Latin America and Europe cement industries

F.N. Stafford & D. Hotza
Federal University of Santa Catarina, Florianópolis, Brazil

J. Labrincha, L. Arroja & A.C. Dias
University of Aveiro, Aveiro, Portugal

ABSTRACT: The cement industry is an intensive consumer of raw materials and energy and is also responsible for great emissions of carbon dioxide. Due to this, many efforts have been made to turn this industry more environmentally friendly. In Europe, cement industries have been practicing the use of wastes as raw material since the 70's. This operation is called co-processing, and substitution rates have achieved 98% as yearly average for single cement plants and over 50% in some countries. In contrast, Latin America started co-processing operation only in the 90's and the highest average substitution rates vary between 7 and 18% according to the country. Thus, Latin America has potential to increase co-processing and many lessons can be learned from the European experience, such as the role of proper waste management, alternatives to landfilling and the importance of popular participation in the project and installation of co-processing units for hazardous wastes.

1 INTRODUCTION

Concrete is the most widely used manmade material and it is composed by water, cement, aggregates and additives. The most well-known form of cement is Portland, which is made of clinker and additives (WBCSD, 2009b). Clinker is produced from natural extracted raw materials such as limestone, clay and marl and smaller amount of other natural minerals, such as sand, bauxite and iron ore. These materials are mixed, originating the raw mix, which is fed to a kiln for pyro-processing at about 1450° C. After a cooling process, the clinker is conveyed to a ball mill for final grinding. In the final mill system, clinker is mixed with a small amount of gypsum in order to finally obtaining Portland cement (Strazza, Del Borghi, Gallo, & Del Borghi, 2011). According to the world business council for sustainable development (WBCSD), the cement production is projected to grow 0.8 to 1.2% per year, reaching between 3,700 and 4,400 Megatonnes in 2050 (WBCSD, 2009a). moreover, the production of cement involves the consumption of large quantities of energy, obtained from a number of different sources (Valderrama et al., 2012). Obviously, the huge amount of energy consumed in the cement production has impacts on the environment. Main emissions are carbon dioxide (CO_2), sulphur dioxide (SO_2) and nitrogen oxides (NO_X) that occur during the pyro-processing. they are mostly related to the chemical reactions and usage of fossil fuels and are responsible for impacts as climate change, acidification and eutrophication, respectively. In fact, 60% of the emission of co_2 is due to the limestone decarbonatation, but other 40% is related to the use of fossil fuels (Pacheco-Torgal, Cabeza, Labrincha, & Magalhaes, 2014).

In this scenario, the use of wastes as an alternative to replace raw materials and fossil fuels is a valid option that provides a solution in terms of reducing fossil fuel dependency as well as is a contribution to achieve lower emissions. this replacement is called co-processing and if carried out

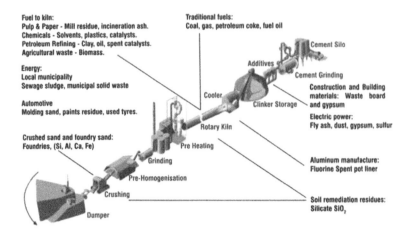

Figure 1.   Production line of cement and examples of alternative raw materials and fuels that can be used (Fonta, 2013).

in a safe and sound manner it should not affect health and safety of workers or neighborhood (cembureau, 2009). Thus, in this paper we analyse the status of co-processing operations in European and Latin American countries. In the first section, co-processing in Europe is discussed as a regular practice. In the following section, we present the efforts made to achieve adequate ways of carrying out co-processing in Latin America. Finally, we made considerations regarding similarities and differences in these two studied scenarios.

2   CO-PROCESSING OPERATION

According to the Cement Sustainability Initiative (CSI), there is substantial evidence that cement manufactured from different types of waste does not change significantly the characteristics of the cement or concrete. However, high levels of some minor components can affect cement performance and it is necessary to assure that specific thresholds are not exceeded (Fonta, 2013). Figure 1 presents a production line of cement and examples of alternative raw materials and fuels that can be used.

The decision on what type of waste can be used in a certain plant cannot be answered uniformly. As a basic rule, a waste accepted as an alternative fuel and/or raw material must give an added value for the cement kiln in terms of the calorific value of the organic part and the material value of the mineral part. However, several factors must be taken into consideration when deciding on the suitability of the materials, including cement chemical composition as well as the environmental impact of the production process. Nuclear waste, infectious medical waste, entire batteries and untreated mixed municipal waste are examples of residues which are not suitable for co-processing in the cement industry (CEMBUREAU, 2009). Some technical criteria have been established for using wastes as raw materials and fossil fuels, such as their physical state, calorific value, physical and grinding properties, moisture content (water content below 20%), compatibility with the current technology or accessible technical changes. Moreover, the cement quality must not be affected and alternative fuels cost must be lower than traditional fuels (Aranda Usón, López-Sabirón, Ferreira, & Llera Sastresa, 2013).

3   CO-PROCESSING IN EUROPE

Several cement producers are present in European Union, totalizing more than 260 cement plants. European business groups members of CSI are CRH (Ireland), Heidelberg Cement (Germany),

Holcim (Switzerland), Italcementi Group (Italy), Lafarge (France), and Secil (Portugal). CEMEX (Mexico) and Votorantim (Brazil) are also CSI members (Aranda Usón et al., 2013; CSI, 2014). According to WBCSD, countries of the European Union produce around 250 megatonnes of cement per year, and this number tends to stabilize (WBCSD, 2009a). The fuels used in cement kilns are fossil fuels as petroleum coke and coal and, since the 70's, alternative fuels started to be used as well (Lamas, Palau, & Camargo, 2013). Nowadays, they include animal meat and bone meat, municipal solid waste (also called refuse derived fuel, RDF), sewage sludge, biomass and end-of-life tyres (Aranda Usón et al., 2013). Typically, in European countries, the average substitution rate is over 50% for the cement industry and up to 98% as yearly average for single cement plants. In 2010–2011, the replacement ratio reached 83% in the Nertherlands, 62% in Germany and 60% in Norway, but in the same period, other countries did not reach a replacement ratio of 10% (Aranda Usón et al., 2013). A successful example is the case of Belgium. In 1999, an urgent solution was needed for the treatment of thousands of animal meal and fat from potentially contaminated animal products. The federal authorities identified the co-processing of the contaminated meat and bone meal in the cement industry as the best way of resolving this crisis. Belgian plants were requested to treat a large amount of animal meal, allowing the complete destruction of the potential contaminants in the kiln, as well as reducing emissions as a result of fossil fuel substitution (CEMBUREAU, 2009). It seems to be consistent with the general statement that Nordic countries and Netherlands have, in general, a much higher use of alternative fuels due to more advanced and efficient waste recycling schemes in place (Josa, Aguado, Heino, Byars, & Cardim, 2004). Besides this, many companies are working on reduce the CO_2 emissions replacing fossil fuels by alternative fuels. Among the main European cement producers some numbers and initiatives must be highlited: the Irish industry CRH used 430,000 tonnes of alternative wastes in 2013, which expresss alternative fuels as 21.2% of the fuel mix. In that year, 54% of alternative fuels were solid recovered fuels and 23% were biomass (CRH, 2013). The German Heidelberg Cement has as aim for 2020 the leadership in the co-processing of alternative fuels and raw materials using the potential of hazardous waste, sorted municipal solid waste and sewage sludge in combination with local opportunities (Heidelberg Cement, 2011). In 2014, the French Lafarge has used 20.7% of alternative fuels instead of fossil fuel, of which 38% was biomass. The goal to 2020 is using 50% of alternative fuels, of which 30% should be biomass (Lafarge, 2014). Another interesting initiative is the mobile sorting lines offered by the Swiss Holcim to companies that collect municipal waste, providing flexibility and extends the waste preprocessing service. In 2014, 14% of Holcim's thermal energy demand was covered by co-processing alternative fuels. By 2030, the company aspire to use 1 billion tons of secondary resources, replacing approximately 25% of primary materials (Holcim, 2015a). Indeed, Holcim also stands out for its initiative Geocycle, which is now, a network of 38 companies developing innovative industrial and municipal waste management services for a wide range of customers, aiming a zero-waste future (Geocycle, 2015). In general, the companies have established programs and targets to replace fossil fuels, by investing in modernization of facilities and programs for waste management. This enabled an increase in replacement rates and reducing operating costs, reasons why these actions are now being extended throughout the world. However, facilities using alternative fuels still continue to generate concern, particularly in the surrounding residential areas. It is especially intense when the facilities are located near populated areas (Rovira, Nadal, Schuhmacher, & Domingo, 2014), which can originate cases of NIMBY (not in my backyard) syndrome. The NIMBY syndrome reflects is an opposition to local siting of hazardous waste facilities and other locally unwanted land uses. Thus, no matter how technically suitable a proposed facility is, there is the possibility that its siting may be obstructed by a NIMBY movement (Kikuchi & Gerardo, 2009). In a characteristic case, a co-processing of hazardous waste was established by the Portuguese government in 2000 as a "National Strategic Plan of Waste Management". Since then a number of discussions regarding health and safety and further legal requirements are happening. Only in 2008, after a number of tests, the first co-processing operation has started. However, since then, it has been subject to legal decisions that sometimes allow the operation and sometimes suspend it (Ambiente, n.d.; CIMPOR, n.d.; Publico.pt/ciencia, n.d.).

4 CO-PROCESSING IN LATIN AMERICA

The annual cement production in Latin America is estimated in 200 megatons of cement per year and projected to grow to 400 megatons in 2050 (WBCSD, 2009a). The biggest cement producers in Latin America are Brazil, Mexico and Argentina. According to CEMBUREAU, those Latin American countries integrate the list of the 20 biggest producers of cement in the world (CEMBUREAU, 2012). Additionally, according to the *Federación Interamericana del Cemento* (FICEM), Colombia is another important producer, which reached in 2012 an equivalent cement production than that of Argentina (FICEM, 2012). Besides this, the market is composed in part by small producers and in part by some of the major world business groups, as the Swiss Holcim and the French Lafarge, and others locally originated as the Mexican Cemex and the Brazilian Intercement and Votorantim (Holcim, 2015b; InterCement, 2015; Lafarge, 2015; Votorantim, 2015; "Worldwide Locations|Cemex," 2013). Data from the International Cement Review, indicates that Latin America have 224 cement factories, of which more than 60% are local producers ("International Cement Review," n.d.). The use of co-processing operations started in the 90's and the most common wastes used nowadays to replace fossil fuels are tires, plastics, textiles, sawdust and wood, wastes from the production of paper, and others such as spent oils or solvents and inks. There is also a promising field to the use of biomass, especially rice husk, peanut and sunflower bagasse. The substitution rates in Argentina, Brazil, Chile, Costa Rica, Colombia, Guatemala, Mexico and Dominican Republic vary between 7 and 18% (FICEM, n.d.). In this scenario, for promoting the responsible co-processing in these and other countries, FICEM instituted a working group on climate change and co-processing with the participation of industry experts from different associated countries. The intent was preparing the Latin American industry to further regulations concerning climate change and supports the development of local regulations to co-processing (FICEM, n.d.). In addition, according to the Intergovernmental Panel on Climate Change (IPCC), many industrial facilities in developing nations are new and include the latest technology with the lowest specific energy use. However, many older, inefficient facilities remain in both industrialized and developing countries so there continues to be a huge demand for technology transfer to upgrade industrial facilities to improve energy efficiency and reduce emissions (IPCC, 2007). An interesting case in Latin America occurs in Brazil. Currently, 1.3 megatons per year of waste are co-processed in Brazil's cement industry, representing about 8% of the fuel matrix; however, the sector has potential to dispose around 2.5 megatons per year, offering potential for additional CO_2 emission reductions (Kihara & Visedo, 2014). In this context, the Brazilians Votorantim and Intercement occupy a prominent position: Votorantim has more than 90% of facilities authorized to co-processing, while Intercement reached 37% rate of thermal substitution in Candiota facility in 2012. In the same year, these two companies coprocessed more than 500,000 tonnes of wastes, each one. The Mexican CEMEX states that they *"put in place corporate guidelines for the introduction and handling of alternative fuels and raw materials in cement kilns to complement local regulation or to serve as a substitute where no regulation exists"*. However, the most intense use of alternative fuels take place in European facilities through the use of Climafuel® refuse derived fuel (United Kingdom) and Enerfuel (Spain) ("Alternative fuels and renewable energy|Carbon Strategy|Cases Studies|CEMEX," 2013).

5 COMPARING SCENARIOS

Opposite to the developed economies of Europe, Latin America is basically constituted by developing economies that are facing many challenges regarding political, social and environmental concerns. Certain countries have standards, regulations and laws for waste management and co-processing, which basically prohibit using untreated urban waste, radioactive, organochlorine, hospital and health services residues, pesticides and other related wastes (FICEM, 2012). Thus, the lack of regulations and properly waste management strategies constitute an obstacle to promoting co-processing. A great potential to encourage the practice comes from the amount of viable waste generated every year and the main cement producers: the presence of multinational companies in

Latin America has been promoting the use of wastes as fuels. However, Cemex recognizes that the use of alternative fuels is highest in Europe. For example, in 2009 they reached substitution taxes of 26% in Spain. The taxes were even better in United Kingdom (40%) and Germany (48%), but the Mexican taxes for the same year were 8% ("Alternative fuels and renewable energy|Carbon Strategy|Cases Studies|CEMEX," 2013; CEMEX, 2013).

6 CONCLUSION

Co-processing operations have been developed in Europe since the 70's. These almost fifty years of experience can teach important lessons of using co-processing as an alternative to landfilling, the role of properly waste management and last, but equally important, the popular participation in the project and installation of an unit that co-processes hazardous wastes. In Latin America, a number of efforts have been made as an attempt of achieve better levels of use of wastes in cement industries. In addition, the presence of multinational companies, such as CEMEX, Holcim and Votorantim has been promoting the use of wastes as fuels. Many countries are celebrating agreements and developing strategies and regulations to improve waste management and stimulate the co-processing, such as the working group started by FICEM in 2010. Despite many problems, a few associated to the incorrectly destination of wastes, Latin America has potential to increase co-processing due the amount of waste generated every year and its cement production.

REFERENCES

Alternative fuels and renewable energy|Carbon Strategy|Cases Studies|CEMEX. 2013. Retrieved June 14, 2014, from http://www.cemex.com/SustainableDevelopment/cases/AlternativeFuels.aspx
Ambiente, A. P. do. (n.d.). Plano Estratégico dos Resíduos Industriais (PESGRI). Retrieved March 10, 2015, from http://www.apambiente.pt/index.php?ref=16&subref=84&sub2ref=108&sub3ref=208
Aranda Usón, A., López-Sabirón, A. M., Ferreira, G., & Llera Sastresa, E. (2013). Uses of alternative fuels and raw materials in the cement industry as sustainable waste management options. *Renewable and Sustainable Energy Reviews*, 23, 242–260. http://doi.org/10.1016/j.rser.2013.02.024
CEMBUREAU. 2009. *Sustainable cement production: co-processing of alternative fuels and raw materials in the european cement industry.*
CEMBUREAU. 2012. *Activity Report – The cement sector: a strategic contribute to Europe's future.*
CEMEX. 2013. Alternative Fuels – CEMEX's position. Retrieved from http://www.cemex.com/MediaCenter/files/CEMEX_POSITION_on_Alternative_Fuels.pdf
CIMPOR. Declaraçao Ambiental 2012 – Centro de Produção de Souselas. Retrieved from http://www.cimpor-portugal.pt/cache/binImagens/Declaracao_Ambiental_2012_-_CP_Souselas-1461.pdf
CRH. 2013. *Sustainability Report.*
CSI. 2014. Members. Retrieved June 17, 2014, from http://www.wbcsdcement.org/index.php/about-csi/our-members
FICEM. Coprocesamiento – una alternativa ambientalmente sostenible.
FICEM. 2012. Co-procesamiento: recuperación de residuos en cementeras. Retrieved from http://ficem.org/brochure-ficem/brochure_coprocesa_oficio-2012.pdf
Fonta, P. 2013. Treating waste productively by the cement industry. *Indian Cement Review*, (September), 30–33.
Geocycle. 2015. Geocycle - News detail. Retrieved from http://www.geocycle.com/about-us/news-detail/article/toward-a-zero-waste-future-geocycles-contribution-in-2014.html
Heidelberg Cement. 2011. *Sustainability Ambitions – 2020.*
Holcim. 2015. Alternative resources. Retrieved June 15, 2015, from http://www.holcim.com/sustainable/environment/alternative-resources.html
Holcim. 2015. Holcim Ltd – Worldwide companies of the Holcim group. Retrieved from http://www.holcim.com/about-us/worldwide-companies-of-the-holcim-group.html
InterCement. 2015. InterCement. Retrieved from http://www.intercement.com/pt/#/mundo/
International Cement Review. Retrieved June 12, 2014, from http://www.cemnet.com/
IPCC. 2007. *Climate Change 2007 – Mitigation of climate change.*

Josa, A., Aguado, A., Heino, A., Byars, E., & Cardim, A. 2004. Comparative analysis of available life cycle inventories of cement in the EU. *Cement and Concrete Research*, 34(8), 1313–1320. http://doi.org/10.1016/j.cemconres.2003.12.020

Kihara, Y., & Visedo, G. 2014. Brazil: a view on the future. *International Cement Review*.

Kikuchi, R., & Gerardo, R. 2009. More than a decade of conflict between hazardous waste management and public resistance: A case study of NIMBY syndrome in Souselas (Portugal). *Journal of Hazardous Materials*, 172, 1681–1685. http://doi.org/10.1016/j.jhazmat.2009.07.062

Lafarge. (2014). *Sustainability report 2014*.

Lafarge. 2015. Onde estamos – Cimento: Lafarge. Retrieved from http://www.lafarge.com.br/wps/portal/br/2_8-Contacts_AND_locations

Lamas, W. D. Q., Palau, J. C. F., & Camargo, J. R. De. 2013. Waste materials co-processing in cement industry: Ecological efficiency of waste reuse. *Renewable and Sustainable Energy Reviews*, 19, 200–207. http://doi.org/10.1016/j.rser.2012.11.015

Pacheco-Torgal, F., Cabeza, L. F., Labrincha, J., & Magalhaes, A. de. 2014. *Eco-efficient construction and building materials: life cycle assessment (LCA), eco-labelling and case studies*. woodhead Publishing.

Publico.pt/ciencia. Co-incineração de resíduos industriais perigosos em Souselas poderá ser retomada. Retrieved March 10, 2015, from http://www.publico.pt/ciencia/noticia/coincineracao-de-residuos-industriais-perigosos-em-souselas-podera-ser-retomada-1413027

Rovira, J., Nadal, M., Schuhmacher, M., & Domingo, J. L. 2014. Environmental levels of PCDD/Fs and metals around a cement plant in Catalonia, Spain, before and after alternative fuel implementation. Assessment of human health risks. *The Science of the Total Environment*, 485–486, 121–9. http://doi.org/10.1016/j.scitotenv.2014.03.061

Strazza, C., Del Borghi, a., Gallo, M., & Del Borghi, M. 2011. Resource productivity enhancement as means for promoting cleaner production: analysis of co-incineration in cement plants through a life cycle approach. *Journal of Cleaner Production*, 19(14), 1615–1621. http://doi.org/10.1016/j.jclepro.2011.05.014

Valderrama, C., Granados, R., Cortina, J. L., Gasol, C. M., Guillem, M., & Josa, A. 2012. Implementation of best available techniques in cement manufacturing: a life-cycle assessment study. *Journal of Cleaner Production*, 25, 60–67. http://doi.org/10.1016/j.jclepro.2011.11.055

Votorantim. 2015. Votorantim: presença internacional. Retrieved from http://www.votorantim.com.br/pt-br/presencaInternacional/Paginas/presencaGlobal.aspx

WBCSD. 2009. Roadmap timeline of carbon emissions reductions. Retrieved from http://www.wbcsdcement.org/pdf/technology/WBCSD-IEA_Cement Roadmap_centre_spread_actual_size.pdf

WBCSD. 2009. *The Cement Sustainability Initiative: Recycling Concrete*.

Worldwide Locations|Cemex. (2013). http://doi.org/10.4135/9781412964289.n147

Wastes: Solutions, Treatments and Opportunities – Vilarinho, Castro & Russo (eds)
© 2015 Taylor & Francis Group, London, ISBN 978-1-138-02882-1

Potential of residues to contribute to the supply of minor metals

S. Steinlechner
Nonferrous Metallurgy, Montanuniversitaet Leoben, Leoben, Austria

J. Antrekowitsch
Christian Doppler Laboratory for Optimization and Biomass Utilization in Heavy Metal Recycling,
Nonferrous Metallurgy, Montanuniversitaet Leoben, Leoben, Austria

ABSTRACT: Within the last decades, mainly the base metals, like zinc, copper and lead were recycled from different industrial residues without taking into account that accompanien minor side-elements are often also present. Indium, silver but also gold or platinum group elements are well known companions of copper, lead and zinc ores, leading to interesting and partly significant content also in residues arising from the corresponding primary industry, like dusts, slags or sludge. Depending on the metal discussed, political unrest, decreasing primary resources leading to supply risks in different countries or also simply economic interests moved these mentioned side elements and with this the residues where they are present in, into the focus of the recycling industry. Therefore, this paper tries to answer the questions how the recycling of these wastes could contribute to the overall supply of for example silver but also other minor elements.

1 INTRODUCTION

Due to the fact that different base- and special metals are not homogenously distributed around the world, raw materials play a crucial role in our life, for the economy but also to maintain and improve the quality of life. As a result a strong discrepancy between supplier and consumer of metals, like for platinum group metals (PGMs) or also Indium, Germanium, etc. can be found. Particularly the areas of the world, which are more industrialized, are consuming high amounts of technologically important metals, like silver or also gold, in for example the electronic industry for contacts.

Based on this situation and the geographical availability of metals the European Union assessed 20 metals/raw materials as critical for the EU region (World Survey Silver, 2014). This includes the PGMs, Indium, Germanium, Gallium, etc. At the moment silver is not declared as critical within the EU, but can be forecasted to steadily grow in its demand due to its main application area in electronic industry after usage for jewelry and coins. A somewhat same situation can be found for the case of gold, for what the consumption per year is increasing continuously at the moment. On top of the demand-list is China, with 1283 tons per year, respectively 28.2% of the global sum (Gold Survey, 2014) in 2013. Especially the growth of the Chinese middle class in the last 10 years beside the government decision to make it easier to buy gold for the population lead to the dominating role of China also in the gold sector. In addition, India and the Arabian countries are famous for investing in gold and are responsible for the forecast of growing gold demand in the upcoming years. This circumstance shifts maybe also the supply of gold in the area of higher risk.

A further crucial aspect in this context is the kind of production technology. In case of critical elements, often the present production is as a side element of other metals (most of the time base metals) due to their low concentration in the ores. The following table 1 summarizes this fact, next to the source of by-product, the share of production in case that there are more than one and its share on the total revenue in the corresponding industry sector.

Table 1. Summary of selected by-products in primary metal production, their sources, recovery efficiencies and contributions to refinery revenues. Critical materials defined by the EU are underlined (European Commission, 2015).

By-product Metal	Sources of Production	Share of Production	Recovery Efficiencies	Max. Share of Total revenues
Gallium	Alumina	90%	10%	≈4%
	Zinc	10%	–	–
Germanium	Zinc	70%	≈12%	≈2%
	Coal	25%	–	–
Gold	Primary	≈90%	–	–
	Copper	≈10%	>99%	≈20%
Indium	Zinc	100%	25–30%	≈3%
Palladium	Platinum	60%	40–60%	≈15%
	Nickel	40%	–	≈15%
Platinum	Nickel	15%	–	≈10%
Silver	Lead-Zinc	≈35%	>95%	≈45%
	Primary	≈30%	–	–
	Copper	≈23%	>99%	≈25%
	Gold	≈12%	–	–

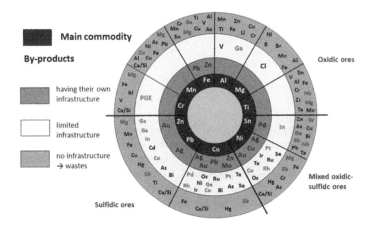

Figure 1. "Wheel of Elements" (Unep Report, 2014) showing the main ore types of metals (dark blue) and their typical by-products, grouped by the infrastructure available for recovery (light blue) or grey and green ring show trace metals (critical metals in red) which have limited or no infrastructure to be recovered.

Table 1 shows that obviously the demand of critical elements does not correlate with the supply of itself. Moreover, the supply is linked to the production amount of the main metal carrying he side element. This circumstance is also illustrated by the so called metal wheel (see figure 1), showing for a wide variety of critical or also non critical side elements, its typical origin ore (carrying metal) and if it is recovered as a by-product or if it is not economically viable.

Especially the metals zinc, lead and copper are carrying a great variety of side elements, leading to possible contents in generated residues of the production route of the main metal. Therefore, this paper summarizes potential generated residues from lead, zinc as well as copper production and their possible contribution to the supply of minor elements, like for instance silver, gold or PGMs. The presentation will further cover the supply and demand situation of selected critical metals.

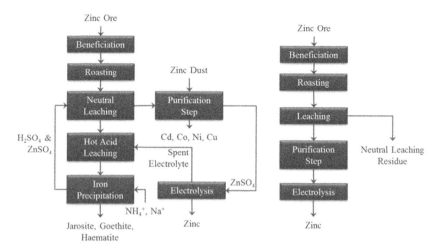

Figure 2. Left flow-sheet: Zinc production by neutral- and hot-acid leaching; Right flow-sheet: One-step leaching for low iron containing ores or oxidic ores (without roasting step).

2 RESIDUES AND BY-PRODUCSTS FROM ZINC, LEAD AND COPPER PRODUCTION

The following chapter describes the production route of lead, zinc and copper including the generated residues/by-products, which are showing potential to contribute to the minor metal supply.

2.1 Leaching residues from hydrometallurgical zinc production

More than 90% of the sulfidic or also oxidic zinc ore concentrates are processed in the hydromet-allurgical primary zinc production. Even though new production technologies were developed and implemented, like the direct leaching or the solvent extraction, in general two ways of hydromet-allurgical zinc winning can be described. Below figure 2 shows the main process steps in the conventional two-step leaching (neutral- followed by hot-acid leaching) and the one-step leaching in primary zinc production. The one step leaching is carried out in case of low iron containing ores or also in case of oxidic ores but then without the roasting step.

In case of sulfidic ores the first step is an oxidation, called roasting before the leaching is carried out. The roasting leads to a zinc ferrite formation between the zinc and iron contained in the ore. This is the reason why the one-step leaching (see figure 2 right) at moderate conditions is only carried out in case of low iron containing zinc ores, respectively low zinc ferrite in the calcine and with this low zinc losses in the neutral leaching residue. Zinc ores always contain certain amounts of silver, lead and sometimes other precious metals, which are not dissolved and remain in the neutral leaching residue (Dutrizac and Jambor, 2000). Depending on the gangue and iron content 600–700 kg neutral leach residue is generated per ton of zinc, with silver contents of up to 0.1% Ag.

If high iron contents in the ore are present, the leaching is carried out in two steps, to minimize the liquid amount for the required iron precipitation step. By this the zinc ferrite is dissolved in the hot acid leaching, keeping the zinc losses low but making an additional iron precipitation mandatory. Sometimes there is no separation of the hot acid leaching residue, also called lead/siler residue, done. In this case the iron precipitates on the remaining material and therefore the lead as well as silver is contained in the iron residue. While the goethite-precipitate can be utilized if a hot-acid-filtration is carried out, the jarosite is worldwide dumped in huge quantities. Nonetheless, also in case that a separation of the lead-silver-residue is carried out, a certain amount of silver and lead anyway is still in the jarosite as a result of the utilized roasted ore (calcine) used for pH adjustment during the precipitation process. Around 500–600 kg jarosite per ton of zinc are produced with up

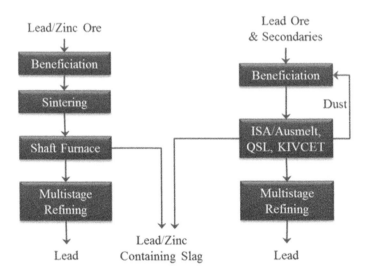

Figure 3. Overview of lead production possibilities.

to 230 ppm Ag (Roca et al, 1999) in this residue next to also present indium. Own investigations showed up to 300 ppm of silver and Indium in investigated dried filter cakes.

2.2 Slags from primary lead production

Different ways for the lead production, depending on the ore type, are available. Figure 3 gives an overview about the general ways of direct production and the lead/zinc ore processing in a shaft furnace in combination with a sintering as a pre-step.

Typically, a lead/zinc concentrate is processed in a shaft furnace, or also called Imperial Smelting furnace. The ratio in the sinter between zinc oxide and the reducing agent has to be kept on a defined level to avoid the unwanted iron oxide reduction. By this, also a certain amount of lead and zinc is lost in the slag. An average generated slag still has 6–9% zinc and up to 2.5% lead. As companion element of lead the silver is known and therefore can be found in the slag as well. Own investigations showed silver contents up to 150-200 ppm per ton of material. The typical ratio between zinc to lead production is Zn/Pb = 2/1, while the generated slag amount is 0.65-1 times the amount of produced zinc.

In case of low zinc containing ores the lead is processed without zinc as a direct by-product in shaft furnaces with sintering before or in QSL-, Ausmelt- or other direct production furnaces. Nevertheless, up to 3% of zinc in the ore is contained in the raw material, which is not recovered and with this has to be removed from the process stream by the slag. This is done by keeping the reduction potential low so that zinc oxide is not reduced. In former years a content of 11% zinc oxide was required for an economical treatment of the slag to recover the zinc (fuming), otherwise the taped slag was dumped without post-treatment (Pawlek, 1983). The amount of generated slag per ton of lead fluctuates from industrial site to site due to different ores, facilities and with this recovery rates between 400–600 kg per ton.

Nevertheless, dumps with zinc contents up to 15% exists next to significant amounts (up to 6%) of lead and with this also potentially silver.

2.3 Residues and byproducts from secondary copper production

Due to the wide range of materials, especially electronic scraps containing gold, silver, PGMs and other minor elements, recycled in the secondary copper industry, process streams arising from this

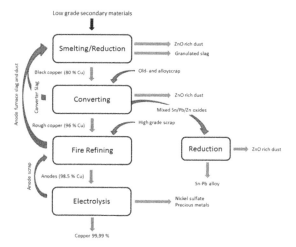

Figure 4.   Flow sheet of a secondary copper production (Ayhan, 2000 and Schlesinger et al, 2011).

sector seems to be potential to carry those valuable elements in particular. By the input of these complex materials, like e-scrap, non-ferrous shredder material, different sludge, etc., an increased input of further elements like zinc, lead, tin, halogens but also precious- and minor metals is the result. With the aim of producing high grade copper, the logical consequence is that those elements (impurities) have to be removed through the slag- and dust phase during smelting, converting or purification step. As a result those elements are enriched in the residues and increase their potential value. The below figure 4 shows the principal flow sheet of a secondary copper plant, with a lead and tin recovery included.

Various materials are charged into the different steps dependent on their grade. Typically, the generated slag is very low in remaining copper or other valuable elements. Nevertheless, in some industrial plants it is processed to recover also these small amounts of e.g. copper droplets. However, a huge amount is dumped without treatment and therefore could be potential for copper, gold and silver recovery.

The anode slime from electrolytic copper refining is processed to recover the precious metals next to other valuable elements already, so that this material forms no new potential. But especially the dust from the first two stages, which is up to 60 kg per ton of produced copper, shows high contents of valuable metals. On the one hand carry-over is responsible for that but the far more important mechanism is the volatilization by for instance halide formation. Halogens are introduced by flame-retardants in the charged PCBs or also plastics. As a result, the quantity as well as chemical composition of flue dust strongly depends on the charged material and process technology. In this context especially the minor metals cannot be predicted exactly.

Zinc, lead, tin and copper are the main valuable fraction in the dust. Beside this, up to 800 ppm silver and gold, sometimes PGMs are present, due to carry-over as well as formed volatile compounds. Although there is a zinc content of up to 50% it is not possible to use this material directly in the primary zinc production. The main reason are the halogen compounds, which would disturb the electro winning process significantly. Moreover, the presence of high lead, tin and copper contents leads to higher effort and additional problems in the zinc winning. Nevertheless, these dusts are showing high potential concerning the secondary production of precious metals (Björkman and Samuelsson, 1996, Hanusch and Bussmann, 1995).

3   SUMMARY AND CONCLUSION

As shown in figure 1 zinc, lead and copper ores carry a wide variety of valuable side elements, like Ag, Au, PGMs or also In and Ge or others. Due to that these side-elements are not always recovered

Table 2. Status of precious metals recycling out of residues from lead, zinc and copper industry.

Base Metal	Mainly recycled	Generally not recycled
Copper	Anode slimes	
		Slags from 1st step
		Dusts from various steps
Zinc	Goethite	
		Jarosite
	Neutral leach residues	Neutral leach residues
Lead		Slag
	Matte	
		Speiss

Table 3. Potential sources of precious metals and estimated amount of silver in untreated residues from copper, lead and zinc industry.

Untreated material	Amount [kt]	Ag [t]	potential for Au	potential for PGMs
Jarosite	4000	800	Yes (low)	
Goethite	400	100		
Neutral leaching residues	200	200	Yes (low)	Yes (low)
Dust from sec. copper industry	100	50	Yes	Yes
Lead slag	3000	150		

or are only inefficiently recovered within the base metal production route, residues from those metallurgical industries show high potential to contribute to the future minor metals demand. Although some of the generated by-products/residues are already treated, some of them are still inefficiently treated or dumped until today. Table 5 summarizes the main generated products/residues and their status of recycling.

Some of those residues are dumped till today because of on the one hand the complexity of the material and with this the costs for a recycling process but on the other hand the fact that different recycling concepts only tried to recover one metal. One example for this is the slag from lead production, which was dumped for years in case of too low zinc contents, although an intelligent combination of metallurgical process steps could recover also the remaining lead and with this could also allow an economic processing in case of lower zinc contents by a mutli metal recovery. In special, those minor metals can contribute to an overall revenue of recycling processes significantly, even though there absolute content per ton of residue is relatively low. Considering this circumstance, materials that were dumped for years and declared as residues until now, become raw material sources for the future. As an example table 6 summarizes this fact for the discussed residues from lead, zinc and copper industry and their potential to contribute to the silver supply.

The average content of Au and PGMs are difficult to estimate, which is the reason why it is only mentioned what residues show potential. However, in case of a realized recovery of silver, very likely also gold and PGMs are collected, for instance by a metal bath process. As an example, gold in jarosite can be around 1.35 ppm in content, which is less than one hundredths of the silver content (~200 ppm) but can contribute nearly with the same monetary value to a possible recycling process if recovered due to a multiple of 70 in its price compared to the silver.

In case of the silver contained in the residues the sum of the potentially recovered amount in is 1 300 tons silver annually, which represents 5.1% of the annual worldwide mine production, showing the huge potential of residues to contribute to the supply of minor metals.

ACKNOWLEDGEMENT

The authors want to thank the Austrian Research Promotion Agency (FFG) and the Federal Ministry of Science, Research and Economy (BMWFW) for the financial support.

REFERENCES

Ayhan, M. 2000. Das neue HK-Verfahren für die Verarbeitung von Kupfer-Sekundärmaterialien. *Intensivierung metallurgischer Prozess.* Heft 87 der Schriftenreihe GDMB 197–207.
Björkman, B., Samuelsson, C. 1996. Dust Forming Mechanisms in the Copper Converting Process. *Second International Symposium on Extractions and Processing for the Treatment and Minimization of Wastes, 1996, TMS* 105–114.
Dutrizac, J. E., Jambor, J. L. 2000. Jarosite and their application in hydrometallurgy. *Reviews in Mineralogy and Geochemistry* 40(1): 405–452.
GFMS Gold Survey 2014. London: Thomas Reuters.
Hanusch, K., Bussmann, H. 1995. Behavior and Removal of Associated Metals in the secondary Metallurgy of Copper. *Third International Symposium Recycling of Metals and Engineered Materials, 1995, TMS* 171–188.
Pawlek, F. 1983. *Metallhüttenkunde.* Berlin, New York: Walter de Gruyter.
Roca, A., Viñals, J., Arranz, M., Calero, J. 1999. Characterization and alkaline decomposition/cyanidation of beudantite–jarosite materials from Rio Tinto gossan ores. *Canadian Metallurgical Quarterly* 38(2): 93–103.
Schlesinger, M.E. et al. 2011. *Extractive Metallurgy of Copper 5th edition.* Elsevier.
Study on Critical Raw Materials at EU Level. Available at http://ec.europa.eu/DocsRoom/documents/5605/attachments/1/translations/en/renditions/native. Accessed 5th March 2015.
UNEP Report. Available at http://www.unep.org/resourcepanel/Portals/24102/PDFs/Metal_Recycling_Full_Report.pdf. Accessed 5th March 2015.
World Survey Silver 2014. London: Thomas Reuters. ISSN: 2372-2312.

Wastes: Solutions, Treatments and Opportunities – Vilarinho, Castro & Russo (eds)
© *2015 Taylor & Francis Group, London, ISBN 978-1-138-02882-1*

Collection of waste batteries in Portugal and Brazil

N. Vieceli & F. Margarido
Center for Innovation, Technology and Policy Research – IN+, Instituto Superior Técnico (Univ. Lisbon), Lisboa, Portugal

F. Durão & C. Guimarães
Instituto Superior Técnico (Univ. Lisbon), Lisboa, Portugal

C.A. Nogueira
Laboratório Nacional de Energia e Geologia, I.P. (LNEG), Lisboa, Portugal

ABSTRACT: The legal aspects governing the disposal of used batteries were established in the nineties of the last century in Brazil and Portugal. In Brazil the law establishes the implementation of a reverse logistics system of this waste, similar to the system of Portugal, in which there is an extended producer responsibility. Portugal, subjected to the targets set by the European Union, has increased collection rates, with emphasis on the role of management entities. In Brazil, progress and the development of strategies to increase the collection rate have been observed, nevertheless, challenges associated with the increase of the coverage of the collection of batteries and also the public information about the disposal are relevant, as well as combating irregularities. Moreover, given the growing concern over the shortage of raw materials, proper management, including the collection and recycling of these wastes, can be seen as an important opportunity for both countries.

1 INTRODUCTION

The appearance of batteries brought considerable advance from the technological point of view, however, their production on an industrial scale and low cost led to an increase in consumption, which resulted in a new paradigm about their disposal because, despite their innocent appearance and small size, batteries are now a serious environmental problem, since in most cases they are disposed of inadequately (Gomes & Melo 2006).

The process of recycling used batteries begins in their collection stage and the difficulties in this step are the main challenges to recycling, particularly in small countries (Nogueira 2001). In this context, the efficiency of the collection process for used batteries and the efficiency of the metal recovery process can be considered factors that will affect the overall environmental and human health impacts of battery systems (Morrow 2001). According to Nogueira & Margarido (2012), the best approach for end-of-life management of used batteries is the recycling, mainly for environmental reasons but also motivated by resources preservation and, in some cases, by economics profits.

The main objective of this work was to study the systems of recovery and management of end-of-life batteries developed in Portugal and Brazil, mainly regarding the collection, once studies on the subject are scarce. The evaluation of the evolution thereof and the identification of the existing challenges and opportunities for improvement, is relevant work for more than just this two countries, since the management of this waste is a global challenge.

2 METHODOLOGY

Information regarding the legal aspects adopted in both countries was consulted in the published laws available on the internet. Data related to the production and disposal of waste in Brazil was obtained from a study by ABRELPE (Associação Brasileira de Empresas de Limpeza Pública e Resíduos Especiais, Brazilian Association of Companies of Public Cleaning and Special Waste) in 2013. Information particularly related to the destination of batteries was obtained mostly with data made available by Abinee (Associação Brasileira da Indústria Elétrica e Eletrônica, Brazilian Association of the Electrical and Electronics Industry), as well as in a study in the Applied Economic Research Institute in 2012 (IPEA, Intituto de Pesquisa Econômica Aplicada). Data on the amount of batteries inserted in the Portuguese market, about the system of management and recycling rates were mainly obtained based on information provided by Ecopilhas. Recently published scientific articles, as well as MSc. (Master of Science) and Ph.D. thesis related to the subject, also assisted in the evaluation of the management panorama, particularly regarding the collection of this waste, in Brazil and Portugal.

3 PANORAMA OF THE MANAGEMENT OF WASTE BATTERIES IN PORTUGAL AND BRAZIL

3.1 *Legal considerations*

The basis of the legal regime in Portugal of batteries and accumulators containing hazardous materials was created with the Decree-Law No. 219/94 of 20 August, which transposed the Directive 91/157/EEC. However, in 1998 the European Commission decided to initiate legal procedure against the Portuguese State by failure to comply with the legislation, which also happened to Greece, France and Italy, since in most cases the reverse circuit of commercialization was rarely executed, being difficult to estimate the flow and collection rates at the time, probably less than 5% (Nogueira 2001).

Currently the management of batteries and accumulators is regulated in Portugal by the Decree-Law No. 6/2009 establishing the placement system of batteries and accumulators on the market and the system of collection, treatment, recycling and disposal of their waste. This Decree transposes into national law the Directive 2006/66/EC of the European Parliament and of the Council (APA nd,a). In addition, special waste flows, such as the batteries and accumulators, which did not have their own management entities or specific strategies set when the PERSU I (*Plano Estratégico de Resíduos Urbanos*, Strategic Plan for Urban Waste) was prepared in 1996/1997, were incorporated in PERSU II, being published specific legislation and created management entities for these flows (Trotta 2011). The flow of batteries and accumulators is still considered relevant in the recent PERSU 2020 (Diário Oficial da República 2014).

Brazil stands out for having been the first country in Latin America to regulate the disposal of batteries, through the Resolution CONAMA 257/1999, which was established by the National Environmental Council (CONAMA, *Conselho Nacional do Meio Ambiente)* on June 30, 1999 (Espinosa et al. 2004).

The CONAMA Resolution 257 was revoked in 2008 by CONAMA Resolution 401/2008, which established a more significant reduction in mercury levels, lead and cadmium of the batteries covered by the Resolution, and defined the criteria and standards for their environmentally adequate management. According to Mantuano (2011), this resolution proposed to give greater effectiveness to the post-consumer responsibility of the manufacturers and importers of batteries, through which they came to be responsible by the full cycle of their products, not only until the products are acquired by consumers. This Resolution also determined to stores selling batteries and authorized technical assistants the creation of collection points suitable for receiving this material from users, for submission to manufacturers and importers, responsible for their destination. This resolution foresees the information on the packaging of the products and advertising about the destination and associated risks, in addition to encouraging the promotion of environmental education campaigns to encourage responsibility and consumer participation in this process. Moreover, the National Solid Waste Policy (PNRS) established in 2010 (Brazil 2010) determined a similar obligation of implementing reverse logistics systems of this type of waste. This law was discussed for more than 20 years in the country, being the post-consumer responsibility one of the points that suffered resistance (Oliveira 2013). The implementation of these systems through sectorial agreements has been discussed in Brazil, but in the case of batteries a system had already been established by CONAMA Resolution 401/2008, as mentioned.

3.2 The management of batteries in Brazil and Portugal

In Portugal, *Ecopilhas*, batteries and accumulators waste management company, performs the management of the Integrated System of Used Batteries and Accumulators (SIPAU) and consists of a non-profit organization, formed by the main producers/importers of batteries and accumulators on the Portuguese market, that use the company's service and transfer it their responsibility, and was licensed in 2002. Its main function is to ensure the functioning of the SIPAU, managing a set of operations that allow the selective collection, temporary storage, sorting and recycling of the batteries and accumulators collected in the country (*Ecopilhas* nd).

According to Gonçalves (2010), by subjecting the management of batteries and accumulators to an integrated system as this, the producers have the obligation to register in the organization and report information about the type and quantity of batteries and accumulators placed on the market and indicate the chosen management system to each type of batteries and accumulators. In 2013 *Ecopilhas* had a membership of 741 partners (*Ecopilhas* 2013), much higher than the initial number of 50-member of the system in 2003. Noteworthy are in this process the *Ecoparceiros*, public or private entities directly involved in the consumption of batteries that collaborate in their collection (Gonçalves 2010).

According to the producer members of *Ecopilhas* in 2013, at least about 1433 tons of portable batteries and 1068 tons of industrial batteries were placed on the market. In 2013 it was recovered 87% of industrial batteries (934 tons) and 29% of portable batteries placed in Portugal by members and the number of portable battery collection points recorded was 7640 (*Ecopilhas* 2013). The minimum collection rates for Member States established by Directive 2006/66 / EC of the European Parliament and the Council are 45% until 26 September 2016 (Official Journal of the European Union 2006). The management of the entire process was carried out since 2004 exclusively by *Ecopilhas*, however, it is highlighted that in early 2010 licenses to conduct the activity of waste management of portable batteries and industrial batteries were granted to more companies in Portugal (Gonçalves 2010), given the potential synergies that derive from the shared management of waste of portable batteries and accumulators and of the waste of electrical and electronic equipment (APA nd,a).

In Brazil, Abinee (Brazilian Electrical and Electronics Industry Association) started in November 2010 the implementation of the program of Reverse Logistics of batteries as provided by CONAMA Resolution 401 (Abinee 2011). From the program, until 2013 it had been collected 420 tons of batteries through 1100 collection points, being the collected batteries forwarded to the company *Suzaquim* Chemical Industry, located in São Paulo, and the costs of the responsibility of the participating companies. All portable batteries sold in Brazil are contemplated, but with different treatments. Regular brands are notified, but in the case of illegal ones, competent bodies are notified to adopt adequate measures (Abinee 2013).

However, according to a study by IPEA in 2012, based on data from the National Basic Sanitation Survey (*Pesquisa Nacional de Saneamento Básico*, PNSB) conducted by IBGE (*Instituto Brasileiro de Geografia e Estatística*, Brazilian Institute of Geography and Statistics) in 2010, from the 5564 Brazilian municipalities evaluated, only 10.99 % had controlled the management of batteries performed by others, a number considered small, and in many of these the batteries were deposited in dumps or landfills.

In 2006, from the initiative of a private institution, the *"Papa Pilhas"* program was created, developed by the *Banco Real* (now Santander), receiving in eight years more than 800 tons of alkaline batteries and cell phones, with 2834 points collection in Brazil in 2010. This material was sent for sorting and subsequent recycling in *Suzaquim*. Nevertheless, with the implementation of PNRS and reverse logistics by the sector itself, this program has been discontinued, since the Law amplifies this initiative for the whole society (Santander 2014, IPEA 2012).

The recycling process carried out by *Suzaquim* has a cost of about US$1000/t, through which pigments made from oxides and metal salts are produced, using as raw material batteries, industrial or cell, fluorescent lamps and galvanic sludge. The product is then sold to the industry for the manufacture of fireworks, ceramic tiles, paints and glass (Menezes 2011, Mantuano 2011).

In Portugal, considering the financial sustainability of the management systems, the concept of *Ecovalor* is an important aspect, being a financial benefit, which is paid for by producers for each product inserted in the market in order to support the various costs of waste management by a Waste Management entity (Portuguese Environment Agency 2015). The *ecovalores* proposed by *Ecopilhas* and approved for the biennium 2014-2015, ranging from 360 euros/t for alkaline batteries, which are the most representative on the market, to 700 euros/t for lithium batteries and others, for example (Diário da República 2014b). After collecting, a sorting is done, due to the great variety of electrochemical systems it becomes difficult to treat them all in a single process (Gonçalves 2010). In Portugal this operation was carried out since 2007 by *Resitejo, Associação de Gestão e Tratamento dos Lixos do Médio Tejo,* through an agreement with *Ecopilhas*, where cells are separated by sieving and manual sorting (Gonçalves 2010) and destined for specialized recycling companies in France and Spain (*Ecopilhas* 2013).

In Brazil, although the *Suzaquim* has a capacity to recycle 950 tons of material, it receives only 330 tons annually (Faleiros 2011). After use, batteries are usually disposed together with household waste in sanitary landfills or dumpsites, thus becoming a serious pollutant. The recycling and proper disposal of batteries in the country is a recent preoccupation, since only 1% of this type of waste is recycled (Mombach 2010). In Brazil 1.2 billion household batteries are sold each year. There is still a long way to go concerning their destination, being required extensive work of awareness and involvement of the population and the associated sectors (Abinee, 2013). It is important noting that when batteries are discarded as unsorted municipal waste, in contact with humidity, heat and other substances they can release their toxic contents, leading to environmental contamination. Thus, it is important to control the concentration of some elements present in these materials (Mombach 2010). The disposal of batteries with household waste is aggravated when considering that in Brazil their inadequate disposal is representative. According to a study of ABRELPE (2013), 41.7% of the waste collected in Brazil in 2013 went to dumps or controlled landfills, which differ little since they do not have the necessary systems for environmental protection. This percentage corresponds to 79000 t/day of waste which is disposed inappropriately. The existence of scavengers on the dumps can increase the risk of contamination.

3.3 *Challenges and opportunities*

The electronics sector recorded in Brazil an increase of 110% between 2003 and 2011, driven by growth in domestic demand due to the increase in employment and wages. Between 2003 and 2010, the production of mobiles increased 125% while the production of computers increased 337%, and this figures should continue to increase (Abinee 2012). These numbers also stand out when considering the end of life of all of these products. Another major challenge is the parallel market of irregular cells, which outside the legal specifications are highly toxic and hazardous to health (Marques & Cunha 2013). In Brazil it is estimated that about a third of the collected material corresponds to batteries of more than 200 irregular marks, being estimated that 40% of the battery market in Brazil is dominated by irregular products. In addition to representing a risk, their marketing unfairly generates costs to the producers that attend the legislation (Abinee 2012). For these products to be withdrawn from the market, in addition to supervision, consumers need to inspect the batteries they buy.

The lack of disposal alternatives and the lack of public information on the risks that the batteries cause when discarded in incorrect locations in Brazil, can result in serious environmental damage (Kemerich et al. 2012, Marques & Cunha 2013). Studies on the perception of the problem of improper disposal of batteries among the population indicated that, although the population was aware of the existence of collection points of these materials, it did not know that they could be recycled, therefore not worrying about their correct destination (Brum & Silveira 2011, Kemerich et al. 2012). Moreover, the mere existence of a law is not enough for it to be known and enforced, in addition to observing the existence of opposing obstacles to communication, such as economic interests (Reidler & Günther 2002).

Although the legal basis on the subject has been inserted in both countries between 1990 and 2000, the development of management systems seems to lie at different stages in them. Initially Portugal also faced obstacles to begin the implementation of the system and the collection rates are still inferior to other countries, being the progressive collection rates imposed by the EU a challenge. The participation of the population is also important in this case, considering not only the increase in collection rates but also the reduction of the generation of such waste, source reduction and sustainable consumption, through the use of rechargeable batteries, for example. In this context, awareness goals are fundamental to the management systems. The targets imposed by the European Union and the associated incentives should also be seen as opportunities to improve the system.

The territorial dimension of Brazil and the great diversity existing between the different regions of the country can also be a challenge in waste management in general. In addition, it is clear the need for a more comprehensive disclosure to consumers about the proper disposal of this waste in Brazil. This is crucial in this process and in the choice of legal products, as well as a more effective supervision, expansion of the collection network and incentives that make the collection system possible. Moreover, given the growing concern over the shortage of the raw materials, proper management, including the collection and recycling of this waste, can be seen as an important opportunity for both countries.

4 CONCLUSIONS

The legal aspects that regulate this subject were set in the nineties of the last century in Brazil and Portugal. In Brazil, the law defines the establishment of a reverse logistics system of the batteries, similar to the system in Portugal, in which there is an extended producer responsibility. Challenges were faced by Portugal in the beginning, nevertheless, given the incentives and goals imposed by EU, an increase in collection rates has been seen. In Brazil, although strategies have been adopted, aspects such as public information, expansion of the collection, monitoring and combating irregular batteries represent major challenges for the management of waste, which are usually inadequately disposed, although some improvements have been observed in the last years.

ACKNOWLEDGEMENTS

The author N. Vieceli acknowledges the doctorate grant ref. 9244/13-1 supplied by CAPES Foundation, Ministry of Education of Brazil.

REFERENCES

Abinee – Associação Brasileira da Indústria Elétrica e Eletrônica 2011. Available at <http://www.abinee.org.br/noticias/com78.htm>. Accessed on 01 abr. 2015.
Abinee – Associação Brasileira da Indústria Elétrica e Eletrônica 2013. Available at <http://www.abinee.org.br/noticias/com28.htm>. Accessed on 01 abr. 2015.
Abinee – Associação Brasileira da Indústria Elétrica e Eletrônica 2012. *A indústria elétrica e eletrônica impulsionando a economia verde e a sustentabilidade.*

ABRELPE – Associação Brasileira de Empresas de Limpeza Pública e Resíduos Especiais 2013. *Panorama dos Resíduos Sólidos no Brasil 2013.*

APA – Associação Portuguesa do Ambiente n.d.a. *Resíduos de Pilhas e Acumuladores.* Available at <http://www.apambiente.pt/index.php?ref=16&subref=84&sub2ref=197&sub3ref=281>. Accessed on 31 mar. 2015.

APA – Agência Portuguesa do Ambiente n.d.b. *Ecovalor.* Available at <http://www.apambiente.pt/index.php?ref=16&subref=84&sub2ref=197&sub3ref=279>. Accesed on 31 mar. 2015.

Brasil 2010. Política Nacional de Resíduos Sólidos. *Lei nº 12.305, de 02 de ago. 2010.*

Brum, Z.R. & Silveira, D.D. 2011. Educação ambiental no uso e descarte de pilhas e baterias. *Revista Eletrônica em Gestão, Educação e Tecnologia Ambiental REGET-CT/UFSM* 2(2): 2005–213.

CONAMA – Conselho Nacional do Meio Ambiente 2008. M.M.A. *Resol. 401, de 04 nov. 2008.*

Diário da República 2014a. Nº 179, 17 de setembro de 2014.

Diário da República 2014b. 2.ª série, N.º 128, de 7 de julho de 2014.

Official Journal of the European Union 2006. *Directive 2006/66/CE of The European Parlament and the Council, of 6 Semp. 2006.*

Gomes, A.C.L. & Melo, S.G. 2006. Pilhas e Efeitos Nocivos. *Pilhas e efeitos nocivos* 10(3): 10-5.

Ecopilhas n.d. Available at <http://www.ecopilhas.pt/portal/>. Accessed on 21 mar. 2015.

Ecopilhas 2013. *Relatório de Atividade 2013.*

Espinosa, D.C.R. 2004 Brazilian policy on battery disposal and its practical effects on battery recycling. *Journal of Power Sources* 137: 134–139.

Faleiros, G. 2011. 52: 22. Available at <http://www.pagina22.com.br/index.php/2011/05/economia-verde-duas-faces-das-pilhas/>. Accessed on 01 abr. 2015.

Gonçalves, J.P.P. 2010. *Valorização de Pilhas Domésticas Esgotadas: Separação e Recuperação de Metais, por extração com solvents.* Master Thesis. Universidade Nova de Lisboa.

IPEA – Instituto de Pesquisa Econômica Aplica 2012. *Diagnóstico dos Resíduos Sólidos de Logística Reversa Obrigatória.* Relatório de Pesquisa. Brasília.

Kemerich, P.D.C. et al. 2012. Descarte indevido de pilhas e baterias: a percepção do problema no município de Frederico Westphaken – RS. *Rev. Elet. Em Gestão e Tecnol. Ambiental* 8(8):1680–1688.

Marques, M.A. & Cunha, E.B. 2013. O descarte inadequado de pilhas e baterias usadas e os impactos sócio-ambientais provocados pela ação do consumidor. *Caderno Meio Amb. e Sustent* 2(2).

Mombach, A. 2010. *Determinação de metais e metalóides em pilhas por ICP OES.* Thesis.

Morrow, H. Environmental and Human Health Impact Assessment of Battery Systems. In Pistoia, J. et al. (eds) 2001. *Used Battery Collection and Recycling* 10:1–32. Amsterdam: Elsevier.

Mantuano, D.P. 2011. Pilhas e baterias portáteis: legislação, processos de reciclagem e perspectivas. *Revista Brasileira de Ciências Ambientais* 21: 1–13.

Menezes, L. 2011. *Como é feita a reciclagem de pilhas e baterias?* Super Interessante, Abril. Available at <http://super.abril.com.br/ciencia/como-feita-reciclagem-pilhas-baterias-667505.shtml>. Accessed on 01 abr. 2015.

Nogueira, C.A. 2001. *Reciclagem de Baterias de Níquel-Cádmio por Processoamento Hidrometalúrgico.* PhD Thesis. Instituto Superior Técnico, Universidade Técnica de Lisboa.

Nogueira, C.A. & Margarido, F. 2012. Battery Recycling by Hydrometallurgy: Evaluation of Simultaneous Treatment of Several Cell Systems. In Salazar-Villalpando, M.D. et al (eds). *Energy Technology 2012: Carbon Dioxide Management and Other Technologies* 227–234. New Jersey: John Willey.

Oliveira, T. 2013. Ministério do Meio Ambiente. *Painel debate desafios da Política Nacional de Resíduos Sólidos.* Available at <http://www.mma.gov.br/informma/item/9743-painel-debate-desafios-da-pol%C3%ADtica-nacional-de-res%C3%ADduos-s%C3%B3lidos>. Accesso on 05 abr. 2015.

Reidler, N.M.V.L. & Günther, W.M.R. 2002. Percepção da população sobre os riscos de descarte inadequado de pilhas e baterias usadas. *Proc. XXVIII Congreso Interamericano de Ingeniería Sanitaria y Ambiental; México, 27–31 October, 2002.*

Santander 2014. *Papa Pilhas.* Available at <http://sustentabilidade.santander.com.br/pt/Praticas-de-Gestao/Paginas/Papa-Pilhas.aspx>. Accessed on 05 abr. 2015.

Trotta, P. 2011. *A Gestão de Resíduos Sólidos Urbanos em Portugal; Proc. II Congresso Nacional de Excelência em Gestão.* ISSN 1984-9354.

Wastes: Solutions, Treatments and Opportunities – Vilarinho, Castro & Russo (eds)
© *2015 Taylor & Francis Group, London, ISBN 978-1-138-02882-1*

The 2ˢDR process – Innovative treatment of electric arc furnace dust

S. Wegscheider, S. Steinlechner & C. Pichler
Nonferrous Metallurgy, Montanuniversitaet Leoben, Leoben, Austria

G. Rösler & J. Antrekowitsch
*Christian Doppler Laboratory for Optimization and Biomass Utilization in Heavy Metal Recycling,
Montanuniversitaet Leoben, Leoben, Austria*

ABSTRACT: The treatment of metallurgical dust, such as Electric Arc Furnace Dust (EAFD), is one main aim for the future to avoid landfilling of residual wastes from the iron and steel industry. A high percentage of valuable metals, which are present in these by-products would contribute to the overall production efficiency if recovered successfully. The 2ˢDR (Two step Dust Recycling) process is a newly developed concept for multi-metal recovery from steel mill dust modelled after the so-called "Zero-waste" – recycling. An EAFD will be reprocessed in a Top Blown Rotary Converter (TBRC) and the generated products zinc oxide, metal alloy and slag can be used in further applications to prevent landfilling.

1 INTRODUCTION

The global production of steel increases every year and therefore the amount of residues rises as well. Most of these residues contain high amounts of valuable metals like zinc. The innovative processing of such wastes helps to protect resources and generate secondary raw materials e.g. for the zinc primary industry. One goal for the near future is to recycle these residues environmentally and cost-friendly without creating additional landfill products.

One of these residues is the steel mill dust which contains a substantial amount of zinc. Through remelting of galvanized steel scrap, metallic zinc is vaporized and collected as zinc oxide in the bag house filter of the steel mill. Globally, the percentage of dumped steel mill dust from Electric Arc Furnaces is about 55%. However, in Europe the recycling rate reaches approximately 90%. An overview of the production and recycled amount of Electric Arc Furnace Dust (abbreviated with EAFD) and the globally zinc recovery capacity is shown in Table 1. As can be seen the main recyclers are located in the NAFTA- as well as in the EU-countries.

In 2013, the annual electric steel production reached about 451 Mio. t with an average amount of 15-20 kg dust per ton of steel (World Steel Association, 2015). This means that up to 9 Mio. t of EAFD are generated each year, theoretically available for reprocessing. About half of this amount

Table 1. Overview regarding EAFD generation and recycling amounts (Rütten, 2011).

	NAFTA	SEAISI	P.R. China	EU 27	Other areas	World
EAF Steel 2009 [Mt/a]	50.7	64.3	48.3	60.3	118.3	341.8
EAFD Generation [t/a]	760,100	1,157,000	724,200	1,024,400	2,129,300	5,795,000
EAFD Recycling [t/a]	810,000	427,000	0	1,071,000	250,000	2,558,000
thereof Waelz	735,000	270,000	0	831,000	210,000	2,046,000
thereof RHF	25,000	62,000	0	40,000	20,000	147,000
thereof other	50,000	95,000	0	200,000	20,000	365,000
Zinc Recovery [t/a]	186,000	98,000	0	246,000	58,000	588,000

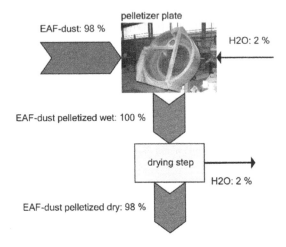

Figure 1. Material preparation for the 2^{S}DR process.

is currently reprocessed. Due to the fact that steel dust contains zinc up to 40%, there is a high potential of recovering pure ZnO out of EAFD. The average content of zinc in the dust reaches 22%, which means that about 1.5 Mio. t of zinc could be produced by recycling (ZincOx Ressources plc, 2015). To obtain a pure marketable end-product, several halides and alkali compounds also present in the dust have to be separated, which displays a metallurgical challenge. Also the utilization of by-products is an increasingly important aspect for the future in order to cope with other recycling processes and to fulfill the environmental requirements. Other considerations for the optimization of the process are CO_2-savings by the use of CO_2-neutral reducing agents as well as the energy recovery by post-combustion of the waste gas.

2   THE 2^{S}DR PROCESS

The newly developed 2^{S}DR process represents a concept for multi-metal recovery from steel mill dust. The main aim is to generate products only to avoid the landfilling of newly generated residues, such as the arising Waelz slag in an alternative steel mill dust recycling process. The 2^{S}DR process basically consists of two main steps, the clinkering and the subsequent reduction.

2.1   *Material preparation*

First of all, a material conditioning – in this case an agglomeration step – takes place due to the required transportability and chargeability. This is done by pelletizing or in an Eirich-mixer. As can be seen in Figure 1, the EAFD is mixed with water for pelletizing on a plate. The present halides in the dust together with the water act as a binder. By varying the speed and the angle of the plate, different pellet diameters can be produced. By enlarging the amount of fine particles to a pellet of about 0.8 cm, a higher weight can be adjusted. Without this step, the fine, low weight dust would otherwise leave the reactor vessel with the high gas stream during charging. Afterwards, the pelletized EAFD enters a drying step to remove the present water.

2.2   *Clinker step*

After the material preparation, the input of the dried and pelletized dust in the clinker step takes place. Impurities like chlorides and fluorides evaporate under oxidizing atmosphere at temperatures of around 1100°C as different chemical compounds – leaving behind a halide- and alkali-free

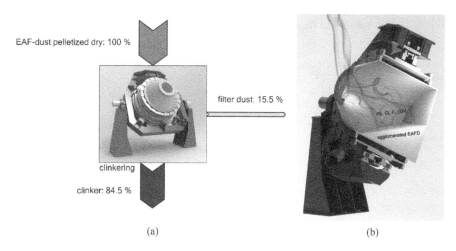

EAF-dust pelletized dry: 100 %

filter dust: 15.5 %

clinkering

clinker: 84.5 %

Pb, Cl, F, (Zn)

agglomerated EAFD

(a)   (b)

Figure 2.   Flow sheet of the clinker step (a) and the cross section of a TBRC while clinkering (b).

agglomerated EAFD. Reactions (1) and (2) describe the principle of the evaporation of chlorides and fluorides.

$$Me_x Cl_y \leftrightarrow Me_x Cl_y(g) \tag{1}$$

$$Me_x F_y \leftrightarrow Me_x F_y(g) \tag{2}$$

The present ZnO stays in the agglomerated dust in the TBRC due to its high evaporation temperature. The removable volatile components mainly include chlorine compounds as only $CaCl_2$ requires a longer treatment time. Fluorides can also be removed easily except CaF_2, because it represents one of the most stable fluorides. A rapid heating of the EAFD has a positive effect on the removal of halides due to the fact that there is no time for any possible interactions that build up undesirable compounds which are difficult to remove. A negative side effect of the removal of these compounds is the resulting zinc loss by the formation of zinc-halide compounds. However, it appears more appropriate to accept a zinc loss and thus obtain a product with high quality than creating a product that shows elevated amounts of halides. (Steinlechner and Antrekowitsch, 2011).

Figure 2 shows a flow sheet of the clinker step (a) and the cross section of a TBRC during this process (b). Instead, there is also the possibility to use a waelz kiln for the clinker step.

2.3   Reduction step

In the second stage of the 2^sDR process, the reduction of the clinker takes place to reclaim zinc as zinc oxide in the bag house filter on the one hand and valuable metals like iron accumulate in the metal alloy on the other. Therefore, a metal bath consisting of pig iron granules which is carburized up to 4.3% by applying a reducing agent, is used in the reduction step. For the carburization it is also possible to take charcoal instead of coke in order to minimize the CO_2-impact. Moreover, the charging of the clinker can be hot or cold, whereas a significant amount of energy can be saved with a hot input of the material. At this stage, the indirect reduction of zinc oxide and iron oxide from the clinker by carbon monoxide (see reactions (3), (4) and (5)) occurs. The iron oxide, which is mainly present as Fe_3O_4 in the clinkered dust is reduced to FeO and further to liquid Fe due to the prevailing high temperatures as well as the high reduction potential.

$$ZnO(s) + CO(g) \leftrightarrow Zn(g) + CO_2(g) \tag{3}$$

$$Fe_3O_4(s) + CO(g) \leftrightarrow 3FeO(l) + CO_2(g) \tag{4}$$

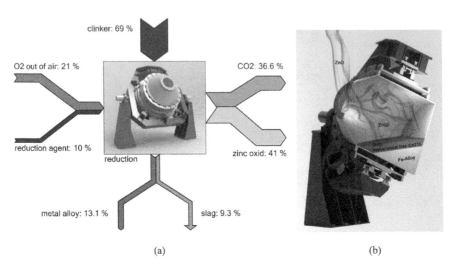

clinker: 69 %

O2 out of air: 21 %

reduction agent: 10 %

reduction

metal alloy: 13.1 %

CO2: 36.6 %

zinc oxid: 41 %

slag: 9.3 %

(a)   (b)

Figure 3.   Flow sheet of the reduction step (a) and the cross section of a TBRC during reduction (b).

$$FeO(l) + CO(g) \leftrightarrow Fe(l) + CO_2(g) \qquad (5)$$

Metallic zinc volatilizes from the converter, reoxidises to zinc oxide (reaction (6)) and can be collected in the filter house.

$$Zn(g) + \frac{1}{2}O_2(g) \leftrightarrow ZnO(s) \qquad (6)$$

The formed carbon dioxide reacts with the available carbon to carbon monoxide according to the Boudouard reaction (7).

$$C(s) + CO_2(g) \leftrightarrow 2CO(g) \qquad (7)$$

Furthermore, carbon monoxide oxidises to carbon dioxide in the exhaust gas system or in the stack (reaction (8)).

$$CO(g) + \frac{1}{2}O_2(g) \leftrightarrow CO_2(g) \qquad (8)$$

During the reduction step the content of carbon in the metal bath sinks and is hence relatively low. Therefore, the metal bath needs to be recarburized by appropriate addition of a reducing agent for a further charge. In Figure 3 (a) the flow sheet of the second process step is shown, while in (b) the cross section of a TBRC during reduction can be seen.

2.4   Advantages of the 2^s DR process

The advantages of the 2^sDR process can be summarized by the following points:

– Simultaneous recovery of Zn (as ZnO) in the bag house filter as well as valuable metals in the metal alloy is possible and verified on technical scale
– Creation of a heavy metal free slag for further application (e.g. for sand blasting or cement industry)
– "Zero-waste"-recycling by usability of generated by-products
– As described in (Griessacher et al., 2012), the utilization of charcoal is possible to avoid payments for carbon credits

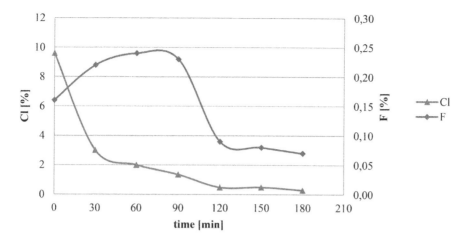

Figure 4.   Behavior of the elements Cl and F during the clinker step.

   – High quality of produced ZnO (compounds of Pb, Cl and F evaporate in the clinker step)
   – No special requirements regarding chemical composition of EAFD

3   EXPERIMENTAL RESULTS

The experimental results are divided into two posts – the clinker step results and the results of the reduction step.

3.1   *Clinker step*

As can be seen in Figure 4, the content of unwanted elements such as chlorine and fluorine decreases during the clinker step. The chlorine content decreases quickly due to the fact that chlorine-compounds are more volatile than those of fluorine. Therefore, the relative fluorine value increases first of all and shows an enrichment during the first 90 min. Subsequently, the fluorine also decreases. After 120 min both elements do not change significantly. This means that clinkering for 120 min is sufficient, which ultimately has a positive effect in terms of time and energy saving.

3.2   *Reduction step*

Three different products are generated in the reduction step which are described below.

3.2.1   *Filter dust:*
The produced filter dust from the 2^sDR process consists of up to 73% Zn while the reached value for chlorine is below 0.52% and below 0.12% for fluorine. The remainder consists of carry-over material, which is not a disturbing material for the primary zinc industry. Therefore, the filter dust can be used as input material for the electrolytic zinc winning process.

3.2.2   *Metal alloy:*
Table 2 shows the chemical composition of the generated metal alloy. As can be observed, the metal alloy is sellable to the iron and steel industry as a scrap substitution because there are no undesirable components in the alloy.

Table 2. Chemical analysis of the metal alloy.

	Fe	C	Zn	Si	S	P	Cu	Mn	Pb
[%]	99.656	0.011	0.004	0.017	0.023	0.026	0.083	0.023	0.0026

Table 3. Chemical analysis of the slag

	Pb	Zn	Femet	CaO	MgO	MnO	SiO2	Al2O3
[%]	0.005	0.063	1.4	5.74	6.4	3.96	9.01	2.19

3.2.3 *Slag:*

Table 3 offers an overview of the analyzed elements in the slag. It shows low contents in metals for which reason it is possible to use the slag for sand blasting or the cement industry.

4 CONCLUSION

The conservation of natural resources and the associated reprocessing of waste material like steel mill dust currently represents an important topic. The available processes show major disadvantages such as poor zinc oxide quality, difficult process control as well as the amount of newly generated residues which have to be landfilled. The development of new recycling strategies for the generation of products that can be supplied for a further use is currently the topic of various research projects. The presented 2^sDR process constitutes a pioneering step towards "Zero-waste" where residuals are fed into the process and products are produced exclusively. The results obtained from the test series symbolize positive findings and are broadly consistent with those from the mass and energy balances from literature (Wegscheider, 2014), where the gained experience serves as potential for the optimization of further calculations. With respect to the ubiquitous CO_2-problem, charcoal can be used as reducing agent. The carbon carrier made out of biomass also leads to satisfactory results. To sum up, the investigations have shown that the described two-step process for the reprocessing of EAFD show a significant potential for the future.

REFERENCES

Griessacher, T., Antrekowitsch, J. and Steinlechner, S. 2012. Charcoal from agricultural residues as alternative reducing agent in metal recycling. *Biomass and Bioenergy* 39: 139–146.

Rütten, J. 2011. Processing EAF Dust Through Waelz Kiln and ZINCEX Solvent Extraction: The Optimum Solution. In Harre, J. (Ed.), *Proc. of 6. European Metallurgical Conference: June 26–29, Düsseldorf, Germany*. GDMB, Clausthal-Zellerfeld. 1673–1687.

Steinlechner, S. and Antrekowitsch, J. 2011. Options for Halogen Removal from Secondary Zinc Oxides. In Harre, J. (Ed.), *Proc. of 3. Seminar Networking between Zinc and Steel, Leoben, Austria, Schriftenreihe der GDMB, Gesellschaft für Bergbau, Metallurgie, Rohstoff- und Umwelttechnik*. GDMB, Clausthal-Zellerfeld. 47–56.

Wegscheider, S. 2014. *Innovative Verwertung von Stahlwerksstäuben zur Generierung sekundärer Rohstoffe, Master thesis*. Leoben: Montanuniversität Leoben.

World Steel Association 2015. *World Steel in Figures 2014*. Available at: http://www.worldsteel.org/media-centre/press-releases/2014/World-Steel-in-Figures-2014-is-available-online.html.

ZincOx Ressources plc 2015. *ZincOx game changing process*. Available at: http://zincox.com/about/game-changing-process.html.

Wastes: Solutions, Treatments and Opportunities – Vilarinho, Castro & Russo (eds)
© 2015 Taylor & Francis Group, London, ISBN 978-1-138-02882-1

Solid sorbents for rare earths recovery from electronic waste

E.M. Iannicelli Zubiani, C. Cristiani, G. Dotelli & P. Gallo Stampino
Politecnico di Milano, Dipartimento di Chimica, Materiali e Ingegneria Chimica "G.Natta", Milano, Italia

ABSTRACT: Activated carbon (AC) and modified AC were tested as solid sorbents for Rare Earths (REs) recovery from Electronic Waste. The modified AC was synthesized by loading Pentaethylenehexamine and the amount of loaded polymer was estimated by COD analysis of the residual amount in solution. The AC and the modified AC were contacted with lanthanum solutions (chosen as representing element of REs family) and the lanthanum adsorbed by the solids was analyzed by ICP-OES of the contacted solution. Finally, release tests were performed on the different samples in order to verify the solids capability not only to capture but also to recover metal ions. The obtained results showed that the experimental procedure was appropriate to load the polymer onto the AC and that the modification of the AC improved both adsorption (from 44% to 100%) and release (from 65% to 91%) with respect to natural AC, ensuring a global recovery efficiency of 90%.

1 INTRODUCTION

REs demand is constantly increasing in the Global Market, since new technological applications such as optical, electronics, ceramics and nuclear (Iannicelli-Zubiani et al., 2014) exploit these materials for their unique properties. Since natural resources are located just in focused areas (mainly China) (Massari and Ruberti, 2013), European Commission highlighted REs as critical raw materials (EU, 2010) and promoted processes for their recovery, separation and substitution (ERECON, 2014). In particular, their recovery from Waste Electrical and Electronic Equipment (WEEE, type of waste widely growing in recent years) represents a big opportunity since it could reduce the supply risk of such raw materials and could transform the waste into an economic valuable product, minimizing the environmental impact (Iannicelli-Zubiani et al., 2012).

Different methods have been proposed for REs recovery (Binnemans et al., 2013) and, among the others, hydrometallurgical method has been reported to be one of the most interesting being: generally applicable to very different compositions and allowing the same processing steps as those for extraction of REs from primary ores, usually realized by solvent extraction.

But in recent years adsorption onto solid sorbents has presented as an alternative approach to solvent extraction systems. The principle is similar, involving a partitioning of solutes between two phases. However, instead of two immiscible liquid phases, adsorption onto solid sorbents involves partitioning between a liquid (sample matrix) and a solid (sorbent) phase. This technique enables both concentration and purification and intends to solve some of the limits of solvent extraction processes, such as: the need of many process steps; the consumption of large amounts of chemicals and the generation of large amounts of waste water (Iannicelli-Zubiani et al., 2013a).

Different solids can be used as matrix: resins (Helaly et al., 2012), nanotubes (Chen et al., 2013, Zakharchenko et al., 2012), clays and modified clays (Iannicelli-Zubiani et al., 2014, Iannicelli-Zubiani et al., 2013b), membranes (Gaikwad, 2012), silica (Wu et al., 2013) and activated carbon (Chen, 2010, Awwad et al., 2010). Among them, activated carbon (AC) has been proven to be an effective adsorbent for the separation of a wide variety of pollutants from aqueous or gaseous media and for this reason it was selected as solid sorbent in the present study.

Table 1. Properties of the used AC.

Sorbent	Humidity (%)	Ash (%)	Iodine	pH	Mesh	Density (kg/m^3)
AC	10	10/15	225/600	8.7	99.8% of powder below 120 μm	600

AC is a porous material with exceptionally high surface area (ranges from 500 to 1500 m^2/g), large pore volume and well-developed internal microporosity. In addition, a chemical modification of AC offers flexibility in their specific physical and chemical properties and enhances their affinity for inorganic and/or organic species present in waters (Marwani et al., 2013). In general, the most successful extractors for metal ions are those sulfur and nitrogen or phosphorous containing compounds (Soliman et al., 2004) and for this reason in the present study the AC was modified with a Pentaethylenehexamine in order to test its adsorption capability towards REs and to compare it with commercial AC.

A NdFeB hard disk magnet was used as example of WEEE rich in REs. The disk was manually disassembled and the magnetic component was removed and leached in a HNO$_3$ solution: as expected, the magnet has a very complex composition and so, in order to test the effectiveness of the proposed solid sorbents, the adsorption capability towards REs was firstly performed on model solutions: lanthanum was chosen as representing element of REs family.

So lanthanum adsorption experiments were carried out using as solid sorbents both AC and modified AC. Since the recovery is the final aim, also the release of the captured lanthanum ions was considered.

2 MATERIALS AND METHODS

2.1 Materials

Powdered AC procured by Torchiani s.r.l. was used as solid matrix. Its main properties are reported in Table 1.

The polymer used as modifying agent is a Pentaethylenehexamine (PEHA in the following), having molecular weight 232 g/mol, an initial pH of 11 and density at 25°C equal to 0.96 g/mL. It was supplied by Sigma-Aldrich. Lanthanum(III) nitrate pentahydrate and nitric acid ACS reagent, both of them from Sigma-Aldrich, were the other reactants used. Deionized water was also used as solvent throughout the whole study.

2.2 AC modification

The polymer modified AC was synthesized by the following method: 2.5 g of carbon powders were mixed, in a closed reactor under vigorous stirring, with 50 mL of aqueous polymer solution for a fixed time of 90 min. Samples prepared with PEHA had an initial polymer concentration of 135 mmol/L. All the intercalation experiments were carried out at 30°C: the temperature was set and controlled with a thermostat.

The pH of the solutions was measured before and after dispersing the carbon powders using a Mettler Toledo FE20/EL20 digital pH meter and it was constant throughout the preparation at about pH 11. After the reaction, the solid phase was separated by the liquid using a centrifuge (HETTICH 32 RotoFix). The solid was dried by one day at room temperature while COD analysis was performed on the liquid in order to calculate the amount of loaded polymer by difference between the initial and the residual amount.

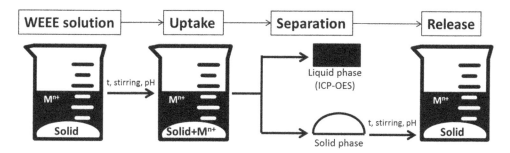

Figure 1.   General scheme of the procedure.

2.3 *Metal ions uptake and release*

A determined amount of AC and modified AC (2 g) was contacted with 50 mL of lanthanum solutions at known initial concentration ($C_0 = 2700$ mg/L for La^{3+}, the same concentration of neodymium in the dissolved magnet) under vigorous stirring. After 90 min, the solid was separated from the liquid using a centrifuge (HETTICH 32 RotoFix). The adsorption experiments were carried out using a solid/liquid ratio of 0.04 g/mL and the pH was monitored and resulted to be constant at about 6; the solutions were analyzed by ICP-OES before and after the adsorption experiments to determine the amount of lanthanum captured by the solid phase.

In particular, the amount of metal ion adsorbed at time t, here termed q_t [mg/g], was calculated from the mass balance equation (Equation 1):

$$q_t = \frac{(C_0 - C_t)V}{W} \tag{1}$$

where C_0 and C_t are the initial and the residual metal concentrations in the solution (mg/L) respectively, V is the volume of solution (L) and W is the mass of AC (g). The uptake efficiency (%) was calculated by Equation 2.

$$\text{Uptake efficiency} = \frac{(C_0 - C_t)100}{C_0} \tag{2}$$

The release tests were performed following previous works (Iannicelli-Zubiani et al., 2013a): by contacting the La-containing AC and modified AC with solutions at acid pH (pH 1), under continuous stirring (magnetic stirrer, 500 rpm). The experiments were performed at room temperature, for 90 min, with a solid/liquid ratio of 0.026 g/mL. The amount of released ions (q_w) was determined by ICP-OES analyses of the liquid phase after separation and calculated from Equation 3:

$$q_w = \frac{(C_w)V}{W} \tag{3}$$

where C_w is the metal concentration in solution (mg/L) after the release step, while V is the volume of solution (L) and W is the mass of AC (g). The release efficiency (%) was calculated from Equation 4:

$$\text{Release efficiency} = \frac{(C_w)100}{C_t} \tag{4}$$

where C_t was the amount of metal ion adsorbed by the solid phase.

A general scheme of the procedure is reported in Figure 1.

Table 2. Adsorbed and released lanthanum in absolute and percentage values.

Sorbent	q_t (mg/g)	Uptake efficiency (%)	q_w (mg/g)	Release efficiency (%)	Global efficiency (%)
AC	29	44	19	65	28
Modified AC	67	100	61	91	91

2.4 Characterizations

2.4.1 COD analyses

The Chemical Oxygen Demand (COD) analyses were carried out using a Spectrodirect Lovibond instrument. In a standard instrumental procedure the unknown sample was oxidized by heating for 120 min at 150°C (ASTM, 2006).

The reaction (Equation 5) that has to be considered for the quantification of loaded polymer, starting from COD expressed in g/L of O_2, is:

$$C_{10}H_{28} + 17O_2 \rightarrow 10CO_2 + 14H_2O \tag{5}$$

2.4.2 ICP-OES analyses

All the metal ions concentrations in solutions were measured by Inductively Coupled Plasma – Optical Emission Spectroscopy (ICP-OES) analyses using a Perkin Elmer Optima 8300 spectrometer.

3 RESULTS AND DISCUSSION

3.1 AC modification

The amount of loaded polymer was calculated by difference between the initial and the residual amount using the COD (Chemical Oxygen Demand) analysis on the liquid solutions. The results showed that 30% of the initial polymer is loaded on the activated carbon, corresponding to a ratio polymer/carbon of 0.2 g/g.

3.2 Uptake and release

The AC and the modified AC were tested in the process of lanthanum recovery and the results are reported in Table 2. It is evident that the modification of the AC significantly enhances the affinity of the solid towards lanthanum uptake (from 29 mg/g of adsorbed lanthanum in presence of AC to 67 mg/g, corresponding to the totality of metal ions in solution, in presence of the modified AC). Also the released lanthanum reflects the improvement caused by the PEHA loading: in fact the desorbed lanthanum is strongly increased in case of modified AC (from 19 mg/g to 61 mg/g) both because the higher uptake both because a more efficient release mechanism. It is supposable that the release phase, realized by changing the pH conditions until values of 1, is more effective in the case of modified AC since the PEHA amino groups result protonated and release the captured lanthanum.

In terms of global efficiencies (calculated multiplying the uptake and the release efficiencies), the chemical modification due to PEHA enhances the system from 28% to 91% of recovered lanthanum.

3.3 Maximum adsorption and desorption capacity

Since the modified AC was able to adsorb all the lanthanum present in the solution simulating the neodymium concentration in a magnet, other experiments with increased lanthanum initial

Table 3. Maximum adsorbed and released lanthanum in present and other studies.

Sorbent	q_t (mg/g)	q_w (mg/g)	References
AC	36	24	This study
Modified AC	88	70	This study
Montmorillonite	57	–	(Bruque et al., 1980)
Bentonite	36	–	(Chegrouche et al., 1997)
Modified clays	60	34	(Iannicelli-Zubiani et al., 2013b)
Iron oxide loaded calcium alginate beads	124	124	(Wu et al., 2010)
Lewatit TP 214 Resin	39	–	(Ferrah et al., 2014)
Polydopamine (PDA) membrane	60	59	(Hong et al., 2014)

concentration were performed in order to evaluate the maximum adsorption capacity both of AC and modified AC. In Table 3 the obtained results are compared with other studies.

While the AC presented a maximum adsorption capacity lower or in the best case comparable to the other solid sorbent considered (clays, modified clays, resins, membranes), the modified AC achieved indeed values in almost all cases higher than the other sorbents. Among the considered solid matrixes, only iron oxide loaded calcium alginate beads are characterized by a maximum adsorbed lanthanum value higher than the one obtained in the present study (124 mg/g and 88 mg/g, respectively) but it is worth to note that the adsorption time of the experiments is pretty different, being 28 h in alginate beads study (Wu et al., 2010) and 1.5 h in the present study.

Concerning maximum desorption capacity (Table 3), it can be remarked the good lanthanum release in case of modified AC (70 mg/g), very often higher than the other considered sorbents.

4 CONCLUSIONS

The following conclusions can be drawn:

– new materials were synthesized starting from a solid matrix and a liquid polymer: the experimental procedure was proved to be appropriate to load the polymer PEHA onto the AC;
– the amount of loaded polymer is 0.2 g/g, equal to 30% of the initial polymer concentration;
– both AC and modified AC were effective in uptake and release processes but with strongly different efficiencies: the modification improved the uptake efficiency from 44% to 100% and the release efficiency from 65% to 91%;
– the maximum adsorption capacity was found 36 mg/g for AC and 88 mg/g for modified AC while the maximum desorption 24 mg/g and 70 mg/g, respectively;
– the modified AC ensured a global process efficiency of more than 90%, meaning more than 90% of lanthanum recovery;
– the high efficiencies in the uptake step suggest also the possibility to use these innovative materials as finishing purifying materials for waste water treatment;

ACKNOWLEDGEMENTS

This work was performed under the project "E-WASTE – Il ciclo intelligente" (ID 40511448, financed by Lombardia Region).

REFERENCES

ASTM 2006. Standard Test Methods for Chemical Oxygen Demand (Dichromate Oxygen Demand) of Water.

Awwad, N. S., Gad, H. M. H., Ahmad, M. I. & Aly, H. F. 2010. Sorption of lanthanum and erbium from aqueous solution by activated carbon prepared from rice husk. *Colloids and Surfaces B: Biointerfaces,* 81, 593–599.

Binnemans, K., Jones, P. T., Blanpain, B., Van Gerven, T., Yang, Y., Walton, A. & Buchert, M. 2013. Recycling of rare earths: a critical review. *Journal of Cleaner Production,* 51, 1–22.

Bruque, S., Mozas, T. & Rodriguez, A. 1980. Factors influencing retention of lanthanide ions by montmorillonite. *Clay Miner,* 15, 413–420.

Chegrouche, S., Mellah, A. & Telmoune, S. 1997. Removal of lanthanum from aqueous solutions by natural bentonite. *Water research,* 31, 1733–1737.

Chen, Q. 2010. Study on the adsorption of lanthanum(III) from aqueous solution by bamboo charcoal. *Journal of Rare Earths,* 28, Supplement 1, 125–131.

Chen, S., Zhu, S. & Lu, D. 2013. Titanium dioxide nanotubes as solid-phase extraction adsorbent for on-line preconcentration and determination of trace rare earth elements by inductively coupled plasma mass spectrometry. *Microchemical Journal,* 110, 89–93.

ERECON 2014. Strengthening the Europesn Rare Earths supply-chain. *ERECON Final Conference.* Milano, Italy.

EU 2010. Critical raw materials for the EU. Report of the Ad-hoc Working Group on defining critical raw materials European Commission.

Ferrah, N., Abderrahim, O. & Didi, M. 2014. Lanthanum (III) Removal onto Lewatit TP 214 Resin in Nitrate Medium: Kinetic and Thermodynamic Study. *IOSR Journal of Applied Chemistry (IOSR-JAC),* 7, 45–52.

Gaikwad, A. G. 2012. *Behavior of transport and separation of lanthanum, yttrium and lutetium metal ions through celluose fiber supported solid membrane.*

Helaly, O. S., Abd El-Ghany, M. S., Moustafa, M. I., Abuzaid, A. H., Abd El-Monem, N. M. & Ismail, I. M. 2012. Extraction of cerium(IV) using tributyl phosphate impregnated resin from nitric acid medium. *Transactions of Nonferrous Metals Society of China,* 22, 206–214.

Hong, G., Shen, L., Wang, M., Yang, Y., Wang, X., Zhu, M. & Hsiao, B. S. 2014. Nanofibrous polydopamine complex membranes for adsorption of Lanthanum (III) ions. *Chemical Engineering Journal,* 244, 307–316.

Iannicelli-Zubiani, E., Bengo, I., Azzone, G., Cristiani, C., Dotelli, G., Masi, M. & Sciuto, D. 2012. Tecnologie di recupero e separazione di terre rare: stato dell'arte e prospettive. *In:* Editore, M., ed. Ecomondo, Rimini.

Iannicelli-Zubiani, E. M., Cristiani, C., Dotelli, G., Stampino, P. G. & Bengo, I. 2014. Lanthanum uptake by clays and organo-clays: effect of the polymer. *Procedia Environmental Science, Engineering and Management* 1, 5.

Iannicelli-Zubiani, E. M., Cristiani, C., Dotelli, G., Stampino, P. G., Bengo, I., Masi, M. & Pelosato, R. 2013a. Rare Earths separation from WEEE by synthetic polymers modified clays. *In:* Wastes: solutions, treatments and opportunities, Braga, Portugal.

Iannicelli-Zubiani, E. M., Cristiani, C., Dotelli, G., Stampino, P. G., Pelosato, R., Bengo, I. & Masi, M. 2013b. Polymers modified clays for separating Rare Earths from WEEE. *Environmental Engineering and Management Journal,* 12, 4.

Marwani, H. M., Albishri, H. M., Jalal, T. A. & Soliman, E. M. 2013. Study of isotherm and kinetic models of lanthanum adsorption on activated carbon loaded with recently synthesized Schiff's base. *Arabian Journal of Chemistry.*

Massari, S. & Ruberti, M. 2013. Rare earth elements as critical raw materials: Focus on international markets and future strategies. *Resources Policy,* 38, 36–43.

Soliman, E. M., Saleh, M. B. & Ahmed, S. A. 2004. New solid phase extractors for selective separation and pre-concentration of mercury (II) based on silica gel immobilized aliphatic amines 2-thiophenecarboxaldehyde Schiff's bases. *Analytica Chimica Acta,* 523, 133–140.

Wu, D., Sun, Y. & Wang, Q. 2013. Adsorption of lanthanum (III) from aqueous solution using 2-ethylhexyl phosphonic acid mono-2-ethylhexyl ester-grafted magnetic silica nanocomposites. *Journal of Hazardous Materials,* 260, 409–419.

Wu, D., Zhao, J., Zhang, L., Wu, Q. & Yang, Y. 2010. Lanthanum adsorption using iron oxide loaded calcium alginate beads. *Hydrometallurgy,* 101, 76–83.

Zakharchenko, E., Mokhodoeva, O., Malikov, D., Molochnikova, N., Kulyako, Y. & Myasoedova, G. 2012. Novel Sorption Materials for Radionuclide Recovery from Nitric Acid Solutions: Solid-Phase Extractants and Polymer Composites based on Carbon Nanotubes. *Procedia Chemistry,* 7, 268–274.

Author index

Printed and bound by CPI Group (UK) Ltd, Croydon, CR0 4YY

24/10/2024

01778293-0006